The 8086 Microprocessor
Programming and Interfacing the PC

The 8086 Microprocessor
Programmimg and Interfacing the PC

Kenneth J. Ayala

Western Carolina University

Delmar Publishers

an International Thomson Publishing company I(T)P®

Albany • Bonn • Boston • Cincinnati • Detroit • London • Madrid
Melbourne • Mexico City • New York • Pacific Grove • Paris • San Francisco
Singapore • Tokyo • Toronto • Washington

PHOTOS
Figure 1.1 A Mainframe Installation Courtesy of International Business Machines Corporation; Figure 1.2 A contemporary Minicomputer Copyright 1994 Hewlett-Packard Company. Reproduced with permission; Figure 1.3 Early Microprocessor CPUs. Photograph by the author; Figure 1.4 The original IBM PC Courtesy of International Business Machines Corporation; Figure 1.5 A modern PC Copyright 1994 Hewlett-Packard Company. Reproduced with permission; Figure 1.6 Weaving Monitor Computer Courtesy of A. B. Carter Company.

Copyediting: David Dexter
Text and Cover Design: Ronna Hammer
Composition: G & S Typesetters, Inc.
Artwork: G & S Typesetters, Inc.
Cover image: Photomicrograph of 8086 microprocessor by Michael Davidson

For more information, contact:

Delmar Publishers Inc.
3 Columbia Circle , Box 15015
Albany, New York 12212-5015

International Thomson Publishing
Berkshire House
168-173 High Holborn
London, WC1V7AA
England

Thomas Nelson Australia
102 Dodds Street
South Melbourne 3205
Victoria, Australia

Nelson Canada
1120 Birchmont Road
Scarborough, Ontario
M1K 5G4, Canada

International Thomson Publishing GmbH
Konigswinterer Str. 418
53227 Bonn
Germany

International Thomson Publishing Asia
221 Henderson Bldg. #05-10
Singapore 0315

International Thomson Publishing Japan
Kyowa Building, 3F
2-2-1 Hirakawa-cho
Chiyoda-ku, Tokyo 102
Japan

02 01 00   10 XXX 8 7 6 5 4

Library of Congress Cataloging-in-Publication Data

Ayala, Kenneth J.
    The 8086 microprocessor : programming and interfacing the PC / Kenneth J. Ayala.
  Stephen L. Herman and Richard L. Sullivan. -- 3rd ed.
        p.   cm.
    Includes index.
      ISBN: 0-314-01242-7   (soft)
      1. Intel 8086 (Microprocessor) —Programming.   2. Computer interfaces
  I. Title.
  QA76.8.I292A99 1994   94-31196
  004.165—dc20   CIP

Dedication

To Anne Thomas Ayala, for thirty-two usually pleasant, often exciting, occasionally irritating, but never dull, years of life together.

Trademarks and Copyrights

CONTENTS

Appendices

PREFACE

Given the number of 8086 assembly language programming books available today, why write another one? The answer is in three parts. First, I could not find a textbook that reflected the current industrial practice of using computer ports and expansion busses to interface computers to external hardware. Second, I could not find a textbook that gave sufficient attention to operating system concepts as an integral part of PC assembly language programming. Third, in order to take my courses in 8086 programming and interfacing, my students must purchase four books: a standard PC assembly language programming tutorial text, a DOS programmer's reference, a PC/ISA bus reference, and an assembler-debugger software manual. What the students need is a single book that can be all of a tutorial, reference, and manual.

I believe this book addresses the three concerns listed above and perhaps a few more. What I have attempted to do is provide the student with a single textbook that includes assembly language programming concepts found in most assembly language texts and operating system topics that are not always covered. Hardware topics include a thorough discussion of disk file organization, interfacing to standard ports, interfacing to the PC/ISA bus, and the design of a single-board computer.

Many who teach assembly language programming today, I suspect, grew up with the original 8-bit microprocessors and their peripherals. Like myself, many are self-taught in formal programming concepts and come from a digital-logic-design hardware background. In the past, I have taught assembly language programming from the standpoint that the microprocessor would be part of a small, single-board computer, and that the microprocessor would execute a single, dedicated program. Our teaching hardware consisted of single-board trainers, usually supplied by the microprocessor manufacture. The crudest trainers had to be programmed manually, in hex code, via a keypad; hardware single-stepping and LED readouts provided us with debugging capabilities. More advanced trainers were equipped with serial ports to which we connected a personal computer; programs could be assembled and downloaded for debugging under a monitor program. Life was simple—programs were small and hardware expensive. Disk drives, large memories, high-level languages, and operating systems were the domain of the Computer Science Department.

No more. In most applications today, the microcontroller has taken the place of the 8-bit microprocessor. It is cheaper today to equip an assembly language programming laboratory with personal computers than with single-board trainers. System programming is done in a high-level language, running on a personal computers under the control of an operating system. Programs and memory capacities are huge; hardware is cheap. Modern microprocessors contain hardware features that are meant to be used *only* by operating system software. Assembly language programming is used for special purposes, such as embedded controllers or as procedures called from a high-level language. Current industrial practice is to treat a computer as a component in an automated system. Indeed, it is now possible to buy a complete PC, less drives and monitor, on a board 4 inches square.

In my opinion, students should be trained so that they view a microprocessor as part of a computer, and the computer as one element in a controlled system. High-level languages and assembly languages should be used as appropriate for the problem at hand. Computer interfacing should be done using standard ports and standard busses. Operating system features should be understood and used as part of the programming process. In modern systems, software concepts have an equal or greater importance than do hardware concerns. The thrust of this book, then, is to introduce the beginner to classical assembly language programming techniques, operating system concepts, and interfacing to a PC using standard ports and busses.

For 4 years, the class notes that form the basis for this book have been in use to teach a two-semester course sequence in assembly language programming and computer interfacing. My students have had two courses in digital-logic design and one course in microcontrollers before beginning the course sequence. I assume that the student who uses this book has studied digital-logic design, or basic binary fundamentals. No prior study of computers or assembly language programming is necessary, however.

There are many ways to organize a sequence of courses using this book. Students who have had a previous microprocessor course might spend a semester studying Chapters 4, 5, 6, 7, and 8, learning how to program the 8086. Students who have no previous microprocessor experience may be best served by including Chapter 3 and deleting Chapter 8 in their studies. A second course on hardware and interfacing should include material from Chapters 10, 11, and 12. Computer science students, who may have limited interest in the hardware designs of Chapters 10, 11, and 12, should find Chapters 8 and 9 useful.

The reviews of this book revealed that no one book will fit everyone's mold. Some reviewers thought there was too much software and not enough hardware. Some thought there was too much hardware and important software issues were ignored. Some preferred that hardware be covered before software, and others liked software before hardware. No matter what the opinion, I wish to thank the reviewers for their comments, and for helping to identify errors of commission and omission, If I ignored some sage advice, it was because there was an equal amount of sage advice on the other side of the argument. The reviewers are:

Farrokh Attarzedeh	University of Houston (TX)
Walter Bankes	Rochester Inst. of Tech. (NY)
William E. Barnes	New Jersey Inst. of Technology
Wayne Dean	Southern Alberta Inst. of Technology
Melvin Duvall	Sacramento City College (CA)

Bruce Hutchinson	Mohawk College (Ontario)
Jeong-A Lee	University of Houston (TX)
Sura Lekhakul	St. Cloud State Univ. (MN)
Michael Lynch	University of Florida
Robert E. Martin	Northern Virginia Comm. Coll.
A. R. Marudarajan	Calif. State Polytechnic Univ.-Pomona
Alan Moltz	Naugatuck Valley Comm.-Tech. Coll. (CT)
Scott Moser	Capitol College (MD)
Gordon Murphy	Northwestern University (IL)
W. Shields Neely	Auburn University (AL)
Ron Nichoel	College of DuPage (IL)
Patrick O'Connor	DeVry Inst. of Technology-Chicago
James A. Resh	Michigan State University
Dennis W. Suchecki	San Diego Mesa College (CA)
Thomas Schultz	Purdue University (IN)
Donald A. Smith	SUNY College of Technology-Morrisville
Wayne J. Vyrostek	Westark Comm. Coll. (AR)
Jean Walls	Mohawk Valley Comm. Coll. (NY)
Michael Walton	Miami-Dade Community College (FL)
George W. Washburn	West Kentucky State Vo-Tech
Jefferey B. Weaver	Pennsylvania College of Technology
James Eris Reddoch	Alara Lee Hildebrand

Finally, my heartfelt thanks to Eric Isaacson, who was my severest, and most helpful, critic. He accurately exposed my errors and attempted to correct many long-held misconceptions. I hope that instructors and students find Eric's A86 assembler and D86 debugger to be as useful and easy to use as I have. But, please remember! The software included with this book is *shareware*. Legally, and ethically, if you use this book, you should buy registered copies of the software, or a site license, from Eric. See Appendix A or B for instructions on how to contact him.

Kenneth J. Ayala
Western Carolina University
Cullowhee, North Carolina 28723

1

A Short

History

of Computing

Chapter Objectives

On successful completion of this chapter, you will be able to:

- Describe the historical development of calculation methods.
- Describe the development of digital computer technology.
- Name the people and companies who have made major contributions to the development of computer technology.
- Understand two ways microcomputers are used and what the employment opportunities are in the field of microcomputer programming.

1.0 Introduction

The history of mankind is usually written with a bias toward seemingly grand events: the rise of nation-states, royalty, politics, war, exploration, and religion. A student of history might easily be able to recite the dates and causes of certain ancient Greek wars but be completely unaware of when the number *zero* was invented.

Nations wax and wane, civilization advances and declines, empires rise and fall, yet one human activity has always increased in power: the ability to calculate. The ability to count (enumerate) is one of the most powerful inventions of human history. Only language and writing might be more important than the concepts of numbers, number systems, and arithmetic.

Numbers

Numbers were invented to help people determine how many things they owned, owed, commanded, built, killed, or begat. The novel idea of owning land led to surveying practice involving rudimentary trigonometry and plane geometry. Because the heavens were thought to influence human fate, the study of the planets and stars drove early astronomers to further developments in arithmetic and geometry. Also, as the science of architecture developed, numbers and means of numeric calculation were needed to handle the challenge of building multistory palaces and shrines.

Number Systems

Number systems were invented in Mesopotamia, Egypt, China, and India to help in the development of early civilized states. Nations that depend on the ownership of land as a basis for wealth have to develop ways to measure and record the extent of land wealth. Trade depends on accounts that show how much was bought or sold. Warfare demands a good estimate of enemy strength, in warriors and horses. Numbers are an essential basis for all civilizations.

The early numbers industry employed countless scribes, whose sole duty was to perform arithmetic and keep records. As time passed, more efficient means were sought by which numbers could be manipulated. The only mechanical aid to calculation was the abacus, probably an invention of the Babylonians around 2000 B.C.

Positional number systems, zero, negative numbers, fractions, and the operations of addition, subtraction, multiplication, and division were all well known to the mathematical elite by the early Middle Ages. Using pen and sheepskin, people from clerks to astronomers were able to deal with the problems of the day. The astronomers, however, were beginning to run into problems with the enormous distances involved in planetary calculations. On earth, the age of exploration opened up new problems for ships' navigators as sailing distances increased.

Logarithms

The invention of logarithms by John Napier and Henry Briggs in the 17th century permitted calculations involving very large and very small numbers to be done much faster than was previously possible. Many hours of labor (most of Briggs's life) were devoted to calculating logarithmic tables and to extending the significant digits in the tables.

Mechanical calculating devices, such as "Napier's bones" developed by Napier, also appeared during the 17th century. The slide rule (based on marking off distances on two sticks in logarithmic lengths) was invented by William Oughtred in 1621. Logarithms and devices based on logarithms continued as indispensable mathematical tools well into the middle of the 20th century in the form of log tables and slide rules.

The invention of logarithms did not, however, solve the problems associated with ever-increasing numeric complexity and civilization's increasing volume of calculations. The industrial age had multiplied human *muscle* power tremendously. Steam, iron, and electricity had mechanized civilization. One worker, with the aid of the new machines, could do the work that previously took hundreds. But calculations continued to be done manually: pencil, paper, slide rule, and log tables. Mechanical adding machines began to appear in the early part of the 20th century, but these were slow and still depended on human operators. Some radical new way was needed to enable humans to mechanize mathematics.

Mechanical Calculators

Early visionaries—such as Ada Byron, Countess of Lovelace (for whom the high-level language ADA is named), and Charles Babbage in 1833—had attempted mechanical "calculating engines." But the technology of the early 1800s could not produce the components needed for the task. Herman Hollerith was more successful; his punched-card method of recording, tabulating, and sorting data proved functional by the 1890s. Hollerith's Tabulating Machine Company would, in 1924, become the International Business Machines Corporation, or IBM. At the same time Hollerith was developing his punched-card system, another inventor, William Burroughs, was perfecting the mechanical printing calculator. The Burroughs Adding Machine Company would come to dominate the office machine business by the beginning of the 20th century.

Punched-card machines and calculators were very adept at sorting and counting data and performing arithmetic. But mechanical devices were inherently speed-limited by their electromechanical nature. By the 1920s, however, a new technology called *electronics* began to develop the necessary component to make a workable high-speed calculating engine: the vacuum tube. And when vacuum tubes became sufficiently reliable, the mathematicians, as usual, had already invented the tools needed to design the calculating machine we now call a *digital computer*.

1.1 Theoretical Foundations of Computer Design

Boolean Logic

In 1847 a British mathematician named George Boole[1] produced a "mathematics of logic." Boole attempted to reduce the process of logical thinking to

---

[1]George Boole, *The Mathematical Analysis of Logic* (Cambridge U.K., 1847).

a system of algebra. He succeeded, and we now call his algebra *Boolean algebra* or *Boolean logic.* Boolean algebra is *not* an algebra of binary systems. Boolean algebra is an algebra of propositions and statements.

Boole's work, as so often happens, was soon forgotten by all but a few specialists and laid to rest in the "attic of mathematics."

Then in 1938, a candidate for a master's degree at the Massachusetts Institute of Technology, Claude Shannon[2], was casting about for a tool to use in the design of relay systems. The electromagnetic relay was, at the time, the primary component used for telephone switches and automatic control systems. Relay circuit design was mainly a discipline of trial and error. Relays are switches, and Shannon wanted to find some way to systematically design switching systems.

Shannon incorporated Boole's earlier work (as updated by Bertrand Russell and Alfred Whitehead in their book *Principia Mathematica*) on the algebra of logic and adapted Boolean algebra as a design tool for relay systems. The Boolean algebra operations of NOT, AND, and OR became the building blocks for relay logic systems. Shannon completed his master's degree and went on to a brilliant career in the field of communication theory. His work on Boolean switching logic, however, did not go unnoticed.

1.2 Early Computers

In 1943, the world was at war. One pressing war need, in the United States, was some method to calculate the trajectories of long-range gun projectiles. Shell trajectories, it turns out, are not the simple ballistic calculations of elementary physics. Air density, wind, temperature, and the earth's rotational effects all serve to complicate the physics of real projectile flight.

ENIAC and Mark I

A revolutionary project to develop an electronic calculating machine was begun in 1943 at the University of Pennsylvania. The computer, dubbed ENIAC (for Electronic Numerical Integrator and Computer), was built of vacuum tubes and relays. ENIAC weighed 30 tons, consumed 174 kilowatts of power, was approximately 80 feet by 80 feet in size, and required 18,000 vacuum tubes to build. ENIAC cost approximately $10,000,000 1992 dollars to build.

Another wartime digital computer, the Mark I, was built at Harvard University. The Mark I, originally backed by IBM, was built of electromechanical counters and relays, was 51 feet long and 8 feet high, and weighed in at a meager 5 tons. One of the project designers was Grace M. Hopper, who later developed the COBOL business programming language. The Mark I, although the first program-controlled automatic calculator, was obsolete at birth; the electromechanical calculator could not keep pace with electronics.

---

[2]Claude E. Shannon, "A Symbolic Analysis of Relay and Switching Circuits," *Trans. AIEE* 57 (1938):713–723.

1.3 Early Mainframes

EDVAC and UNIVAC

After World War II ended, the ENIAC designers, J. P. Eckert and J. W. Mauchly, worked on a successor to ENIAC called the EDVAC. They formed the Eckert-Mauchly Computer Corporation to develop the first successful commercial computer, the Universal Automatic Computer or UNIVAC. In 1950, Remington Rand acquired the firm, and Remington Rand was, in 1955, acquired by Sperry Corporation.

Mauchly and Eckert, along with John Von Neumann[3], are generally credited with the development of the electronic stored-program digital computer.

During the late 1940s and early 1950s, other fledgling computer operations were formed, including Engineering Research Associates, whose founder, William Noris, went on to found Control Data Corporation (CDC). The federal government was also actively engaged in developing computers for national defense, including project Whirlwind at MIT, and LARC at Livermore Labs.

Control Data

It was at CDC that a brilliant computer designer, Seymour Cray, first gained national recognition. CDC became a principal supplier of large, high-speed computers in the 1960s. Cray later formed the supercomputer company that still bears his name.

The first generation of postwar computers were large, expensive, and remarkably unreliable. With all these problems, however, the early computers were better than any manual counterpart.

The Transistor

At the same time as the early computer companies were forming, another important and related event took place at Bell Laboratories in 1948: the invention of the solid-state transistor by John Bardeen, Walter Brattain, and William Shockley. Astonishingly, Bell did not retain sole rights to the new invention. Commercial models of the new transistor were soon being manufactured, under license from AT&T, by GE, RCA, Westinghouse, Raytheon, and Western Electric.

The transistor would prove to be the technology that would change the bulky, expensive, vacuum-tube based digital computer into a reliable, faster, smaller, and ultimately, cheaper solid-state machine. By the end of the 1950s IBM, Philco, and Sperry had all brought out cheaper and more reliable computers based on the new transistor technology.

IBM

IBM, which had made its mark in office machines and punch-card sorting equipment, was not part of the pioneering electronic computer effort. IBM caught up quickly by participating in the Whirlwind project at MIT and later

---

[3]John Von Neumann, *First Draft of a Report on the EDVAC,* Applied Mathematics Panel, 1945.

in the huge air-defense computer system named SAGE. In 1953, IBM brought out its first vacuum-tube computer model, the IBM 701. Some IBM managers thought they might sell a half-dozen or so 701 models; they sold 19.

The next IBM model, the 650, was predicted by many inside IBM to be as disappointing as the 701. Two hundred and fifty sales of the 650 were predicted by one IBM optimist, C. C. Hurd; 1,500 were sold over a production run of 15 years. The IBM model 650 launched the age of mass-produced commercial computers and helped place IBM in the forefront of the *mainframe computer* business.

The commercial world welcomed the new computers. By the 1950s, governments, insurance companies, and banks were drowning in a sea of paper and paperwork. The defense industry was also finding itself hampered by a lack of computing power needed to design ever more sophisticated weapons. Telephone traffic was straining the bounds of mechanical switching centers. Manual computing means could not keep pace with the demands of the day, calculations had to be automated, and the new computer companies were ready to fill the need.

By the early 1960s the mainframe computer was a mature and established part of modern corporate equipment. Companies such as Burroughs, GE, Honeywell, NCR, RCA, CDC, Philco, IBM, and Sperry had products available to meet the needs of industry, education, and government.

Computers began to become a subject of discussion in academe and on Wall Street. Digital computers began to be noticed too by the general public, through the popular media of science-fiction movies and the more prosaic "computer numbers" inked on bank checks. Soon the term *computer error* appeared in the mass media. Prestigious universities began to acquire computers, computer centers, and computer operators. The study and employment fields of computer science and computer engineering began to emerge from departments of mathematics and electrical engineering. Digital computers, based on discrete transistor and, later, integrated circuits, established the mainframe computer as essential to business, industry, and education.

The mainframes of the 1960s and 1970s were expensive, powerful (for the times), large, and required a small army of computer operators, maintenance personnel, and administrators to produce results. The average person had no access to computing power—nor seemed to need such access.

The Integrated Circuit

Meanwhile, in 1958, at a small company named Texas Instruments located in Dallas, Texas, the first integrated circuit was built by Jack Kilby. The discrete resistor-transistor circuit was on the way to becoming a single integrated circuit. Integrated circuits made an immediate impact on electronics as digital logic gates and operational amplifiers were reduced from a boardful of discrete components to a single component. Integrated-circuit digital logic gates would lead to the minicomputer, and then to the microcomputer.

1.4 Minicomputers

DEC

In 1965, a small, unknown company named the Digital Equipment Company (DEC) introduced a new type of digital computer designed under the leadership of Gordon Bell—the DEC PDP-8. The DEC PDP-8 was unique because of its relatively small size and low cost. DEC, founded in 1957 by Kenneth Olsen, had become known in digital logic design circles as a manufacturer of digital logic circuit cards. DEC had also been engaged in the design of a special-purpose computer for the U.S. space exploration program, the Programed Data Processor, or PDP-1. The PDP-1 required no special room or army of personnel to run and sold for the rock-bottom price of $120,000.

Building on its experience with later PDP models 2 through 6, DEC produced the PDP-8 as a small, "inexpensive" computer aimed at users in the engineering and scientific world. The original PDP-8 was made up of discrete transistors and sold for the then unheard-of low price of $18,000. Unlike the mainframes, the PDP-8 was intended for single users running a single program. The PDP-8, and the succeeding integrated-circuit based PDP-11, soon made DEC a scientific and engineering computing household word. The PDP-11 had more power and sold for less than the PDP-8, thus opening the computer world to many users who had not been able to afford the pricey offerings of the established mainframe names.

DEC—and later minicomputer firms such as Data General, Hewlett-Packard, Gould, Texas Instruments, and Computer Automation—began producing relatively low cost computers that opened up computing to more people. Mainframe makers ignored the new market for small computers until DEC had a commanding lead in the marketplace.

The Role of Minicomputers and Mainframes

To differentiate between computer types, the term *minicomputer* was coined at DEC. Mainframe computers are expensive and massive machines that require special air-conditioned rooms and constant operator and maintenance attention. Figure 1.1 shows a picture of a typical mainframe, or "large-scale" computer. Figure 1.2 shows a photograph of a typical minicomputer.

Mainframe computers are the steam shovels of computers and typically are used to process massive amounts of data, such as bank statements, airline reservations, air-defense coordination, payroll checks, college records, and complex engineering and scientific calculations. Mainframes can also be operated in a *multi-user* mode, that is, many people can access the computer at the same time, each running a different program.

Most of the tasks that were formerly reserved for mainframes have only now been taken over by minicomputers and microcomputers. One reason for the decline of IBM's mainframe business in the early 1990s has been the replacement of mainframes by minicomputers and microprocessors.

Mainframe computing is also characterized by extensive use of *off-line data*, in a *batch mode*. Off-line means that the data processed by the computer has been stored before the computer program is run. Batch mode means that only one type of data and one program, for instance printing payroll checks, is run at a time. Mainframe computers have the memory and printer peripheral resources to handle computing chores that require processing large amounts of data and paper.

The reverse of off-line is *real time*. Real-time computing means that the processed data is gathered at the time it is needed by the computer program. Some mainframe applications are run in real time, as for airline reservation systems, air defense, and telephone switching.

Figure 1.1 ■ A Mainframe Installation

Figure 1.2 ■ A Contemporary Minicomputer

Minicomputers—because of their relatively low cost, unattended operation, and ruggedness compared to mainframe computers—began to be used in real-time applications as well as off-line applications. Computer control of chemical processes, paper-making machines, robots, spacecraft, milling machines, and other automated activities became commonplace as minicomputers became cheaper and more powerful. By the mid 1970s, minicomputers, based on integrated-circuit logic gates, were the dominant small computer. The price of a minicomputer, however, at around $5,000 or so, still kept a lot of people out of the computer marketplace.

1.5 The Microcomputer Revolution

Intel

Minicomputers dominated the engineering and small scientific application fields until the middle of the 1970s. Then, in the year 1971, another unknown electronics company, Intel Corporation, introduced the first integrated-circuit central processing unit, the 4004, designed by Marcian E. Hoff, Jr.

A computer *central processing unit,* or CPU, is the core of a computer system. Up until Intel's offering, CPUs were built from hundreds of integrated-circuit logic gate packages (chips) that were connected to form CPU circuits. Intel's integrated CPU put all the circuits into a single chip package using a new solid-state technology, MOS. Older integrated-circuit logic designs used bipolar TTL, which was faster than MOS but required more power, space, and, most importantly, money.

Integrated circuit technology had progressed steadily since the first Texas Instruments circuit. Using photolithography techniques that continue to make circuit details smaller and smaller, integrated circuits progressed from single transistors to complete circuits. The circuits became more and more complex, and soon, entire electrical functions could be found inside one package. Intel placed 2,000 transistors on a chip to make the 4004 possible.

Very complex integrated circuits began to appear in the late 1960s and early 1970s, as markets opened up for electronic CRT terminals, data transmission circuits, instruments, and the newest craze: calculators. It was a short step from a calculator to a CPU, and Intel made it first.

The 8080

Intel soon followed the 4004 with a more powerful model, the 8008, in 1972. The 8008 and a later version, the 8080, helped make the new term *microcomputer* recognizable to the general public. Soon Motorola, RCA, Texas Instruments, Zilog, and Rockwell International had competing microprocessors in the market. Figure 1.3 shows a photograph of several early microprocessor CPUs.

Microcomputers were incredibly inexpensive, rugged, and small. Early CPU models sold for as little as $100, and a complete computer could be built for under $2,000. Interestingly, small start-up computer companies, computer hobbyists, and game and toy manufacturers were the first to adopt the new

Figure 1.3 ■ Early Microprocessor CPUs

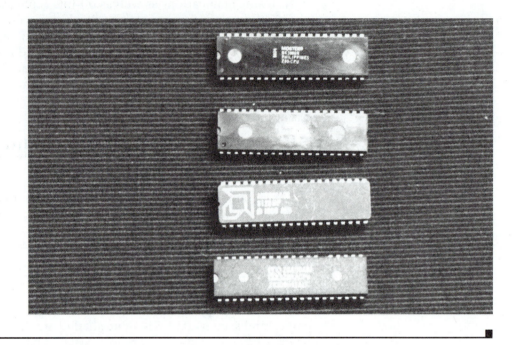

microprocessors—and to use them. The established mainframe and minicomputer companies ignored the potentials of the microprocessor.

Computer hobbyists, or *hackers* as they became known, were a vital driving force in the early days of microcomputer technology. The first microcomputer was marketed by Ed Roberts's MITS company as a kit, the Altair, in 1975. A surprisingly vigorous market developed for microcomputers, and small companies such as IMSAI, Cromemco, North Star, and Ohio Scientific appeared to serve the hacker community. Another start-up company, Apple Computer, offered a sophisticated (for the times) business-oriented microcomputer named the Apple II in 1977.

The 8086 and the IBM PC

The microcomputer business boomed, and in 1978 Intel introduced a much improved CPU, the 8086. Three years later IBM tentatively entered the microcomputer fray with the first IBM personal computer, the PC. IBM managers, in a virtual replay of the 1950s, thought they might sell a few hundred thousand PCs. They sold 300,000 the first year, and 600,000 the next. Figure 1.4 displays a picture of the first IBM PC.

IBM's successful entry, and the rapid growth of Apple, established an entirely new kind of computer and computer business. Computing was now open to millions of customers. Software became an everyday topic of conversation. Tiny, unknown software companies, such as Microsoft and Lotus, became "overnight successes." Personal computer brands began to proliferate; soon personal computers became a commodity, sold in department stores and by mail. Personal computers that are much more powerful than mini-computers of 10 years ago can now be purchased for under $300.

Microprocessors enable almost everyone to have access to very inexpensive computing power. Small businesses, grammar school children, office workers, and many others have benefited, and continue to benefit, from the era of cheap computing. A typical contemporary PC microcomputer is shown in Figure 1.5. Today's PC has more power than did most mainframes of the 1960s—at one-thousandth of the cost and space.

1.6 Microcomputer Applications

Computers may be designed to be used for *general-purpose* or *special-purpose* applications. General-purpose computers are equipped to run many different types of programs. Special-purpose computers are optimized to run a single program.

General-Purpose Computers

General-purpose computers are the types that the computer-buying public associates with the word *computer*. General-purpose computers are

Figure 1.4 ■ The Original IBM PC

Figure 1.5 ■ A Modern PC

equipped with various input-output peripheral devices, such as keyboards for input and printers for output. Mass storage devices, such as floppy and hard disk drives, provide a medium by which different programs may be placed into a general-purpose computer and run. General-purpose computers range in size and power from the personal desk computer to giant super computers that occupy entire buildings. The most popular general-purpose computer is the desktop personal computer that is found in most offices and classrooms.

General-purpose computer usage requires operators that are skilled in using *application programs.* Application programs—which include word processors, spreadsheets, computer-aided drafting (CAD), computer-aided manufacturing (CAM), and desktop publishing—can only be used effectively by persons skilled in their use.

Special-Purpose Computers

Special-purpose computers go unnoticed by the public, because they are often "buried" inside a larger entity, such as a microwave, a robot, an automobile, or a milling machine. Special-purpose computers are equipped to run a single program and often do not have printers, CRT screens, or keyboards. Figure 1.6 shows a picture of a special-purpose computer used to monitor the operations of cloth-weaving looms in a large textile mill.

Special-purpose computers are often used to operate automatic machinery—such as in robots, spacecraft, radios, VCRs, airplanes, metal-working machines, chemical processes, printing presses, check sorters, weaving looms, cigarette makers, sewing machines, video games, motion-picture cameras, scientific experiments, automatic bank tellers, telescopes, submarines, bakery ovens, laser printers, guided missiles, and almost any application in which automated intelligent control is cost-effective.

Special-purpose computers can range in size from 1-chip *microcontrollers* found in appliances and automobiles, to minicomputers and personal computers used to control industrial machinery and processes.

Typically, no special computer or program skills are needed to operate a special-purpose computer. One reason that a machine becomes equipped with a special-purpose computer is so that its operation by the user may be made simpler.

Figure 1.6 ■ Weaving Monitor Computer

1.7 Employment Opportunities in Small Computer Applications

The name *small computer* encompasses both the desktop-style personal computer and the breadbox-sized special-purpose computer. Because of the ever-increasing computing capabilities of microprocessor CPUs, however, one cannot equate "small" size with reduced computing power.

Employment in the area of general-purpose computers is divided between those who use application programs and those who write application programs. The number of users of application programs far outweighs the number of persons who write application programs.

One trend in application programs has served to diminish the number of persons needed to write them: all-purpose application programs. At the start of the personal computer revolution, users had to write, or have written for them, their own programs to fit their particular needs. Many programmers found employment in writing great numbers of unique programs for many different customers.

All-purpose application programs, however, can be tailored by the customer to fit almost any conceivable need. Once written, the same all-purpose program can be sold to many different types of customer. A spreadsheet is a good example of an all-purpose application program. Spreadsheets allow many, interrelated computations to be done in any manner specified by the user. Originally intended for financial analysis and reporting, spreadsheets have also been adapted to perform engineering calculations and other uses not thought of when the spreadsheet was written. The popularity of general-purpose, one-size-fits-all application programs has reduced the need for general-purpose programmers.

Special-purpose computer applications, however, continue to require increasing numbers of programmers who can write machine-control application programs. Because machine-control applications are tailored to the machinery involved, special-purpose application programs are also unique and cannot generally be re-used in another type of machine. The world continues to automate, and demand is increasing for persons who are skilled in applying special-purpose computers to solve automation problems.

1.8 Assembly Code Programming

Solving problems using special-purpose computers often requires that portions of the computer program be written in what is known as CPU *assembly code*. Every CPU is designed to use its own unique assembly code, yet all assembly code languages share common features. Assembly code programming is the subject of this book, using the 8086 CPU as an example of a modern microprocessor.

You can use a general-purpose computer, such as a PC-compatible model, to learn how to program the 8086 CPU in assembly code. To help in the mechanics of assembly code programming, this book includes two application programs called an *assembler* and a *debugger*, which greatly speed up the learning process (see the appendixes). Assembly code programming skills you acquire from using a general-purpose computer can be transferred to special-purpose computer applications with ease.

Bibliography

Augarten, Stan. *Bit by Bit: An Illustrated History of Computers.* New York: Ticknor & Fields, 1984.

Ayala, Kenneth. *The 8051 Microcontroller.* St. Paul, MN: West, 1991.

Bergamini, David. *Mathematics.* New York: Time-Life, 1963.

Evans, Christopher. *The Making of the Micro.* New York: Van Nostrand Reinhold, 1981.

Moore, Doris. *Ada, Countess of Lovelace.* New York: Harper & Row, 1977.

Peak, Martha, ed. "Reflections—An Oral History of the Computer Industry." *Computer System News*, November 1988.

Rotsky, George, ed. "Fifteen Years That Shook the World." *Electronic Engineering Times*, November 1987.

Rotsky, George, ed. "The 30th Aniversary of the Integrated Circuit—Thirty Who Made a Difference." *Electronic Engineering Times*, September 1988.

Shurkin, Joel. *Engines of the Mind.* New York: W. W. Norton, 1984.

2

A REVIEW

OF BINARY

ARITHMETIC

Chapter Objectives

On successful completion of this chapter you will be able to:

- Describe the differences between symbolic and positional number systems.
- Understand how positional number magnitudes are formed.
- Calculate the decimal equivalent of a nondecimal base number.
- Convert numbers from decimal to binary and hexadecimal, and the reverse.
- Understand the concept of signed complementary number systems.
- Structure a complementary signed number system of any size.
- Perform addition and subtraction operations using signed complementary numbers.
- Understand the meaning of an overflow in signed number mathematics.
- Perform multiplication and division of unsigned binary numbers.
- Describe some of the uses for the ASCII, BCD, Gray, and Parity error binary codes.

2.0 Introduction

The human need to count things goes back to the dawn of civilization. How many sheep did one possess? More ominously, how much land was owned, and how much tax was due? Most urgently, what was the size of the opposing army? To answer the questions of "How much" or "How many," people invented number *systems*. A number system is any scheme used to count things. There are two general types of number systems, *symbolic* and *positional*. Positional number systems are the most popular number systems in use today and have been for hundreds of years.

2.1 Symbolic Number Systems

The symbolic number system uses *distinct* symbols, called *numerals*, to construct larger numbers. The most famous symbolic number system, in the Western world, is the Roman numeral system. The Roman system has seven distinct numerals from which all larger numbers can be constructed. These and their decimal number equivalents are shown in Table 2.1.

A symbolic number system, such as the Roman, expresses large numbers by adding together a string of numerals. For instance, to express the decimal number 27 in Roman numerals results in the Roman number XXVII, or X + X + V + I + I. The Roman number for decimal 6,000 is obtained by placing together six Roman symbols for 1,000, MMMMMM, or M + M + M + M + M + M.

The Roman system forms large numbers from seven numerals by placing numerals of the same or increasing size to the left of each other, as in our previous examples. If a smaller number were placed to the left of a larger number, however, then the smaller is subtracted from the larger. For instance, the Roman number for decimal 9 is IX (10 − 1), not VIIII (5 + 1 + 1 + 1 + 1).

Clearly, the Roman numbering system is not convenient for expressing large numbers and certainly must have been a nightmare for students learning the Roman numeral multiplication tables. Moreover, there was no Roman symbol for zero, a much later invention of clever Arabic mathematicians. The Arabs, in turn, had gotten a very innovative numbering system

Roman Number	Decimal Number
I	1
V	5
X	10
L	50
C	100
D	500
M	1,000

Table 2.1 ■ Roman and Decimal Numbers

from India that uses the *position* of a numeral to express its *size*. The Indians had gotten the idea of a positional numbering system from the Mesopotamians, who had also decided to use *sixty* numerals! The Indians pared the number of numerals down to nine, the Arabs added the symbol for zero, and the positional decimal number system was introduced to Western Europe, via Moorish Spain, around the 11th century. The Arabic system became the *decimal* number system, from the Latin name for *tenth*, "decima."

The computational advantages of a decimal positional number system slowly gained acceptance in the Western world, starting with commercial companies located in Italy and Germany. The positional decimal number system has now entirely supplanted the symbolic Roman numerals in our society except for use in books, buildings, and the Superbowl.

2.2 Positional Number Systems

A positional number system is one in which a set of numerals form the basic counting symbols, or *digits*. The decimal number system, for instance, has 10 numerals: 0, 1, 2, 3, 4, 5, 6, 7, 8, and 9. Numbers larger than a single digit are expressed by arranging the numerals from right to left in a row. The position of a digit, in the number, is counted from right to left with the first position counted as position zero.

The total number of numerals used for a given positional number system is the *radix* (r) of the system. The radix of the decimal numbering system is 10, the total number of decimal numerals. The radix is used to make up numbers in the system that are greater than a single digit. The total value of a positional number is found as follows:

Count the position of a digit, starting at the right and numbering each position toward the left as position number 0, 1, 2, 3, and so on.

Raise the radix of the system to a power equal to each position number, as r^0, r^1, r^2, r^3, and so on.

Multiply the digit found at each position by its associated raised radix.

Add the resulting numbers together.

For instance, let us investigate the positional decimal numbering system, which has 10 numerals. Ten seems like a goodly number of digits, because humans are equipped with 10 fingers, and 10 numeric symbols are easy to learn. The 10 numeric symbols chosen for the Arab-Indian system are, in increasing order of size, 0, 1, 2, 3, 4, 5, 6, 7, 8, and 9. Ten digit symbols results in a radix of 10, which is a number made up of numerals 1 and 0. There is *no* numeral *10* in our system. Now, we construct a small table of the radix 10 raised to positional powers, as shown in Table 2.2.

Note that the first position is position number 0, *not* position number 1. Any number (except 0) raised to the power of 0 is a 1. Many grammar school children learn to express large decimal numbers by memorizing a few of the powers of 10 from Table 2.2, such as the "1's place," "10's place," "100's place," and so on.

We may now construct numbers in the decimal system using the 10 deci-

Position	Power of 10
0	1
1	10
2	100
3	1,000
4	10,000
5	100,000
6	1,000,000
7	10,000,000
8	100,000,000

Table 2.2 ■ Powers of the Radix 10

mal numerals and Table 2.2. For instance, the decimal number 654,321 is constructed as follows:

Numeral	Position	Numeral × Radix^{Position}
1	0	$1 \times 10^0 = 000001$
		+
2	1	$2 \times 10^1 = 000020$
		+
3	2	$3 \times 10^2 = 000300$
		+
4	3	$4 \times 10^3 = 004000$
		+
5	4	$5 \times 10^4 = 050000$
		+
6	5	$6 \times 10^5 = \underline{600000}$
		654,321

The decimal number system has succeeded because very large numbers can be expressed using relatively short series of easily memorized numerals. Systems with a radix larger than 10 require additional numeric symbols but do result in more compact numbers than the decimal system. Systems with a radix smaller than 10 do not require as many numerals but result in longer numbers. Humankind also uses number systems based on a radix of 20 ("Four score and seven years ago . . ."), 60 (60 seconds in a minute, 60 minutes in an hour), and, since about 1940, a radix of 2.

Positional numbers permit the number system *inventor* to choose any radix and create any desired *symbol* for each numeral in the system. It is usually most convenient to pick from the decimal numeral symbols for number systems with a radix up to 10. For systems with a radix larger than 10, new symbols have to be invented, the most convenient being letters of the alphabet.

For instance, suppose that a system with a radix of 5 is chosen. A radix of

5 means that there must be five digits in the system, equivalent to decimal 0 to decimal 4. If we use decimal digits for our radix-5 system, then the number 12344 radix 5 has the following decimal number system equivalent:

Position	Numeral	$5^{Position}$	Decimal Equivalent
4	1	$\times \quad 5^4 \quad =$	$1 \times 625 = 625$
			+
3	2	$\times \quad 5^3 \quad =$	$2 \times 125 = 250$
			+
2	3	$\times \quad 5^2 \quad =$	$3 \times \; 25 = \; 75$
			+
1	4	$\times \quad 5^1 \quad =$	$4 \times \quad 5 = \; 20$
			+
0	4	$\times \quad 5^0 \quad =$	$4 \times \quad 1 = \quad \underline{4}$
			974 radix 10

To convert from nondecimal radix number A to decimal radix number B, multiply each digit of number A by its radix raised to the appropriate positional power. The product of each multiplication is expressed as a *decimal* number, and the resulting individual decimal numbers are added together to form the final, equivalent decimal number.

The reverse process, going from a decimal radix number back to a radix 5 number, involves division. From the previous example, 974 decimal is converted back to a 12344 radix 5 as follows:

Dividend ÷ 5	Quotient	Remainder	
$\dfrac{974}{5}$	= 194	Remainder of 4	(least significant digit)
$\dfrac{194}{5}$	= 38	Remainder of 4	
$\dfrac{38}{5}$	= 7	Remainder of 3	
$\dfrac{7}{5}$	= 1	Remainder of 2	
$\dfrac{1}{5}$	= 0	Remainder of 1	(most significant digit)
		12344	

The process of converting decimal radix number B to a different radix number A is to divide decimal number B by the radix of number A and keep the remainder as a digit of the new number. The quotient from the first division is divided by the A radix, and the process repeated until the quotient is zero. The first division produces the *least significant digit* (LSD) of number A, and the last remainder is the *most significant digit* (MSD) of number A.

2.3 Integer Binary Numbers

Sixty years ago, the practice of performing mathematical operations using binary numbers would have been thought absurd. To be sure, the mathematics of positional number systems had been well developed since the 9th century, but a positional number system based on a radix of 2 would likely have been thought an odd curiosity 60 years ago.

There is nothing new, or "modern," about binary numbers or binary mathematics. Binary numbers follow the same rules for arithmetic as does the more familiar decimal number system. Performing numerical calculations using binary numbers is no more difficult than performing calculations using decimal numbers. The challenge of binary arithmetic is to become comfortable with a system that has only two numerals—and to become comfortable with another system, called *hexadecimal*, that has 16 numerals.

Binary Digital Computers

So, the question might be, what happened 60 years ago that requires us to deal with binary numbers? The answers to that question are: the digital computer and the integrated circuit. Binary numbers and binary math are used today because humans have invented the digital computer and developed very *reliable* semiconductor electronic switches with which to build digital computers.

To date, reliable semiconductor switches must operate either turned on or turned off; they are inherently binary devices. Binary has evolved as the electronic digital system of choice because it is fairly easy to build reliable electronic switches that are either turned fully on or turned fully off. It is very difficult to build *reliable* electronic switches that can be on, half-on, and off, for instance. Electronic technology drives digital computer development, and electronic technology has progressed to the point at which extremely reliable and inexpensive binary electronic solid-state switches can be made.

The ability to make low-cost, reliable solid-state switches has, in turn, led to the development of very *inexpensive* digital computers. The availability of inexpensive digital computers has led to *increasing* employment demand for individuals who know how to use computers to solve society's problems. The ability to use a digital computer, particularly for engineering applications, requires a knowledge of binary numbers. Digital does *not* mean binary. Every number system has digits, and a computer based on a four-digit number system would also be a digital computer.

We normally study number systems in grammar school. After some confusing exposure to the "new" math, we settle down to using one number system—the decimal number system. Decimal numbers become so familiar to us that we begin to think they are the only numbers possible. Thus we find ourselves, many years past grammar school, beginning all over again with the "new" math.

The Binary Number System

A binary number system uses two decimal numerals, 0 and 1. A binary number is made up of a collection of binary numerals using a radix of 2. Binary digits are also called *bits*, which is a contraction of the words *Binary* and *digITS*.

In an electronic circuit, 1 and 0 might correspond to a switch that is off or on, or to a circuit that has a high voltage or a low voltage output. The assignment of binary 1 to off, for example, or binary 0 to a high voltage is arbitrary. A binary 1 may correspond to a switch that is off, or one that is on. Standards for the computer industry, however, have generally followed the practice that a circuit that is in a high-voltage state is a binary 1, and one that is in a low-voltage state is binary 0.

Conversions Between Decimal and Binary Numbers

The decimal equivalent of a binary number is formed by multiplying each bit in the binary number by the binary radix of 2 raised to the position's power. The result of each multiplication is expressed as a decimal number. The individual decimal numbers are added to obtain the decimal equivalent of the binary number. For instance, the binary number 111110100 can be converted to a decimal equivalent number as follows:

Digit	Position	$2^{Position}$	Decimal Equivalent
1	8	256	$1 \times 256 = 256$
1	7	128	$1 \times 128 = 128$
1	6	64	$1 \times 64 = 64$
1	5	32	$1 \times 32 = 32$
1	4	16	$1 \times 16 = 16$
0	3	8	$0 \times 8 = 0$
1	2	4	$1 \times 4 = 4$
0	1	2	$0 \times 2 = 0$
0	0	1	$0 \times 1 = \underline{0}$
			500

The reverse process, that of converting a number with a decimal radix to a number with a binary radix, is done by repeated division of the decimal number by the binary radix of 2. The decimal number is repeatedly divided by 2, and the remainder becomes a bit of the equivalent binary number. Any decimal number divided by 2 must have a remainder of 0 if the decimal number is even, or 1 if the decimal number is odd.

The remainder of the first division operation yields the *least significant bit* (LSB) of the binary number. The number left after the first division by 2 (the first quotient) is again divided by 2, yielding a second remainder and a second quotient. The second remainder is the next most significant bit of the binary equivalent number. The process of dividing the quotient by 2 and keeping the remainder as a bit of the binary equivalent number is continued until the *quotient* is zero. The final remainder is the *most significant bit* (MSB) of the binary equivalent number.

To demonstrate conversion from a decimal number to an equivalent binary number, we shall take the decimal 500 number from the last example and reconvert it to binary.

Dividend ÷ 2	Quotient	Remainder

$$\frac{500}{2} = 250 \text{ Remainder} \longrightarrow 0 \quad \text{(LSB)}$$

$$\frac{250}{2} = 125 \text{ Remainder} \longrightarrow 0$$

$$\frac{125}{2} = 62 \text{ Remainder} \longrightarrow 1$$

$$\frac{62}{2} = 31 \text{ Remainder} \longrightarrow 0$$

$$\frac{31}{2} = 15 \text{ Remainder} \longrightarrow 1$$

$$\frac{15}{2} = 7 \text{ Remainder} \longrightarrow 1$$

$$\frac{7}{2} = 3 \text{ Remainder} \longrightarrow 1$$

$$\frac{3}{2} = 1 \text{ Remainder} \longrightarrow 1$$

$$\frac{1}{2} = 0 \text{ Remainder} \longrightarrow 1 \quad \text{(MSB)}$$

$$\overline{111110100}$$

Hexadecimal Numbers

The *larger* the radix of a number system, the more *compact* is the expression for any number in the system. Because each digit is multiplied by a larger radix raised to the positional power of the digit, the value of the number grows rapidly as digits are added.

A radix 5 number is more compact than a binary number, and a decimal radix number is more compact than a radix 5 number. For example, the number of digits required to express decimal 57 in systems of radix 2, 3, and 5 are shown next.

$$111001 \text{ radix } 2 = 57 \text{ decimal}$$
$$2010 \text{ radix } 3 = 57 \text{ decimal}$$
$$212 \text{ radix } 5 = 57 \text{ decimal}$$

Radix 12 numbers, in turn, are even more compact than decimal numbers. But, a problem arises when we begin to use numbers that have a radix larger than 10: we must invent *new* symbols for the numerals greater than decimal 9. One popular method used to invent new numerals is to borrow other well-known *symbols from the alphabet.*

Alphabetic letters can be adapted for each new numeral, such as A for decimal 10, B for decimal 11, C for decimal 12, and so on until all the new numerals have a symbol. Number systems that use a radix greater than 10 have not proved useful to date, except one, known as *hexadecimal.*

Hexadecimal numbers have a radix of *16, with numerals equivalent to decimal number 0 to decimal number 15.* Hexadecimal numbers are useful for binary work because *one* hexadecimal numeral is equivalent to a 4-bit binary number. Hexadecimal numbers then, may be thought of as a type of binary shorthand. Each hexadecimal numeral can be replaced by its equivalent binary number, or the reverse. (Hexadecimal is also commonly referred to as *hex* by most programmers.)

A hexadecimal radix of 16 requires 16 symbols for the 16 numerals used in the system. The decimal number system supplies numeral symbols 0–9 for the first 10 hexadecimal digits, and letters A–F are borrowed from the alphabet for the remaining 6 hexadecimal numerals. The hexadecimal numeral symbols and their binary and decimal equivalents are shown in Table 2.3.

Hexadecimal Numeral	Binary Equivalent	Decimal Equivalent
0	0000	0
1	0001	1
2	0010	2
3	0011	3
4	0100	4
5	0101	5
6	0110	6
7	0111	7
8	1000	8
9	1001	9
A	1010	10
B	1011	11
C	1100	12
D	1101	13
E	1110	14
F	1111	15

Table 2.3 ■ Hexadecimal Numerals

Conversions Between Hexadecimal, Binary, and Decimal Numbers

Because each hexadecimal *numeral* is equivalent to a 4-bit binary *number,* conversion between hexadecimal and binary numbers involves no calculations. Converting from a hexadecimal number to a binary number means simply replacing each hexadecimal numeral with its 4-bit binary equivalent. Converting from a binary number to a hexadecimal number involves replacing each *group* of 4 binary bits with their equivalent hexadecimal numeral.

CONVERTING FROM BINARY TO HEXADECIMAL

Table 2.3 can be used to convert binary numbers to hex numbers by dividing the binary number into groups of 4 bits—*starting at the LSB and grouping to the left.* Table 2.3 then shows how to replace each 4-bit group with its equivalent hex digit. *Leading zeroes* can be added to the leftmost bit (MSB) as needed to fill out the final (leftmost) group of 4 bits.

As an example, take the binary number for 500 decimal that was just computed above and convert it to hex as follows:

500 decimal = 111110100 binary

Third Group of Four	Second Group of Four	First Group of Four
0 0 0 1	1 1 1 1	0 1 0 0
| 1 |	| F |	| 4 |
hex 1	hex F	hex 4

111110100 binary = 1F4 hexadecimal

The binary number 111110100 is converted to the hexadecimal number 1F4 by forming groups of 4 binary bits, beginning at the LSB, and grouping to the *left.* Note that a binary 000 was added to the third group to fill it out to 4 bits.

CONVERTING FROM HEXADECIMAL TO BINARY

To convert from a hex number to a binary number requires that a 4-bit binary number from Table 2.3 be inserted in place of each hex numeral. Each numeral of the hex number becomes a 4-bit binary number. For example, hex 3DEC is converted to binary as shown here:

	3	D	E	C
Original hex number				
Binary equivalent from Table 2.3	| 0011 |	| 1101 |	| 1110 |	| 1100 |

Table 2.3 should be committed to memory. The language of digital computers is filled with hexadecimal numerals and their binary number equivalents.

CONVERTING HEXADECIMAL NUMBERS TO DECIMAL NUMBERS

Conversions between hex numbers and decimal numbers proceed as in previous discussions involving binary numbers. Compact hex numbers expand into large decimal numbers because of the large radix size of a hex number. Conversion from hex to decimal is begun by converting each hex numeral of the hex number into its decimal equivalent found in Table 2.3. Each decimal number is then multiplied by hex radix 16 raised to its positional power. As an example, converting the hex number 5DECA to a decimal number is done as shown next:

Hex Number	Decimal Equivalent of Hex Numeral	$16^{Position}$	Decimal Equivalent
5×16^4 =	5	$\times$ 65,536 =	327,680
$D \times 16^3$ =	13	$\times$ 4096 =	53,248
$E \times 16^2$ =	14	$\times$ 256 =	3,584
$C \times 16^1$ =	12	$\times$ 16 =	192
$A \times 16^0$ =	10	$\times$ 1 =	10
			384,714

CONVERTING DECIMAL NUMBERS TO HEXADECIMAL NUMBERS

Converting a decimal number into a hexadecimal number involves repeated steps whereby a decimal quotient is divided by hex radix 16. At each step, the decimal remainder is replaced with its equivalent hex numeral. The process begins by dividing the original decimal number by 16 and obtaining a decimal quotient and a decimal remainder. The remainder is converted into a hex numeral using Table 2.3, and the quotient is again divided by 16. Each division step produces a quotient, which is fed to the next division step, and a remainder, which is converted to a hex numeral; the conversion ends when the quotient reaches zero. The remainder of the first division step is the LSD of the hex number; the remainder of the last division step (the one that results in a quotient of zero) is the MSD of the hex number.

For example, converting the decimal number 384,714 back to a hex number is shown next:

Dividend ÷ 16	Quotient Remainder (decimal)	Hex equivalent	$16^{Position}$
$\dfrac{384,714}{16}$ =	24,044 Remainder of 10 = A		1 (LSD)
$\dfrac{24,044}{16}$ =	1,502 Remainder of 12 = C		16
$\dfrac{1,502}{16}$ =	93 Remainder of 14 = E		256
$\dfrac{93}{16}$ =	5 Remainder of 13 = D		4,096
$\dfrac{5}{16}$ =	0 Remainder of 5 = 5		65,536 (MSD)
		5DECA	

Hexadecimal numbers are so convenient to use in place of binary numbers that almost all binary numbers longer than 2 or 3 bits are usually expressed in hex. *This book uses hexadecimal numbers* in place of binary numbers almost exclusively, in keeping with standard practice.

The only other popular nondecimal number system still in use is *octal (radix 8)*. Exercises involving octal number conversions will be left for the end of the chapter.

2.4 Fractional Binary Numbers

Integer (whole) number binary-decimal conversions are readily handled by the conversion techniques of multiply by raised radix or divide by radix discussed in previous examples. Fractional (less than unity) numbers are converted from one number system to another in exactly the same manner, except that the positions of the digits are numbered as negative, not positive, numbers.

Before beginning a discussion of fractional binary-decimal number conversions, however, the concept of the decimal point (binary point) has to be investigated. The point (.) is used to *separate* the positive radix positions of a number from the negative radix positions. To the left of the point are increasingly positive position digits, and to the right of the point are increasingly negative position digits.

Converting Binary Fractions to Decimal

Converting from a binary number that has a fractional part to a decimal number is done by multiplying each bit by the binary radix 2 raised to the appropriate positional power. But, remember that positional powers to the *right* of the binary point are *negative,* not positive, numbers. The individual results of each multiplication step are then added to obtain the final decimal number. As an example, a fractional binary number, 1101.1011, is converted to decimal as shown below.

$$
\begin{array}{llll}
2^{\text{Position}} & 8\ 4\ 2\ 1\ . & .5 & .25 & .125 & .0625 \\
\text{Position} & 3\ 2\ 1\ 0\ . & -1 & -2 & -3 & -4 \\
\text{Bit} & 1\ 1\ 0\ 1\ . & 1 & 0 & 1 & 1
\end{array}
$$

Integer Part	*Fractional Part*
$1 \times 8 = 8$	$1 \times .5 \quad = \quad .5000$
$1 \times 4 = 4$	$0 \times .25 \quad = \quad .0000$
$0 \times 2 = 0$	$1 \times .125 \quad = \quad .1250$
$\underline{1 \times 1 = 1}$	$\underline{1 \times .0625 = \quad .0625}$
$13 \quad +$	$.6875 = 13.6875$

As the fractional example shows, as one moves to the right of the binary point, the power of the radix, 2, is raised to the power -1 to yield $\frac{1}{2}$, to -2 to yield $\frac{1}{4}$, to -3 to yield $\frac{1}{8}$, and so on.

Converting Hexadecimal Fractions to Decimal

Hexadecimal to decimal fractional conversions are done using the hex radix, 16, raised to negative powers. The preceding binary number,

1101.1011, convertsd to hex number D.B, which is converted to decimal as shown next.

$$1101.1011 = D.B$$
$$D = 13$$
$$.B = B \times 16^{-1} = \tfrac{11}{16} = .6875$$
$$D.B = 13 + .6875 = 13.6875$$

When converting from hex fractional numbers to decimal numbers, the negative powers of the hex radix 16 begin with $\tfrac{1}{16}$ followed by $\tfrac{1}{256}$, $\tfrac{1}{4096}$, and so on. The last example again demonstrates how binary-decimal conversions are quickly performed when a hex number is used in place of its binary equivalent.

A second example, converting the hex number 3E.4FC to a decimal number, follows next.

Integer Part

3×16			$= 48$
$E \times 1$	$= 14 \times 1$		$= \underline{14}$
			62

Fractional Part

4×16^{-1}	$= \tfrac{4}{16}$		$= .2500000000$
$F \times 16^{-2}$	$= \tfrac{15}{256}$		$= .0585937500$
$C \times 16^{-3}$	$= \tfrac{12}{4096}$		$= \underline{.0029296875}$
			$.3115234375$

$$3E.4FC \quad = 62 + .3115234375 = 62.3115234375$$

Converting Decimal Fractions to Binary and Hexadecimal

Decimal *fractional* numbers are converted to fractional binary or hex equivalents by repeated *multiplication* of the decimal fraction by 2 (binary) or 16 (hex). Every multiplication step will result in a product that has an integer part and a fractional part. The integer part of each product is saved as a binary (hex) numeral. The fractional part of each product is again multiplied by 2 (binary) or 16 (hex) until the fractional part of a product is zero, or it becomes clear that the *process will never end*. The first multiplication yields the MSD of the fractional part of the binary or hex number.

There is one important fact to note when converting from decimal to binary fractional numbers: *exact conversions may not be possible*. Exact binary equivalents of decimal fractions will only be exact if the multiplication product, at *some* stage of the conversion process, ends as an integer product with *no* fractional part. Should repeated multiplication by 2 (or 16) of the decimal fraction result in a *repeating* pattern of decimal remainders, the binary fraction is *infinite in length*.

For instance, the decimal number 21.1, which has a fractional part of decimal .1, will be shown to have *no* exact binary equivalent. The decimal number, 21.1, when converted to binary, is processed as shown next.

Convert integer part of decimal number to binary by repeated division:

$$\frac{21}{2} = 10 \qquad \text{Remainder of 1 \quad (LSB)}$$

$$\frac{10}{2} = 5 \qquad \text{Remainder of 0}$$

$$\frac{5}{2} = 2 \qquad \text{Remainder of 1}$$

$$\frac{2}{2} = 1 \qquad \text{Remainder of 0}$$

$$\frac{1}{2} = 0 \qquad \text{Remainder of 1}$$

Integer Part = $\overline{10101}$

Convert fractional part of decimal number to binary by repeated multiplication:

Fractional Part = .0 0011 0011 0

(MSB)

.1 × 2 = 0.2 Integer = 0

First pattern is 0011
.2 × 2 = 0.4 Integer = 0
.4 × 2 = 0.8 Integer = 0
.8 × 2 = 1.6 Integer = 1
.6 × 2 = 1.2 Integer = 1

Repeat pattern is 0011
.2 × 2 = 0.4 Integer = 0
.4 × 2 = 0.8 Integer = 0
.8 × 2 = 1.6 Integer = 1
.6 × 2 = 1.2 Integer = 1

And so on forever
.2 × 2 = 0.4 Integer = 0

Using the binary digits of the preceding decimal-binary fractional example yields the *inexact* equivalent of 21.1 decimal as 10101.000110011 binary.

As can be seen in the example of repeated multiplications of the decimal fraction .1, a repeating pattern of integers is established that will continue for as long as the multiplication continues.

The accuracy of the binary equivalent of a decimal fraction may, of course, be as precise as may be desired. Increased accuracy is obtained by carrying the calculations out to as many binary places as needed. For the prior example of converting .1 decimal to 9 binary places, the conversion from binary back to decimal becomes the following:

$$
\begin{aligned}
.000110011 = 0 \times \tfrac{1}{2} \ \ &= .00000000 \\
0 \times \tfrac{1}{4} \ \ &= .00000000 \\
0 \times \tfrac{1}{8} \ \ &= .00000000 \\
1 \times \tfrac{1}{16} \ \ &= .06250000 \\
1 \times \tfrac{1}{32} \ \ &= .03125000 \\
0 \times \tfrac{1}{64} \ \ &= .00000000 \\
0 \times \tfrac{1}{128} &= .00000000 \\
1 \times \tfrac{1}{256} &= .00390625 \\
1 \times \tfrac{1}{512} &= \underline{.00195312} \\
&\ \ \ \ .09960937
\end{aligned}
$$

The accuracy of the decimal fraction to binary fraction conversion, from .1 decimal to 9-bit binary places, is seen to be 99.61%.

The same fractional decimal, .1, when converted to 3 hex digits is processed by repeated multiplication by 16, as follows.

$$
\begin{aligned}
&\qquad\qquad\qquad\qquad \text{(MSD) .199 hex} \\
.1 \times 16 &= 1.6 \ \ \text{Integer} = 1 \underline{\quad\quad} |\ |\ | \\
.6 \times 16 &= 9.6 \ \ \text{Integer} = 9 \underline{\quad\quad\ } |\ | \\
.6 \times 16 &= 9.6 \ \ \text{Integer} = 9 \underline{\quad\quad\quad} |
\end{aligned}
$$

.1 decimal = .199 hex (.0001 1001 1001 binary)

The decimal .1 fraction to hex conversion shows the appearance of the repeating integer (9) pattern, which signals that the process will have no finite end.

To check the accuracy of the conversion of .1 decimal to 3 hex digits, the hex fraction is converted back to decimal as shown next.

$$
\begin{aligned}
\tfrac{1}{16} \ \ &= .06250000 \\
\tfrac{9}{256} \ \ &= .03515625 \\
\tfrac{9}{4096} &= \underline{.00219726} \\
&\ \ .09985351
\end{aligned}
$$

The accuracy of converting .1 decimal to 3 hex places is 99.85%.

Note that hex conversion routines proceed much *faster* than do binary conversion routines, because of the larger size of the hex radix.

2.5 Number System Notation

Normally, in everyday life, numbers do not have to have any *notation*. Notation involves some special written clues as to the radix base of the number system in use. Everyone *assumes* that all numbers used in conversation, or seen in any publication, are decimal. Decimal numbers are so common that a special "decimal" notation does *not* have to be used. This book, however, uses *three* number systems: decimal, binary, and hexadecimal (hex). To help you distinguish among the three number systems, the following notation is used throughout the text:

■ Decimal numbers are written with *no* special notation. For instance, decimal 123,456, is written 123,456, with no identifying notation.

■ Hexadecimal numbers are written with an *h, after* the number, to denote hexadecimal. For example, hex ABCD will be written ABCDh, or abcdh. Hex numerals may be written using capital letters for the digits greater than 9 or using lowercase, as, for instance, 12ach. In this book, capital letters will normally be used in *explanatory* text and lowercase letters in *programs*.

■ Binary numbers will be written with a small *b* or *xb* after the binary number. For instance, binary 1011 will be written 1011b or 1011xb. The *xb* notation for binary numbers is a feature of the A86 assembler, which *ensures* that the assembler does not confuse 101b (binary) with 101b (hex). Binary numbers written as, for instance, 1011xb remove any doubt as to the binary radix of the number.

2.6 Binary Addition and Subtraction

Binary math, when done by a digital computer, involves some ordinary, and extraordinary, concepts. Binary arithmetic calculations are extremely simple because there are only two numerals, 0 and 1. Interesting problems arise, however, when dealing with positive and negative binary numbers. Additional challenges may also appear because of the fact that every binary number has a finite size inside a computer.

Binary Number Addition and Subtraction

There are four possible results when two binary numerals are added together or one numeral is subtracted from the other.

Addition of two binary numerals may have one of the four results shown next.

```
    Carry           Carry           Carry           Carry
      |               |               |               |
     1)1             0)0             0)1             0)0
  add  1          add  1          add  0          add  0
  ──────          ──────          ──────          ──────
      0               1               1               0
```

As shown when adding 1 to 1, the *possibility* of a carry from one bit position to the next higher order bit position must be included in the addition process. (*Carries* are shown using a right parenthesis to indicate the carry bit.)

Adding one 8-bit number to another 8-bit number, for instance, may produce a number of carries from each bit to the next, as the following example illustrates.

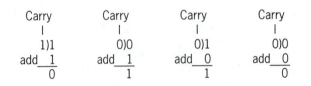

```
Carry  =  1)1)1)1)1)1)1)1)
  +95  =  0 0 1 0 1 1 1 1 1 b
add +189  =  0 1 0 1 1 1 1 0 1 b
 +284  =  1)0 0 0 1 1 1 0 0 b = 256) + 28 = +284
```

Note that the *final* carry, from bit position 7 of the sum to bit position 8, must be included as part of the total, or the sum is in error.

The last example began by adding two 8-bit numbers but obtained a result that requires 9 bits to hold the total sum. In general, every time an *n*-bit number is added to an *n*-bit number, there is the possibility of an *n+1* bit sum. But, if the computer is not *designed* to hold an *n*+1 bit number, then errors can occur.

Binary subtraction, involving 2 bits, also has four possible outcomes, as shown next.

```
   Borrow          Borrow          Borrow          Borrow
     |               |               |               |
     0}1             1}0             0}1             0}0
sub   1          sub   1         sub   0         sub   0
   ─────          ─────          ─────          ─────
     0               1               1               0
```

A borrow, from the next higher order bit position to the lower order bit position, when 1 is subtracted from 0, is a possibility that must be considered during the subtraction process. (*Borrows* are shown using a right curly bracket.)

Subtraction of one 8-bit number from another 8-bit number also may generate borrows from higher order bits. For example, by reversing the previous addition problem, subtracting binary equivalent 95 from binary equivalent 189, proceeds as shown next.

```
Borrow  = 0}1}0}1}1}1}1}0}
  +189 =    1 0 1 1 1 1 0 1 b
sub +95 =   0 1 0 1 1 1 1 1 b
  ──────────────────────────
  +94 = 0}0 1 0 1 1 1 1 0 b = 0} + 94 = 94
```

Subtracting a smaller number from a larger number presents no problems, because the result is less than either of the two original numbers. However, reversing the numbers in the previous subtraction problem so that +189 is subtracted from +95 poses some difficulties when the subtraction is done in binary, as shown next.

```
Borrow  = 1}0}1}0}0}0}0}0}
   +95 =    0 1 0 1 1 1 1 1 b
sub +189 =  1 0 1 1 1 1 0 1 b
  ──────────────────────────
  − 94 = 1}1 0 1 0 0 0 1 0 b = 1} + 162
```

The result of a borrow plus 162 as the answer can be converted to −94 by using the borrow. If the borrow into the 8th binary place is understood to be -2^8, or −256, then −256 + 162 is −94. As was the case for 8-bit addition, however, subtracting two 8-bit numbers can result in a 9-bit answer.

To subtract +189 from +95, manually in the decimal system, we *actually* subtract +95 from +189 and then affix a negative *sign* to the answer to arrive at −94. By inspection, we know that +189 is larger than +95, and we subtract the smaller number from the larger, adjusting the sign of the answer as needed. The same process (subtracting the smaller from the larger number) in a computer binary circuit works just as well as manual decimal subtraction, but a *negative sign* for the answer is needed.

The possibility of generating negative and positive binary numbers, as well as the formation of *n*+1 bit numbers from *n*-bit numbers, leads to the need for a system of *signed* binary numbers of *finite* size.

Signed Binary Numbers

Arithmetic defines both positive and negative numbers, as well as the operations of addition, subtraction, multiplication, and division. Addition, multiplication, and division operations with positive numbers involve no awareness of signs; all such positive number operations have positive results. Subtraction of a large positive number from a smaller positive number, however, introduces the need for numbers that are less than zero, the so-called *negative* numbers. The introduction of signed numbers requires that the result of any arithmetic operation be adjusted to reflect the sign.

Not only must the signs of results be adjusted, but some symbol must be invented to indicate the sign of numbers. Pencil-and-paper arithmetic carries the sign of a number as a preceding positive (+) or negative (–) symbol. Usually, if a number is positive, *no* sign is attached. It is often assumed that a number is positive unless preceded by a negative sign.

Positive and *negative* sign symbols should *not* be confused with the math operations of *addition* and *subtraction*. It is unfortunate that we tend to use the same symbols for addition (+) and subtraction (–) as we do for positive (+) and negative (–). Fortunately, computer programs can spell out arithmetic operations to be done, such as ADD or SUB, and so signs are never confused with math *operations*. To prevent confusion in this book, the words *add* and *subtract* (or *sub*) will be used to differentiate between a mathematical *operation* and a magnitude *sign*.

Note that binary numbers do not *have* to be signed. The computer *programmer* decides, when the program is *written*, if the binary numbers involved should be signed. The programmer may decide, for instance, that the numbers used in a program are to be unsigned 8-bit integers. Each program number is then an 8-bit positive binary number that ranges from 00h to FFh. Some examples of 8-bit binary, positive, unsigned integer numbers, and their decimal equivalents, are shown next.

Decimal		Unsigned Binary
000	=	00000000b
080	=	01010000b
123	=	01111011b
128	=	10000000b
255	=	11111111b

All the bits of an unsigned binary number are used as part of the *magnitude* for the number, and the number is *assumed* to be positive.

Numbers in a program may be unsigned positive integers (such as those just listed), signed integers, signed or unsigned fractions, packed BCD, unpacked BCD, or ASCII. The choice of what type of numbers to use, and how to code the numbers, is *entirely up to the programmer*. The programmer can also decide how *large* the numbers are to be in the program. Binary numbers extend away from both sides of the binary point. Large numbers, and exact fractions, require more bits to represent them than do small numbers and rounded fractions. The *programmer, and the problem at hand*, determine to how many significant binary places program numbers will be extended.

SIGN-MAGNITUDE BINARY NUMBERS

There are two ways to express signed numbers: *sign-magnitude* signed numbers and *two's complement* signed numbers. Both sign-magnitude and two's complement numbers use the most significant bit of the number as a *sign* bit. The sign bit is a 0 for a positive number and a 1 for a negative number.

Sign-magnitude numbers are the type of numbers taught in grammar school. To make a sign-magnitude number, simply place the *sign* of the number in front of the *magnitude* of the number. For example, sign-magnitude numbers −123 and +123 both use the *magnitude* 123 and affix the proper sign. Some examples of 8-bit binary sign-magnitude numbers, and their decimal equivalents, are listed next.

Sign-Magnitude Decimal		Sign-Magnitude Binary
+000	=	00000000b
+096	=	01100000b
+127	=	01111111b
−000	=	10000000b
−001	=	10000001b
−087	=	11010111b
−127	=	11111111b

To make an 8-bit sign-magnitude number, 7 bits are used for the magnitude of the number, and the MSB is the *sign*. The magnitude of any such 8-bit number is thus limited to 127. Note that sign-magnitude binary numbers may define both a positive *and* negative 0. Sign-magnitude numbers are the type of numbers we deal with in everyday decimal number systems. Decimal systems use subtraction of sign-magnitude numbers, and adjust the sign of the answer to that of the largest signed number.

Sign-magnitude binary numbers, although perfectly feasible for use in computer programs, are *not* employed to any great extent. One difficulty with using sign-magnitude numbers in a program is excluding the sign bit from arithmetic operations. To use an 8-bit sign-magnitude number, for instance, the sign bit must be removed, bit 7 made a binary 0, the operation performed, the sign of the result calculated, and then bit 7 of the result re-inserted.

Most computer programs that use negative numbers use *complementary negative* numbers. Complementary numbers automatically adjust the sign of the result for addition and subtraction operations. Complementary numbers can be used for arithmetic in all number systems, including decimal, but are not usually taught in grammar school.

Complementary Numbers

Complementary numbers enable subtraction to be done using *addition*. Complementary numbers can do subtraction by addition, because subtraction is done when a complementary negative number is *formed*.

Complementary number theory became popular because early computers could only perform one mathematical operation, that of addition. Complementary numbers and complementary arithmetic became highly developed to serve the arithmetic needs of the first computers. Complementary math, although it appears awkward at first inspection, is actually much simpler than conventional "grammar school" math.

Complementary numbers are also *finite* numbers, that is, the *programmer must decide how large the largest number in the program needs to be* and then use complementary numbers based on the size of the largest number. If it turns out that the programmer erred, and larger numbers are needed, then *all* of the numbers in the program must be re-sized.

TEN'S COMPLEMENT NUMBERS

We begin our study of complementary numbers using the familiar decimal radix 10 number system.

First, a decision as to the *largest possible* number needed for our discussion must be made. *Assume* that the largest number needed for our purposes is positive 49, and the smallest number needed is negative 50. Each number in the system will use the most significant decimal digit of the number to indicate the sign of the number. All numbers that are 49 or *less*, down to 00, are *positive*. All numbers that are 50 or *greater*, up to 99, are *negative*.

A partial listing of some complementary decimal numbers in our system is shown next.

Positive Numbers	Negative Numbers
+00 = 00	−01 = 99
+01 = 01	−05 = 95
+10 = 10	−10 = 90
+20 = 20	−20 = 80
+40 = 40	−30 = 70
+49 = 49	−50 = 50

Positive numbers, in our system, are formed by simply writing the numbers from 00 to 49. The smallest positive number is 00, and the largest is 49, as expected. Positive numbers are said to be in *true* form.

Negative numbers, in our system, are formed by *subtracting* the *magnitude* of the negative number from 100. The number formed by subtracting the magnitude of the negative number from 100 is called the *ten's complement* of the negative number. Unexpectedly, our largest (in magnitude) negative number is 50, and the smallest is 99. Ten's complement negative numbers are said to be in *complementary* form.

The ten's complement form of the example negative numbers is formed as follows.

$$-01 = 100 \text{ subtract } 01 = 99$$
$$-05 = 100 \text{ subtract } 05 = 95$$
$$-10 = 100 \text{ subtract } 10 = 90$$
$$-20 = 100 \text{ subtract } 20 = 80$$
$$-30 = 100 \text{ subtract } 30 = 70$$
$$-50 = 100 \text{ subtract } 50 = 50$$

Negative and positive numbers carry *no* sign attached. The most significant digit of each number determines its sign.

Negative numbers are formed by *pre-subtracting the magnitude* of the negative number from the system radix raised to a power equal to the number of digits in the negative number. For instance, there are 2 digits in negative 01 to negative 50 in our ten's complement example, so all negative magnitudes are subtracted from 10 raised to the 2nd power, or 100.

To see the advantages that complementary numbers enjoy over "standard" sign-magnitude numbers, consider the following *conventional* subtraction of number *B* from number *A:*

1. The sign of *B* is changed.
2. The signs of both *A* and *B* are compared.
3. If the signs are the *same*, the numbers are added, and the sign of the result is the same as the signs of *A* and *B*.
4. If the signs of *A* and *B* are *different*, then the smaller *magnitude* is subtracted from the larger, and the sign of the difference is the sign of the larger magnitude number.

Conventional addition follows the same process, except the first step, changing the sign of *B*, is omitted.

Complementary subtraction proceeds in a more direct manner, because no consideration is given to the relative sizes of either *A* or *B*. Complementary subtraction proceeds as follows:

1. *B* is complemented (or is already in complementary form).
2. *B* is added to *A*.

Complementary addition involves simply adding *B* to *A*. Complementary math is simpler than conventional math because complementary operations of addition and subtraction automatically *adjust* the sign of the result.

The *downside* to complementary numbers is a condition called *overflow* (discussed later). Complementary number systems always involve the concept of a *finite* number limit. Pencil-and-paper math, however, always assumes that numbers can be made arbitrarily large, or arbitrarily small, simply by writing more digits.

SUBTRACTION BY ADDITION OF TEN'S COMPLEMENT NUMBERS

Several examples of adding ten's complement numbers together using complement numbers that vary from 00 to 99 demonstrate how subtraction can be done by addition.

Example 1 ■ Subtract Positive 12 from Positive 32

Subtracting +12 from +32 is the same as adding 32 to the ten's complement of 12.

$$\begin{array}{cc}
\textit{Standard Math} & \textit{Complementary Math} \\
+32 & 32 \\
\text{subtract } \underline{+12} = (100-12) = & \text{add } \underline{88} \text{ (complementary form)} \\
20 & 1)20 \\
 & \uparrow \\
 & \text{discard}
\end{array}$$

The result of the complementary subtraction has a carry of 1, and result of +20.

Adding positive 32 to the ten's complement of 12 is the same as subtracting 12 from 32, as shown here:

$$32 + (100 - 12) = 100 + 32 - 12 = 100 + (32 - 12) = 1)20$$
$$\hspace{7cm} \uparrow$$
$$\hspace{7cm} \text{discard}$$

The result is 32 subtract 12, the desired operation. As the largest number in our system is 49, the carry of 1 into the hundreds place in the answer is, conveniently, thrown away. Subtraction by addition "works" for complementary numbers because the negative form of a number has been "pre-subtracted" when the negative number is formed.

Example 2 ■ Subtract Positive 32 from Positive 12

The reverse of Example 1 will yield a negative number, in ten's complement form. The subtraction proceeds, for standard and ten's complement numbers, as shown next.

$$\begin{array}{cc}
\textit{Standard Math} & \textit{Complementary Math} \\
+12 & 12 \\
\text{subtract } \underline{+32} = (100-32) = & \text{add } \underline{68} \text{ (complementary form)} \\
-20 & 0)80 = (100-20) = \text{negative } 20
\end{array}$$

There is no carry from the operation, therefore the result is the ten's complement negative number for 20 (100 − 20 = 80). Note that the size of the most significant digit, 8, correctly shows that the result is *negative.*

Example 3 ■ Add Positive Numbers and Overflow

Complementary number mathematics will work correctly as long as the sum of any addition does not *exceed* the largest finite number allowed when the system is set up.

Examples 1 and 2, above, will work correctly as long as no result is smaller than −50 (50) or larger than +49 (49). Should any operation result in a number out of the allowed *bounds,* an *overflow* is said to have occurred. An overflow condition is a serious *error* on the part of the programmer: A number has been generated that is larger than planned for by the *programmer.*

Usually, when an overflow occurs, the programmer must rewrite the program or, at least, resize all of the numbers in the program. Overflows are the result of using signed numbers in a program. Unsigned numbers cannot overflow, because there is no finite limit to their size, except the total memory capacity of the computer.

To see an overflow occur in our example ten's complement system, we shall add +30 to +21, and get +51, which exceeds the largest positive number, 49, *allowed* in the system. The overflow occurs as shown next.

Complementary Math
```
        +30
add    +21
       0)51
```

The resulting number is 51, with no carry. A 51 is supposed to be a negative 49 in our ten's complement system, therefore the answer is in *error*. We have added two legal-size positive numbers and gotten an illegal negative result, because our largest legal positive number is 49.

Overflow can also occur if two *negative* numbers are added together and a *positive* result is reached. For example, adding a negative 30 to a negative 21 will result in a positive number, and an overflow condition, as shown next.

Complementary Math
```
      -30 = (100 - 30) = 70 (complementary form)
add  -21 = (100 - 21) = 79 (complementary form)
                       1)49
```

The result has a carry and a sum of 49, or positive 49. An overflow error has occurred because the sum of two legal-size negative numbers has resulted in an illegal positive number.

Overflows can occur only when two positive or two negative numbers (in ten's complement form) are added, yielding numbers larger than those allowed by the system. Additions involving one negative ten's complement number and one positive number will always result in a correct answer, because the result must be smaller than either of the parts.

To solve the overflow problem of the preceding examples, a *larger* signed number system that uses three digits, based on numbers from 000 to 999, could be used. The new system would then have positive numbers from 000 to 499, and negative numbers from 999 (−001) to 500 (−500). Negative numbers in this larger system are formed by subtracting positive magnitudes from 10 raised to the 3rd power, or 1,000.

Our preceding overflow examples would then be legal, as shown next.

Complementary Math Addition
```
       +030
add   +021
      0)051
```

The answer, 051, is within the allowable range of positive numbers that range from 000 to 499.

Complementary Math Subtraction
```
      -30 = (1000 - 30) = 970
add  -21 = (1000 - 21) = 979
                        1)949 = -51
```

The answer, 949, is −51 in ten's complement form for a system with three digits. No overflow has occurred, because the result fits into a legal-size answer.

TWO'S COMPLEMENT NUMBERS

Negative *two's complement* numbers are formed by subtracting the binary *magnitude* of the negative numbers from 2 raised to a power equal to the number of bits used in the system. For example, if the *programmer* decides that the largest binary number in a program is to be positive 01111111, or negative 10000000, then an 8-bit two's complement number system has been decided on. The two's complement of any number in an 8-bit system is found by subtracting the number from 2 raised to the 8th power, or 100000000, a 9-bit number.

The two's complement of a binary number may also be found by inverting every bit in the number (also called the *one's complement*), and adding 1 to the least significant bit of the inverted number. For example, the two's complement of an 8-bit number equal to 80 decimal is formed using both methods, as shown next.

$$80 \text{ decimal} = 50h = 01010000b$$

$$
\begin{array}{r}
\textit{Invert and add} \\
01010000b \\
\text{complement} \quad |\,|\,|\,|\,|\,|\,|\,| \\
10101111b \\
\text{add} \quad \underline{1b} \\
10110000b = B0h = -80 \text{ decimal}
\end{array}
$$

$$
\begin{array}{r}
\textit{Subtract from } 2^8 \\
1\ 00000000 \\
\text{subtract} \quad \underline{01010000} \\
0\ 10110000 = B0h = -80 \text{ decimal}
\end{array}
$$

Several numbers in an 8-bit two's complement number system are shown next, in true and two's complement form.

	Decimal Value	Binary Value	Hex Value
	+127	0111 1111	7F
Range of	+080	0101 0000	50
positive	+050	0011 0010	32
numbers	+001	0000 0001	01
	+000	0000 0000	00
	−001	1111 1111	FF
Range of	−050	1100 1101	CE
negative	−080	1010 0000	B0
numbers	−127	1000 0001	81
	−128	1000 0000	80

Using 8 bits to represent all numbers in a program results in numbers that vary from as small as −128 (10000000b) to as large as +127 (01111111b).

Note that, for an 8-bit number system, there are 128 positive numbers (0 to +127) and 128 negative numbers (−1 to −128). All 256 possible binary numbers, from 00h to FFh, are used to represent positive and negative numbers.

All negative numbers, in two's complement form, *begin* with a binary *1* in the *most significant bit* position of the number. All positive numbers *begin* with a binary *0* in the *most significant bit* position. The *most significant bit* of a *signed* binary number is named the *sign bit*. However, if the number is *not designed* to be a signed number (by the programmer), then the most significant bit of the number is a magnitude bit, not a sign bit.

The programmer chooses to use signed numbers, or not. If signed numbers are chosen then the programmer must also choose a range of signed binary numbers as large as may be needed by the program. The programmer chooses as follows.

■ A *group* of bytes is chosen by the programmer to represent the largest number to be needed by the program.

■ Bits 0 to 6 of the most significant *byte* (MSBY), and any other lower order bytes, express the *magnitude* of the number.

■ Signed numbers use a 1 in bit position 7 of the *MSBY* as a negative sign and a 0 as a positive sign.

■ All negative numbers are in two's complement form.

When doing signed arithmetic, the programmer must *know in advance* how large the largest number is to be; that is, how many bytes are needed for the largest *possible* number the program can generate.

Two's Complement Mathematics

ADDITION

Signed numbers may be added two ways: the addition of like (same sign) signed numbers, and the addition of unlike signed numbers. An example of adding two 8-bit signed positive numbers that does not produce an overflow condition is as follows:

```
      +045 = 00101101b
add   +075 = 01001011b
      +120   01111000b = 120
```

If two 8-bit positive numbers are added, there is always the *possibility* that the sum will exceed +127 and an overflow will occur. An overflow condition is demonstrated in the example below when 50 is added to 100:

```
      +100 = 01100100b
add   +050 = 00110010b
      +150   10010110b = −106 (two's complement form)
```

There is an overflow because the sum of the positive numbers (150) exceeds the largest positive number (127) allowed in an 8-bit system.

If *unlike* sign 8-bit numbers are added, then it is not possible that the result can be larger than −128 or +127, and the sign of the result will be correct. For example:

```
          −001 = 11111111b              −128 = 10000000b
    add   +027 = 00011011b        add   +127 = 01111111b
          +026   00011010b = +026        −001   11111111b = −001
```

The result of adding two negative numbers together for a sum that does *not* exceed the negative limit is shown in this example:

```
          −030 = (100h − 1Eh) = E2h =   11100010b
    add   −050 = (100h − 32h) = CEh =   11001110b
          −080 = (100h − 50h) = B0h = 1)10110000b = −080
```

The carry is discarded, and the result is correct.

On adding two negative numbers whose sum *does* exceed −128, we have the following overflow condition:

```
          −070 = (100h − 46h) = BAh =   10111010b
    add   −070 = (100h − 46h) = BAh =   10111010b
          −140  Exceeds allowed range    1)01110100b = Carry and +116
```

An overflow error has occurred by adding two negative numbers and receiving a positive result.

By changing our numbers from 8-bit to 9-bit, or decimal values from 000 to 511, the overflow conditions generated by our previous examples may be avoided. In a 9-bit system, positive numbers are those that start at 0000000000b (000 decimal) and extend up to 011111111b (255 decimal). Negative numbers, in a 9-bit system, extend from 111111111b (−001 decimal) to 10000000b (−256 decimal).

Most programmers do not size numbers to the nearest bit needed, however, but to the nearest byte.

MULTIPLE-BYTE ADDITION

Number size is increased by using more bytes to express the number. Multiple-byte addition involves adding carries from low-order sums to higher-order sums.

Using multi-byte numbers in *unsigned* addition means that carries between bytes are propagated from the highest-order bit of a low-order byte to the lowest-order bit of the next-higher-order byte. Carries are propagated from byte to byte by the simple technique of adding the carry bit to the next higher byte.

For example, a pair of 2-byte unsigned numbers is added as shown next.

```
              High-order byte           Low-order byte
    Carry   1)1)0)1)1)1)1)1)1) ←      1)1)1)1)0)0)0)0)
              0 1 1 0 1 0 1 1 1         0 1 1 1 1 0 1 0 0 b
    add   0 0 1 0 0 1 0 0 1           1 0 0 1 1 1 0 1 0 b
              1 0 0 1 0 0 0 0 1         0 0 0 1 0 1 1 1 0 b
```

As can be seen in the example, the carry out from the most significant bit of the low-order byte sum is added to the lowest-order bit of the next higher byte. Should there have been a carry out of the high byte of the example, it could be added to a higher-order byte, and so on. The process of adding the carry to the next-higher-order byte is continued until the last (highest-order) byte in the number is reached.

Signed numbers appear to behave as unsigned numbers until the highest-order byte is reached. Signed numbers use bit 7 of the *highest*-order byte in the number as the sign. If the sign bit is negative, then the *entire* number, from the lowest-order byte upward, is in two's complement form. For example, the smallest and largest positive and negative numbers of a 2-byte (1-word) signed number are expressed as follows:

$$+32767 = 01111111\ 11111111b = 7FFFh$$
$$+00000 = 00000000\ 00000000b = 0000h$$
$$-00001 = 2^{16} - 00000000\ 00000001b = 11111111\ 11111111b = FFFFh$$
$$-32768 = 2^{16} - 10000000\ 00000000b = 10000000\ 00000000b = 8000h$$

Note that the *low*-order bytes of the numbers 00000 and −32768 are *exactly alike*, as are the low-order bytes for +32767 and −00001. The difference between positive and negative numbers appears at the sign bit of the high-order byte for each number.

For multi-byte signed number arithmetic then, the lower bytes are treated as unsigned numbers. All checks for overflow are done only for the *highest*-order byte, which contains the sign bit.

SUBTRACTION

Binary subtraction *can* be done, as noted in the section on two's complement addition, by taking the two's complement of the number (the *subtrahend*) to be subtracted. The subtrahend is *added* to the number (the *minuend*) from which the subtrahend is to be subtracted. Subtraction using addition is an obsolete concept, however. Most modern computers can carry out subtraction operations just as they carry out addition operations.

As was the case for addition, the programmer may choose to use signed or unsigned numbers, according to the needs of the program. An overflow condition, if signed numbers are in use, indicates that an error has occurred when two numbers of *unlike* signs are *subtracted*. Overflows cannot occur when unsigned numbers are subtracted, because all unsigned numbers are, essentially, positive.

UNSIGNED SUBTRACTION

Unsigned subtraction begins with a positive minuend and subtrahend, but the possibility of a negative result is always present.

The result of an unsigned subtraction operation will be in *true* (positive) form, with no borrow, if the subtrahend is *smaller* than the minuend. The result of unsigned subtraction will be in two's complement form, with a borrow, if the subtrahend is *larger* than the minuend. A negative result in two's complement form is *not* a signed number, as all bits of the number are used for magnitude, and there is *no* sign bit.

For example, using 8-bit unsigned numbers, 100 is subtracted from 15 as shown next.

$$
\begin{array}{r}
015 = 00001111b \\
\text{subtract } \underline{100} = \underline{01100100b} \\
-085 = 1\}10101011b = \text{Borrow, ABh} = \text{Borrow, 171}
\end{array}
$$

The borrow signifies that the 8-bit magnitude result of the subtraction is in two's complement form, with a sign bit of 1. (The 2^8 two's complement of 85 is 171.)

The same result, with no borrow, could have been reached by subtracting 1 from 172, as demonstrated in the next example.

$$
\begin{array}{r}
172 = 10101100b \\
\text{subtract } \underline{001} = \underline{00000001b} \\
171 = 0\}10101011b = \text{No Borrow, ABh} = 171
\end{array}
$$

No borrow signifies the result is in true, positive form. Note the first bit of the unsigned result is a 1, yet the number is *unsigned positive.*

SIGNED SUBTRACTION

As was the case for signed addition, there are two possible combinations of signed numbers when subtracting. One may subtract numbers of like or unlike signs.

When numbers of *like* sign are subtracted, it is impossible for the result to exceed the positive or negative magnitude limits of the signed number system in use. The result must be smaller than either the minuend or subtrahend. For example, subtraction involving two 8-bit signed positive numbers is shown next.

$$
\begin{array}{r}
+100 = 01100100b \\
\text{subtract } \underline{+126} = \underline{01111110b} \\
-026 = 1\}11100110b = \text{E6h} = 2^8 - (+026) \\
| \\
\text{discard}
\end{array}
$$

The result, –26 in two's complement form, is smaller than the minuend, +100, or the subtrahend, +126. Note the sign bit of the result is a 1, signifying the result is a signed negative number in the system.

The next example, using two 8-bit signed negative numbers, shows a correct result when numbers of like signs are subtracted.

$$
\begin{array}{r}
-061 = 2^8 - 00111101b = 11000011b \\
\text{subtract } \underline{-116} = 2^8 - 01110100b = \underline{10001100b} \\
+55 0\}00110111b = 37h = +55
\end{array}
$$

The answer, after subtracting –116 from –61, is +55, which is numerically smaller than the minuend or subtrahend. Note that the sign bit of the result is a 0, signifying that the result is a signed positive number in the system.

An overflow becomes possible when subtracting numbers of opposite sign because the situation becomes one of adding numbers of like signs. This can be demonstrated in the following example that uses 8-bit signed numbers.

```
          −099 =   10011101b
subtract  +100 =   01100100b
          −199   1}00111001b = 39h = +057
                   |
                 discard
```

The result, −199, exceeds the size of the maximum negative number (−128) allowed in an 8-bit signed number system. An overflow has occurred, because two legal-size numbers, that should have generated a negative result, have generated an illegal positive result.

An example of an overflow resulting from subtracting a negative 8-bit signed number from a positive 8-bit number is shown next.

```
          +087 =   01010111b
subtract  −052 =   11001100b
          +139   1}10001011b = 8Bh = −117
                   |
                 discard
```

The result, +139, exceeds the maximum positive number (+127) allowed in an 8-bit signed number system. An overflow has occurred because two legal-size numbers, that should have generated a positive result, have combined into an illegal negative result.

Again it must be emphasized: When an overflow occurs in a program that uses signed numbers, an *error* has been made in the *estimation* of the largest number needed to successfully operate the program. Theoretically, the program could stop and re-size every signed number. Re-sizing every number in a program, as the program operates, is rarely feasible.

RECOVERING FROM AN OVERFLOW

The remedy for an overflow error is to make the range of signed numbers larger. Increasing the range of signed numbers means increasing the number of bits used in the number system. For instance, examples of the same signed numbers in an 8-bit and a 16-bit system are listed below.

Number	8-Bit Form	16-Bit Form
+52	00110100	00000000 00110100
+120	01111000	00000000 01111000
+870	Too big	00000011 01100110
−52	11001100	11111111 11001100
−120	10001000	11111111 10001000
−870	Too big	11111100 10011010

Note that numbers, such as 52 or 120, may be expanded from 8-bit size to 16-bit size by *extending* their sign bits. Sign extension is done from a low-order byte by copying the sign bit of the low-order byte into every bit of succeeding higher-order bytes. Thus, as shown in the last example, −52 in 8-bit form (11001100) becomes −52 in 16-bit form (11111111 11001100) by copying the negative sign bit (1) into the high-order byte.

Expanding the size of signed numbers is easily done before the computer program begins to run. Once in operation, however, it is very difficult for the program to correct the effects of an overflow. If an overflow occurs, the programmer has made a *mistake* and has made no provisions for a number as large as the one that generated the overflow. Some error acknowledgment procedure, or user notification, should be included in the program if an overflow is a possibility. Most pocket calculators, for instance, will show some kind of error message in the display should the user try to generate a number larger than the calculator can hold.

2.7 Binary Multiplication and Division

Binary multiplication and division proceed as in other number systems. Multiplication can be carried out as a repeated series of shift-and-add sequences. Division can be carried out as a sequence of trial subtractions.

Overflow conditions are generally not a possibility for multiplication or division operations, unless *division by 0 is attempted.* Overflows occur when two legal-sized signed numbers are added or subtracted, generating an illegal result that is of the *wrong* sign. Multiplication and division operations handle sign and magnitude separately, and the sign of the result can be made to be correct.

Binary Multiplication

The multiplication table for binary numbers involves four entries, as shown next.

Binary Multiplication

$0 \times 0 = 0$
$0 \times 1 = 1$
$1 \times 0 = 0$
$1 \times 1 = 1$

Binary numbers can be multiplied together by a repetitious shift-and-add process. If the number is signed, the sign bits can be separated from the magnitude bits, and the multiplication carried out. The sign of the result may be found by XORing the sign bits.

An example of multiplying two 8-bit unsigned binary numbers is shown as follows.

```
        130 = 10000010b
      × 240 = 11110000b
                00000000
                00000000
                00000000
                00000000
                10000010
               10000010
              10000010
             10000010
        0111100111100000 = 79E0h = 31,200
```

Every multiplication operation must make allowances for a result that is up to twice as large as each component part. Note that multiplying two 8-bit numbers together, as in the example, yields a result that can be as large as 16 bits in length. Multiplying the two largest possible 8-bit unsigned numbers (255 by 255) yields an answer of 65,025, or FE01h.

Multiplication (and division), in primitive computers, is done using *software* programs to perform the shifting and adding of each multiplication step. Modern computers, such as the 8086, do multiplication and division in *hardware*.

Binary Division

Binary division is a repeated process of *trial* subtractions. If the divisor will not divide into the partial dividend by 1, then the quotient bit is 0, another bit is *added* to the partial dividend, and the trial subtraction re-tried.

An example of dividing an unsigned 16-bit number by an unsigned 8-bit number is shown next.

```
                              0000000011101010 = EAh = 234 (Quotient)
  32780                      ┌────────────────
  ───── = 10001100 │ 1000000000001100
   140                        10001100||||||||
                              ─────────||||||||
                              011101000||||||
                              10001100||||||
                              ─────────|||||||
                              010111000|||||
                              10001100|||||
                              ─────────|||||
                              010110001|||
                              10001100|||
                              ─────────|||
                              0010010110|
                              10001100|
                              ─────────|
                              00010100 = 14h = 20 (Remainder)
```

Every division operation results in a quotient and a remainder. Note that the quotient can be as large as the dividend (when the divisor is 1) and that the remainder can contain as may bits as the divisor. Division by 0 is undefined, and many computers contain circuits that detect and act on division by 0.

2.8 Binary Codes

A *code* is any arrangement by which one set of items, say numbers, are uniquely assigned to members of another set of items. Human beings are assigned code numbers in many ways. We all have our social security codes, telephone codes, home address codes, and driver's license codes, for instance. The word *code* usually brings to mind spies, secrets, and intrigue, but there are many more common uses for codes, particularly in the binary world.

Character Codes

Each letter, number, space, and punctuation mark, as well as each drawing seen in this book, has been *coded* as a set of binary numbers and stored on a magnetic disk. Things such as letters, numbers, and punctuation marks are all human-invented symbols called *characters*. Characters let us communicate with each other using standard symbols.

Clearly, a binary digital computer cannot internally store a character in its written form, say the letter *a*. The computer must store some binary number made up of 1s and 0s that is the *code* for the letter *a*. In order to store all possible characters, a unique code number for each letter, number, punctuation mark, and any other symbol humans might use, must be established. Keyboards, for instance, are often the means by which humans communicate with computers using keys that have been imprinted with common human characters. Each key placed on a keyboard has a unique binary code assigned to it; therefore the keys are said to be *encoded*.

In order to avoid confusion in the computer industry as to how each character is to be encoded, standards have been agreed on that define which binary number is assigned to which character. The character code used in the United States, and most other countries, is the *ASCII* character code. Appendix E lists the standard binary ASCII code for characters.

The ASCII Character Code

The acronym *ASCII* stands for the American Standard Code for Information Interchange. Every industry attempts to set standards for its products. There are standards for tires, paper, thread, print fonts, milk, shirts, fishhooks, pedigree dogs and cats, beer, doorknobs, gasoline, and just about anything that is produced and consumed in great quantities. About the only thing that has not been standardized is human beings.

Various industries band together in voluntary associations to set standards for their products. A United States national standards body exists, whose name is the American National Standard Institute or ANSI. ANSI, in collaboration with the electronic communications and computer industry, has established the ASCII standard. Most standards are *voluntary*, but only the largest and financially strongest manufacturer can choose to ignore industry standards.

Standards basically exist in order to expand markets. Consumers will not tolerate mass consumption products that are "one-of-a-kind." Most of us, for instance, would not buy a car that had to use a certain gasoline that was made only by the car manufacturer.

The original ASCII character standard assigned a 7-bit binary number (00h–7Fh) to code 127 characters. The original 7-bit 127 character ASCII code has been expanded to 8 bits (80h–FFh) to code international and graphics characters. Most of the ASCII characters are the familiar symbols seen on keyboards, such as letters, numbers, and punctuation marks. Some of the ASCII characters are called *control* characters, or *nonprintable* characters. Control character symbols are not seen on key caps, computer monitor screens, or the printed page, thus they are *nonprintable*.

Two of the most common ASCII control characters are the *carriage return* (CR) and *line feed* (LF) characters. Each line of visible text characters stored in the disk file for this book ends in invisible control characters CR and LF. Every time the Enter key is struck on the computer keyboard used to write this book, a CR, LF character sequence is generated to the word processor program. The word processor program then knows that it is time to start a new line on the computer screen. When the word processor program stores visible book text characters to the disk file, many control characters, such as tabs and the CR, LF control characters are stored also. Word processors may also *add* other ASCII control characters (or "happy faces" as they sometimes appear on the screen) in order to add features such as underlines, superscripts, boldface, and the like. A *pure* ASCII file will contain nothing but printable ASCII characters and CR, LF control characters. Most word processor files are *not* pure ASCII but may be saved on a disk as pure ASCII as a user option.

The ASCII code was originally designed to handle a typewriter-style keyboard. As computer keyboards added new keys, such as function keys and cursor control keys, new key codes have been defined. The new codes are often called *extended* codes. Extended codes use an ASCII 00h byte *followed* by a second byte to represent the new computer keyboard keys. The extended codes have not yet been standardized by ANSI. The dominance of IBM in the PC marketplace established the IBM PC and AT class keyboards as de facto industry standards. Personal computer keyboards assign a *scan code* to each key on the keyboard, and Appendix E also lists PC keyboard scan codes for each key on the keyboard,

Table 2.4 lists the ASCII codes for some common keyboard characters. ASCII is a 7-bit code for characters from 0000000b (NUL) to 1111111b (DEL).

Note that the decimal numerals 0–9 differ from their true form only by the number *30h*. The decimal numeral 8, for instance, is ASCII 38h, and decimal numeral 3 is coded as ASCII 33h. ASCII decimal numeral codes can easily be converted to the binary code for the decimal numeral by subtracting 30h from the ASCII code.

ASCII Code (hex)	Keyboard Character
0A	LF
0D	CR
20	(SPACE)
30	0
31	1
32	2
33	3
38	8
39	9
41	A
42	B
61	a
62	b
63	c
77	w

Table 2.4 ■ Examples of ASCII Codes

Numeric Codes

BINARY CODED DECIMAL (BCD) CODE

Decimal numerals play such an important part in computer programs that several *codes for decimal numerals* have evolved. The ASCII code for decimal numerals, shown in Table 2.4, adds a 30h to the binary equivalent number for each decimal numeral. Another popular code for decimal numerals is named the *Binary Coded Decimal* or BCD code. Table 2.5 lists the BCD code for the 10 decimal numerals, 0 through 9.

In essence, the BCD code is the radix 2 binary numbers for decimal numerals 0 to 9. In the BCD code shown in Table 2.5, 4 binary bits (1 nibble) are used to express each decimal numeral. The BCD code stops at decimal 9, or 1001b. Decimal 17, for instance, requires two BCD numbers of 0001b (1) and 0111b (7). The BCD code "wastes" the remaining 4-bit binary numbers greater than 1001, so that *none* of the six nondecimal hex numerals A–F are used in a BCD scheme.

A BCD number is a binary number in which each group of 4 bits codes a decimal numeral from 0 to 9. Two BCD numerals require two 4-bit codes (a byte), three BCD numerals require three 4-bit codes, and so on. The inefficiency of the BCD code can be seen by noting that the largest BCD byte-sized number is 99 decimal, whereas the largest byte-sized hex number is FFh, or 255 decimal.

BCD numbers, although not as bit-efficient as pure binary or hex numbers, have a place in computer programs that must deal with decimal-oriented humans.

PACKED AND UNPACKED BCD NUMBERS

ASCII numbers and BCD numbers are closely related. The BCD code for decimal 8, for instance, is 1000b; the ASCII code for decimal 8 is 00111000b (38h). Adding 30h (00110000b) to any *unpacked* BCD number yields the equivalent ASCII character for the decimal number. Unpacked BCD numbers are those numbers that occupy an *entire* byte. The 4 *least* significant bits of the byte hold a binary number ranging from 0000b to 1001b, and the 4 *most* significant bits are set to 0000b. For example, an unpacked BCD 5 is 05h or

BCD Code (binary)	Decimal Numeral
0000	0
0001	1
0010	2
0011	3
0100	4
0101	5
0110	6
0111	7
1000	8
1001	9

Table 2.5 ■ Binary Coded Decimal (BCD) Code

00000101b. When 30h is added to 05h, the ASCII character for decimal 5, 35h, is generated.

Packed BCD numbers are those BCD numbers that use *both nibbles* of the byte to hold a BCD numeral. Packed BCD numbers range from 00h to 99h, with no nibble greater than binary 1001. To convert from packed BCD to unpacked BCD requires that each nibble of the packed number be put into a single byte that begins with a 0000b nibble. For example, the packed BCD number 87h becomes the unpacked bytes 08h and 07h.

Conversions from packed BCD numbers to unpacked BCD and ASCII numbers are often useful for programs that deal primarily with decimal numbers.

Other Binary Codes

Several specialized binary codes have been formulated for use with position sensors known as *encoders*. Encoders serve to convert rotary motion in degrees, for example, into binary codes. One degree of rotation might be encoded as a single binary number, so that 360 binary numbers would encode a complete revolution of a shaft. Many encoders use optical pickups to read a circular pattern of tracks. Each track is made up of black-and-white patches that represent binary 0 and 1. There is one track for each significant bit of the encoder binary number.

A popular encoder code is known as the *Gray* code. Table 2.6 shows a Gray code for the first 16 hex numerals.

A careful inspection of Table 2.6 reveals that, as the count progresses from 0 hex to F hex, only one bit at a time changes between *adjacent* code numbers. For instance, when the count goes from 8 to 9, the code changes

Digit	Gray Code
0	0000
1	0001
2	0011
3	0010
4	0110
5	0111
6	0101
7	0100
8	1100
9	1101
A	1111
B	1110
C	1010
D	1011
E	1001
F	<u>1000</u>
	0000

Table 2.6 ■ Encoder Gray Code

from 1100b to 1101b. Only a single bit, the LSB, changes from 1 to 0 as the count changes.

Gray codes permit slight misalignment of encoder readout pickups and make changes between a 1 and a 0 on each encoder disk track at least 2 bits long. For instance, assume that the LSB track pickup is 1 bit ahead of the MSB track pickup, and the encoder goes from 7 to 8. The Gray code correctly changes from 0100 to 1100, as only the MSB changes. If the encoder used a natural binary code, the numbers would incorrectly change from 0110 to 1001.

Other standard industrial binary codes include Universal Product Codes (UPC) bar codes, excess-three codes, and Hamming error-correcting codes.

Error Correction and Detection Codes

Many coding schemes exist for detecting and correcting errors in binary numbers, particularly numbers that are subject to distortion when they are transmitted over great distances by communication systems. Discussion of the most common error-detection and correction code system, Hamming codes, is beyond the scope of this book. Hamming codes, in essence, using *redundant* bits, enable the receiver to detect and correct errors in a binary message.

PARITY

One very simple error-detection code is the *parity* code concept. Parity involves adding a single bit to a string of coded bits. The bit that is added, called the *parity bit*, makes the total *number* of 1 bits in the string, *including the parity bit, odd* or *even*. If the parity bit makes the total number of bits in the string, *including* the parity bit, odd, then the code is called an *odd parity* code. If the parity bit added to the string makes the total number of bits in the string, *including* the parity bit, even, then the code is called an *even parity* code. The parity bit is referred to in each case as an *odd* parity bit or an *even* parity bit. Parity is used extensively in binary data transmission schemes and memory systems in computers.

Parity bits are usually added at the end of a string of bits to be parity coded during transmission. The value of the parity bit is calculated based on the preceding bits and added at the end. ASCII parity codes, for instance, make the parity bit the most significant bit of the character byte, bit 7. ASCII transmissions send each ASCII byte low-bit first, so that the last bit transmitted is the parity bit.

Parity bits enable a receiver to determine if one bit in the bit string is in error. For instance, if the ASCII character 9 is transmitted, using even parity, then the ASCII character is encoded by the sender as:

```
ASCII 9, even parity = 0011 1001b = 39h
                          |
                       Parity bit
```

The ASCII code for 9 is 0111001b or 39h using 7-bit ASCII code. There are four 1 bits in 39h, so there are an even number of bits in the character. A

0 parity bit is added at bit position 7 to keep the number of 1 bits an even number.

Assume that an error occurred in bit position 0 of a character 9 transmitted in ASCII, and the character was received as:

Received character, even parity = 0011 1000b = 38h

The received character is ASCII 8, but there are only three 1 bits in the character. The receiver knows that the parity is supposed to be even, counts the number of 1s, and detects that an error has been made. The receiver can request that the character be retransmitted.

Parity will *not* always detect errors involving more than 1 bit. For instance, had 2 bits of the ASCII 9 been changed during transmission to yield an ASCII 3, the received character would be:

Received character, even parity = 0011 0011b = 33h

The receiver would count four 1 bits, determine that the number of 1 bits are an even number, and be unaware that an error had taken place.

ASCII character codes may be coded even, odd, or *no parity*. No parity ASCII means that bit 7 of the character is always a binary 0. The ASCII code shown in Appendix E is for no parity ASCII.

2.9 Summary

The binary number system is a positional number system that consists of two digits, 0 and 1. The radix of the binary system is 2. Binary numbers are expressed as:

$$b_n \times 2^n + b_{n-1} \times 2^{n-1} \ldots b_1 \times 2^1 + b_0 \times 2^0 . b_{-1} \times 2^{-1} + b_{-2} \times 2^{-2} + \ldots b_{-m} \times 2^{-m}$$

Binary numbers may be treated as pure numbers, or they may be used to encode other quantities. Common binary codes include the following:

Binary Code	Use
ASCII	Codes decimal, alphabetic, and control characters
BCD	Codes the decimal numerals 0–9 in binary
Gray	Used in position sensors
Parity	Detects errors in binary data

Binary math may be done using signed or unsigned numbers. Unsigned numbers are assumed to be positive numbers. Signed numbers consist of positive numbers in true form and negative numbers in two's complement form. The two's complement of a binary number is found by subtracting the number from 2 raised to a power equal to the number of bits in the number.

Signed numbers use the most significant bit of the most significant byte of the number as the sign bit. Positive numbers are signified by a leading 0 bit, and negative numbers by a 1 bit.

Forming two's complement negative numbers by subtraction from 2 raised to a finite power results in signed numbers that are limited to programmer-defined magnitudes. Any operation involving signed numbers that exceeds the programmer-defined maximum number size results in an error condition called an overflow. Overflow errors are usually serious programming blunders.

Most contemporary microprocessors perform the arithmetic operations of addition, subtraction, multiplication, and division using signed or unsigned binary numbers.

2.10 Problems

1. Express the following decimal numbers as separate numerals multiplied by 10 raised to the positional power of the numeral.
   a. 123   b. 124,678   c. 3,204   d. 4,096   e. 1,896,573   f. 1,024

2. Convert each number of Problem 1 to a number based on a radix of 6. Check your answer by converting the radix 6 number back to decimal.

3. Convert each number of Problem 1 to a number based on a radix of 8 (the octal number system). Check your answer by converting the radix 8 number back to decimal.

4. Convert the following decimal numbers to binary. Check your answers by converting the binary answer back to decimal.
   a. 127   b. 89   c. 356   d. 16   e. 3,289   f. 487,941   g. 255

5. Convert the following binary numbers to decimal. Check your answer by converting the decimal answer back to binary.
   a. 101101011   b. 111111011   c. 10101   d. 10101100   e. 111001

6. Convert the decimal numbers of Problem 4 to hexadecimal. Check your answer by converting the hex answer back to decimal.

7. Convert the binary numbers of Problem 5 to hexadecimal, and then to decimal. Check your answers by converting the decimal answers back to hexadecimal, and then to binary.

8. Convert the following binary fractions to decimal. Check your answers by converting the decimal fractions back to binary.
   a. .10110   b. .11111   c. .00101   d. .01011   e. .10001   f. .1101

9. Convert the following decimal fractions to binary. Limit the binary fraction to 5 binary places. Check your answers by converting the binary fractions back to decimal.
   a. .123   b. .4   c. .999   d. .125   e. .034   f. .078

10. Convert the binary fractions of Problem 8 to hex, and the hex fractions to decimal. Check your answers by converting the decimal fractions back to hex, and then to binary.

11. Convert the decimal fractions of Problem 9 to hexadecimal fractions.

Limit the hex fractions to 3 places. Check your answers by converting the hex fractions back to decimal.

12. Write the first 10 and last 10 positive and negative numbers of an 8-bit signed number system.

13. Add together the following sets of unsigned binary numbers.
    a. 10110110110   b. 1100000101   c. 101010100011
       10000011101   0011111011   100001111001

    d. 111111111111   e. 0111111111   f. 10100011110
       000000000001   1000000000   10000000011

14. Add together the numbers of Problem 13 as sets of signed numbers. Indicate which sums result in an overflow error.

15. Subtract the numbers of Problem 13 as unsigned numbers. Subtract the top number from the bottom number, and the reverse.

16. Subtract the numbers of Problem 13 as signed numbers. Subtract the top number from the bottom number, and the reverse. Indicate the results that generate an overflow error.

17. Sign extend the negative number 10111000 to 2 bytes in length.

18. Sign extend the positive number 011111011 to 3 bytes in length.

19. Multiply the numbers of Problem 13 together.

20. Divide each number of Problem 13 by 1011 binary.

21. Convert every letter of this sentence to no parity ASCII.

22. Decode this message in ASCII: 20h 48h 65h 6Ch 6Ch 6Fh 20h.

23. Convert 99d to ASCII, then to unpacked BCD and packed BCD.

24. Convert packed BCD 54H to ASCII.

25. Take all of the no parity ASCII codes for lowercase letters *a* through *d* and convert them to even parity ASCII codes.

3

THE
8086 FAMILY
ARCHITECTURE

Chapter Objectives

On successful completion of this chapter you will be able to:

- Describe digital computer system organization and operation.
- Explain the function of the CPU and memory.
- Explain the difference between code and data memory.
- Describe the structure of a CPU instruction.
- Explain why assembly language programming is used.
- Write assembly language programs for the PAL CPU.
- Describe the 8086 CPU register architecture.
- Describe the functions of the 8086 CPU registers.
- Explain how memory is segmented into 64K blocks.
- Understand the function of the data memory stack.
- Explain how a 20-bit memory address is formed using a 16-bit segment register and a 16-bit offset quantity.
- Explain the difference between a physical and a logical memory address.
- Explain the purpose of an instruction pipeline in speeding up CPU operations.

3.0 The Forest and the Trees

The next five chapters of this book focus on programming, in *assembly language,* the 8086 family of microprocessors.

Programming is both a *science* and an *art* and is similar in concept to writing fiction. The science of writing involves rules of grammar, punctuation, spelling, and the structure of a story; the art of writing is how the words are arranged by the author. To write fiction you must first have a set of words to use in writing your story. You must learn how to spell the words correctly and how to grammatically arrange the words. Then you learn to write acceptable fiction by studying common literary techniques, by reading many works of fiction by successful authors, and by much *practice* writing.

You learn to program in the same way you learn to write fiction. First, you must have a set of words, grammar, and punctuation to use in writing your program. Then you study common programming techniques, analyze example programs, and write *many practice* programs. As with fiction, the chances are that the more you practice writing programs, the better you will become. In one sense, programming is a set of rules that are followed in order to get a computer to accomplish some objective. Rules of programming are similar to rules of grammar and punctuation for an author. The art of programming is in how you *arrange* the words.

Programming any computer implies that some method can be found that will enable us, the humans, to get it, a lump of semiconductor, to do what we "tell" it to do. How to tell a lump of sand what to do is the purpose of this book. But, instead of getting bogged down in the root, trunk, branches, limbs, twigs, and leaf details of programming the 8086, let us step back from the tree and view the forest.

Before we get into the exact details of programming a particular computer, such as the 8086, we need to develop the concept of a *generic* computer. Generalization will carry us only so far, however. Computers are all alike, in that they all can produce the same results, and all different in the *exact* way they produce those results. Eventually we must pick a real example of the general model, such as the 8086, in order to do actual programming.

3.1 A Generic Computer

Question: What is a computer? Answer: It depends on whom you ask. Everyone's concept of a computer is not the same. Technically oriented persons tend to forget that, to the "man or woman in the street," a computer is mysterious and unknowable.

In its most basic sense, a digital computer is any machine that can be made to perform a series of binary operations. Common computer *operations* include storing and retrieving binary numbers, addition, subtraction, and Boolean logic. Computer operations may be done in any *sequence.* The arrangement of the *sequence of operations* that the computer is to perform is called a computer *program.*

Computer operations are carried out in a special binary circuit named the *Central Processing Unit* or CPU. Binary numbers, which are needed by the CPU for the operations, are stored in other circuits called the *Memory.*

Contemporary computers are made of silicon, steel, glass, and plastic and are powered by electricity. Most computer circuits operate using only two levels of electrical voltage because this has been found to be the only way to make the circuits operate reliably. Dual-voltage computer circuits operate as either "on or off," "high or low," or "1 or 0"—that is to say, in a binary mode.

Hardware Concepts: The Central Processing Unit

A microprocessor CPU is made of semiconductor material (usually silicon) that is arranged to form hundreds of thousands of transistors. The transistors are connected together (as an integrated circuit) as gates that perform the Boolean logic operations of AND, OR, and NOT on voltages applied to them. Flip-flops are also made from the transistors and arranged into counters, latches, and registers. Signals applied to the CPU logic circuits are applied in synchronism with internal pulses, called a *clock*. The clock frequency is determined by an oscillator that normally uses a synthetic mineral crystal as a reference.

A CPU is an enormous collection of combinational and sequential logic circuits similar to those studied in logic design courses—it is both the heart and brains of a computer. The scale of the number of CPU circuits is staggering, involving hundreds of thousands of hours of human design work. The CPU provides the *timing pulses* that regulate the rest of the computer and the logical operation circuitry that carries out the program. The CPU is "told what to do," or programmed, by a *sequence of binary numbers*. The *total sequence* of the binary numbers is called, collectively, the *computer program*. The program, however, is not generally stored inside the CPU but is located in the computer system memory. The CPU is designed to automatically *get* the program sequence from memory. The programmer is responsible for the CPU's getting the *right* sequence in memory.

Hardware Concepts: Memory

Creating a complete computer system involves connecting the CPU to many other types of integrated circuits, most notably circuits that are solely designed to *store* binary data in electrical form. Binary bits may be stored as charges in a capacitor, in switched transistors, or in metal fuse patterns. Circuits that are meant to store binary data in electrical form are known as *solid-state* memories and are made up of millions of storage devices, or *cells*. Each cell holds a bit of a binary number and can interchange that bit with the CPU under the *control of the CPU timing circuits*. Typically, bits are transferred between the CPU and memory as positive-logic voltages.

ROM

Memory circuits may *permanently* store binary numbers and are known as *read only memory* or ROM. ROM memory cell contents may not be changed (written to) by the CPU, but they may be used (read) by the CPU. ROM is also called *nonvolatile* because it does not depend on electrical power to store the numbers.

RAM

Memory circuits that store numbers that may be both read *and written to* by the CPU are given the name of random access memory or RAM[1]. RAM is also called *volatile* memory, because its contents are lost when power is removed. RAM types include static RAM (SRAM) and dynamic RAM (DRAM).

Software Concepts: Code and Data Memory

The function of memory is to store binary numbers for use by the CPU. The numbers in ROM and RAM are of two types: *program code* numbers and *program data* numbers.

CODE NUMBERS

The bits at a memory address might be used by the programmer to instruct the CPU as to what *program operation* it is to perform. Another name for a program operation is an *instruction*. If a bit pattern in memory is intended to be used as a CPU instruction, it is generally made up of two parts. The first part of the instruction is the operational code, or *opcode*. The opcode is the "verb" part of the instruction and contains the binary code for a CPU *action*. The second part of an instruction is the *operand(s)* code. The operand code is the "noun" part of the instruction and describes *what* is to be acted on by the CPU. A *complete* CPU instruction contains an opcode and, generally, operands. The complete sequence of program instructions *placed in memory by the programmer* are numbers known, collectively, as *code* and are stored in the area of memory known as *code memory*.

THE CPU FETCHES CODE NUMBERS

The process of getting an instruction from code memory into the CPU is called *fetching* the instruction. Once an instruction is fetched by the CPU it is then *executed*, that is, the CPU does whatever the instruction tells it to do. The operation of fetching and executing instructions from code memory is *a completely automatic* CPU action, much like the involuntary actions of our own bodies. We have many automatic functions in our bodies, such as digestion, heartbeats, and hearing. We do not have to program our bodies' automatic actions; they are "transparent" to our thought processes. In a similar manner, the CPU will carry out the operation of fetching and executing program code with no direction from the programmer. All other CPU actions, other than fetching and executing instructions, are controlled by the *sequence of instructions* stored in code memory. The original code number fetched from code memory on start-up is the first opcode of the computer program. The CPU executes the operation that is indicated by the first instruction number and then fetches the next instruction. The process of fetching and executing program code continues until the computer is turned off or is otherwise halted.

[1]ROM may be accessed (read) in a random manner also, but the names ROM and RAM are common industrial usage. There are also hybrid electronic memory circuits that may be written to by the CPU but retain bits when power is removed. The hybrid memories are generically named *electrically erasable and programmable read only memories* or EEPROMS.

We, the programmers, are responsible for making sure that the sequence of program instructions work toward some goal. The CPU will execute (run) the program *exactly* as we write it. We hear, via the media, that a "computer error" has caused some disruption of normal life. Computers rarely make errors. Computer programmers, however, often make errors.

DATA NUMBERS

Some addresses in memory may be used by the programmer to be accessed by the CPU under program control. Addresses in memory that are meant to be read by the CPU, or written to by the CPU, are called *data* addresses. The contents of a data address is a data number.

Data numbers may be *variable,* meaning they may be changed as the CPU executes the program, or they may be *constant* and not change during program execution. In general, data is stored in RAM, because RAM can be altered by the CPU as the program runs. Fixed data, however, may also be stored permanently in ROM as a read-only source of data for the program.

The CPU has *no* way of discerning between code and data numbers. If, by some mistake on the programmer's part, or by a rare mishap of nature, the CPU should *fetch data* when it was expecting code, it will decode and attempt to execute *whatever* operation is coded by the data number. Such accidents involving mistaking data for code are usually *attributable to the programmer.* Occasionally nature, in the form of electrical noise, will cause an erroneous fetching of data instead of code. Nature is very *rarely* to blame for program mistakes, particularly in well-designed personal computers. Your instructor will not be noticeably moved to pity if you claim, "The *computer* messed up *my* program."

Computer Concepts: A Computer Model

A fundamental model of a computer is made up of a CPU and a ROM and RAM memory, as shown in Figure 3.1. Inspection of Figure 3.1 shows that the CPU obtains numbers stored in memory by *addressing* memory locations that contain numbers.

MEMORY ORGANIZATION

Memory is organized as a group of *locations* where binary numbers may be *stored.* Each memory location is assigned a unique binary number, called the *address* of the location. The CPU uses addresses to find code and data numbers in memory. Memory addresses are assigned by the computer system designer who connects the CPU and memory together. Memory address numbers may begin at zero and go up to as large a number as the CPU can physically generate. Figure 3.1 illustrates a memory made up of over 1 million addresses, from address 00000h up to address FFFFFh.

Each individual storage element in a memory is a bit. A bit is one digit of a binary number and may be a 0 or a 1. (Binary numbers, and binary math, are discussed in Chapter 2.) Bits stored at each memory location can be organized into binary numbers of any length. The names given to binary numbers of various lengths are shown in Table 3.1.

Any group of bits, typically a byte, that can all be accessed by the CPU *at one time* are the *contents* of an address in memory.

Figure 3.1 ■ A fundamental computer model.

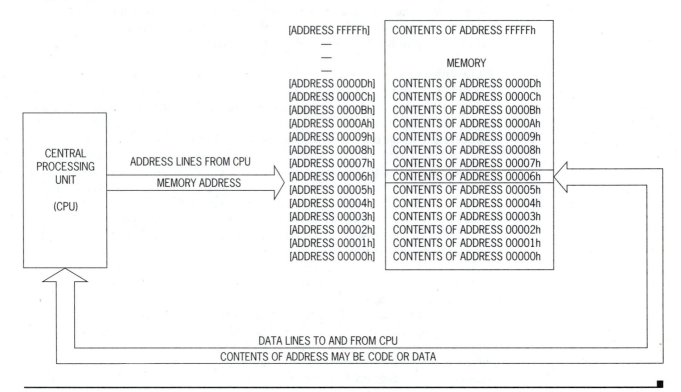

Memory can be thought of as a huge college dorm building with hundreds of thousands of rooms. Each room (group of people) has a unique address in the building. Any group of people, in any room, can be accessed at the address of the room in which they are placed. Another allegory for computer memory are post office boxes. Each P.O. box has an address number. The contents of each box is mail. We access our mail by addressing our P.O. box.

The CPU is physically connected to the memory by a number of electrical connections, called *address lines* and *data lines*. The CPU places signals on the address lines so that a unique address in memory can be accessed by

Name	Length, in bits
Nibble	4
Byte	8
Word[2]	16
Double Word	32
Quad Word	64

Table 3.1 ■ Binary Number Names

[2]Strictly defined, a word's length is the *maximum* number of bits a specific computer can access in *data* memory *during the execution of a single instruction*. The names used in Table 3.1 apply to the 8086 CPU and are used in this book.

the CPU. Information stored at that address is then accessed (fetched, read, or written) using the data lines. The name *data* lines does *not* mean that only data numbers may be accessed by the CPU. The contents of *any* memory address (code or data) are accessed using the data lines.

If the memory is RAM, then the CPU may obtain data bits from an address (read the data at the address) or place new data bits in an address (write data to the address). The CPU can also fetch program code instructions from RAM memory.

If the memory is ROM, then the CPU may only read data bits or fetch code bits at an address, because of the read-only nature of ROM.

Keep in mind that reading or writing data at a memory address means that the CPU has been instructed *by the program* to access a certain memory address. Fetching a code number from an address in memory is an *automatic* CPU function.

The concept of code memory and data memory is central to programming a computer. In one sense, a computer can be thought of as a machine that transforms data numbers in memory under the control of code numbers in memory. The totality of code in memory is the computer program. All other addresses in memory can contain data for use by the program. *Very few programs use all of the available data memory*, however, and the unused data space contains whatever random bits (also known as *garbage*) are formed when the computer is powered up.

Computer Concepts: Starting Up

Since we, as programmers, are responsible for getting code and data properly organized in memory, how does the computer find the very first opcode, when it is first turned on? To find the first opcode, the CPU is *designed* to do so by the manufacturer.

When the CPU is first energized (also known as *resetting* or *booting*), it will *automatically* generate a *reset code memory address* that is designed into the CPU hardware and fetch the code bits at that address. The bits from the first address will be interpreted by the CPU as an instruction that the CPU is to perform. The CPU decodes the opcode, performs the instruction, and then obtains the next instruction from another code memory address.[3]

A computer manufacturer will normally supply a start-up program in ROM with the computer. The start-up program initializes the computer system and then may load the operating system program from disk. Small computers, such as single-board types, may not have an operating system and the program begins at the CPU reset address. The programmer *must* place the first instruction at the reset address.

Computer Concepts: Machine Language

A computer instruction is a group of bits that controls the CPU so that it will perform a desired CPU operation. Most CPUs are designed to perform hundreds of different operations. Instruction codes are located in the code

---

[3]Other microprocessor designs, notably those of Motorola, fetch the *address* of the first instruction from the reset address and *then* fetch and execute the first instruction from that address.

section of memory and are *placed there by the programmer*. Every CPU instruction requires a *unique* bit pattern to distinguish it from another instruction. The CPU instructions are said to be *coded*, because one unique instruction can be distinguished from another unique instruction only if each instruction is encoded as a different pattern of bits.

Another name for code is *machine language*, because the codes are the only bit patterns (language) to which the CPU (a machine) can respond. *Every* computer program must be coded, ultimately, in machine language. Machine language is understood perfectly by the CPU but is *not* suitable for human use.

Software Concepts: Assembly Language Mnemonics

Computer operations are directed by binary-coded machine language instructions fetched from code memory. Although the code bits fetched may make perfect sense to the CPU, they are not readily understandable by humans. The gap between humans and computers is bridged by human language. All computers, from calculators to supercomputers, are programmed using some sort of human *computer language*. A computer language is whatever scheme is used by the programmer to get the computer to produce results. *Assembly language* is the human language most closely tied to machine code.

A human language that uses a *mnemonic* scheme is used for machine code programming. A mnemonic is defined[4] as some plan (a word, rhyme, or the like) designed to aid the (human) memory. As applied to assembly language programming, mnemonics are words that "sound like" an equivalent machine code operation.

Assembly language is usually written in instruction mnemonics that have been invented by the CPU *manufacturer*. Once a program has been written in mnemonics, it is translated into machine code by a process known as *assembling* the program. Assembling machine code, which was once done manually, is now done using a computer. Using the computer to assemble a program leads to the need for a text editor program, an assembler program, and a program that operates the computer.

Software Concepts: Operating Systems and Application Programs

An operating system program—such as DOS, VMS, or UNIX—is a program that *manages* the overall computer *system*. Computer *assets*—such as RAM memory, disk drives, cathode ray tube (CRT) monitors, keyboard, printer ports, and serial ports—are under the general control of the operating system program. All other programs, such as the ones we write, are called *application programs*. Application programs are placed in computer RAM, from a disk, by the operating system. The operating system program then lets the application program have access to computer system assets so that the application program can do its job. Text editors and assembler programs are the application programs we shall need in order to program the 8086.

---

[4]*Computer Dictionary* (Redmond, WA: Microsoft Press, 1991), p. 229.

3.2 The Mechanics of Programming

Getting Instructions into Code Memory

The task facing a computer programmer is, fundamentally, how to place the right set of instructions in code memory in order to get the CPU to solve the problem at hand. Some way must be found to translate human thoughts into bits inside of semiconductor code memory.

Personal computers are equipped with ROM code memory that has been formed and inserted permanently into the computer so that the computer is equipped with some minimum program when it is first turned on. The CPU, under the direction of ROM memory code, can then read the operating system program from a disk and place it in RAM. The operating system program then is given control of the computer and awaits orders from the computer programmer. Using the operating system program, the programmer can direct that an application program (*user code*) be placed in RAM and executed. The application program may help the programmer write programs, in which case it is often called a *utility program*, or it may be the programmer's own creation.

Many types of utility programs exist to help get human thoughts converted into code memory. The only fundamental difference between the humblest pocket calculator and the mightiest mainframe is the means by which their software and hardware convert thoughts, and language, into code bits.

Computer Languages

Humans think and communicate using languages (both verbal and more subtle means). Computers communicate using voltages and currents. An ideal way to convert from thoughts to voltages would be to issue instructions to the computer by speaking to it exactly the way one would speak to another person. A less desirable method, but the one most used today, is to type instructions to the computer on a keyboard.

The complexity of the language used to instruct the computer establishes the level of communication between human and computer. The simplest possible computer language is that used to get results from an inexpensive pocket calculator. We can program the calculator using no more than a few keys that have the programming language printed on separate keys (+, −, =, 0–9).

HIGH-LEVEL LANGUAGES

At a more advanced level than a pocket calculator, we can program a mainframe computer in a high-level language using a keyboard and a CRT. Interestingly, we need no knowledge of the internal workings of the calculator or the mainframe computer to use either one. High-level languages are said to be *machine independent* because the internal circuits of the computer used to program in a high-level language are not part of the high-level language. Examples of high-level languages include BASIC, FORTH, Pascal, C, and FORTRAN.

Note that the high-level language names are generic. Many versions of each high-level language exist, depending on what brand of computer they will program, but the language is the same. A high-level language is not tied to any particular processor type and no details of CPU construction are needed by the programmer. For instance, a Pascal program can execute on a mainframe, a PC, or an Apple with equal results. The Pascal programmer does not have to be concerned with what particular computer will be used to execute the Pascal program. Clearly, some utility programs must take the Pascal program and convert it into the machine language of the particular CPU in use, but such conversion is not the concern of the Pascal programmer.

High-level languages are often called *transportable,* that is, the same high-level language programs may be used on many makes of computer, all of which have different processors.

ASSEMBLY LANGUAGES

Assembly language programming, unlike high-level language programming, is tied very closely to the *physical* makeup of the CPU. Assembly language programs use the internal circuits (registers) of the computer as part of the assembly code language. The programmer must have a very *detailed* knowledge of the arrangement of a particular CPU's internal circuits in order to program that CPU in assembly language code. Examples of assembly-level languages include those associated with the 8051, 8086, 6502, 68000, and Z80 CPUs.

Assembly languages are *specific* to a unique microprocessor CPU type and are intimately bound up with the *exact construction* details for that particular microprocessor. An assembly language programmer *must know* the internal makeup of the CPU in order to program it. CPUs are not interchangeable. For instance, assembly language programs written for a Zilog Z80 will not work for an Intel 8086. Assembly code programs are bound to a particular CPU type and cannot be used for a different CPU type.

Assembly-level languages are not transportable; programs written for one type cannot be used on another. In general, it is quicker to program in a high-level language than in assembly language because detailed knowledge concerning the hardware arrangement and operation of the processor is not needed for a high-level language. Detailed knowledge of the CPU, however, is essential for programming in assembly language for that processor.

Why Use Assembly Language?

There are at least five reasons to write computer instructions in assembly language:

1. To speed computer operation. Programs written in assembly language can be stored compactly in code memory, and less time is spent fetching the code. High-level languages are converted to code by utility programs named *compilers.* Because of the general nature of high-level languages, the compilers often produce excess or *overhead code.*

2. To reduce the size of the program. Assembly language requires no extra overhead code because the assembly language programmer is aware of the exact needs of the program for any given situation.

3. To write programs for special situations. Often, particularly when dealing with machine control, no standard programs (named *drivers*) exist. Robot arms and antilock brakes, for instance, have no standard drivers. It is generally more efficient to write nonstandard driver programs in assembly code.

4. To save money. Small computer systems, such as those that are embedded inside other machines, are often produced in large numbers. Reducing code size also reduces the cost of associated ROM chips.

5. To better understand how computers operate. In order to fully understand what is going on "under the hood" of the CPU, you should learn to program the CPU in assembly language.

Speed, size, and uniqueness are advantages assembly language programs offer over high-level languages.

Programs written in high-level languages are converted to machine code by a type of assembler named a compiler. Compilers take the high-level language and convert it to the machine code for the particular CPU in use. The resulting machine code program often requires large amounts of memory storage because each high-level instruction is compiled into many, often redundant, machine code instructions. High-level languages, after they are compiled into machine code, usually take longer to run and occupy more memory space than do assembly language programs. The excess high-level language machine code instructions take extra time to be fetched and executed by the processor. The time needed to run a high-level program is greater than the time needed to run a nonredundant assembly code program.

Execution time and program size may be of little interest if the application is, for instance, a word processing program. Word processing programs running on a personal computer have enormous amounts of memory at their disposal and the speed of program response far exceeds the response time of the user. It makes sense to write a word processing program in a high-level language.

Execution time and program size is of great interest, however, if the application is an automatic braking system on an automobile. Response time of an ABS program is in milliseconds, and the program must fit into a small ROM. Machine control programs are often written in assembly code where speed of response, and compact program size, are needed.

Cost can also be a factor in choosing assembly code over a high-level language. Pocket calculators do not require fast response time from internal code. Cost is crucial, however, in the calculator market, and the size of the program used to operate the calculator is costly in terms of memory chips. Program size restraints apply to any application in which cost is paramount, which includes most home appliances, computer peripherals, automotive electronics, and toys. Finally, size sometimes plays an important part in the decision to use assembly language rather than a high-level language. Applications that require the ultimate in miniaturization, such as portable phones, space vehicles, and many weapons, require that the controlling computer be as physically small as possible. Again, to shrink program size and memory chip requirements, assembly code is needed.

3.3 The Assembly Language
Programming Process

Assembly language programming requires the use of application programs, known as *utilities* or *tools*, to get our assembly language programs converted into code memory bit charges. Assembly code utility programs can range from very crude board-level *monitors* to very sophisticated personal computer *assembler* programs. We shall assume that our assembly language programming is to be done using a personal computer. The personal computer enables us to speedily write and test our assembly language programs.

To get your thoughts converted into code memory charges, you must know how to use a set of programs that have been written to help ease the way. The following programs are essential to your success:

1. An *operating system* program, which controls the operation of the personal computer used for the entire programming process. DOS (PC-DOS or MS-DOS) is the operating system program used in this book.

2. A *word processing*, often called a *text editor*, program. Programs in assembly language mnemonics are written using the text editor and stored on a disk as files that, normally, end in the extension *.ASM.* The .ASM files are intended for the use of another program called the assembler program. Any text editor that can produce an ASCII (text) file is suitable for writing assembly language .ASM files.

3. An *assembler* program, which takes the .ASM assembly language program file and *converts* it to a machine code *.COM* file (the type we shall use most often). The assembler converts the ASCII mnemonic text file into a .COM file that contains machine code instructions to the CPU, in binary form.

    The assembler may also produce .BIN files that contain the same machine codes as a .COM file but omit some .COM features. Another output file type is an .OBJ file for input into a *linking* program. Linking is usually done when an assembly code file is included (linked) with a high-level programming language, such as Pascal. Linkers produce a resultant .EXE file that, for instance, is a combination of the Pascal and assembly code programs.

4. A *linking* program, which converts an .OBJ file(s) to a single .EXE file. Linking programs, such as the DOS version's LINK, may be purchased from many vendors. High-level languages, such as Pascal, often have their own linkers included. We shall seldom need a linking utility because our .COM files may be run without linking.

5. A *testing* program, which lets you run and test your program under controlled conditions. Testing your program is the *most important step* of the programming process. To be able to test your program you must be able to execute each instruction and see the results. Utility programs that allow the user to test programs are called *debuggers*.

To summarize, the programming process used in this book proceeds as follows. The name of your file is up to you; we shall use the generic name *anyname* to identify the example file as it goes through the process.

- Your thoughts are edited, using a word processor or text editor, into your *anyname.asm* file and stored on disk as the pure ASCII text file *anyname.asm*. A pure text (ASCII) file is one that can be viewed on the CRT using the DOS TYPE anyname.asm command. Many word processors will produce pure ASCII text files if requested to do so.

- The assembler program supplied with this book, A86, is loaded and run. A86 converts the anyname.asm file on the disk into an anyname.com file on the disk. Anyname.com can then be loaded by DOS into the semiconductor memory of the computer at the direction of the programmer.

- The debugging program supplied with this book, D86, can be loaded by the programmer and used to test anyname.com. After testing, anyname.com may be run, full speed, by the programmer.

In order to help understand the fundamentals of assembly language programming, we shall write our first programs using a "practice" CPU.

3.4 The PAL Practice CPU

Let us, for the sake of practice, invent a very small computer CPU called the *Practice Assembly Language* or PAL. We shall invent a set of binary coded PAL instructions.

We shall use the PAL for practice programming by using only very crude (manual) means of converting PAL assembly language instructions into machine code. At the dawn of the computer age, assembly was done manually. Manual assembly methods have the distinct advantage of providing the user with a clear picture of exactly what happens when source (.asm) files become machine code. Manual assembly also provides the impetus to learn to use an assembler program such as our assembler, A86.

Figure 3.2 shows the details of the *internal registers* of the PAL. As shown in Figure 3.2, PAL contains three internal registers:

Register A: Performs all operations in the CPU

Register B: Stores numbers temporarily

Register IP: Holds the address of the next instruction to be executed in code memory

Registers A and B can hold 8 bits (1 byte) of binary data, and register IP (the *Instruction Pointer*) can hold a 16-bit address number. PAL is called an *8-bit* computer because the *working* registers A and B can hold a 1-byte number. PAL register IP determines the memory address range of the CPU to be 64K because it can address code from address 0000h to FFFFh. Memory can contain *any* mix of code and data bytes. For instance, there might be 12K of code and 10K of data, with the rest of memory unused. Code is stored in the

Figure 3.2 ■ PAL CPU and memory.

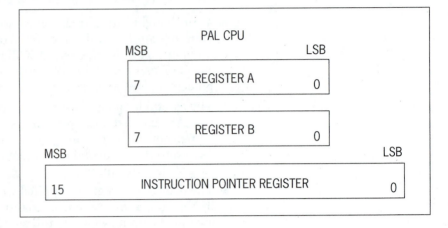

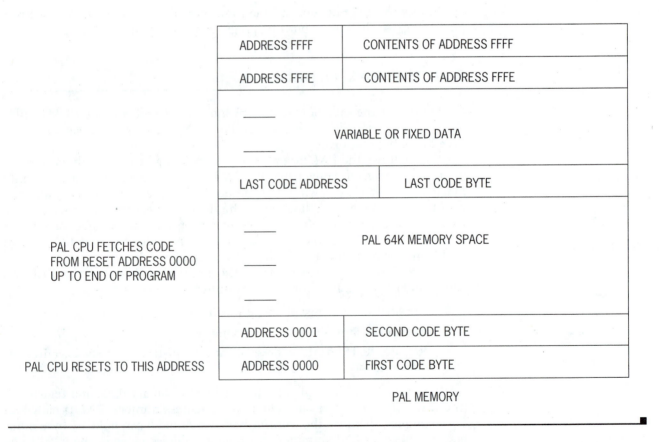

PAL CPU FETCHES CODE
FROM RESET ADDRESS 0000
UP TO END OF PROGRAM

PAL CPU RESETS TO THIS ADDRESS

code area of memory, and data in the data area of memory. It is up to the *programmer* to make sure that the CPU *fetches* from code memory and *reads and writes* data memory.

When PAL is reset, the IP register is set to 0000h by internal circuits. Every time an instruction is fetched from code memory, IP is indexed (counts up) to point to the address of the *next* instruction. Note that IP points to the next instruction *before* executing the current instruction just fetched. PAL fetches the first instruction at code memory address 0000h and proceeds as directed by the programmer.

PAL Instructions

PAL has been equipped with nine instructions. These are the only instructions possible for PAL. The instruction mnemonics are shown in Table 3.2. In the table, the instruction mnemonic and data location are listed on the left; the opcode for the mnemonic (in hexadecimal) with the instruction operand and the CPU action for the instruction are listed on the right.

Note that the *single* instruction mnemonic MOV can be encoded into one of *five* opcodes, depending on the *location of the data* to be moved.

We now have a set of instructions for PAL that is capable of performing some common computer operations. Our example program will consist of placing the numbers 1, 2, and 3 in data memory locations 1000h to 1002h. Because we know that PAL resets to 0000h in memory, our program must start with the first opcode at memory location 0000h.

The first step is to write our program using the instruction mnemonics,

Instruction Mnemonic	Data Location	Machine Code (hex)	CPU Action
MOV	A,num	3Dh,num	Move the byte *number* num to A. num means any *1-byte* number from 00H to FFH.
MOV	A,[addr]	3Eh,addr	Move the *byte contents* of *address* addr to A. addr is any *2-byte address* number in memory. Note the use of *brackets*, [], in the instruction to mean *address contents*.
MOV	[addr],A	3Fh,add	Move the *byte* number in A to the *contents* of address addr.
MOV	B,A	4Ch	Move the contents of A to B.
MOV	A,B	4Dh	Move the contents of B to A.
AND	A,B	F3h	Logical AND the contents of A and B. Results of the AND in A.
OR	A,B	9Ah	Logical OR the contents of A and B. Results of the OR in A.
NOT	A	67h	Logical complement the contents of A.
GO	addr	77h,addr	Set IP equal to the 2-byte number addr. Note there are *no* brackets, so that the *number* addr is placed in IP, *not* the contents of the address.

Table 3.2 ■ PAL Opcodes

with appropriate comments beside each instruction as a reminder of what each is to do:

Instruction	Comment
mov a,01h	;put the number 1 in A
mov [1000h],a	;store in memory location 1000h
mov a,02h	;put the number 2 in A
mov [1001h],a	;store in memory location 1001h
mov a,03h	;put the number 3 in A
mov [1002h],a	;store in memory location 1002h
go 0000h	;set IP back to start of program

Note that we do *not* have to write our program in capital letters. Instruction mnemonics are often printed as uppercase letters in manuals, but the assembler (in this case, ourselves) will accept lowercase letters. Numbers are all expressed in hexadecimal by using the letter *h*. If we do not use the *h*, then we may forget and use decimal numbers when we wish to use hex numbers. Our comments all *begin with a semicolon* to remind us that they are comments and not part of the instruction.

Now that the program is written in assembly language, we must assemble it into pure hexadecimal machine code and place each instruction in the right place in code memory. To assemble our program we must convert each line of our program from assembly language to its equivalent hex code. Our program is assembled as follows:

Code Memory Address	Contents of Code Memory Address	Address Instruction
0000h	3Dh	mov a,01h
0001h	01h	
0002h	3Fh	mov [1000h],a
0003h	00h	
0004h	10h	
0005h	3Dh	mov a,02h
0006h	02h	
0007h	3Fh	mov [1001h],a
0008h	01h	
0009h	10h	
000Ah	3Dh	mov a,03
000Bh	03h	
000Ch	3Fh	mov [1002h],a
000Dh	02h	
000Eh	10h	
000Fh	77h	go 0000h
0010h	00h	
0011h	00h	

We can see, from our manually assembled program, that a very short program takes 18 bytes of code memory. Several items of importance are illustrated by our manually assembled program:

■ Manual and automatic assembly (done using assembly programs) both use a text file as an input and produce a hex file as an output.

■ The program must begin at 0000h and proceed sequentially. We must make sure that the correct hex number is in *each* code address.

■ Instructions are stored with the opcode first; then the data operands (*constant* numbers, or addresses) are stored next in code memory. Note that the register names do *not* have to be stored, because the opcode contains that information.

■ Data addresses, which are *2 bytes* in length, are stored *low* byte first in code memory, then *high* byte next in code memory. Low-high is one of two ways to store 16-bit numbers in *sequential* byte locations in code memory. One way to remember how 2-byte numbers are stored is to note that we go from 16-bit number low byte to 16-bit number high byte as we go from the lower to higher address in the code memory.

There must be some instruction that lets the programmer limit the code addresses available to the program. PAL uses the GO instruction to limit the example program to occupy code addresses 0000h to 0011h. If we did not have a GO instruction at locations 000Fh–0011h in code memory, the CPU would merrily keep on going by fetching every sequential byte in memory up to address FFFFh. The CPU has no way of knowing what is code, data, or just random garbage numbers. The *programmer* has the responsibility to make sure that the CPU stays *within code memory* and cannot possibly find its way to data or garbage.

Manually assembled programs suffer from two basic problems. First, the assembly process is very tedious and prone to many mistakes when the human assembler puts in the wrong opcode, or forgets to number code memory addresses correctly. Second, should any changes be required in the program, the entire program must be re-assembled. To see the impact of the second point, take the original program and make this small change:

Instruction	Comment
mov a,01h	;put the number 1 in A
mov b,a	;store the number in B *** change ***
mov [1000h],a	;store in memory location 1000h
mov a,02h	;put the number 2 in A
mov [1001h],a	;store in memory location 1001h
mov a,03h	;put the number 3 in A
mov [1002h],a	;store in memory location 1002h
go 0000h	;loop back to the beginning

Now, re-assemble the program. Every byte, after the first two, must be assigned new addresses in code memory!

Manual assembly, for even a small computer such as the PAL, is informative but soon becomes tedious. Manual assembly of programs for sophisticated CPUs, such as the 8086, is totally impractical.

3.5 A Complete Computer System

It is most important that the assembly language programmer have a clear picture of each *part* of a complete computer *system*. Let us reinforce our understanding of a computer by examining a complete computer system and *reviewing* earlier concepts of computer operation.

In its most basic form, a computer system consists of three distinct physical parts, as shown in Figure 3.3—a CPU, memory, and *peripheral* devices.

The CPU controls the operation of memory and peripherals. The CPU places memory and peripheral addresses on the address lines and generates the control signals that access memory code and memory and peripheral data. The actions of fetching code and accessing data are referred to as CPU *cycles*. Reading data from memory or a peripheral is a read cycle. Writing data to

Figure 3.3 ■ A basic computer system.

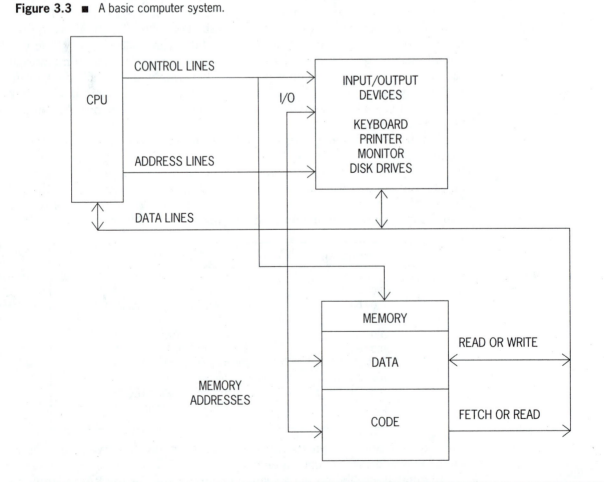

memory or a peripheral is a write cycle. Fetching code from memory is a fetch cycle.

The second part of the computer system is solid-state memory, made up of ROM and RAM. Memory has the sole function of storing binary numbers for use by the CPU. Memory holds two kinds of numbers, coded program instructions and program data, and its circuits respond to address and control signals from the CPU. Memory supplies instruction code to the CPU during a fetch cycle and data to the CPU during a read cycle. Memory stores data from the CPU during a write cycle.

The third component of a computer system are circuits and devices that behave much like memory but are more commonly known as *input and output* (I/O) peripherals. The function of I/O is to enable data to be interchanged between the CPU and the "outside world." Peripherals—such as disk drives, printers, keyboards, and CRT terminals—make the computer useful.

The CPU reads data from or writes data to a peripheral much as it does with data memory. But there are significant differences between memory and I/O. One difference between memory and I/O is speed of response. Memory can supply or store a number in nanoseconds. Peripherals may take many milliseconds to react to CPU control. A second difference is the fact that the system may have millions of memory addresses, but only a few I/O peripheral addresses.

Computer Operation

When the CPU is first energized (or reset), it begins to function as a computer by automatically fetching a binary number from memory. The CPU treats the first number fetched from memory as a CPU instruction opcode. An opcode is a binary number in memory that is part of a coded CPU instruction. The CPU is designed to decode and execute one of its many functions based on the instructions that are fetched from memory. The CPU decodes the first instruction code to determine what the next CPU action should be, executes the first instruction, and automatically fetches the next instruction. The CPU continues to fetch and execute instructions until the computer system is shut off or otherwise disabled.

An area of memory that holds a sequence of CPU instructions is called *code memory*. Memory areas that do not hold code numbers may be used by the program as *data memory*.

Peripheral I/O devices let the CPU exchange data with the computer system operator (the user). The user can type on a keyboard to supply data to the program. Printers and CRT monitors may be used by the program to output data to the user. Disk drives can be used to permanently store programs and program data.

There is a hidden fourth part of a computer system, which is not shown in Figure 3.3. The invisible part is the *operating system software*. Clearly, if the computer fetches its first instruction from memory when power is applied, someone must have written the instruction. If the computer continues to fetch and execute code, then someone must have written a program that controls the computer system from the instant it is turned on. The "someone" is the operating system programmer, and the program that controls the computer is the operating system. Operating systems may be simple or complex, but they are an essential part of a computer system.

Operating system software is partially stored in ROM memory so that it is instantly available when the computer system is turned on. The remainder of the operating system software is stored on a disk and placed in memory after the computer system is initialized by the ROM program.

CPU Design

To be able to function as part of a computer system a CPU must have the means to address memory and I/O, fetch and execute instructions, and retrieve and store data. Fetching and executing instructions are operations that are completely automatic. Everything else the CPU does is under our program's control. Our program, in turn, is made up of the opcodes that the CPU can decode and execute.

The CPU design—that is, how the various internal registers are arranged—determines the types and variety of opcodes the CPU can execute. CPU power, such as the number of bits it uses and the amount of memory it can address, is a function of the CPU *architecture*. The term *architecture* calls to mind the stylistic features of a building rather than the materials used to physically build the building. CPU architecture focuses on the computing features of the CPU, not the integrated circuit technology used to construct the CPU. A CPU architecture is an abstract *model* of circuit reality. The number, bit size, and types of CPU internal registers and the instructions that control those registers are interdependent. Let us now investigate how the architects of the 8086 CPU met the challenge of designing an industry standard.

THE X86 FAMILY

The 8086 family of central processing units consists of the 8086, 8088, 80186, 80188, 80286, 80386, 80486, and Pentium names. The original 8086 and 8088 CPUs were introduced by the Intel Corporation in 1978. Subsequently, the 8088 gained tremendous commercial acceptance when it was chosen by IBM to power its first personal computer, the PC, in 1981.

One generic numbering system for the 8086 family is to simply use the notation 80X86, or, as is done in this book, the X86. The X86 family features an internal CPU architecture that makes each new model compatible with the family members that preceded it. Each new X86 family member, as one progresses from the 8086 to the Pentium, offers new instructions, on-chip integration of more system functions, and increases in computing speed over previous models. The family is still evolving. Intel announced the *Pentium* 64-bit CPU in 1993. Other vendors, notably AMD and Cyrex, continue to improve the X86 family.

For instructional purposes, we must choose one CPU as our general model for the X86 family. Because more advanced models are *backwards compatible* to the beginning 8086 model, the 8086 CPU is chosen to represent the entire X86 family. Backwards compatible means that programs written for the 8086 will run on any "PC compatible" computer that uses an 8086 or a more advanced X86 family member. The reverse is not true; programs written to take advantage of the new instructions of the 80386 CPU will *not* run on an 8086 CPU.

This chapter concentrates on the structure of the 8086 CPU. More advanced X86 family members are discussed in Appendix G, but in a more abbreviated format. The student, new to microprocessors, will find the 8086 sufficiently advanced for learning purposes!

Now that the 8086 has been chosen as our generic CPU, one question remains: Why should anyone choose one of the X86 family members for computing? There are two good reasons for considering one of the X86 central processing units for use in a computer system. First, the X86 family has been, and continues to be, produced in such quantities that the prices of an X86 CPU, and products based on the X86 CPU, are very inexpensive. Second, and more important, a vast quantity of software has been written for the X86 family. Software developed for the X86 family includes application programs such as word processors, spreadsheets, databases, publishing, graphical user interfaces, CAE, and CAD. Program development software includes not only the familiar high-level language compilers such as FORTRAN, Forth, Pascal, BASIC, and C, but also assemblers and debuggers for writing programs in the X86 machine language. In addition, operating system programs include the most popular operating system in the world, DOS, as well as the highly regarded UNIX system.

Many can argue, and successfully, that the 68000 family, the SPARC family, and the 29K family of CPUs are technically superior to X86 models. Such arguments miss the point. When you choose a family of CPUs to address your computational problems, you must look at total system cost, including hardware and software. You must also consider maintenance costs, personnel costs, component availability, and manufacturer support. Choosing the most advanced CPU does not automatically guarantee that the overall system will be superior to one based on a less sophisticated model. The sheer number of X86-based computers currently in use, the ongoing development of new family member CPUs, and the base of software now available for the X86 family makes a strong argument for choosing one of the X86 CPUs for a computing task.

And one last point. From an educational standpoint, all CPUs basically operate in the same manner. The knowledge gained when learning how to use one particular CPU family applies, generally, to any family. An individual who understands the fundamental operating and programming principles of an 8086 will be equally able to use any other popular CPU type. The 8086 standard, by which all others are judged, is a good starting point.

3.6 Organization of the 8086 CPU

For *programming* purposes, the 8086 CPU is composed of a *set of registers* as shown in Figure 3.4. Figure 3.4 shows each register as a rectangular box containing letters and numbers. The *name* of each register is shown in large capital letters, such as AX, SP, or CS.

The number of bits in each 8086 register is indicated by the LSB number 0 on the right and the MSB number on the left of the box. For instance, register AH is an 8-bit register containing numbers made up of bit positions 7, 6, 5, 4, 3, 2, 1, and 0. Register AX is a 16-bit register containing a binary number made up of bit positions 15 to 0. Figure 3.4 also spells out the names of certain registers and lists additional register features that are explained in this and other chapters.

Figure 3.4 ■ 8086 CPU registers.

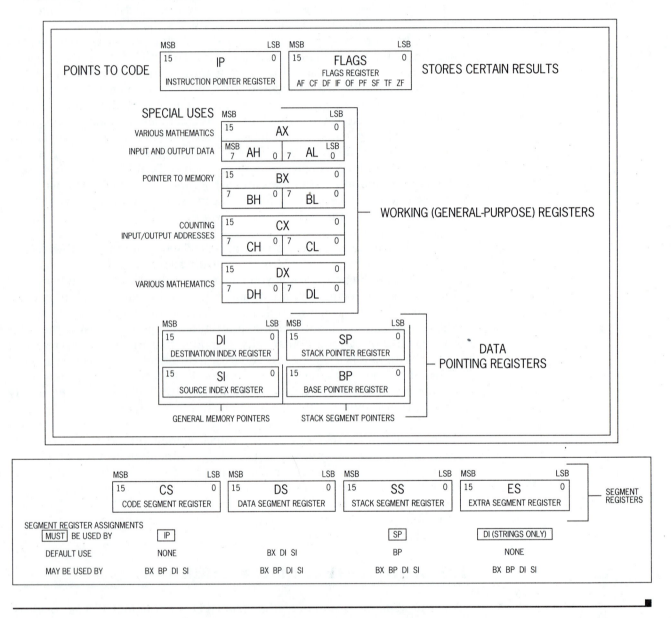

The registers of Figure 3.4 are the *only* ones we may use in a program. The 8086 register set is also called the programmer's *model,* because the CPU can be thought of as just these registers, and nothing else. Many, many other circuits exist inside an X86 CPU integrated circuit. Most X86 CPU circuit details are hidden from us and are the proprietary property of the manufacturer. Knowledge of the internal circuit details of the CPU, although such knowledge may be of intellectual interest, does not add to our ability to utilize the 8086. For our purposes, the 8086 CPU *register set,* and the *instructions that control that register set,* are the computer.

We shall now investigate the functions of each internal register, beginning with the most important, the Instruction Pointer.

The Instruction Pointer: IP

The CPU must be able to specify the addresses of the program opcodes that it is to fetch and execute. This is done using a register that holds the address of the next instruction to be fetched. This register is named the *Instruction Pointer* or IP *register* by the 8086 designers. The IP is 16 bits in length and can hold numbers from 0000 to 65,535 (0000h to FFFFh).

Each addressed location in an 8086 memory holds a 1-byte number. After an instruction is fetched from memory, and before it is executed, the IP is incremented to point to the next instruction. Unless the IP is changed by an instruction, it counts *up* as the program is executed. The counting action of the IP is completely automatic.

Do not assume that the IP contents are absolutely sequential—such as 1, 2, 3, 4, and so on—*after* a complete instruction is fetched and executed. A *single* instruction may occupy *several* consecutive code address locations because most 8086 instructions take *more* than 1 byte to encode. The resulting IP contents may change as 1, 3, 5, and so on, as multi-address (multi-byte) instructions are fetched for execution from code memory.

Flags

Certain computer operations require that the outcome of the operation be "remembered" for use by later instructions. Digital circuits remember events by storing unique binary numbers in dedicated memory locations for later recall and inspection.

Frequently, the outcome can be stored as a single bit. A binary 1 might mean that a certain result did occur, whereas a binary 0 might mean that the result did not occur. Both pieces of information are valuable. If a single bit stores a result, then that bit is named a *status flag* bit, because it acts as a signal flag for the CPU to use in later operations. A status flag bit stores the result of (the status of) an operation.

Flag bits are stored inside the CPU in bit locations with names that specify the type of result the flag bit signifies. For instance, the *Zero flag bit* is set to a 1 if the outcome of certain operations results in a register containing all binary 0s. Should the result of an operation be nonzero, then the Zero flag bit will be cleared to 0. Not all instructions affect the flags, and many instructions affect only a few of the flags. One must refer to an instruction reference, such as Appendix F, to find what flags are affected by which instructions.

Status flags are set automatically by the CPU when it executes an instruction. Other flags, named *control flags,* may be set or reset by the programmer using opcodes designed for that purpose. Control flags are used by the programmer to choose the way in which certain opcodes operate on data or to control CPU behavior. Some status flags may also be set or reset by the programmer in order to get them to a known condition in the program.

Because they are only a single bit, and the CPU is organized as a set of 16-bit registers, the flags are grouped into a register named the *Flag register.* The Flag register is 16 bits long. The Flag register may also be referred to as the *processor status word* or PSW, because it stores the latest results of CPU operations as a 16-bit, or word length, number. Other than being grouped into the Flag register, the PSW flags have no connection with each other and are set or reset independently of each other. Several of the PSW bits are not used for any purpose, and their contents are not documented for any given CPU operation.

Each flag in the Flag register is interpreted by flag decision opcodes in one of two ways: either the flag is set to 1, or it is cleared to 0. How the flag came to be in its present state is of no interest to the flag instruction. Flag opcodes operate on the state of a flag at the instant the flag instruction is executed. The programmer should follow flag-setting instructions with flag-decision instructions, to prevent any intervening instruction from changing the flags in error. Many, very-hard-to-find, problems are caused by programmer inattention to the flags.

The meaning of each flag contained in the Flag register is shown in Table 3.3. Status flags will have an *S* in brackets below the flag name and control flags a *C*. Flags that may be set by the programmer will also have a *P* in brackets under the flag name.

Not all the flags are of equal importance. You will find that the Carry flag and the Zero flag tend to be used most often in your programs, whereas others, such as the Parity flag, Sign flag, and Overflow flag are rarely needed.

At this point in our studies, when all is new and unfamiliar, we should not become overly concerned with the meaning of each flag. In Chapter 5 we make use of the Carry flag, CF, and the Zero flag, ZF, and Chapter 8 demonstrates some uses for the Auxiliary Carry flag, AF, and the Direction flag, DF. The Interrupt flag, IF, comes into play when we discuss hardware concepts in

Flag	Meaning of the Flag if Set to 1
Auxiliary Carry flag (AF) [S]	A carry out of the low nibble of a result or a borrow into the low nibble of a result.
Carry flag (CF) [S] [P]	A carry out of the most significant bit of a result or a borrow into the result.
Direction flag (DF) [C] [P]	Forces certain opcodes to process data in memory from high to low addresses.
Interrupt flag (IF) [C] [P]	Allows the CPU to be interrupted by an externally generated interrupt.
Overflow flag (OF) [S]	Applies only to signed numbers used in a program. Indicates that the result is too large for the signed number range chosen.
Parity flag (PF) [S]	The operation generated a result with even number of 1's in the least significant byte.
Sign flag (SF) [S]	The result of an operation contains a 1 bit in the most significant bit position.
Trap flag (TF) [S]	Performs a software interrupt at the end of the next instruction.
Zero flag (ZF) [S]	The last CPU operation resulted in a quantity that contains all binary 0s.

Table 3.3 ■ Flag Results

Chapters 11 and 12. The meanings of the Parity flag, PF, Sign flag, SF, and Overflow flag, OF, were discussed in Chapter 2. The remaining flag, the Trap flag, TF, is primarily used by software debugging programs and will not concern us further.

The A Registers: AX, AH, and AL

If the computer is to be of any practical use, it must be capable of performing basic mathematical and logical operations on binary numbers. Computer mathematical and logical operations were originally carried out in a circuit named the *accumulator,* which accumulated the results of the mathematical and logical operations. Most early computers had only one register, the accumulator register, that held the results of all mathematical and logical operations. The accumulator was normally named the *A register.*

Contemporary CPU designs, such as the 8086, usually have an A register and many other registers that can hold the results of the operations formerly assigned to the accumulator register. The A register does not play as important a part in computations for modern designs as it did in more primitive CPUs. Arithmetic is performed in the *arithmetic logic unit* or ALU. In the 8086 design, the ALU is a separate (hidden) circuit and can place results in most of the other working registers.

There still exist, however, some reserved mathematical operations for the A register in the 8086, such as multiplication and division, and several specialized mathematical conversions. The register that holds a result of multiplication and division in the 8086 is called the *AX register.* The AX register is 16 bits long and is divided into two 8-bit parts named *AH* (the high byte of AX) and *AL* (the low byte of AX). Either part of AX, or the whole register, may be used by 8086 programs. We shall find AX to be one of the most often used registers in the 8086 CPU.

The Stack Pointer: SP

Frequently, the computer must suspend the orderly fetching and executing of sequential instructions and change the number in the IP to one that is unrelated to the next sequential instruction address. An abrupt replacement of the next instruction address by an entirely new instruction address often happens when some result forces the computer to change what it is doing.

When the computer abruptly changes from one program sequence to another it is said to have *jumped, called, interrupted,* or *branched* from one part of the program in order to execute another part of the program. Program changes can be caused by the program itself, using software instructions for that purposes, or by hardware signals generated in the CPU circuitry.

Branching from one part of the program to another implies that the CPU will resume the original program at the instruction where the branch took place. To be able to resume the original program at exactly the right instruction requires that the address of the next instruction of the original program be saved prior to changing the IP to the new program address. The CPU "remembers" where it was in the original program by storing the next instruction address in an area of memory named the *stack.* The CPU accesses the stored address on the stack and places it back in the IP when it "returns" to the original program sequence. The CPU is able to return to its place in the original program because it stores a "return address" on the stack.

The stack is an area of *data memory* that typically holds program numbers that are meant to be used and then *discarded*. The CPU has a dedicated register, the *Stack Pointer* or SP, that is used to address the stack data area. The SP register holds the *last* stack address used in the stack area of data memory. The SP is 16 bits long and is not divided into 8-bit parts.

In general, the stack is a "scratch pad" type of memory where the program can *temporarily* store information for later use. The concept of a stack of data is analogous to the concept of coins in a car coin holder. The last coin pushed into the holder is at the top of the stack of coins in the holder. The first coin we can retrieve from (*pop off* of) the coin stack is the last one we *pushed onto* the stack of coins. Such a stack is called a *last in first out* or *LIFO stack.*

We could, if we wished, assign an address to each coin on our stack and keep track of each coin's address. But, as long as we know that we have a LIFO stack, then we can get to any coin simply by *popping* from the stack as many coins as have been *pushed* on the stack above our desired coin.

The 8086 stack is made of words (2 bytes) of data. The Stack Pointer register keeps track of the address in RAM where the last word was pushed. Each time a word of data is pushed onto or popped off of the stack, SP adjusts to point to the next remaining word of data on the stack.

Physically, the Stack Pointer counts *down* in memory as data is pushed onto the stack. As data is popped back from the stack, the Stack Pointer counts *up*. To prevent the SP from "hitting bottom" in the stack area of memory, it is usually set to begin at a *high* address in the stack area of memory. Figure 3.5 shows how the SP decrements when pushing and increments when popping the stack area of memory. As shown in Figure 3.5, the SP decrements and increments by 2 as 2 bytes (one word) are pushed or popped in the stack.

SP is the second *pointing register* we have discussed, the first being the IP. A pointing register is one that is primarily used by the programmer to hold the address of a location in memory. The register is said to "point to" the address, much in the same way that a telephone directory points to our home address. A telephone directory can be considered to be a pointing register to a city.

Pointing registers are convenient places to store an address of some data. Often we do not have to know the *exact* number that is in the pointing register. We only need to know that the pointing register has a valid address number in it that we can use.

In our branching example, when the original program changes to a second program, the CPU automatically stores the return address *on the stack, wherever* SP points at that *instant*. The address SP points to is called the *top of the stack*. On completion of the second program, the CPU pops the return address from the top of the stack back to IP. The actual value of SP is not important as long as we are careful to make sure that no other data was pushed on the stack, and none popped off, prior to the return.

We shall see more examples of stack operations when program branching to *procedures* are discussed in Chapter 6.

Figure 3.5 ■ The stack.

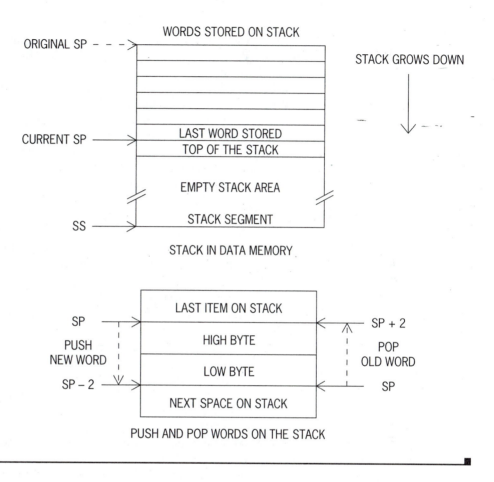

PUSH AND POP WORDS ON THE STACK

General-Purpose Registers

The registers that have been identified thus far are all the registers that are absolutely necessary to permit the 8086 to perform all standard CPU functions. With an IP, Flag register, AX, and SP, the CPU can fetch instructions from memory using IP, operate on data using AX, remember operation results in the Flag register, and store return addresses on the stack using SP. In fact, many early (1950s) computers contained similar registers and nothing more.

The number of internal registers contained in any microprocessor is, in large part, a function of just how many transistors can be squeezed onto an integrated circuit. The more dense the integrated circuit can be made, the more internal registers the designers can employ.

Many forces are at work when the designers begin with a supposed "clean sheet of paper" for a new CPU design. One force is the desire to design and implement an innovative and clear advance in the state of the art of computing. A balancing constraint that confines the designer is customer desire for compatibility of the new design with earlier models. Compatibility issues drive the CPU designers to try to maintain some sort of resemblance of the new CPU with older CPUs. Marketing forces also dictate that there be lots of "bells and whistles" on any new design. Typical marketing desires include the ability of the new CPU to easily handle high-level languages and advanced operating system software.

The designers of the 8086 family, in response to creative, customer, and market forces, went far beyond the minimum number of registers needed for a basic CPU and added many more. The addition of these extra registers greatly enhances the usefulness of the 8086 family. Even though the term *general-purpose* or *working register* may be applied to the next set of registers we shall discuss, there are still some opcodes that may work only if certain members of the set of general-purpose registers are used for the instruction. Specialized working registers are indicated by the heading *Special Uses* in Figure 3.4.

Instructions that *must* use a certain register are said to be *register-specific*. Because of the register-specific nature of some instructions, not all of the general-purpose register functions are interchangeable.

THE WORKING REGISTERS: BX, CX, AND DX

The precursor to the 8086 CPU was the Intel 8080 family of CPUs, which included the Intel 8080 and 8085 models, and Zilog's Z80. The Intel 8080 family is an 8-bit design that enjoyed popular acceptance by the technical community until it was overshadowed by the Zilog Z80. The 8080 family contained seven 8-bit registers named A, B, C, D, E, H, and L. The A register was the main math and logical operation register, and the remaining six registers were *working* registers. A working register is one that is interchangeable with other registers of the same type. Many 8080 instructions could use the registers as 8-bit types or combine the 8-bit registers into 16-bit pairs known as BC, DE, and HL.

The designers of the 8086, in an effort to maintain some compatibility with the 8080, retained the concept of the six 8080 working registers. The 8086 designers kept the 8080 working register set by including in the 8086 CPU three 16-bit general-purpose, or working, registers named BX, CX, and DX. Each working register is divided into two 8-bit registers in a manner similar to that of AX. Registers AL, CH, CL, DH, DL, BH, and BL of the 8086 correspond to the earlier A, B, C, D, E, H, and L 8-bit registers of the 8080.

The addition of working registers makes a computer design much more flexible. Working registers can be used to perform mathematical and logical operations, point to memory, take part in repetitive operations, or hold temporary data generated by the program. The ability to store temporary results in working registers greatly improves program execution speed. Any time data must be interchanged with memory, the CPU must produce numerous time-consuming control signals in order to access the data in the memory chips. Memory circuits are also much slower in responding to CPU control signals than are internal CPU circuits. Data in the internal CPU registers can be accessed much faster than data in memory. A program executes quicker when general purpose registers are provided in the CPU for storing temporary results as opposed to storing everything in memory.

Pointing Registers

The address of the next instruction that the CPU is to execute is contained in the Instruction Pointer. The IP is said to "point to the address of the next instruction." IP operation is automatic, with no programmer effort needed to oversee it. The IP is a register that points to addresses of code bytes in code memory. Every time a *byte* of code memory is read into the CPU, IP is incremented by *1* to point to the next byte address in code memory.

The SP register is a pointing register, and like the IP it is automatically used by the CPU when stack operations take place. SP, however, points to *words* in the stack and is incremented or decremented by *2* when accessing the stack area of memory. Other pointing registers—DI, SI, and BP—may be found in the working registers.

POINTING REGISTERS DI AND SI

A register that can point to addresses of data in memory, rather than addresses of code in code memory or data in the stack, can be of great convenience to the programmer. Registers that point to data addresses in memory (not instruction addresses) are very useful in fetching and storing large amounts of data. A second advantage of using data pointing registers is that the pointing register contents can be *changed as the program executes by the CPU*. By repetitively changing the data pointers, a program can access huge amounts of memory with a few program opcodes.

The 8086 contains two 16-bit registers that are primarily intended to hold data addresses, that is, point to data bytes located in addresses in data memory. The two 16-bit data pointing registers are named the *Source Index* or SI *pointing register* and the *Destination Index* or DI *pointing register*. An *index* register is the name given any register that is primarily used to point to data addresses, much like the index of a book points to subjects in the book. In addition, index registers usually can be automatically incremented or decremented (indexed) by various opcodes in order to quickly point to new data addresses.

Registers SI and DI are *not* divided into two 8-bit parts, as are the AX, BX, CX, and DX registers. SI and DI are used solely as 16-bit registers because they are not primarily intended to be used as working registers. The pointing registers SI and DI may be used, however, to temporarily store 16-bit data words just as may be done using the working registers.

There are no counterparts to SI and DI in the 8080 design. SI and DI were added to the 8086 design to facilitate moving large amounts of data from a source of data in memory to a destination for data in memory. Several opcodes can cause the DI and SI pointing registers to change their contents repetitively as data is accessed in memory, much as the IP register contents are automatically incremented as each code byte is fetched. Repetitive opcodes enable the programmer to access up to 64K of data addresses at a time with just a few opcodes.

THE BASE POINTER REGISTER BP

The 16-bit register SP points to the stack area of memory. Another 16-bit pointing register, named the *Base Pointer* or BP, also points to addresses in the stack area of memory. If we consider the SP as a pointer to the top of the stack, then the BP can be considered to point to the base (bottom) of an area of data.

Actually, the most significant difference between the SP and the BP is the fact that the CPU can change the SP *automatically* for certain stack operations. For instance, the programmer does not have to be conscious of the SP, or its operation, when using the branching capability of the CPU. The BP, however, has no automatic features, and the programmer must make a conscious effort to use the BP in the program. The inclusion of a BP, which has no counterpart in the 8080 CPU, adds to the overall flexibility in using the stack area of memory. Chapter 8 discusses some uses for the BP for processing data configurations named *tables*.

3.7 How the 8086 Addresses Memory Locations

Memory is anything that is not contained within the CPU. CPU registers, such as BX or DI, are not "addressed" because they have names, and using the name of a register lets the program access the register. Memory, however, is one large mass of byte locations that are all the same. The only way to get to any particular memory location is to assign it an address. Memory addresses can vary from address 00000h to address FFFFFh, or 1 million bytes of memory, each with a unique address assigned to it. How the 8086 CPU "generates" or makes up an address number is fundamental to understanding 8086 machine programming.

Segmented Memory

Figure 3.6 shows a portion of memory that has been arranged in groups of addresses called *segments*. Memory segments are pointed to by special 8086 registers called *segment registers*. Segmented memory is a unique feature of the X86 CPU family and has no counterpart in earlier 8-bit CPUs such as the 8080 or 8085.

Memory is, physically, a linear sequence of addresses, each address holding a single byte. The lowest memory address is address 0, the next address is address 1, and so on until the highest memory address is reached. The highest memory address for the 8086 is address FFFFFh. Although the physical memory is composed of a million sequential address locations, *how* each address is generated inside the CPU is completely up to the CPU designers.

Two kinds of memory addresses are generated by the CPU—code addresses and data addresses. Code address size determines how many bytes of code memory can be used for programs. Data address size limits the number of data bytes that can be used by the program. Usually, code address size and data address size are the same.

One choice for address generation is to use registers that all contain as many bits as are needed to specify any address in memory. For the 8086, this would require registers that are all 20 bits long (00000h to FFFFFh). The 8086 does *not* use 20-bit registers, but many CPU designs that compete with the 8086 make the register size equal to, or greater than, the number of bits needed to address memory. For instance, the 8080 CPU has an IP that is 16 bits

Figure 3.6 ◼ Segment arrangements.

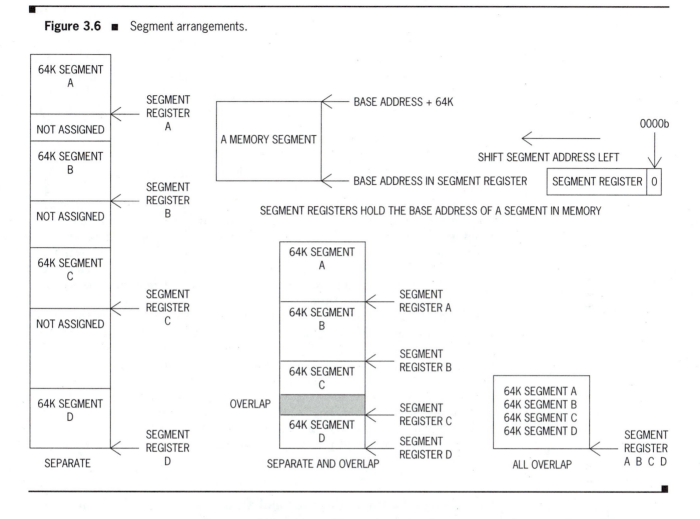

long, and thus can address 64K of code memory. The 8080 data addresses are limited to the same 64K. The 68000 CPU has an IP that is effectively 24 bits long, and thus can directly address 16M of code memory. The 68000 data addresses are also limited to 16M although the 68000 data address registers are 32 bits long.

Both the 8080 and the 68000 use the IP *alone* to address program code memory. The 8086 designers, however, chose *another* way to form code and data addresses, and their way uses 16-bit registers called *segment registers* to form the 20 bits needed to address memory.

Segment Registers: CS, DS, SS, and ES

The key to understanding the operation of the 8086, and the DOS operating system, is the concept of memory *segments*. As previously discussed, everything external to itself is considered to be memory by the CPU. Addressing memory is the primary task of the CPU, and most CPU time is spent accessing memory. Memory addressing flexibility, then, is very important if the CPU design is to be successful in the marketplace. The 8086 makes use of segmentation of memory to meet the goal of memory addressing flexibility.

Segment registers are 16-bit registers that hold the *beginning address* of some memory segment. The address in the segment register is the lowest (smallest) memory address for the segment. All other memory addresses in the segment are higher (greater than) the segment register address. A memory segment starts at the address in the segment register, and ends 64K *above* the address in the segment register. A segment is thus defined as any continuous memory area of 64K, which is the total size of the original 8080 CPU memory addressing range.

The 8086 segment registers are named *Code Segment* or CS, *Data Segment* or DS, *Stack Segment* or SS, and *Extra Segment* or ES. The names of the segment registers imply the nature of the memory area that each segment register is used to address. At any time, segmented memory is divided into as many segments as there are segment registers: code segment, data segment, extra segment, and stack segment. *Many* segments of the same type may be used by a program as it operates, simply by having the program change the number in the segment register. But, only *one set* of segment addresses can be *actively* in use at any one time.

Segmenting a monolithic 1M memory into 64K pieces reflects the earlier heritage of the 8080. The 8080 programs could not exceed 64K, and 8086 code segments are the same size: 64K. The significant difference between the 8086 and the 8080 is that the 8086 can address 1M of memory.

Figure 3.6 shows that segments, since they start at the addresses contained in the segment registers, can occupy any *mix* of areas in memory. The programmer can choose the addresses placed in the segment registers. Segments can be *separated* physically in memory, *overlap* in memory, or occupy exactly the *same* physical part of memory—as desired by the programmer.

Combining a Segment Register and an Offset to Address Memory

To form segmented memory addresses, the designers of the 8086 used a concept of *combining two short memory address numbers to generate a longer memory address number.* A complete segmented memory address is formed when one 16-bit register, a segment register, is combined with a second 16-bit *quantity* called an *offset quantity.* An offset quantity is *not always* a register. Many data memory addressing opcodes allow a fixed *number* (called a *displacement*) to be added to a pointing register to form an offset, which is then added to a segment register to make a physical address.

Keep in mind that the CPU *automatically* adds the offset address to the segment register to form a physical address in memory. The programmer instructs the CPU what offset and which segment register to use when accessing memory. After the CPU fetches the instruction, however, it automatically adds the programmer-specified offset to the programmer-specified segment register to access memory.

Offset means the *distance*, in bytes, from the start of the segment to a memory address. All offsets are *relative* to the beginning address of the segment, and offsets can range in size from 0000h to FFFFh bytes.

We are all very familiar with the concept of an offset. If you are in a building that happens to be in the mountains, the floor you are on is an offset, in feet, above the particular ground level on which the building sits. Ground level is analogous to the start of a segment in memory. The *absolute* height of the floor of the mountain building, above sea level, is quite high. Should an identical building be built near the sea, the same floor is at the same offset above ground level, but the *absolute* height of the floor above sea level has changed. In the same way, a segment of memory can begin at a very low or a very high address in memory. Similar offsets into different segments are identical when referenced to the beginning of the segment. A memory segment is the "ground level" reference for offsets *above* the beginning of the segment.

For example, "sea level" in the memory is at address 00000h. A high-level address could be at memory address F0000h. A segment could begin at address 00000h, with an offset of 1000h. The combination of segment and offset is then absolute memory address 01000h. The same offset, in the high address segment, would give us an absolute memory address of F1000h. The offset is still 1000h in both cases, only the beginning address of the segment has changed.

How a Segment Register Is Combined with an Offset

An absolute 20-bit (1,048,576-byte) memory address is formed by the 8086 CPU when a 16-bit segment register is *combined* with a 16-bit offset quantity. The offset quantity could be the contents of a 16-bit *pointing register*, a 16-bit data *number*, or some *combination* of register contents and data numbers. We have already studied the pointing registers; they are any of the previously mentioned pointing registers IP, SP, BP, SI, or DI and the working register BX. Offset data numbers can be any 16-bit constant *number* used by the programmer.

One megabyte of physical memory requires 20 address bits in the address number. The internal registers are all 16 bits in length, and thus can only address, by themselves, 65,536 bytes of memory. If two 16-bit registers are *directly* added together, the result cannot exceed 17 bits, or a total address range of 131K. The problem of getting 20 bits by adding together two 16-bit registers is solved, in the 8086, by *automatically shifting* a segment register 4 bits (1 nibble) to the *left* before adding it to an associated offset quantity. Shifting the segment register by 4 bits makes the segment register essentially 20 bits long.

As the segment register is shifted left, zeroes shift into the bits vacated in the least significant nibble of the segment register. The least significant nibble of a left-shifted segment register is thus *always* 0h. The sum of the 20-bit left-shifted segment register and the *unshifted* 16-bit offset number generates a 20-bit physical address number.

The automatic shifting and adding of a segment register and an offset takes place in a special CPU circuit named the *bus interface unit* or BIU. The BIU takes care of generating all 8086 addresses to the memory address lines, which are collectively called the memory *bus*. Again it must be emphasized that address generation is an automatic CPU operation similar to fetching code. The programmer can specify the segment register and the segment offset in the computer code. The BIU forms the 20-bit memory address when the code is fetched and executed.

As an example of segmented memory addressing, say the IP contains F12Ch and the CS contains 5ABEh. The actual 20-bit physical code memory address that is generated to fetch the next instruction is automatically formed in the BIU as follows:

```
IP  = F12Ch unshifted =   F12Ch
                        +++++ by the BIU
CS = 5ABEh shifted    = 5ABE0h
                        69D0Ch = 20-bit code address
```

As another example, let DS contain 4000h and BX contain 5000h. A 20-bit address in the data segment pointed to by DS is automatically formed in the BIU as follows:

```
BX = 5000h unshifted =   5000h
                       +++++ by the BIU
DS = 4000h shifted    = 40000h
                        45000h = 20-bit data address
```

Instead of thinking of *shifting* the segment registers 1 nibble to the left, we can also think of them as 20-bit registers that *increment by a count of 16, or 10h*. Thus, if CS contains 1000h and DS contains 1001h they point to areas of memory separated by 10h bytes, *not* 1h bytes. This is shown below.

```
DS = 1001h, points to address 10010h
CS = 1000h, points to address 10000h
DS – CS pointing difference is  00010h
```

The 8086 can address 1,048,576 bytes of memory. One megabyte of integrated circuit memory was considered to be beyond the most extravagant needs of most users when the 8086 first appeared on the scene. Early PCs might have had as little as 16K of internal memory; 128K or 256K was reserved for the legendary "power user." But, the eternal story of more for less continues to be valid in electronic manufacturing. Integrated circuit memory is now so cheap that a megabyte of memory is considered the minimum amount for any serious programmer. Unfortunately, programs have expanded, in a sort of bureaucratic manner, to fill the memory allotted to them. One suspects that there will never be enough memory.

Segmented memory, although it takes some getting used to, is a very handy way to divide total memory into well-defined pieces. Code segments, for instance, let us place many programs in memory, each in a different segment. By keeping track of which program is in which segment, we can insure that one program will not interfere with another. Operating systems, such as DOS, can use the segments to keep everything organized in memory.

The segment concept introduces the whole subject of memory addressing and some definitions of memory address nomenclature are in order here.

3.8 Physical and Logical Memory

Physical Memory Addresses

As discussed previously, the fundamental model of a computer is a CPU and associated memory devices. Memory is physically realized in the form of integrated circuits that are interconnected to the CPU in order to receive electrical signals from and to interchange data with the CPU. (Chapters 11 and 12 will discuss the necessary electrical connections between CPU and memory.) For the present, we shall view memory as some number of addresses, each address containing a 1-byte binary number. The integrated circuit memory is the physical memory available to the CPU. Each memory address generated by the CPU will ultimately end at a single physical location within one of the memory circuits. The addressed locations that receive the electrical signals generated by the CPU are the real, physical memory addresses.

The physical memory arrangement, and the means by which address signals reach the physical memory, are normally the concern of the computer system circuit designer. Programmers tend to view memory in a more abstract manner as *logical* addresses.

Logical Memory Addresses

Another name for the segment registers might well be *base registers*, except the term *Base Pointer* has been taken by the BP register. Segment registers hold what is known as the segment, or *base*, address of the final, physical memory address. The programmer who uses an operating system, such as DOS, is *rarely* aware of the actual segment base address numbers, because DOS assigns the segment register values when the program is loaded into RAM.

Physical addresses are formed when the left-shifted segment base address is added to the offset address. The combination of segment register base address and offset address is the *logical* address in memory. The programmer will not know the actual, physical address in memory because that requires knowing the actual segment register base address contents. But, the programmer will know the logical address, which is some *known* segment register *name* added to a known offset.

To illustrate the difference between physical and logical addresses, say the CS register is set to a base address of 5678h in memory and the offset is 1234h. The programmer could, *if* the contents of CS were known, calculate the *physical* address of 579B4h by adding the shifted segment base address to the offset. But, if the programmer knows only that CS will be added to an offset of 1234h, then the *logical* address is CS + 1234h. The actual contents of CS is usually not known by the programmer because the number in CS is assigned by DOS when the program is loaded from disk to RAM. The programmer knows the logical address is in the code segment, with an offset of 1234h, but the absolute, physical address is generally unknown. The programmer trusts DOS to make sure the code segment is placed where no harm will come to the program.

Offset Address

We shall use the term *offset address* when discussing memory addressing to mean the *final* number that is added to a segment register to yield a physical memory address. An offset address is an unsigned (positive) 16-bit number that is *added* to a left-shifted segment address to form a physical address. The offset address may be a number found in one of the nonsegment registers, a combination of two nonsegment registers, or a number added to a nonseg-

Figure 3.7 ■ Offset address calculations.

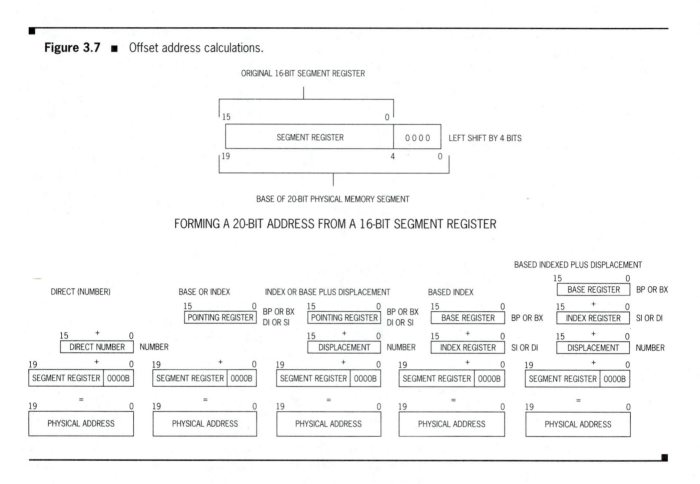

FORMING A 20-BIT ADDRESS FROM A 16-BIT SEGMENT REGISTER

ment register(s). Figure 3.7 shows how offset addresses may be added to segment registers to form physical addresses.

As can be seen in Figure 3.7, an offset address can be a *number*, a *pointing* register, a *number added to a pointing* register, or *two pointing registers added together, and then to a number*. Chapter 4 discusses, in greater detail, exactly how numbers and pointing registers may be combined to form segment offsets.

Legal Segment Register–Offset Register Combinations

The 8086 CPU uses one of the four available segment registers to form a physical address depending on what sort of program action is taking place. For instance, the Code Segment register, CS, is used with offset addresses found in the IP when opcodes are to be fetched from the code segment. The Stack Segment register, SS, is used with the offsets found in the SP or BP when the stack segment is accessed. The Data Segment register, DS, may be used when data areas of memory are pointed to by the offset addresses in BX, DI, or SI.

Another name for the offset address in memory is the *effective* address. An effective address is the final number, when all is said and done, that is added to a segment register to generate a physical address in memory. Usually, effective addresses are offsets into data memory.

Matching a segment register with an offset register is *not* random. Certain offset registers are *always* used with certain segment registers, unless the *programmer specifically requests* another segment register in the program. Figure 3.4 lists, below each segment register, the pointing registers that *must* be used, are used by *default,* and *may* be used with a particular segment register.

The four segment registers—CS, DS, SS, and ES—are automatically left-shifted and added to one of the pointing registers whenever the pointing register is used in an instruction. The programmer *may,* in some cases, *override* the automatic segment register selection by *naming* another segment register *as a prefix to the instruction.*

The *automatic* association of segment registers with offset registers is as follows:

- ◼ The IP offset register is *always* added to the CS segment register when addressing code memory. There is *no* override available to the programmer.

- ◼ The BP or SP offset register is automatically added to the SS segment register when stack operations are in progress. The programmer may *not* override the automatic selection of SS with SP. The programmer may override the automatic selection of SS with BP and choose segment register CS, DS, or ES.

- ◼ Any offset quantity using registers BX, DI, and SI is automatically added to the DS base register. The programmer may override the automatic selection of the DS register and use any one of the other three segment registers CS, ES, or SS.

- ◼ The offset register DI is automatically used with the ES base register when *string operations* (see Chapter 8) are taking place. The programmer may *not* override the selection of DI with ES when string operations are done.

The pointing register–segment register pairings are not as arbitrary as might first appear. Clearly IP must pair with CS, and SP with SS, because these registers are involved in automatic CPU operations. It seems reasonable to pair the general-purpose *data pointing* registers BX, DI, and SI with the *data* segment register DS.

A Word on Notation

Addresses in memory are numbers that may be expressed using two methods of notation. The first method uses a hex number to specify the actual physical address in memory. The physical address is the actual hardware location in memory where the addressed data resides. Physical addresses are expressed as *five* hexadecimal numbers. As an example, F123Ah is a physical address.

The second memory notation method uses a combination of the segment register number and the associated offset to specify the address in memory. The segment register is given first, then the pointing register, separated by a colon. For example, the notation F000h:2000h specifies segment F000h and an offset of 2000h into the segment. Physical addresses are expressed using actual segment register *contents.* For example, A000h:1000h represents physical address A1000h. Logical addresses are expressed using the *symbol* for one of the segment registers, for example, CS:0200h specifies offset 200h into the code segment.

3.9 Grouping 8086 Registers into a BIU and EU

As previously noted, memory addresses are generated in an 8086 CPU circuit called the BIU. Figure 3.8 shows the segment registers and working registers that we have discussed divided into two larger groups—the *execution unit* or EU, and the *bus interface unit* or BIU. Although these groupings are properly the domain of hardware discussions, a short introduction to these concepts is appropriate.

Memory Access and Program Speed

Computers are advertised by the *millions of instructions per second* or MIPS, much the same way that cars are sold by horsepower. Generally, the more instructions that a computer can execute in a given time, the better the computer brand is able to compete for customers. Instruction execution time is a complex mix of CPU clock speed, instruction coding, and internal CPU

Figure 3.8 ■ CPU partitioned into an EU and a BIU.

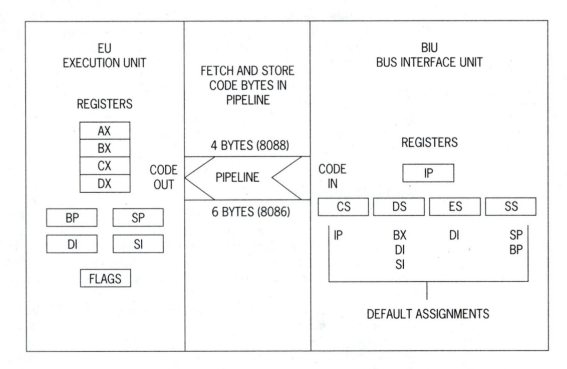

circuits. A significant factor in execution speed is, however, the time it takes for the CPU to fetch instructions from external code memory. If the CPU is limited to fetching an instruction before it can execute it, then there is much "dead time" while the CPU waits for the complete instruction to be fetched from memory.

Almost all early 8-bit CPUs used the approach of "fetch-fetch-fetch-execute-fetch-fetch-fetch" when running a program from code memory. During the time that an instruction is fetched, the CPU is waiting, wasting time. A better approach is to let the computer be composed of two parts that operate asynchronously. One part, called a BIU in the 8086, fetches instructions from code memory whenever the memory is free. Another part, called the EU in the 8086, executes instructions on a continuous basis. The BIU gets instructions bytes and stuffs them in an internal memory, called an instruction *queue* or *pipeline.* The EU gets the instructions out of the queue and executes them as fast as it can. The BIU puts instructions in one end of the pipeline, and the EU takes them out at the other end of the pipeline. Thus, a BIU-EU combination can fetch and execute code simultaneously.

There are times, of course, when the EU needs to access memory for data. When the EU needs access to memory, the BIU stops filling the instruction queue and attends to the needs of the EU. Then, while the EU is busy with other things, the BIU keeps filling the pipeline. The 8086 has an instruction queue of 6 bytes (enough to hold one or more opcodes). The 8086 queue is very modest by current standards.

We now are equipped with a software model of the 8086 CPU. The reader, if desired, may now begin to write programs for the 8086 by advancing to Chapter 4. Those who are curious to know how the 8086 family has evolved may wish to browse through Appendix G.

3.10 Summary

A fundamental computer model can be divided into two parts, a central processing unit (CPU) and memory. The CPU fetches binary instruction numbers from memory and performs binary operations on other numbers as directed by the opcodes. The sequence of CPU opcodes in memory is called a computer program. Results of the binary operations are stored in memory as data. Fetching opcodes is an automatic function of the CPU; all other CPU actions are controlled by the program in memory. Memory may be further divided into high-speed storage named solid-state memory and low-speed storage devices named peripherals.

The 8086 CPU consists of the following 16-bit registers:

Instruction Pointer, IP

Code Segment Register, CS

Data Segment Register, DS

Stack Segment Register, SS

Extra Segment Register, ES

Working Registers AX, BX, CX, and DX

Stack Pointer Register, SP

Stack Base Pointer Register, BP

Index Registers SI and DI

Flag Register, FLAGS

Up to 1M of physical memory can be addressed using a combination of segment register contents added to an offset number. The offset may be the contents of register or a fixed number. Segments may be up to 64K in size.

Further discussion of the X86 family members can be found in Appendix G. The X86 family consists of the 16-bit 8088, 8086, 80188, 80186, and 80286 CPUs and the 32-bit 80386, 80486, and Pentium CPUs. Each family member can be equipped with a math coprocessor, either as an addition to the CPU or integrated on the same chip with the CPU. The 80286 and subsequent CPU models are capable of use in multitasking operating system computers.

3.11 Questions

1. Name two basic components of a computer.
2. Computers deal only with binary numbers. Name two types of computer numbers.
3. Identify six binary number lengths, starting with the smallest.
4. What is the function of an address?
5. What is the difference between fetching a number from memory and reading a number from memory?
6. What is the function of an operating system program?
7. What is software?
8. What is hardware?
9. Name three high-level languages.
10. Why are high-level languages machine-independent?
11. Why must the programmer know about the CPU in order to program in assembly language?
12. Name three assembly languages.
13. Identify two reasons to program in a high-level language.
14. In general, which takes more space in memory, an assembly-coded program or a high-level language program? Why?
15. List four reasons to program a CPU in assembly language.
16. List four types of utility programs.
17. What type of file goes into an assembler program?
18. What type of file comes out of an assembler program?
19. What utility is used to test a program?
20. Name the programs that are used to convert from human thoughts to binary-coded computer instructions.
21. List the names, and functions, of the utility programs included with this book.

Write, and manually assemble, the following programs for the PAL CPU. All memory addresses include the starting and ending addresses. Begin your code at memory address 0000h.

22. Invert all the data from memory locations A000h to A003h.

23. Place 00h in all memory locations from 2000h to 200Ah.

24. Move the data in addresses 1000h to 100Ah to addresses 2000h to 200Ah. For example, move the data at address 1000h to address 2000h, and so on until data at address 100Ah is moved to address 200Ah.

25. Make every odd bit (bit 1, 3, 5, and 7) in memory addresses C000h and D000h a 0. Do not change any other bit at each address.

26. Make the low-order nibble of the byte at address 1234h a 1. Do not change the high-order nibble.

27. Determine what you must do to assemble the program of Exercise 24 beginning at memory address 0100h.

28. Add the instruction mov b,a to the beginning of the program of Exercise 24 and re-assemble the program.

29. Show how PAL stores the word *ABCDH* in memory locations 1234h and 1235h.

30. Explain the differences between memory and I/O.

31. What determines the address of the first instruction in memory?

32. Explain why you cannot tell the difference between a code number and a data number, if viewed randomly in memory.

33. Name two automatic (not the result of a program code instruction) actions a computer performs.

34. How many bytes are addressable by the 8086?

35. What register holds the address of the next instruction?

36. Name all of the segment registers.

37. List the names of the index registers.

38. Name all of the registers that are divided into 2 bytes.

39. List the flags, and the situation that exists if a flag is a 0.

40. Describe the stack area of memory, and how it is addressed.

41. If the CS register contains the number B000h, and the IP contains the number ABCDh, what is the address of the instruction?

42. If the SS register contains 1234h, the CS register contains 4567h, the IP register contains 1000h, and the SP register contains 2000h, what is the address of the next instruction? The next stack operation?

43 What is the maximum amount of memory (in segments) that can be addressed at any instant?

44. What is the minimum amount of memory (in segments) that can be addressed at any instant?

45. Why are memory segments limited to 64K?

46. Determine which of the following statements are true.
    a. Two segments can overlap.
    b. All four segments can overlap.

c.   No segment can overlap another segment.

d.   All four segments can be identical.

47. How many complete 64K segments can be placed in memory?

48. How many segments, of any size, can be placed in memory?

49. How does a logical address differ from a physical address?

50. How does an offset address differ from a logical address?

51. List the default segment register–offset register pairs.

52. Name the registers that can be paired with no optional segment register.

53. Explain the functions of the BIU and the EU.

54. Why does a pipeline speed up program execution?

55. Are data bytes placed in the pipeline by the BIU?

Appendix G

56. List all of the X86 family members that you find in this section, and that you read about in outside reading.

57 How does the 80186 differ from the 8086?

58. What are the two operating modes of the 80286?

59. A selector register points to what type of data?

60. The 80286 is designed to support what type of operating system?

61. What is a virtual memory?

62. How much physical memory can the 80286 address in real mode?

63. In what important way does the 80386 differ from the 80286?

64. How does the 80486DX differ from the 80486SX?

65. Explain the function of a math coprocessor.

4

ASSEMBLY
LANGUAGE
PROGRAMMING

Chapter Objectives

On successful completion of this chapter, you will be able to:

- State the steps taken to write an assembly language program for an X86-based personal computer.
- Understand how an assembler uses a directive.
- Understand the 8086 assembly language syntax.
- Understand the register addressing mode.
- Understand the immediate addressing mode.
- Understand the direct memory addressing mode.
- Understand the indirect memory addressing mode.
- Assemble and debug a program supplied in the text.
- Write and debug your own simple programs.

4.0 Introduction

All contemporary digital computers, no matter how powerful, operate using binary circuits under the control of binary, machine-language codes.

We begin in this chapter by discussing the concepts necessary to understand how to program any computer using assembly language instructions. Next, we introduce some rules of grammar for the 8086 assembly language and study exactly how data addresses are expressed. In Chapter 5, a limited set of 8086 instructions is introduced. A small set of 8086 instructions is sufficiently powerful, however, for the problems at the end of each chapter.

All computers, in the final analysis, are identical in the methods they employ to produce useful computing results. Because all computers are fundamentally similar, the way in which they are programmed in assembly language is the same no matter what particular model is programmed. Once you have learned to program the 8086 in assembly language, you can use the same techniques, with allowances for architectural differences, to program any computer.

We tend to become focused on the assembly language of the particular central processing unit (CPU) at hand. Keep in mind the generality of the actions that take place when assembly language instructions are used. A common error, for beginning programmers, is to learn one CPU and assembly language and then resist changing to another CPU type. Closed-minded behavior soon leads to technical obsolescence and stifles personal growth. You may be sure that the industry will produce new and vastly improved microprocessors about every 5 to 10 years. Embrace the new CPUs and learn to use them; for, no matter how advanced they may seem to be, the basic functions will remain the same.

4.1 The Programming Process

As noted in Chapter 3, a digital computer is a sequential circuit that fetches a program instruction from code memory, decodes the instruction, and then executes it. The computer instructions are placed in code memory by the programmer with the aid of utility programs that have been written by others to facilitate this chore.

The challenges facing a beginning programmer are extensive because of a complete unfamiliarity with any of the procedures and tools that must be mastered before the first, simple program is run. A sequence of learning and action items found on the road to the first programs is listed in Table 4.1.

The list appears to be daunting, but the difficult parts are only the last three items.

Programming can be reduced to procedural steps that can be applied to any problem by a well-trained individual. There is still much art in the process though, and to master the art requires more than formal education and rote learning. Some individuals possess a unique programming talent that lets them "see" ways of doing things that are not obvious to most of their peers. People who possess extraordinary programming skills, whether through natural insight or (as is more often the case) by dint of long, hard work, are often

> a. Learn about CPU architecture (internal register arrangement).
>
> b. Learn a small, but complete, set of CPU assembly language instructions.
>
> c. Learn how to use a computer and its host operating system.
>
> d. Learn how to use a text editor program.
>
> e. Learn how to use an assembler program.
>
> f. Learn how to use a linker program (optional).
>
> g. Learn how to use a debugger program.
>
> h. Practice writing and running small programs until proficient.
>
> i. Add new assembly language instructions and master their use.
>
> j. Continue to write larger and more complicated programs.

Table 4.1 ■ Learning Process for Assembly Language

referred to as *wizards*.[1] I cannot guarantee that you will become a wizard, but I do guarantee that you can become a competent programmer.

Understanding the Computer Architecture

Before any program instructions can be written, the programmer must have a knowledge of what computational *hardware* has been provided by the manufacturer of the computer CPU.

When a design team sets out to build a new CPU, the team has some understanding of what market the new CPU will serve. Microprocessors can be built to operate toys, appliances, printers, robots, automobiles, and guided missiles. Microprocessors are built to be used in microcomputers that range in power from limited-function calculators to multi-tasking *servers* that can execute millions of instructions per second and address huge amounts of memory. Microprocessors can be used to display 8 digits on an LCD screen or generate three-dimensional graphics and process video signals for special visual effects. For each application there is a "best" microprocessor design for that application. *Best* meaning that the microprocessor can do the job in the most cost-effective manner.

The 8086 is a general-purpose CPU that can fill many roles from machine control to personal computing. The 8086 family is best known, by the general public, for its use in personal computers. Many people would be surprised, however, to learn how many of the X86 family are hidden away as embedded controllers in other machines.

Assembly language programming is very machine-specific, that is, assembly language instructions written for an 8086 microprocessor cannot be used on a 68000 microprocessor. The same results can be obtained using an 8086 or a 68000, or any other CPU, but the assembly language instructions will be different in each case. We find the same situation regarding human languages. One may write a poem in French or Turkish; each poem may say the same thing, but it will not be using the same words as the other.

---

[1]*Computer Dictionary* (Redmond, WA: Microsoft Press, 1991), p. 368.

The first task at hand, then, is to gain some understanding of what internal registers are built into the CPU by the manufacturer. The internal architecture of the 8086 is discussed in Chapter 3. A model of the internal registers from Chapter 3 is repeated in Figure 4.1.

Review the 8086 CPU architecture, shown in Figure 4.1, before proceeding. Once you are comfortable with the main structure of the 8086 CPU, proceed to the next step from the list in Table 4.1, which is to learn how to use 8086 assembly language instructions.

Figure 4.1 ■ 8086 CPU registers.

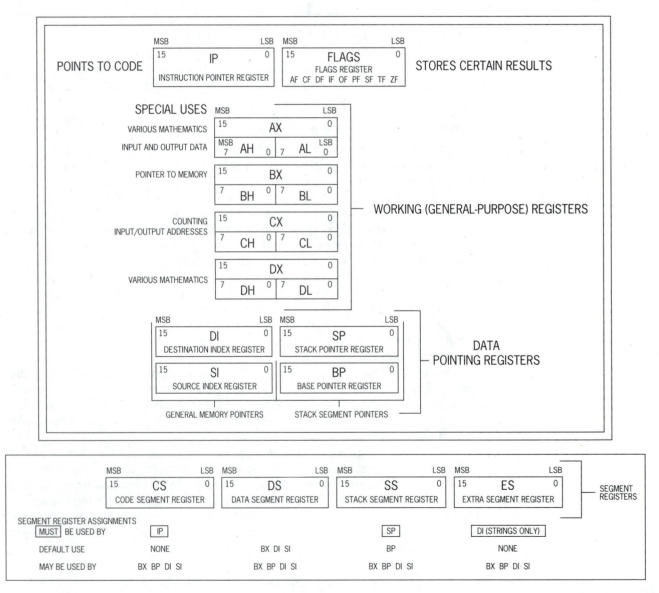

```
mov r0,r1          (Intel 8051)
ld a,b             (Zilog Z80)
move.w a1,d1       (Motorola 68000)
mov ds,ax          (Intel 8086)
txa                (Rockwell 6502)
```

Table 4.2 ■ Examples of Register-to-Register Data Moving Instructions

Understanding Assembly Language Syntax

8086 assembly language is a subset of the English language, but the rules of spelling, grammar, and punctuation are more *rigidly* defined: There are no alternate spellings in assembly language, and no leeway in how a comma is placed. Assembly language is rigid because *assembler programs* cannot tolerate any deviation from the way instructions are defined. The *rules* that apply to a particular assembly code language, and a particular assembler, are grouped under the heading *syntax*. Assemblers are very picky about syntax.

For instance, we might say to a friend, "I ain't got no money," or "I'm broke," or "I find myself temporarily financially embarrassed." In all cases we should be understood. Not so with assembly language. Each CPU assembly language, and each assembler program for that language, admits only one way of expressing a particular action. For example, the assembly language instructions that move data from one internal register to another in several popular microprocessors are written as shown in Table 4.2.

Clearly, each instruction, which has the same result for each computer architecture used, is expressed in very different ways. One CPU uses the term mov and another uses the term ld to get the same result: copy data from one register to another. Each instruction listed in Table 4.2 uses a different mnemonic (memory aid) for the same CPU action. Each mnemonic reminds us of what the instruction does when it is executed by the computer CPU. For example, the Z80 "loads" (ld) data, the 68000 "moves" (move) data, and the 8086 "moves" data also (mov) but with a different mnemonic spelling.

Different manufacturers have adopted different mnemonics for the same CPU action for various competitive reasons. Originally, when almost all programs were written in assembly language, it was thought that the programmer might become "married" to the mnemonics, and thus would be loath to change over to a competitor. Legal questions also arose, because instruction mnemonics could be copyrighted and could not be used by another company without permission. The industry has matured. Instruction mnemonics are no longer subject to copyright law, and most programming is now done in semi-generic, high-level languages. Microprocessor manufacturers have remained with the original forms of their early mnemonics, however, and thus we have 8086 mnemonics that can trace their roots to the first Intel primitive 4-bit microprocessor, the 4004.

Understanding the Assembler Program

Assembly language programming, in the bad old days, was done entirely in a manual fashion, as examined in Chapter 3 for the fictitious PAL CPU. An outline of a typical manual assembly code process is listed in Table 4.3.

a. Write the program, on a pad of paper, using the instruction mnemonics.

b. Convert each instruction to its hexadecimal code, using a code book that lists each instruction mnemonic and its hexadecimal code.

c. Assign each code byte an address in code memory.

d. Use a paper tape punch to record the program code bytes.

e. Enter the program into computer code memory via a paper tape reader under the control of a reader utility program.

f. Find the mistakes, re-write and re-code the program, and re-enter it.

Table 4.3 ■ Manual Assembly Process

The process listed in Table 4.3 is very tedious, especially when new lines of code have to be inserted into the program, and all of the addresses of each line of code reassigned. Today, with the availability of inexpensive personal computers and computer software, the process has been greatly speeded up. The process for writing assembly language programs is now as shown in Table 4.4.

Writing the program is no easier using automated means. But the process of converting assembly language instructions to hex codes, assigning addresses, loading the program, and changing the program is much, much faster using automated assembly.

The increase in the pace at which programs can be edited and assembled has not come without a price, however. Now, in addition to having to learn the instruction mnemonics for a CPU, the budding programmer must also learn to use a personal computer, a text editor, and an assembler program.

Assembler programs that were used with first-generation 8-bit CPUs were fairly easy to use. Assemblers for 8-bit CPUs generally dealt with data in byte sizes, had relatively small memory address ranges of 64K or less, and were not concerned with the needs of complicated operating systems. The advent of 16- and 32-bit CPUs, such as the X86 and 68000 family, with their greatly expanded data sizes and addressing capabilities, has resulted in equal-

a. Write the program in instruction mnemonics, using a computer and text editor.

b. Store the edited assembly language program in a disk file.

c. Use a program, called an *assembler*, to convert the assembly language file stored on disk to a second disk file. The second disk file contains binary equivalents of the assembly language file that the computer can execute (machine language instructions).

d. Load the assembled machine-language program from disk into computer RAM and test the program using a debugger program.

e. Find the mistakes, re-edit the program, re-assemble, load, and test again.

Table 4.4 ■ Automated Assembly Process

ly more powerful assemblers. (Equally more confusing assemblers might be more accurate.)

ASSEMBLER DIRECTIVES

An assembler program is a program like any other, one that has its own programming language that must be understood before any programs can be written. The complexity of assemblers used for modern 16- and 32-bit CPUs is caused by two things.

First, data can exist in byte (8-bit), word (16-bit), double word (32-bit), and, perhaps, quad word (64-bit) size. There must exist, therefore, corresponding special assembler terms for all of the various data sizes.

Second, the range of addresses in memory, for code and data, can easily exceed a million bytes. Moving within this address space requires assembler terms that indicate how far (in bytes) it is between one location and another in memory. For instance, an address could be a "short" distance away, or "near," or very "far" away.

Other assembler program concerns deal with such things as setting the IP contents, assigning names to numbers, and defining data, stack, and code areas of memory. The title given to an assembler program language term is a *directive*.

The 8086 architecture adds a twist to addressing memory through the use of segment registers to help form absolute physical addresses in memory, as discussed in Chapter 3. Because of the segmented nature of 8086 memory, the 8086 has two fundamental ways of measuring distance in memory. One address category includes all memory addresses that are in the *same* segment. Addresses located in the same segment of memory are called *intrasegment* addresses. The second address category includes all memory addresses in any *other* memory segment. Addresses in another segment are *intersegment* addresses. Addresses in the same segment may be a *short* way away, or at least *near*. All intersegment addresses are *far* away, in another segment. Segmented memory results in the need for additional 8086 assembler directives that indicate short, near, or far addresses.

Appendix A discusses the assembler used in this book (A86) and some of the more commonly used A86 directives. The use of many of the A86 assembler directives will be demonstrated and explained in the example programs of this and following chapters. We examine two very common data size directives next.

TWO ASSEMBLER SIZE DIRECTIVES

We shall need two directives to the assembler at the very beginning of our studies. One is the directive byte ptr or b, and the other is the directive word ptr or w.

These two directives are needed because of the way bytes and words of data are addressed in memory by the CPU. Memory is organized so that the *smallest* piece of data that has a *unique* address is a byte. When we wish to access a byte of data, we only need to specify the address where that byte resides. A word of data occupies *two successive* byte addresses, with the low order byte of the word in the smaller (lower) of the two successive byte addresses. A word of data is addressed by specifying the address of its *low* order byte. Thus, an address in memory could be the address of a byte, or the address of a word.

How is the assembler to know which size of data the programmer wishes to use? Sometimes the data used in an instruction indicates the size of the data to be accessed in memory. When a 16-bit register is specified, it is clear that the data size is word-length. An 8-bit register half indicates byte-length data. Whenever the instruction cannot make clear which data size is to be used, then the programmer must *tell* the assembler by including the keywords byte ptr (b) to mean a byte of data, or word ptr (w) to indicate a word. Keywords are intended to leave no doubt as to the programmer's intentions.

We shall use as few assembler directives as possible when we first begin to write programs. Later, after you are comfortable with your programming skills, we can examine what the assembler directives do, and why they are necessary.

Understanding the Problem to Be Solved

The problem at hand, whether assigned by the teacher as a classroom exercise or presented by an actual application, must be *fully* understood by the programmer before any programs are written. At the problem *definition* stage, it is not important to know which computer is to be used; any computer type could do the job. But the most powerful, multi-byte, multi-megahertz computing wonder will do a poor job if the program does *not* solve the problem. There are no shortcuts here; you must persist in asking questions of the teacher or the customer, or nature, until both you and the end user are in agreement as to *what* is to be done.

Many times, especially for complex and expensive applications, a formal, written definition of the problem (a contract) is formulated and agreed on by all parties. The problem definition phase is the *most important* part of the programming process.

Designing the Program

Once the problem is known, then the programmer can begin to lay out a sketch or overall plan of how to solve the problem. The plan is called an *algorithm*. An algorithm is any scheme, such as a list of actions or a diagram, by which the programmer is guided in solving the problem. It can be very precise or quite general. For instance, the algorithm for finding the square root of a 4-word number can be exactly specified, and in great detail. The algorithm for adding together several sets of numbers in memory can be quite general.

A common technique used to *document* an algorithm are diagrams called *flow charts*. The history of flow charts goes back to the dawn of the computer age, when they were considered essential in the programming process. Flow charting very large programs soon became unproductive however, because of their ever-increasing complexity, so large programs were divided into independent pieces called *modules*. Each module is then written using standard programming techniques named *structures* and is documented using written *comments*. Almost all of the example programs in this text are short, and comments will be used to document them.

Writing and Testing the Program

Once the program algorithm has been designed, the actual assembly language instructions for the type of CPU that is to be used are written, assembled into instruction codes, and placed in the computer. The means by which the code is placed in the computer, and tested, varies with the computer hardware and software.

TESTING PROGRAMS USING A PERSONAL COMPUTER

If the program is to be run on a personal computer that has the same processor as that for which the program is written, then the text editing, assembly, and execution of the program proceeds without leaving the keyboard. Personal computers are equipped with a master program, the operating system, which operates the assets of the computer under user command. Personal computers based on the 8086 family of processors generally use an operating system called the disk operating system or DOS. The 8086 programmer, using DOS as a tool to load and run programs, proceeds to test and debug an 8086 program as shown next.

1. A text editor is loaded from disk into RAM, and the programmer edits the program, saving it on a disk as a pure ASCII file. The saved program is referred to as the *source* file.
2. The assembler program is loaded from disk to RAM and assembles the source program as directed by the programmer. The assembler saves the assembled program on disk as a machine-language file. Any errors found during assembly are reported by the assembler program to the programmer on the CRT monitor or system printer.
3. The syntax errors in the source file are corrected by the programmer and the program re-assembled. A debugging program is loaded from disk into RAM and tests the machine-language file under user control.
4. Any conceptual errors (also called *run-time* errors) are found during debugging, and the entire process repeated until the program is seen to run as intended.

Debugging programs usually permits the programmer to execute the test program one instruction at a time. After each step, the programmer can visually inspect CPU register and system memory contents for proper test program operation. The debugger can also run the program full speed, stopping only at programmer-specified places in the program, called *breakpoints*.

TESTING PROGRAMS ON A SINGLE-BOARD COMPUTER

If the test program is to be exercised on a small, board-level computer that has no disk drives or operating system, then the debugging procedure is more cumbersome. Programs are written and assembled using the PC, and then transmitted, via a serial data link, to the computer board using the PC as a terminal. The process of transmitting the program to the board is called *downloading* the program.

The board will have a small debugging program, called a *monitor* program, in a ROM memory on the board, which controls the downloading and subsequent communications with the PC. The monitor will generally have the same general debugging abilities as the DOS-based debugging program. DOS-based debuggers tend to be menu-driven and easy to use because the instructions are always shown on the screen. Monitors tend to require that the user remember certain keystrokes for each desired result. Board-level monitors are very inexpensive and are usually supplied by the board manufacturer.

More advanced board-level debuggers are called *in-circuit emulators* or ICE debuggers. ICE debuggers involve plugging special hardware into the PC and also into the CPU socket on the board. ICE debuggers tend to be costly, in the range of $5,000 to $15,000.

Using a debugger, such as D86 used in this book, becomes natural after a short exposure to their command structure. Testing for (and finding) all of the errors in the program is generally the *real* challenge to the programmer. Small programs, such as are found in this and other books, are relatively easy to test with some assurance that the programs really do work.

THE IMPORTANCE OF TESTING PROGRAMS

Testing a program is the most *crucial* programming step after defining the problem and causes the most grief to novice and experienced programmers alike. Simple programs can be tested by exhaustive means, that is, all possible combinations of conditions can be put into the computer, and the simple program run for each condition. Large, complex programs cannot be exhaustively tested, because the combinations of possible inputs are so numerous that they cannot be known. Large, complex programs can sometimes be tested using test programs that are specifically written to simulate actual use. Then the final bugs are found by using the program in actual situations until, hopefully, all of the problems have been identified. Even then, it may be that in fixing one bug, another is inadvertently introduced. This has led to the cynical observation by jaded programmers that one has "the latest fix—and the latest bug."

4.2 Programming the 8086

We begin our study of the 8086 family of microprocessors by investigating the general structure of 8086 syntax. We shall conclude by discussing how many ways the 8086 has to identify where to locate and store data.

Lines of Code

Assembly language programs are written as a sequence of text *lines*. A text line is nothing more than a line of alphanumeric characters that are stored in a disk file. Each line ends when a carriage return character (or carriage return–line feed) is stored by the text editor on the disk. The file is created by using a word processor or text editor. The only condition placed on the file is that it must be stored on the disk in pure ASCII (American Standard Code for Information Interchange) format. Using ASCII format for the file ensures that unusual characters, used by most word processors for special functions (such as underlining or tabs), do not appear in the final file.

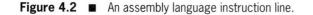

Figure 4.2 ■ An assembly language instruction line.

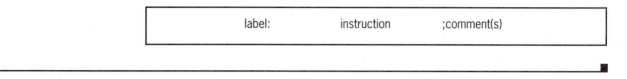

| | label: | instruction | ;comment(s) |

Each line of text is an instruction to the CPU, a directive to the assembler, or a combination of the two. Many names have been used for a single text line in a source file, such as a *statement*, or a *line of code*, or an *instruction*.

8086 Instruction Syntax

Almost all computer instructions involve taking one piece of data from a location somewhere inside the computer and "operating" on that data together with another piece of data located somewhere else inside the computer. Data originates at the *source* location and ends up at the *destination* location. (We will use the terms *location* and *address* interchangeably when discussing the physical place that binary data is stored.)

Instructions commonly start with an action mnemonic that reminds us of what the instruction does. The instruction mnemonic[2] is the *first* word of the instruction and indicates in what manner, or how, the data are to be combined or otherwise manipulated by the CPU.

Each code line may also contain comments and a name for the instruction address in code memory, called the *label* of the code address. A label is a *name* given to the address *number* where the first byte of the labeled instruction is stored in code memory. We have seen that instructions are assembled into code memory with each byte of code located at a unique address number. *Labels* let us *convert* address *numbers in code memory* into *names* of our choosing and let the assembler worry about exactly which address number has each name. If the address number of a line of code should happen to change as a result of adding a new instruction, for instance, the name remains the same, only the address number for that name will change. The assembler can keep track of address numbers for each name, and we can keep track of the names.

The overall syntax of a line of 8086 program code is shown in Figure 4.2.

Labels

A label can be any combination of up to 127 letters (A–Z) and numbers (0–9). Certain punctuation marks (@ _ $?) may also be used.

Labels *must* begin with a letter from A(a) to Z(z) or the characters _, @, $, and ?. Do *not* begin a label with any decimal number from 0 to 9, or the

---

[2]The instruction mnemonic is sometimes called an *opcode* because early microprocessors had one *unique* code per instruction type. Contemporary microprocessors may have many different codes for the same mnemonic. For example, there are six codes for the 8086 mov mnemonic, depending on the address mode used.

assembler may assume it is a *number*. Examples of code address labels are the following:

```
square_root:
tabletwo:
_curses:
setmode:
_transmit:
receive?:
ad1234h:
```

Although you may legally use labels up to 127 characters in length, experienced programmers recommend names of 7 to 14 characters. Try to keep label names *short* and *related* to the program. Do *not* use label names that could be considered *offensive* to a reasonable person (such as your instructor).

Labels are often placed in the first text column, on the far left of the source file page, so that we can see, quickly, that it is a label and not an instruction. Labels may be placed, however, anywhere on a line, as long as they precede the rest of the instruction. Labels may also be used alone on a line, with no following instruction. The assembler will associate the label with the first instruction line *after* the label.

Labels, *when used to assign a name to a location in the program*, always end with a colon (:) to indicate they are labels and not instructions. Any other use of a label, such as part of an instruction operand, *always omits* the colon.

Labels can*not* be the name of any register or instruction used by the 8086 such as ax, mov, or jne. Such names are called *reserved words*, or *keywords*, because they have *already been defined and reserved for assembler* use by the A86 assembler author, Eric Isaacson. Appendix A has a list of A86 assembler program reserved words, at the end of the appendix.

Instructions, without a label preceding them on a line, can begin in any text column including Column 1. Programs are more readable, however, if instructions are indented a space or two from the left margin.

INSTRUCTIONS

An 8086 instruction is any of the *coded set that have been defined by the manufacturer* of the 8086 CPU. There are 122 instruction actions (25 of which are redundant) that have been designed into the 8086 CPU and can be assembled by A86. Every instruction can be converted into a unique machine-language binary code that can be acted on by the 8086 internal circuitry. Undefined (not one of the 97) binary codes will cause the CPU to perform erratically. Examples of defined 8086 instructions include the following:

```
mov destination,source
adc destination,source
cmp destination,source
xchg destination,source
```

Each instruction is generally made up of three distinct parts, as shown in Figure 4.3.

Part 1, the instruction mnemonic, is intended to jog our memories by sounding like the operation to be performed by the CPU. For instance, the instruction shown in Figure 4.3 will move data from one location in the computer (the source) to another location (the destination).

Figure 4.3 ■ An 8086 instruction.

Instruction Part	1	2	3
	mov	destination,	source
	(mnemonic)	(operands)	

Part 2 is called an *operand* or *data address,* because it specifies the destination for the data that is being copied from the source.

Part 3 is also an operand and contains the address of the source location in the computer that is providing data to be copied to the destination.

The names of the destination and source address are separated by a comma (,) so that the assembler will know when one operand name ends and the other begins. Do not leave out the comma, for it is as important as any other part of the instruction.

Note that I have used the term *copied* rather than *moved* in the description of a data mov instruction. This is usually the case for most computer data transfer operations. Data is copied from the source to the destination, not physically moved and leaving nothing behind.

There is no reason that the destination address should be listed before the source address except that doing so is a *convention* established for the 8086 assembler. A programming convention, unlike a law of physics, is purely a human invention.

There are many ways to specify the operand addresses of the destination and source data. One of the attractions of powerful CPUs such as the 8086 is the varied ways in which the addresses of the operands can be identified.

COMMENTS

Comments are included by the programmer as simple, terse explanations of exactly what the instruction is doing to make the program function. Beginning programmers often omit comments, a very *dangerous* practice that can lead to sloppy and erroneous programs. Commenting each program line is encouraged, even when the meaning of the line is obvious. Comments *begin with a semicolon* (;) to indicate that they are not one of the operands. Anything can be typed after the semicolon with impunity, even reserved words. Examples of comments include the following:

```
;this is a comment
;and so is this
;mov ax,bx
;hello mom!
;and so on
```

One rather handy way to use the semicolon when debugging a program is to place a semicolon in front of a line of code that you may think (but are not absolutely sure) needs to be deleted. Placing a semicolon in front of the line of suspect code will ensure it is not part of the assembled program. You can easily restore the line in your program by removing the semicolon, should your suspicions prove wrong.

Segment Register	Segment Type
CS	Code Segment
DS	Data Segment
ES	Extra Data Segment
SS	Stack Segment

Table 4.5 ■ Segment Registers

4.3 8086 Addressing Modes

We have seen that instructions may have operands that are destination addresses and source addresses of data. The *way* in which these addresses are *specified* is named the *addressing mode* for the source or destination address. Addressing modes may be mixed in one instruction so that the source address is specified using one addressing mode and the destination address using a different addressing mode. A CPU design that has many different types of addressing modes is considered to be commercially desirable because high-level languages—such as BASIC, C, and Pascal—can run very quickly and efficiently on which talented CPUs. In this section we shall inspect the modes that are employed in specifying source and destination addresses by 8086 instructions. We shall find that a few addressing modes (immediate, register, memory direct, and memory indirect) are used in our programs most often. The 8086 addressing modes are not unique, and similar modes are used by most CPU designs.

Register Addressing Mode

The 8086 architecture provides four segment registers that hold the base addresses of the active memory segments. The segment registers are listed in Table 4.5.

The 8086 CPU also contains eight general-purpose or "working" registers, which are shown in Table 4.6. Table 4.6 also shows those working registers that can be divided into two parts: a high byte part (H) and a low byte part (L).

Working Register	High Byte	Low Byte
AX	AH	AL
BX	BH	BL
CX	CH	CL
DX	DH	DL
BP		
DI		
SI		
SP		

Table 4.6 ■ 8086 Working Registers

The letters, such as AX and SS, are *names* for the CPU registers and these register names can be used in an instruction as the operands for the destination or source of data. When the name of a register is used to identify the location of data, the addressing mode is formally called the *register addressing mode*.

Figure 4.4 diagrams moving data from one register to another using register addressing modes for both the source and destination of the data.

Several examples of using the register names for the source address and destination address of a data move are shown here.

```
mov ax,cx          ;moves word from register CX to AX
mov ds,si          ;moves word from register SI to DS
mov al,dh          ;moves byte from register DH to AL
mov dh,dl          ;moves byte from register DL to DH
```

- ■ *Caution:* You can move data from any register to any other register except that you may not move data into register CS. Doing this will immediately change the address of the next instruction fetched by the CPU, with unknown (but usually disastrous) results.

- ■ Data from the source register must fit exactly into the destination register. You may not move from a byte-sized address into a word-sized address or the reverse. Thus,

```
mov ax,bl          ;won't work
mov cl,si          ;won't work either
```

What happens if you type a mnemonic, or use an operand in a program that is not defined for the 8086? Nothing, really. The assembler will recognize that the instruction is *illegal* and refuse to assemble it into machine code. The assembler will place an error message at the offending line of code and continue on through the remainder of the program. Illegal instructions are simply those that use undefined mnemonics, use improper operands, or misplace a comma or some other punctuation mark.

Addressing modes are used to tell the CPU where the location of the addressed data is to be found. Register addressing tells the CPU that the data is found in one of the internal CPU registers. The register name is an address, not the data. What is inside the register is the data. The accepted term for data bits stored at an address is the *contents* of an address. Often, as a shortcut, we may forget to say "the contents of a register" and just use the register name. For instance, both of the following comments discuss copying the data contents of register BL to register AL:

```
;move the contents of BL to AL
;move BL to AL
```

We shall see that we may also assign names to data memory locations just as has been done for labels in code memory. Names and labels allow us to focus on program concepts and let the assembler take care of address details.

Immediate Addressing Mode

All addressing modes, except one, use the convention that the operand is the *address* of the data, not the data itself. The exception to normal usage is

Figure 4.4 ■ Register addressing mode.

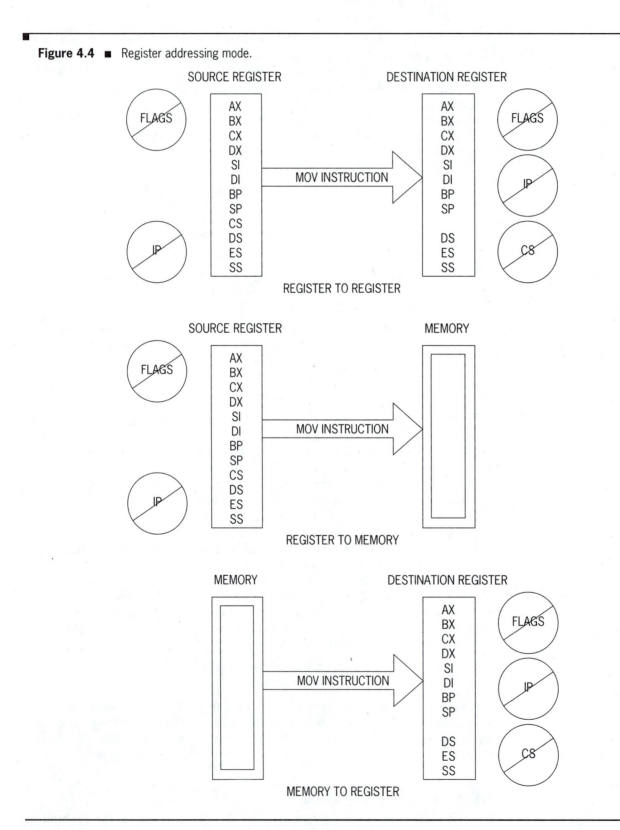

the *immediate* data addressing mode. Immediate addressing for the *source* operand lets the programmer specify the *actual* data number that is to be copied into the destination address. When the immediate addressing mode is used, the number is written as an unchanging fixed *part of the instruction* fetched by the CPU from code memory.

Immediate numbers are called *fixed* data numbers, because they are assembled as part of the program code and cannot change as the program runs. Other data in the program, which can change as the program runs, are called *variable* data or, simply, *variables*.

The register move instructions shown in the previous section on the register addressing mode tell us nothing about what is in the registers. For instance, the instruction mov ax,cx tells us only that whatever data is in register CX is copied into register AX. The actual data could be any number from 0000 to FFFFh. By putting the actual number in the source operand, the program contains the exact number that is copied to the destination address. This mode of addressing the source data is named immediate because the actual data number becomes part of the assembled code. Immediate data is available to the CPU as the number that follows the instruction in *code* memory. Immediate data can be moved to registers by simply specifying what data is to go to which register, as the following examples demonstrate.

```
mov ax,0123h        ;put a 0123h in AX
mov si,3bcdh        ;put a 3BCDh in SI
mov sp,0ffffh       ;put FFFFh in SP
```

Immediate data becomes part of the instruction code and never changes as the program executes. Every time the program is run the same immediate number is used. Figure 4.5 diagrams the immediate addressing mode concept.

- *Caution:* As was the case for register-to-register data moves, the source data must fit the destination exactly. Thus:

```
mov al,5678h        ;will not work
mov ax,32h          ;you will get AX = 0032h
                    ;not AX =3200h
```

- The assembler will supply leading zeroes if the immediate number is *smaller* than the register length.

- Be sure you specify the number base you wish the number to have. If you are using hexadecimal numbers, say so by placing an *h* after the number.

- All numbers must start with a *decimal* digit 0 through 9. 0CADh is a number, CADh is a name.

- Immediate data may not be moved to any segment register. If you wish to put an absolute number into a segment register, put it in a working register first. Then move it as in this example:

```
mov ax,1234h        ;put 1234 in AX
mov ds,ax           ;move 1234 to DS
```

- *Remember:* No moves to CS allowed.

- You cannot have an immediate number as a destination. The instruction mov 1000h,cx will not work.

Figure 4.5 ■ Immediate number addressing mode.

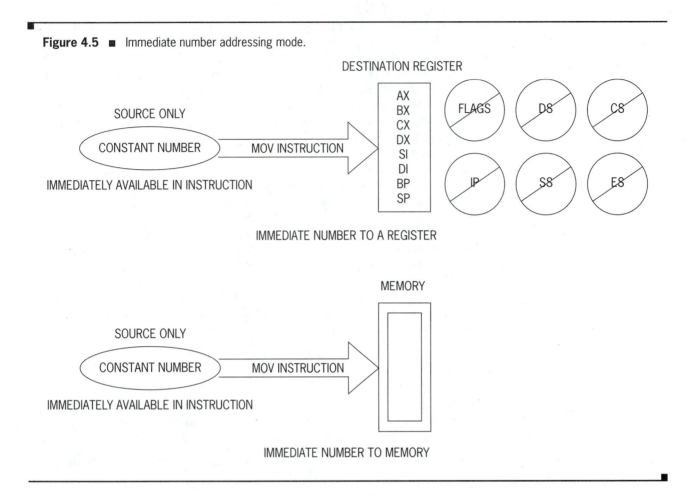

Register and immediate addressing modes confine the source and the destination addresses to the interior of the CPU. This is very advantageous for high-speed computing because the CPU has everything it needs inside itself and can move data very quickly from one internal location to another or from code memory to an internal register. A program to preload all of the interior registers with the number 1234h could be written as shown in the next example.

```
mov ax,1234h        ;move 1234h to AX
mov bx,ax           ;move data to next register
mov cx,ax           ;move data to next register
mov dx,ax           ;move data to next register
mov si,ax           ;move data to next register
mov di,ax           ;move data to next register
mov sp,ax           ;move data to next register
mov bp,ax           ;move data to next register
mov ds,ax           ;move data to next register
mov es,ax           ;move data to next register
mov ss,ax           ;move data to next register
```

Most useful programs, however, must access external memory to retrieve or store data that is generated by the program. Indeed, it has been the advent of cheap dynamic rams (DRAMs) that has made the multi-megabyte personal computer common and unremarkable. Part of external DRAM memory is used to store the application program code that is copied from disk to memory. The remainder of the DRAM memory (or that part not used by the operating system) is available to store the application program's variables.

Memory Addressing Modes

External memory (any memory location outside of the CPU is *external*) must be accessible to the CPU. Three external memory addressing modes are defined for the 8086: *direct, indirect,* and *based-indexed*. All three modes are generically named *memory* addressing modes. In Chapter 3 we discussed the fact that any number combined with a segment register to form a physical address is called the *offset* into the segment. Offsets *into memory* are also called *effective addresses* or EA addresses.

DIRECT MEMORY ADDRESSING MODE

One addressing mode that is used to address external memory is called the *direct* addressing mode, so named because the address in external memory is specified as a *fixed* address number. This is very similar to the immediate addressing mode except that the fixed number specified in the program is an *address* in external memory, *not* immediate data.

Either of the operands may be the direct memory address of a memory location, but *not both.* Moving data from one memory location to another is *not* permitted using direct addressing.

The syntax used to specify direct addressing is to enclose the address of the memory operand in brackets ([]). This has the advantage of letting us differentiate between an immediate number such as 1234h and the direct memory address location [1234h].

The brackets used to indicate a memory address tell us that the quantity inside the brackets is an offset into a memory segment. The brackets do not, however, tell us which of the four segment registers will be used to form the absolute physical address. If the programmer does not specifically specify which segment register to use, then the assembler will make a *default* decision for the programmer. Segment register-offset default parings are discussed in Chapter 3 and summarized in Table 4.7. The default segment register, for most variables, is *normally the data segment* register DS.

Offset Register	Default Segment	Optional Segment
IP	CS	None
SP	SS	None
BP	SS	CS, DS, ES
Any Other Offset	DS	CS, ES, SS

Table 4.7 ■ Default Segments for Memory Addressing

The syntax for specifying a direct address in external memory will result in instructions that look like those shown in the following example.

```
mov [1000h],ax       ;mov AX to location DS:1000h
mov [1000h],1000h    ;move the number 1000h to memory address DS:1000h
mov ax,[2000h]       ;move the data at DS:2000h to AX
```

Direct addressing lets us move data between the CPU registers and any location in memory. Immediate data may be moved to a direct memory address but, clearly, not the reverse. Figure 4.6 diagrams moving data using the direct addressing mode.

OVERRIDING THE DEFAULT SEGMENT REGISTER

If you wish to make sure there is no doubt as to the final absolute memory location segment, the desired optional segment register name may be *added* to the operand. (Refer to Chapter 3 for a review of how absolute addresses are formed.) Use the initials of the segment register *before* the mnemonic, as the following examples demonstrate.

```
ss mov [1000h],1234h    ;move 1234h to SS:1000h
es mov [0abcdh],bx      ;move BX to ES:ABCDh
cs mov ax,[2000h]       ;move the word at CS:2000h to AX
```

- *Caution:* Only one of the operands can be a direct memory address. Data moves from one direct memory location to another will not work.
- Numbers can be moved from working registers to every segment register except CS.
- Note that a segment override is added before the instruction mnemonic is typed.

HOW TO SPECIFY A KNOWN SEGMENT BASE ADDRESS

All of the instructions, just listed in our last example, identify an address in external memory that is found when the bus interface unit (BIU) combines the segment register and the offset. Inspection of the instructions does not reveal the final, absolute physical address of the data address, however. For instance, the instruction ss mov [1000h],1234h tells us only that an offset of 1000h is added to whatever number is in SS, and the immediate number 1234h is placed at that address. The address SS:1000h is an example of a *logical* address.

We do not normally know the absolute, *physical* address of the memory location because we do not know what number is in segment register DS, SS, ES, or CS when our program runs. The exact numbers placed in the segment registers are determined by DOS, when DOS loads the program into DRAM memory. The numbers DOS uses depend, in turn, on the version of DOS, the make of the computer, and what other programs are already loaded into DRAM.

The next program example solves the unknown segment dilemma for us by loading DS with a known number using immediate and register addressing.

Figure 4.6 ■ Direct memory addressing mode.

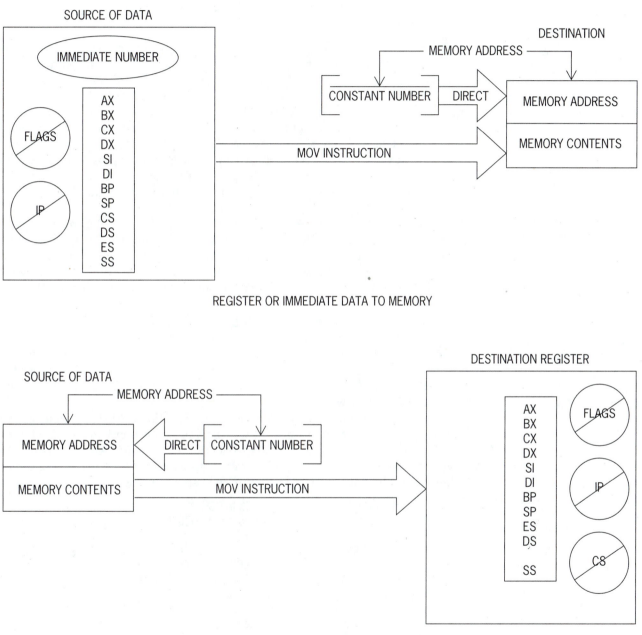

REGISTER OR IMMEDIATE DATA TO MEMORY

MEMORY DATA TO REGISTER

Using this program we can control the absolute physical data addresses in external memory.

```
mov ax,6000h        ;load 6000h into DS via AX
mov ds,ax           ;DS = 6000h
mov ax,1234h        ;AX = 1234h
mov [1000h],ax      ;location 61000h = 1234h
```

If we are writing a program that will be loaded from a disk into the internal memory of a personal computer, we will usually not be concerned with the contents of the segment registers. Operating system software, such as DOS, will assign values to the segment registers depending on what other programs already exist in the DRAM. If the programs we write are to be used on a small 8086-based computer that does not have a disk and DOS, however, the segment register contents *must* be specified in the program. Many of the exercises at the end of each chapter will require that your programs place data in *absolute* external memory addresses. This is done solely to facilitate debugging the programs and to allow the instructor to *grade* the programs quickly.

One word of *warning.* Although the CPU can address 1M of memory, not all of the memory is usable by the application program. DOS reserves almost half of the 1M of memory for its own dark purposes, and your application program may easily "trash" a crucial DOS location. So, until you know exactly what is safe, stay within the memory bounds listed in the problems.

Indirect Memory Addressing Mode

The direct memory addressing mode previously discussed has one serious drawback: Each direct address (a fixed number) cannot be modified by the CPU *as the program executes.* For example, let us assume that we must fill an area of memory with the byte AAh such as might be done to test the memory for read or write errors.

If the area of memory is large, say from 60000h to 61FFFh, then we are faced with the daunting prospect of writing a program that is over 8000d lines long! We could begin by setting DS to 6000h, placing an AAh in register AL, and then, laboriously, putting each byte into a memory address using direct addressing. The outline of such a program is shown next.

```
mov ax,6000h        ;get 6000h to DS
mov ds,ax           ;DS now points to 60000h
mov al,0aah         ;AL = AAh
mov [0000h],al      ;location 60000h = AAh
mov [0001h],al      ;location 60001h = AAh
mov [0002h],al      ;location 60002h = AAh
mov [0003h],al      ;location 60003h = AAh

      .             ;good grief!!
      .

mov [1ffeh],al      ;location 61FFEh = AAh
mov [1fffh],al      ;location 61FFFh = AAh
                    ;finally done!!!!
```

Clearly, it will take many hours to write this program, which is a complete (and expensive) waste of time for the programmer. Not only is writing the program inefficient, but many bytes of internal code memory space will be taken up by the 8195-line program.

But, wait! A computer is very good at doing repetitive tasks of this kind, so an addressing mode that lets the CPU change the address *as the program runs* will save programmer time and code memory space. The CPU can easily change the contents of an internal working register by, say, simply adding a 1 to the number found there. Using an internal register to hold the address of the memory location then lets the CPU, under program direction, change the register contents in microseconds. Sequential memory addresses can be generated automatically, and the problem is solved.

The new addressing mode is dubbed *register indirect* or *indirect memory* because any one of the working registers BX, BP, SI, or DI can be used to hold the address number of the memory location.

Direct memory addressing works by having the CPU combine the operand number (number in brackets) to the segment register to form an absolute memory address. Indirect addressing works by having the CPU combine the *contents* of the register (name in brackets) to the segment register to form the absolute memory address. It is called indirect because the address number is indirectly found *inside* the working register, not directly found as a fixed number in the operand.

Register indirect syntax uses the bracket ([]) around the name of the register that holds the indirect address to indicate that register indirect mode is chosen to specify the operand address.

Not all registers can point to memory. Only working registers BX, BP, SI, or DI may be used to move data using register indirect addressing. A constant (an immediate number) can also be added to the indirect addressing register, inside the brackets. The constant number is referred to as the *displacement* part of the offset. Displacement numbers are the same as immediate numbers—a fixed number we put in our programs to get things where we want them.

As was the case for direct memory addressing, the segment register used to form the physical address can be assigned by default action of the assembler or by the programmer. Data Segment register DS will be assigned, by default, any time registers BX, SI, or DI are used to indirectly address memory. Stack Segment register SS will be the default choice of the assembler whenever register BP is used to indirectly address memory. The programmer may override these default segment register assumptions by specifying any other segment register shown in Table 4.7.

The only exception (there are, alas, always exceptions) to overriding the default segment register occurs when ES is used with DI during *string* operations. String operations are repetitive instructions that move entire groups of data. String operations are examined in Chapter 8.

Some authorities further subdivide indirect addressing into two subspecies: base and index. *Base addressing* takes place whenever any of the *base* registers, BP or BX, are used for indirect addressing. *Index addressing* takes place whenever one of the *index* registers is used for indirect addressing. The results are the same for any of the four registers: The register in use holds the indirect address.

There might be some slight advantage in choosing BX for indirect addressing, because it may be divided into two, 8-bit parts (BH and BL) that could be individually altered to form an indirect address. However, we shall use any of the four indirect addressing registers as functional equals.

Several examples of instructions that use register indirect operands are shown next.

```
          mov bx,1000h        ;set BX to 1000h
          mov [bx],1234h      ;memory location DS:1000 = 1234
   es     mov cx,[bx]         ;register CX = contents of ES:1000h
          mov [di],dx         ;mov the contents of DX to memory DS:[DI]
          mov [bp],ax         ;mov contents of AX to memory SS:[BP]
          mov [bx+100h],ax    ;mov contents of AX to DS:[BX+100h]
```

The absolute address that finally is used in external memory can only be known by knowing exactly what numbers are contained in the segment register and in the indirect register.

Note that we are becoming more and more unconcerned about where, exactly, in physical memory our variables are placed. We are content to know that a variable is at DS:[BX] and let DOS keep track of DS while we make sure BX stays in a safe area.

With the distinct advantage of having indirect addressing modes, our original problem of filling memory locations 60000h to 61FFFh becomes quite easy. Borrowing a few instructions from Chapter 5, we now write the *looping* program shown below to accomplish the same task as the 8,000-line program that uses only direct memory addressing.

```
          mov ax,6000h        ;Use DS to hold the base address in memory
          mov ds,ax           ;DS = 6000H
          mov al,0aah         ;AL = AAH
          mov bx,2000h        ;BX = 2000H
   more:
          dec bx              ;BX = BX–1 (1FFFH the first time)
          mov [bx],al         ;location 6000:[BX] = AAh (no flags changed)
          jnz more            ;if BX is not 0000 then go to label "more"
                              ;and continue to load AAh to DS:[BX]
          nop
```

The looping program shows the power of a computer to use a few instructions to do a great deal of data processing. The CPU can count the contents of BX down from 2000h to 0000h and store AAh in each memory location using BX as an indirect addressing register. The JNZ command lets the program jump (loop) to the instruction found at the program address label named more by testing to see if BX has reached zero. If BX is not zero, then the jump to more is taken, and the next address is loaded with AAh. Finally, BX will reach zero, and the NOP (No Operation) instruction is done.

The operation of the indirect memory addressing mode is shown in Figure 4.7.

- ■ *Caution:* As was the case for direct memory addressing, you may not move data from one memory location to another memory location using indirect addressing.
- ■ Use only registers BX, BP, SI, or DI for indirect addressing.
- ■ DS is the default segment register for BX, DI, and SI, except for string moves, when DI defaults to ES.
- ■ SS is the default segment register for BP

Figure 4.7 ■ Indirect memory addressing mode.

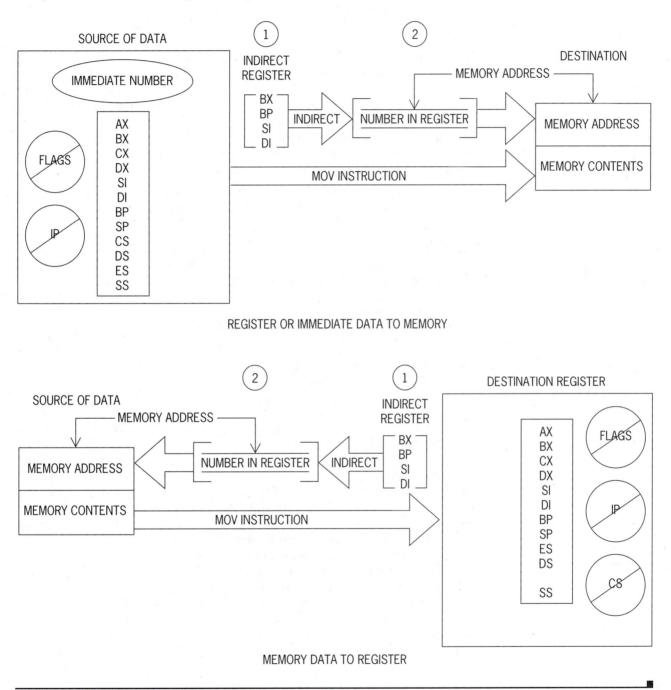

REGISTER OR IMMEDIATE DATA TO MEMORY

MEMORY DATA TO REGISTER

BASED-INDEXED MEMORY ADDRESSING MODE

There remains one more addressing mode available to the programmer, and that is a mode that uses a base register (BX or BP) and one of the index registers (SI or DI). Generally referred to as *based-indexed* or *base-index* addressing, it is a specialized addressing mode normally reserved for advanced programming techniques. Base-index addressing uses *one* of the *base* registers, BX or BP, and *one* of the *index* registers, DI or SI, to form an indirect memory address.

An indirect register address is formed by adding the contents of one of the base registers to one of the index registers. A constant *displacement* number may also be added to the base-plus-index register sum to form the final, effective address. Typical mnemonics that employ base-index addressing appear as follows.

```
mov [bp+si+100h],ax        ;move the contents of AX to DS:EA
mov ax,[bx+di]             ;move DS:EA to AX
mov [bp+di+1000h],1000h    ;move 1000h to DS:EA
```

BP or BX may be added to SI or DI. BP may not be added to BX, nor DI to SI. The displacement number is optional. Data may be interchanged between registers and memory locations, and immediate numbers may be loaded into memory, using base-index addressing.

Note that DS is the default segment register for base-index addressing, and the other segment registers may override DS.

POINTERS TO MEMORY

Before leaving the subject of addressing mode concepts, we need to define one of the most often-used terms in programming, and that is *memory pointers* or *pointers to memory*. When an address in memory is specified, whether using direct or indirect addressing modes, the mode used is said to *point* to memory. The address operand contains a number that is combined with a segment register that yields an absolute, physical address in memory. Thus the operand identifies, or "points to," a unique memory location. Indirect addressing is considered to be the most general way to point to memory because it is such a flexible and powerful method of addressing huge quantities of memory. In fact, the BP register is named the *Base Pointer* to reinforce the concept of using registers to point to memory.

Clearly, we may simply point to memory by using direct addressing. If we want memory location 2000h, we may just specify that location in our bracket. If we wish to access a huge amount of memory, then indirectly using BX, BP, DI, or SI lets the CPU do the work of generating a series of sequential memory addresses. Should we need more complicated means of accessing memory, base-index addressing lets us have the CPU adjust two indirect registers, base and index, to generate sequential and nonsequential addresses.

An Address Mode Program

As an introduction to programming, assembling, and debugging in 8086 assembly code, let us take our single instruction mov and use all of the 8086 addressing modes. Chapter 5 will discuss 17 more 8086 instructions that are sufficient for some serious programming.

Our challenge now is to write a program that will use all of the addressing modes. We cannot possibly use all of the possible memory-register combinations, not with 1 million memory addresses! Even trying all possible register-register combinations will yield a huge program! The addressing mode program address.asm is shown next.

```
;a program, named "address.asm", that shows some of the many
;addressing modes possible on the 8086.
;
code segment                    ;"code segment" is an assembler directive
        mov ax,1234h            ;immediate moves to 16-bit registers
        mov al,65h              ;and 8-bit registers
        mov ah,87h
        mov bx,1234h
        mov bl,76h
        mov bh,98h
        mov cx,4321h
        mov cl,34h
        mov ch,12h
        mov dx,1234h
        mov dl,43h
        mov dh,21h
        mov sp,1111h            ;only 16-bit immediate moves possible now
                                ;note the stack display line changes

        mov bp,2222h
        mov di,3333h
        mov si,4444h            ;all immediate moves to registers done
        mov bx,ax               ;try a few register-register moves
        mov cx,ax
        mov dx,ax
        mov sp,ax
        mov bp,ax
        mov si,ax
        mov di,ax
        mov ds,ax               ;try register to segment register moves
                                ;note D86 memory display changes
        mov es,ax               ;every segment register but CS.
        mov ss,ax               ;note the stack display line changes
        mov ax,0000h            ;try segment register to register
        mov ax,ds
        mov ax,0000h
        mov ax,es
        mov ax,0000h
        mov ax,ss
        mov ax,cs               ;it is OK to move FROM CS.
        mov ds,ax               ;get original DOS segments back
        mov es,ax
        mov ss,ax
        mov al,11h              ;new test data in AL
        mov ah,al              ;try some 8-bit register-register moves
        mov bl,al
        mov al,22h              ;new test data in AL
        mov bh,al
        mov cl,al
        mov ch,al
```

```
            mov al,33h                ;new test data in AL
            mov dl,al
            mov dh,al
            mov [1234h],ax            ;try some memory direct moves to DS:[ ]
            mov [1236h],bx
            mov [1238h],cx
            mov [123ah],dx
            mov [123ch],bp
            mov [123eh],sp
            mov [1240h],di
            mov [1242h],si
            mov [1244h],cs
            mov [1246h],ds
            mov [1248h],ss
            mov [124ah],es
            mov ax,6000h              ;load registers with 6000h.
            mov bx,ax                 ;we will keep DS for D86 debug window,
            mov cx,ax                 ;and make sure all the segment registers
            mov dx,ax                 ;return to their original DOS values.
            mov sp,ax
            mov bp,ax
            mov di,ax
            mov si,ax
            mov ss,ax
            mov es,ax                 ;BUT NOT CS!!!!
            mov ds,[1246h]            ;test to see if we get them back
            mov es,[124ah]            ;get the others back
            mov ss,[1248h]            ;keep same segments for debugging
                                      ;purposes
            mov si,[1242h]            ;CS MAY NOT be moved to
            mov di,[1240h]
            mov sp,[123eh]
            mov bp,[123ch]
            mov dx,[123ah]
            mov cx,[1238h]
            mov bx,[1236h]
            mov ax,[1234h]
            mov bx,1234h              ;try a few indirect moves
            mov ax,9999h              ;put test data in AX
            mov [bx],ax               ;put AX at DS:1234h
            mov cx,[bx]               ;check it out by getting it back
            mov [bx+10h],ax           ;try a displacement
            mov dx,[bx+10h]           ;test it by getting it back
            mov [bx],8888h            ;try immediate data to memory
            mov ax,[bx]               ;test it by getting it back
            mov di,0200h              ;try base-index
            mov [bx+di],ax            ;put it in DS:[BX+DI]
            mov bp,[bx+di]            ;test it by returning it to BP
            mov [bx+di+200h],ax       ;base-index addressing and displacement
dun:        mov si,[bx+di+200h]       ;test to see if it comes back to SI
code ends                             ;another assembler directive
```

We have not exhausted all of the combinations of memory and register addressing possible, but we find that our program is several pages in length.

Two assembler *directives*, code segment and code ends, have been used in the addressing mode program. These directives are not necessary for a program that has only a code segment, and A86 will assemble the program without them. A code segment, as noted in Chapter 3, is an area of memory that contains CPU instructions. Code segments can also be used to hold programmer-defined data. See Appendix A for details on placing data in a code segment.

4.4 Assembling and Debugging a Practice Program

A very good way to learn to use A86 and D86 is to write, assemble, and debug the addressing mode program address.asm listed in the previous section. Any mistakes made while typing the program into the source file will demonstrate how A86 identifies errors.

Assembling the Program

Appendix A contains detailed instructions on how to use the A86 assembler. Input into A86 consists of one file, filename.asm, where filename is any legal name assigned to the assembly code source file.

There are three possible output files from A86: filename.com, filename.bin, and filename.obj. Each type of output file has a different purpose. Refer to Appendix D, "DOS and DOS Interrupts," for a detailed discussion of .COM and .EXE type programs.

Filename.com is a single-segment program that causes DOS to assign the same value to all the segment registers—CS, DS, SS, and ES—when the program is loaded from disk into DRAM. We shall use .COM files almost exclusively in this book, because of their simplicity and the belief that very few assembly-language programs need to exceed 64K in size. Programs that do exceed 64K will, almost always, be written in a high-level language containing, perhaps, some assembly-code parts.

Filename.bin files are machine-code programs that are much like .COM programs but are missing some information needed by DOS to load them into memory. Files of type .BIN are normally burned into EPROMS for operation in a computer that does not have a DOS operating system.

Filename.obj files are assembly-code programs that can be combined with other .OBJ type programs, or with high-level language programs, using a linker program. The output of the linker will be files of the .EXE type.

A86 will produce .COM files by default, that is, if you do not specify some other output type, A86 will produce .COM type output files automatically.

Debugging the Program

Once A86 has assembled the addressing mode program, call up D86 and test the program using the following procedure. (If D86 brings up a window full of NOP instructions, then D86 could not find your assembled program. Quit D86 and be sure to specify the *complete* path including the drive letter to your assembled file.) Refer to Appendix B for more details on how to use D86.

a. Set the memory window of Line 1 to view data memory by typing the command 1,w,1234h on the command line. Note that the memory display line will change whenever DS is changed.

b. Use Ctrl-N to inspect the data memory above address DS:1234h.

c. *Slowly* single-step the program, and make sure that every command works as it should.

d. Try your own numbers, by filling the registers with mov register, immediate commands typed on the D86 command line.

WARNING: Changing DS and SS may get you to parts of memory that are "off limits" for application programs.

e. When satisfied that all of the code works, quit D86.

The warning concerning changing DS and SS is not trivial. DOS will load your program into a "safe" area of DRAM memory. Other areas of DRAM are used by DOS for very important data, and "trashing" the DOS data may mean you will lose control of the computer. Should loss of control occur, a quick Ctrl-Alt-Del usually will re-boot the system. Try to leave DS and SS as set by DOS unless specifically requested to change them by a homework problem.

Note that the label dun is used at the end of the program. A label at the end of the program is very handy to use as a breakpoint address, so the program may be run full speed to the end. But, you should only run to the end of the program after the single-step testing is done.

See Appendix B, which discusses how to use D86, for more information about D86 features.

4.5 Summary

Assembly language programming consists of writing programs using instruction mnemonics that are specified by the manufacturer of the microprocessor. An instruction consists of a label, an instruction mnemonic, operands, and optional comments. The instruction mnemonic specifies the action to be taken by the CPU, and the operands specify the addresses of data used in the action. Labels are names given to address numbers in program code memory.

Operands may specify the address of data using one of five modes of address. These modes are as follows:

Register Data Addressing	The data is found in one of the working registers in the CPU.
Immediate Data Addressing	The data is included as part of the instruction.
Direct Memory Addressing	The address of the data is included as part of the instruction.
Indirect Memory Addressing	The address of the data is held in one of the registers BX, BP, DI, or SI.

Based-Indexed Memory Addressing   The address of the data is formed by adding BP or BX to SI or DI.

Direct and indirect memory addresses use DS as the default segment register, except that SS is used with the BP register. The programmer may override the default segment register by specifying the desired segment register.

Data may be accessed as a byte (8 bits) or a word (16 bits). Data is stored in memory as bytes or words. Memory addresses identify the location of bytes. Words are addressed by addressing the low order byte of the word at the address of the low order byte. When an instruction cannot uniquely identify that the data size to be used is a byte or a word, the keywords byte ptr (b) or word ptr (w) should be used to specify the desired data size.

Assembly language programs are written using a computer and a text editor program. The resultant assembly language programs are converted to machine hexadecimal code (opcodes) by programs called assemblers. Single segment programs of the .COM type are featured in this book.

Programs are tested using special debugging programs that allow the programmer to execute the program one instruction at a time and observe the results. The debugging program may be run on a personal computer under an operating system, or included in ROM as part of a smaller, board-level computer.

4.6 Questions and Problems

1. Name the 8086 internal registers.

2. List the steps involved in getting a program from your mind into a computer.

3. What is the function of an assembler?

4. What is the function of a debugger?

5. Name the parts of a line of code.

6. Name the parts of an instruction.

7. List the 8086 addressing modes, and give an example of each.

8. Explain why CS cannot be a destination.

9. Explain why CS can be a source.

10. Determine which of the following instructions are illegal, and state why.

```
mov al,cx
mov 1234h,ax
mov dx,al
mov cs,1234h
mov [1234h],[5678h]
mov [cx],ax
mov ax,[bp+bx]
mov [si+di],cx
mov 1234h,[bx]
mov cs,[si]
```

Exercise Problems

Write programs, using the mov instruction, to produce the following results in Problems 1–15. Assemble and debug the programs.

1. Move the contents of register AX to registers BX, SI, and DS.
2. Move the contents of register DL to registers AH, BH, and CL.
3. Move the word contents of memory location DS:0FFC1h to register DI.
4. Move the byte contents of memory location DS:1234h to register DL.
5. Move the contents of register BX to memory location DS:0A000h.
6. Move the contents of register CL to memory location DS:4CA0h.
7. Move the word contents of memory location DS:1234h to memory location DS:5678h.
8. Move the contents of memory location DS:1ACDh to register ES.
9. Interchange the contents of registers DI and SI.
10. Interchange the byte contents of memory location DS:1234h with the contents of memory location DS:5678h.
11. Store the number 5678h in memory location DS:2000h using indirect addressing only.
12. Store the contents of memory location DS:1234h in register BX using indirect addressing only.
13. Store the contents of register AX at the word memory location contained in AX. Thus, if AX = 1234h, the number 1234h would be stored at location DS:1234h in memory.
14. Store the number 1234h in absolute memory address 60000h.
15. Move the number at absolute memory address 60000h to DX.
16. Create the following source program and assemble it. Use the debugger to find the contents of register BX.

    ```
    code segment
                mov ax,cs       ;note this is a comment that follows a ;
                mov si,ax
                mov bx,si
    stop_here:  nop             ;do not single-step beyond this instruction
    code ends
    ```

17. Modify the program in Problem 16 so that the contents of SP end up in register BX.

18. Create the following source program and assemble it. Use the debugger to find the contents of register CX.

    ```
    code segment
                mov bx,1000h
                mov ax,1234h
                mov [bx],ax
                mov cx,[bx]
    stop_here:  nop             ;do not single-step beyond this instruction
    code ends
    ```

19. Create the following source program and assemble it. Use the debugger to find the contents of register CX.

```
code segment
            mov ax,1234h
            mov [2000h],ax
            mov cx,[2000h]
stop_here:  nop             ;do not single-step beyond this instruction
code ends
```

20. Create the following source program and assemble it. Use the debugger to find the contents of register CX.

```
code segment
            mov bp,4000h
            mov ax,4567h
            mov [bx],ax
            mov [bp],ax
            mov cx,[bp]
stop_here:  nop             ;do not single-step beyond this instruction
code ends
```

5

A Useful Set of Instructions

Chapter Objectives

On successful completion of this chapter you will be able to:

- Move data inside the computer.
- Do logical operations on data.
- Add and subtract data.
- Count registers up and down.
- Change program flow.
- Diagram program flow using charts.
- Understand how to write sequential programs.
- Understand how to write looping programs.
- Understand how to use label, variable, and constant names in programs.
- Understand how assemblers handle constant, label, and variable names.
- Write successively more complicated programs.

5.0 Introduction

In Chapter 4, we discussed a set of addressing modes that can be used to access data at locations inside the CPU and in external memory. One instruction mnemonic, MOV, was used to demonstrate how data may be accessed by the various addressing modes. We lack, however, a complete set of instructions with which to create and manipulate data to accomplish some computing task.

All computers are equipped with instructions to meet the needs of the market for which the computer is designed, restrained only by available technology. Early CPUs were equipped with a limited set of instructions because of the high cost of hardware. Today, some microprocessors are still designed to execute relatively few basic instructions, but the rationale is to increase speed, not to reduce hardware cost. Limited instruction CPUs are called *Reduced Instruction Set* computers (RISC). Another design option is to equip the CPU with many instructions, some highly specialized. Instruction rich computers are said to be *Complex Instruction Set* computers (CISC). RISC designs gain speed at the cost of increased programming effort; CISC designs simplify programming at the cost of speed. Both RISC and CISC design philosophies are valid, depending on the intended market. General-purpose CPUs, such as the X86 family, tend to be CISC. The designers of the 8086 family have provided us with a catalog of some 122 CPU instructions that can be used to move data, do arithmetic, perform logical operations, make decisions, and interface the CPU to the outside world.

One hundred and twenty-two instructions! Actually, 25 are redundant, which reduces our load to "only" 97 instructions. The beginning programmer, when faced with an awesome array of 97 instructions, is very likely to get bogged down trying to learn each one before attempting any programming. Learning each opcode is a mistake, not only because it delays the learning process but because it is not necessary to know *all* of the instructions, in detail, to write acceptable programs. One does not need to know the *entire* vocabulary of a foreign language, for instance, before using it, and one does not have to know many instructions to write programs. Experienced programmers memorize some instructions they use most often, through sheer repetition. They know that there are other, less-used instructions that can be looked up in the book, if needed.

An old adage pronounces the 80/20 rule of life. As applied to programming it becomes: "You will use 20% of the instructions 80% of the time." This is true, and, in keeping with the saying, I will pick out 18 instructions from the 97 distinct 8086 instruction mnemonics that are available. We shall make extensive use of our 18 instructions for writing programs in this and following chapters. When we need one of the other 79 instructions, we can retrieve it from opcode Appendix F.

5.1 18 Useful Instructions

Moving Data

Computers spend more time just moving data around than doing perhaps any other activity. We have already been introduced to the most often used instruction of all in Chapter 4, the MOV instruction. We shall not need any other data move instructions until we study string moves in Chapter 8.

INSTRUCTION NOTATION

Let us take this opportunity to introduce some notation so that a quick glance at an instruction tells us exactly how it behaves. The notation will indicate the mnemonic for the instruction; operands; the data size it will operate on, byte (B) or word (W) sized; and a summary of the opcode operation. In addition, any of the Flag register flags that are affected by the instruction are identified.

Addressing modes for the operands are also identified. Finally, some usage examples and any special cautionary notes are included to remind the programmer what is legal, and illegal, for the instruction. The MOV instruction is listed next, as an example of the instruction format.

MOV   Move Source to Destination

Mnemonic	Operands	Size	Operation	Flags
MOV	DESTINATION, SOURCE	B/W	(DST) ← (SRC)	None

Addressing Modes
(Destination, Source)

register, immediate
register, register
register, memory
memory, immediate
memory, register

Examples:

```
mov ax,1234h        ;MOV the number 1234h to register AX
mov bx,ax           ;MOV the contents of register AX to register BX
mov [1234h],ax      ;MOV the contents of AX to memory offset 1234h
mov [1234h],5678h   ;MOV the number 5678h to memory offset 1234h
mov cx,[0abcdh]     ;MOV the contents of memory offset ABCDh to CX
mov cx,[bx]         ;MOV the contents of memory offset pointed to
                    ;by BX to CX
```

Operation:

MOV copies data from the source address to the destination address. Byte- or word-length data may be moved.

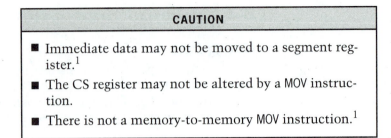

Each instruction heading indicates the following:

Mnemonic: The mnemonic for the instruction action.

Operands: The location(s) affected by the instruction action.

Size: The quantity of binary bits that may be contained in a *memory* location operand. These may be 8 bits (Byte;B) or 16 bits (Word;W), or both (B/W).

Operation: The action produced in the CPU when the instruction is performed. In this case, the contents of the source address, denoted by (SRC), will be copied to the destination address, denoted by (DST) ←.

Flags: Every instruction has the potential to alter one of the result flags in the Flag register. These are most often the Zero (ZF), Carry (CF), Sign (SF), or Parity (PF) flags. If the changed state of the flag is known, it will be listed. For example, if the Carry flag is always reset to 0 as a result of the instruction, then the notation will list *CF = 0*. If the setting of the flag depends on the result of the operation, then only the name of the flag will be listed; it may be a *1* or a *0*. If the state of the flag is unknown (random), then the notation *Flag = ?* is used. For the MOV instruction, no flags are changed, so the word *None* appears. Remember, if a flag is *not* listed, it means that the flag is *not* changed from its previous state. If a flag is listed, then it is set or reset depending on the conditions that exist in the CPU when the instruction is executed. See Chapter 2 for discussions of carry, borrow, overflow, and sign.

Addressing Modes: These are the legal addressing modes for the instruction operands, consisting of immediate, register, and memory addressing modes. Memory address mode can be any one of the direct, indirect, or based-indexed memory addressing modes.

Notation Notes: Parentheses ( ) are used to indicate "the contents of." (DST) means the contents of the destination operand; (SRC) means the contents of the source operand. Brackets [ ] mean the "contents of the contents of" and are used to indicate register or memory indirect addresses.

---

[1]Some assemblers, including A86, *will assemble* an immediate number move to a segment register and a memory-to-memory move by "patching in" code that accomplishes these actions.

Logical Operations

A computer is a binary machine. The common binary operations of AND, OR, XOR, and NOT can be done on data using mnemonics of the same name. Data may also be rotated and shifted left or right.

Logical instructions are very handy when doing work that involves a single bit of information. For example, in many industrial control applications, the computer must sense the status of switches and relay contacts and then operate lights, valves, or motors based on the contact states. An open switch can be represented by a binary 0, for instance, or a motor can be turned on by a binary 1 that controls the motor circuit. A certain class of computers used extensively for the type of industrial control just mentioned are called *programmable logic controllers* or PLCs. They do not appear to be computers, because they often lack keyboards and CRT screens, but buried within almost all PLCs is a general-purpose CPU similar to the 8086 family of microprocessors.

Logical instructions are unusual in that they really operate on each individual bit of the data. That is, each bit is operated on completely independently of the rest of the bits in the data byte or word.

AND   Logical AND Source to Destination

Mnemonic	Operands	Size	Operation	Flags
AND	DESTINATION, SOURCE	B/W	(DST) ← (DST AND SRC)	CF = 0
				OF = 0
				AF = ?
				PF SF ZF

Addressing Modes
(Destination, Source)

register, immediate
register, register
register, memory
memory, immediate
memory, register

Examples:

```
and dh,0fh        ;AND each bit of the number 0Fh with DH
and ax,cx         ;AND each bit of CX with the same bit in AX
and bl,[1234h]    ;AND each bit of the byte at offset 1234h with BL
and [5000h],cx    ;AND each bit of CX with word at memory offset 5000h
and ax,[bx]       ;AND each bit of memory offset [BX] with AX
```

Operation:

AND logically ANDs together each corresponding bit of the source binary quantity and destination binary quantity, leaving the result in the destination address. A destination bit will be a 1 only if the destination and source bit were both 1 before the operation.

AND operations are useful for "masking out" certain 1 bits while keeping others intact. For instance, the instruction and dh,0fh in the examples above will make bits 4–7 of DH all 0 while keeping bits 0–3 of DH in their original state. Every 0 in an AND source operand will "mask out" (set to 0) every corresponding bit in the destination register.

CAUTION
■ No segment register may be source or destination.

NOT   Logical NOT Destination

Mnemonic	Operand	Size	Operation	Flags
NOT	DESTINATION	B/W	(DST) ← (NOT DST)	None

Addressing Modes
(Destination)
register
memory

Examples:

```
not dh                ;NOT each bit of the number in register DH
not ax                ;NOT each bit of register AX
not byte ptr [1234h]  ;NOT each bit of the byte at address 1234h
not word ptr [1234h]  ;NOT each bit of the word at address 1234h
not byte ptr [di]     ;NOT each bit of the byte at address [DI]
```

Operation:

NOT inverts each bit of the destination address contents. This operation is also called the *one's complement* because each bit can be considered to be subtracted from a binary 1. Thus 1 minus 1 is 0, and 1 minus 0 results in a 1.

CAUTION
■ No segment register may be the destination.

ASSEMBLER DIRECTIVES BYTE PTR (B) AND WORD PTR (W)

Note that the assembler keywords byte ptr and word ptr have been used for the first time in our last examples. The reason is plain. A memory address, such as DS:[1234h], can address the *byte* contained at offset address 1234h or the low order byte of the *word* of data that starts at address 1234h. The same is true for indirect addressing using index register DI as a pointing register to memory. The instruction not [di] confuses the assembler. A86 cannot determine if you wish to invert the byte pointed to by DI or the word pointed to by DI.

Keywords byte ptr (pointer to a byte) or word ptr (pointer to a word) tell the assembler program what, exactly, you the programmer wish to use—a byte of memory or a word of memory. A86 will accept the abbreviations B (for byte pointer) and W (for word pointer) to save some typing. We shall use B(b) or W(w) in this book.

Sometimes A86 can assume the data size, as when you execute the instruction mov ax,[1234h], because A86 knows AX is a word-length register. A86 will also guess correctly if you type mov [1234h],5000h, because 5000h is clearly a word-length operand. But A86 cannot guess mov [1234h],0000h because all zeroes could be 00h or 0000h. If in doubt, specify the data size as b or w. For the last example, use mov w[1234h],0000h if you wish to fill the word at offset 1234h with 0000h, or mov b[1234h] if you only wish to zero the byte at offset 1234h.

OR   Logical OR Source to Destination

Mnemonic	Operands	Size	Operation	Flags
OR	DESTINATION, SOURCE	B/W	(DST) ← (DST OR SRC)	CF = 0
				OF = 0
				AF = ?
				PF SF ZF

Addressing Modes
(Destination, Source)

register, immediate
register, register
register, memory
memory, immediate
memory, register

Examples:

```
or dh,0fh          ;OR each bit of the number 0Fh with DH
or ax,cx           ;OR each bit of CX with the same bit in AX
or bl,[1234h]      ;OR each bit of the byte at address 1234h with BL
or b[1234h],33h    ;OR each bit of 33h with the byte at address 1234h
or [5000h],cx      ;OR each bit of CX with word at memory address 5000h
or ax,[bp]         ;OR each bit of memory word location SS:[BP] with AX
```

Operation:

OR logically ORs together each corresponding bit of the source binary quantity and destination binary quantity, leaving the result in the destination address. A destination bit will be a 1 if either the destination bit or source bit, or both, was a 1 before the operation.

OR operations are useful for masking out 0 bits in the destination location. The instruction or dh,0fh in the examples above will set bits 0–3 of DH to 1, while bits 4–7 of DH are left in their original state. Every 1 in the OR source operand will "mask on" (set to 1) every corresponding bit in the destination operand.

CAUTION
■ No segment register may be source or destination.

RCL   Rotate Destination Left with Carry Flag

Mnemonic	**Operands**	**Size**	**Operation**	**Flags**
RCL	DESTINATION, 1	B/W	CF ← MSB...LSB ← CF	OF
RCL	DESTINATION, CL			CF

Addressing Modes
(Destination)

register
memory

Examples:

rcl ax,1	;rotate AX 1 bit to the left, with carry
rcl ax,cl	;rotate AX (CL) bits to the left, with carry
rcl b[bx],1	;rotate the byte at [bx] 1 bit to the left with carry
rcl w[bx],cl	;rotate the word at [bx] (CL) bits to the left ;with carry

Operation:

RCL rotates the destination operand, to the left, using the Carry flag, CF, as part of the operation. CF is stored in an internal CPU temporary location; the MSB (most significant bit) of the destination operand is shifted into CF; the MSB–1 bit of the destination operand is shifted into the MSB; and so on until the LSB (least significant bit) of the destination operand is shifted into the LSB+1 bit. The old CF bit is shifted from the temporary location into the LSB of the destination operand. The practical result of RCL is to create a 9-bit (byte destination) or a 17-bit (word destination) rotation, because the CF bit is the MSB of the rotated quantity.

The number of rotates to be done is indicated as 1 or the contents of register CL. The rotate operation may be done once, if the number 1 is the rotate counter, or any number of rotates from 0 to 255 times using the contents of register CL. (Rotates greater than 16 are repetitive.)

The CF becomes the contents of the MSB of the destination operand every time a rotate is done. If the number 1 is used as a rotate counter, then the OF flag is set equal to the MSB of the destination operand XORed with CF, after the rotate is done. If the CL register is used as a rotate counter, then the OF flag is undefined. The OF flag can be used for signed arithmetic operations.

CAUTION
■ No segment register may be rotated.
■ If CL = 00h, then no rotation will take place.
■ Rotations involving more than the number of bits in the destination operand, although possible, simply repeat the same operation.

As an example of a rotate operation, assume that AX contains the number 1234H and that CF is 1. Then,

AX = 0 0 0 1 0 0 1 0 0 0 0 1 1 0 1 0 0   and   CF = 1

before a rotate is done, and

AX = 0 0 1 0 0 1 0 0 0 1 1 0 1 0 0 1   and   CF = 0

after one rotate is done, and

AX = 0 1 0 0 1 0 0 0 1 1 0 1 0 0 1 0   and   CF = 0

after the next rotate is done.

XOR   Logical Exclusive OR Source to Destination

Mnemonic	**Operands**	**Size**	**Operation**	**Flags**
XOR	DESTINATION,SOURCE	B/W	(DST) ← (DST XOR SRC)	CF = 0
				OF = 0
				AF = ?
				PF SF ZF

Addressing Modes
(Destination, Source)

register, immediate
register, register
register, memory
memory, immediate
memory, register

Examples:

```
xor dh,0fh        ;XOR each bit of the number 0Fh into DH
xor ax,cx         ;XOR each bit of CX into the same bit in AX
xor bl,[1234h]    ;XOR each bit of the byte at address 1234h into BL
xor b[1234h],33h  ;XOR each bit of 33h into the byte at address 1234h
xor [5000h],cx    ;XOR each bit of CX into word at memory offset 5000h
xor b[si],3eh     ;XOR each bit of byte at offset location [SI] into
                  ;3Eh
```

Operation:

XOR logically exclusive ORs together each corresponding bit of the source and destination contents. The result is left in the destination. A destination bit will be a 1 if the destination bit and source bit are different before the operation. A destination bit will be a 0 if the destination bit and the source bit are the same before the operation.

XOR operations are useful for inverting selected bits in the destination operand. For example, xor dh,0fh in the examples listed above will invert bits 0–3 of DH while bits 4–7 of DH are unaffected. Every 1 bit in the XOR source operand will invert a corresponding bit in the destination operand.

CAUTION
■ No segment register may be source or destination.

Mathematics

Computers are often used to perform mathematical calculations. The calculations can range from simple problems, such as calculating payroll checks, to very complex problems, such as computing the trajectory of a missile using trigonometric functions, square roots, and exponentials. Early digital computers could basically add and shift binary numbers. Using the concept of two's complement numbers, subtraction could be implemented on such primitive computers, and multiplication and division were done using add and shift operations.

Most modern CPUs have built-in hardware, an arithmetic logic unit (ALU), that performs the four fundamental mathematical operations. The addition of ALU arithmetic hardware speeds up the operation of the computer for ordinary arithmetic calculations. Complex arithmetic calculations involving trigonometric and other transcendental functions can be done using software.

If speed is of the essence when using transcendental or other special functions, a special chip (a *math coprocessor*) can be added to the computer. Math coprocessors perform involved math calculations in hardware, relieving the CPU of the task.

ADD   Arithmetic Add Source to Destination

Mnemonic	Operands	Size	Operation	Flags
ADD	DESTINATION, SOURCE	B/W	(DEST) ← (DEST)+(SRC)	AF CF OF PF SF ZF

Addressing Modes
(Destination, Source)

register, immediate
register, register
register, memory
memory, immediate
memory, register

Examples:

```
add ax,0f123h       ;ADD the number f123h to register AX
add dx,si           ;ADD register SI to register DX
add ax,[5678h]      ;ADD the contents of word address 5678h to AX
add w[3215h],8adch  ;ADD the number 8ADCh to word location 3215h
add [3598h],di      ;ADD register DI contents to word address 3598h
add bx,[bp]         ;ADD the contents of address SS:[BP] to BX
```

Operation:

ADD performs the arithmetic addition of the source operand to the destination operand, leaving the result in the destination address. The operands may be used in the program as signed or unsigned numbers. The result of an ADD operation will be the correct signed or unsigned number result. If the programmer is using unsigned numbers in the operands, CF = 1 signals a carry out of the result. See Chapter 2 for a discussion of carries, signed numbers, sign bits, and signed number overflows.

CAUTION
■ No segment register may be an operand address.

ADC   Arithmetic Add Source and CF to Destination

Mnemonic	Operands	Size	Operation	Flags
ADC	DESTINATION, ADDRESS	B/W	(DEST) ← (DEST)+(SRC)+CF	AF CF OF PF SF ZF

Addressing Modes
(Destination, Source)

register, immediate
register, register
register, memory
memory, immediate
memory, register

Examples:

adc ax,0f123h	;ADC the number f123h to register AX
adc dx,si	;ADC register SI to register DX
adc ax,[5678h]	;ADC the contents of address 5678h to AX
adc w[3215h],8adch	;ADC the number 8ADCh to word location 3215h
adc [3598h],di	;ADC register DI contents to address 3598h
adc dl,[bp]	;ADC the byte at memory address SS:[BP] to DL

Operation:

ADC performs the arithmetic addition of the destination operand to the source operand. The state of CF *before* the addition (1 or 0) is then added to the result. The final sum is placed in the destination operand address. The operands may be used in the program as signed or unsigned numbers. The result of an ADC operation will be the correct signed or unsigned number result. If the programmer is using unsigned numbers in the operands, CF = 1 signals a carry out of the result. See Chapter 2 for a discussion of carries, signed numbers, sign bits, and signed number overflows.

CAUTION

■ No segment register may be an operand address.

When adding together multi-byte (or multi-word) unsigned number operands, the possibility of a carry out of the most significant bit of the result is a distinct possibility. ADC permits any carry from a previous addition to be added to the next pair of higher-order operands.

DEC   Decrement Destination

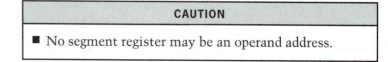

Mnemonic	Operand	Size	Operation	Flags
DEC	DESTINATION	B/W	(DEST) ← (DEST)–1	AF OF PF SF ZF

Addressing Modes
(Destination)

register
memory

Examples:

```
dec ax          ;DEC the contents of register AX
dec dl          ;DEC register DL
dec b[89abh]    ;DEC the byte at address 89ABh
dec w[6754h]    ;DEC the word at address 6754h
dec w[bx]       ;DEC the word at address [BX]
```

Operation:

DEC can be thought of as an instruction that subtracts an immediate 1 from the contents of the destination operand and leaves the result in the destination operand. The operand contents will underflow from 00h (for a byte of data) to FFh, or from 0000h (for a word of data) to FFFFh. Decrementing an operand could be done by actually subtracting a binary 1 from the operand (see SUB below), but DEC encodes more compactly.

CAUTION

■ Unlike a true subtraction operation, the Carry flag is *unchanged* by a DEC operation; in particular, CF is *not* set when the destination underflows.

■ No segment register may be the destination address.

Remember that an assembler keyword of byte(b) or word(w) ptr is required when using a memory address that can be interpreted either as a byte or word address.

INC   Increment Destination

Mnemonic	Operand	Size	Operation	Flags
INC	DESTINATION	B/W	(DEST) ← (DEST)+1	AF OF PF SF ZF

Addressing Modes
(Destination)

register
memory

Examples:

```
inc ax              ;INC the contents of register AX
inc dl              ;INC register DL
inc byte ptr [89abh] ;INC the byte at address 89ABh
inc word ptr [6754h] ;INC the word at address 6754h
inc b[bp]           ;INC the byte at address [BP]
```

Operation:

INC adds a 1 to the contents of the destination operand and leaves the result in the destination operand. The data that is incremented overflows from FFh (for a byte) to 00h and from FFFFh (for a word) to 0000h. Incrementing an operand could be done by adding a binary 1 to the operand (see ADD above), but INC encodes more compactly.

CAUTION
■ Unlike a true addition instruction, INC does *not* change CF; in particular, INC does *not* set the Carry flag when the destination operand overflows.
■ No segment register can be a destination operand.

SUB   Arithmetic Subtract Source from Destination

Mnemonic	Operands	Size	Operation	Flags
SUB	DESTINATION, SOURCE	B/W	(DEST) ← (DEST)–(SRC)	AF CF OF PF SF ZF

Addressing Modes
(Destination, Source)

register, immediate
register, register
register, memory
memory, immediate
memory, register

Examples:

```
sub ax,0f123hh     ;SUB the number F123h from register AX
sub dx,si          ;SUB register SI from register DX
sub ax,[5678h]     ;SUB the contents of address 5678h from AX
sub w[3215h],8adch ;SUB the number 8ADCh from word location 3215h
sub [3598h],di     ;SUB register DI contents from address 3598h
sub ax,[bx]        ;SUB the number at address [BX] from register AX
```

Operation:

SUB subtracts the contents of the source operand from the destination operand and puts the result in the destination operand address. The operands may be used in the program as signed or unsigned numbers. The result of a SUB operation will be the correct signed or unsigned number result. If the programmer is using unsigned number operands, CF = 1 signals a borrow condition into the result. See Chapter 2 for a discussion of borrows, signed numbers, sign bits, and signed number overflows.

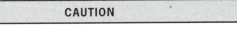

CAUTION
■ No segment register may be an operand.

SBB   Arithmetic Subtract Source and CF from Destination

Mnemonic	Operands	Size	Operation	Flags
SBB	DESTINATION, SOURCE	B/W	(DEST) ← (DEST)–(SRC)–CF	AF CF OF PF SF ZF

Addressing Modes
(Destination, Source)

register, immediate
register, register
register, memory
memory, immediate
memory, register

Examples:

```
sbb ax,0f123h       ;SBB the number F123h from register AX
sbb dx,si           ;SBB register SI from register DX
sbb ax,[5678h]      ;SBB the contents of address 5678h from AX
sbb w[3215h],8adch  ;SBB the number 8ADCh from word location 3215h
sbb [3598h],di      ;SBB register DI contents from address 3598h
sbb [658eh],[bx]    ;SBB the contents of memory location [BX] from
                    ;the contents of memory location 658Eh
```

Operation:

SBB subtracts the contents of the source operand from the destination operand. The *original* Carry flag (1 or 0) is then subtracted from the intermediate result. The final result is placed in the destination operand address. The operands used on the program may be signed or unsigned numbers. The result of an SBB operation will be the correct signed or unsigned number result. If the programmer is using unsigned number operands, CF = 1 signals a borrow condition into the result. See Chapter 2 for a discussion of borrows, signed numbers, sign bits, and signed number overflows.

CAUTION
■ No segment register may be an operand address.

The Carry flag lets the programmer take into account the results of previous unsigned number subtraction operations when performing multi-byte or multi-word subtraction. For SBB, however, the Carry flag is actually a borrow from a higher-order byte or word when the source data is larger than the destination data.

Jumps

Computers cannot yet think, despite the efforts of artificial intelligence and fuzzy logic adherents. But, computers can make *decisions*. A decision can be made by a computer through the simple process of altering the *sequence* of instructions that are to be executed. We make decisions constantly: what to eat or wear, whom to date, which course to take, or what degree for a major. Decisions involve following one path, or another: eat pizza or yogurt, study English or engineering.

A computer makes decisions by following one program path, or an entirely different program path. Decisions are made possible by the automatic action of the CPU in fetching code instructions. The CPU always fetches the *next* instruction to be executed in code memory from the instruction address contained in the Instruction Pointer register. Thus, it is only necessary to change the number in the Instruction Pointer to have the program sequence altered from the next *sequential* instruction to any other instruction address in the program.

Instructions that can alter the contents of the Instruction Pointer are called *jumps* or *program branches*. Both names suggest that program execution suddenly goes from one location in program code memory to another location. The address in program code memory to which the CPU jumps is normally designated by a *label* that the programmer assigns. The assembler keeps track of each label and knows the code address number that each label represents. When the programmer writes an instruction that jumps to a label, the assembler inserts the actual code address number for that label in the assembled program code. If a jump is made, then the jump address number is placed in the Instruction Pointer when the program actually executes.

The conditions that will cause a jump to take place usually depend on the states of the various Status flag bits. Status flags (refer to Chapter 3 for a discussion of flags) are included in a computer as "reminders" to the program of past results (the status) of certain instructions, much as we leave little notes to ourselves to remember to do certain things in the future. Most CPU instructions alter the Status flags so that future instructions can jump on the flag states. There is no way to know if the flags will actually be used by future instructions, but they are set by the CPU "just in case."

A jump instruction examines the state of a particular Status flag or flags. The jump will replace the sequential contents of the Instruction Pointer with a new jump address if the *condition* of the flag used by the jump is *valid*. Condition, for a flag, means the jump will happen *if* the flag is set to the state the jump is looking for. In the vernacular of programmers, the flag is "tested" by the jump instruction, and the jump is "taken" *if* the condition tested by the jump instruction is "true." If the condition tested by the jump is not true, then the jump to the new address in the program is *not* taken. If the Status flag condition is false, the instruction that *immediately* follows the jump instruction is fetched. The jump is said to "fall through" to the next instruction if the jump is not made. To summarize, if the Status flag is true for the jump then the IP is changed to the new jump address. If the Status flag is false for the jump, then the IP is unchanged, and the CPU fetches the next sequential instruction in the program.

Jumps that depend on the state of the flags are called *conditional jumps* because the jump will be done only *if* the conditions of the Status flags are correct for the type of jump in question. If the flag for the jump condition is not set to the state tested by the jump, then the jump is not made, and the instruction that follows the jump is fetched and executed.

We should note here that the programmer decides whether or not to make use of the Status flags. If a program needs to jump on certain conditions, then the programmer includes instructions that set the Status flags. The flags are altered by most instructions and can be used or ignored as the programmer wishes. We should also note that the programmer should be very careful to ensure that the flag condition used in a jump is the one that is desired. Many mistakes are made by beginning programmers who note that a flag is affected by a certain instruction but do not jump on the state of that flag until many instructions *later* in the program. Intervening instructions may alter the very flag that is to be used as a criterion for the jump, and the jump is made (or not made) in error. Try to place your jump instructions as close as possible to the instruction that alters the jump flag. If the jump op-code cannot be placed nearby, make sure that no instruction between a flag-setting instruction and the flag-testing jump instruction affects the flag in question.

You also need to keep in mind that conditional jumps can only branch a *short* distance in the program code. Take care to keep jump address labels close to conditional jump instructions.

Jumps can also be *unconditional*, that is, *no* flags affect the operation of the jump. Unconditional jumps are *always* taken by the program. Unconditional jumps are used by programmers to "jump over" parts of a program or to jump to locations in the program that are beyond the short range of conditional jumps.

THE RANGE OF JUMPS

How *far* in the program can you jump? By far we mean how many code bytes can there be between the jump instruction and the address label that is the destination of the jump instruction. All jumps have a *range*, or maximum number of code bytes between jump instruction and jump address. The 8086 can address up to 1M of memory, so some types of jump can be as far as 1 megabyte. But, as most programs are arranged as segments of 64K each, other types of jumps range within a particular code segment. Or, as is the case with conditional jumps, the range is limited to only several hundred bytes.

Jump instructions for the 8086 are grouped by jump ranges into three distinct categories. Each category depends on the instruction code range that may be taken by jumps within the category. The jump ranges are listed in Table 5.1 and shown in Figure 5.1.

A *short* range jump is thought of as a "local" jump and is the most *often* used jump range. *Near* range jumps are also called *intrasegment* jumps, and *far* range jumps to anywhere in memory are also named *intersegment* jumps. Keywords short, near, and far are used in the program syntax to inform the assembler of the expected range to any label so that it may set aside code

Jump Range	Jump Distance in Bytes
Short	Forward 127, Backward 128
Near	Anywhere within a 64K segment
Far	Anywhere within memory

Table 5.1 ■ Byte Range for Jumps

Figure 5.1 ■ Direct and indirect jump ranges.

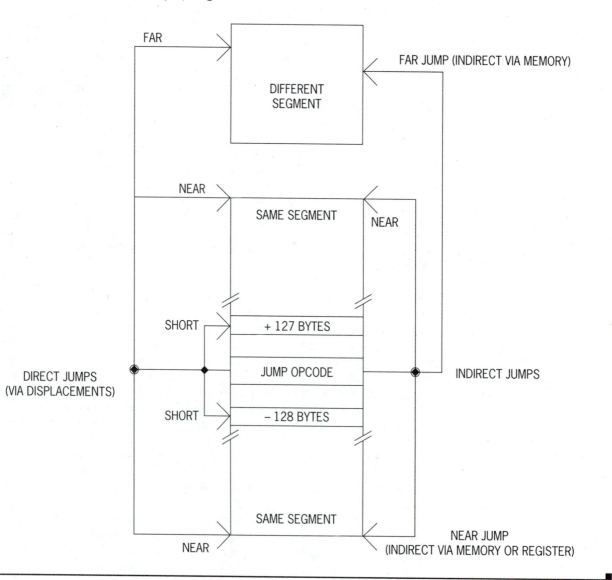

bytes to hold the jump address. Note that jump ranges are relative to the instruction that *follows* the jump instruction. The CPU increments the IP to point to the *next* (following) instruction before executing the present instruction. If the present instruction is a short- or near-range jump, the IP, which points to the next instruction code address, is the reference for the jump range.

Conditional jumps are short-range jumps. It might seem that the range restrictions on short jumps are overly restrictive. Experience will show that most programs jump locally and that longer ranges (beyond +127 or −128 bytes from the instruction *after* the jump) are rarely needed. Moreover, short-range jumps take fewer bytes to code.

The jump address for a short-range jump is formed by the CPU, as the program runs, by adding a single-byte signed number to the current IP. The address of a short jump is *encoded by the assembler* as the two's complement number of bytes from the current IP to the jump address *label*. The two's complement number added to the IP is called the *displacement*, because it is not really an address number but the number of bytes that separate the IP from the jump address. Because the *current* IP points to the *next* instruction, short-range jump addresses can be up to +129 or −126 bytes from the jump instruction. Should it be necessary to extend a short jump further than the range allowed, the programmer can jump to a longer range jump instruction.

Near intrasegment jumps are coded using a 16-bit displacement number. Near intrasegment jumps can jump anywhere within a 64K segment.

THE COMPARE INSTRUCTION

Conditional jumps test Status flags set by prior instructions in the program. One of the most popular flag-setting instructions for use by conditional jumps is the compare instruction, which is discussed next.

CMP   Compare Source and Destination

Mnemonic	Operands	Size	Operation	Flags
CMP	DESTINATION, SOURCE	B/W	(DEST)–(SRC)	AF CF OF PF SF ZF

Addressing Modes
(Destination, Source)

register, immediate
register, register
register, memory
memory, immediate
memory, register

Examples:

```
cmp ah,0f1h        ;CMP register AH with number F1h
cmp dx,si          ;CMP register DX to register SI
```

```
cmp ax,[5678h]        ;CMP AX to the contents of address 5678h
cmp w[3215h],8adch    ;CMP word at location 3215h with number 8ADCh
cmp w[si],1234h       ;CMP the word at address [SI] with 1234h
cmp [3598h],di        ;CMP contents of address 3598h to register DI
```

Operation:

CMP subtracts the contents of the source operand from the destination operand contents but *does not change* the destination or source address contents. CMP does what is known as a *trial subtraction,* and the *real function of the CMP opcode is to determine the relative sizes of the operands.* CMP can be used to determine the relative sizes of *both signed and unsigned* numbers.

The Status flags settings used by *signed* number comparison jumps are set as follows.

Condition	Flag Conditions
(DEST)>(SRC)	SF = OF, ZF = 0
(DEST)<(SRC)	SF <> OF, ZF = 0
(DEST)=(SRC)	ZF = 1

The Status flags settings used by *unsigned* number comparison jumps are set as follows.

Condition	Flag Conditions
(DEST)>(SRC)	CF = 0, ZF = 0
(DEST)<(SRC)	CF = 1, ZF = 0
(DEST)=(SRC)	ZF = 1

The states of the AC and PF flags depend on the nature of the exact results of the CMP instruction. Refer to Chapter 3 for a discussion of the result conditions that set or clear a particular flag.

The CMP instruction is often used with the conditional jump (decision making) instructions of the next section to alter program flow depending on the relative size of the two operands.

Note that the CMP instruction sets most of the *Status* flags, and it is up to the programmer to use the *proper* conditional jump instruction based on the nature of the numbers (signed or unsigned) used in the program. In this book, we shall use five of the *unsigned* number conditional jump instructions. Signed number conditional jump instructions may be found in Appendix F.

CAUTION
■ No segment register may be an operand address.

CONDITIONAL JUMP INSTRUCTIONS

Note that the notation for jump instructions differs from that used previously. The Operand heading becomes Displacement and the Size heading becomes Range.

JC   Jump on Carry
JB   Jump on Below
JNAE   Jump on Not Above or Equal

Mnemonic	Displacement	Range	Operation	Flags
JC	SHORT-LABEL	Short	(IP) ← LABEL IF CF=1	None

Examples:

```
jc _addit      ;as _addit must be short range, no "short" keyword needed
jc lp4         ;lp4 must be a short range away
```

Operation:

JC will replace the IP with the address of the short-range address label if the Carry flag is set to 1. If the Carry flag is 0 then the jump will not be taken, and the instruction that follows the JC instruction will be executed.

CAUTION

■ Note that labels, when used as destinations for jump instructions, do *not* use a colon after the label name. The colon is used with a label only when the label is used to mark an address for a line of code. The assembler will not assemble any short-range jump if it detects that the code address specified by the label is not within +129 or −126 bytes of code from the jump opcode.

ALTERNATE MNEMONICS FOR THE SAME JUMP OPCODE

The JC/JB/JNAE jump instruction is our first experience with several instructions that do exactly the same thing. JC, JB, and JNAE are, in fact, the very same opcode when assembled. Why does one opcode have three different instruction mnemonics? The reason is that many of the jump instructions can be used in various ways by the programmer, depending on the *sense* of how the jump is being used in the program at that particular point.

When an alternate form of a jump instruction is used, then the mnemonic used indicates the relative size of the *destination* versus the *source*. An alternate mnemonic is one other than a jump on the flag that is actually tested by the jump instruction, JC in this case.

For instance, perhaps a CMP has been done, and the programmer is interested in the relative sizes of two numbers. Should the number in the destination be smaller than the source, then the programmer may wish to jump to some other part of the program. The programmer should then use the JB form of the instruction to indicate what condition (destination *below* source) is causing the jump to be taken. Conversely, the programmer may wish to take the jump only if the destination is not greater than or equal to the source. The programmer might then use the JNAE mnemonic.

At another point in the same program, the programmer may wish to change program flow if the action of some instruction left a carry as a result. The programmer should then use JC for the jump mnemonic to indicate that the jump is made because of a carry. When the program is read, the reader understands, by the jump mnemonic used, what the programmer is doing. In all cases the jump command tests the Carry flag and the jump is taken if the flag is 1.

JE   Jump on Equal
JZ   Jump on Zero

Mnemonic	Displacement	Range	Operation	Flags
JE	SHORT-LABEL	Short	(IP) ← LABEL IF ZF=1	None
JZ	SHORT-LABEL			

Examples:

        je same   ;jump to the label "same" if two things are equal
        jz empty   ;jump to the label "empty" if quantity is zero

Operation:

JE (or JZ) will replace the IP with the address of the short-range label if the Zero flag is set to 1. If the Zero flag is 0 then the jump will not be taken, and the instruction that follows the JE/JZ instruction will be executed.

Note that the JZ form of the opcode means that the jump is taken if the Zero flag is set to 1. The flag condition tested by the JZ instruction: Is it true that the Zero flag is 1? If the flag is set to 1 (Yes, it is true that ZF = 1), then JZ takes the jump. If the condition is false (ZF = 0), then JZ goes on to the next instruction.

The JE form of the instruction would normally be used after a CMP instruction if the programmer wishes to jump on equality. If CMP found that the two operands are equal, then the ZF is a 1 (source − destination = 0), and JE takes the jump. Otherwise, JE "falls through" to the next instruction. The JZ mnemonic might be used, for instance, when a counter is decremented, and the jump is to be taken when the counter reaches zero.

JNC   Jump on No Carry
JAE   Jump on Above or Equal
JNB   Jump on Not Below

Mnemonic	Displacement	Range	Operation	Flags
JNC	SHORT-LABEL	Short	$(IP) \leftarrow LABEL$ IF $CF = 0$	None

Examples:

```
jnc up              ;jump to the short-range address "up" if Carry flag = 0
jnc down            ;jump to short-range address "down" if CF = 0
```

Operation:

JNC will replace the contents of the Instruction Pointer with the address of the short-range label if the Carry flag is cleared to 0. Note that the jump is taken if the NC(flag) condition is true, that is, the CF is 0. JNC/JAE/JNB is the reverse of JC/JNAE/JB.

Reverse (or negative) jumps are included in an instruction set so that the programmer can indicate the inverse of the positive jump. For instance, after a CMP instruction, the programmer may wish to make a jump if the source is equal to or greater than the destination. The programmer could use the JB and JE instructions to imply that the destination is smaller than or equal to the source. Or, the programmer may use the JAE mnemonic to show the source is larger than or equal to the destination.

If we were in a pinch to limit our number of instructions below the magic 20%, we could get the same results by using only the positive type followed by an unconditional jump. For example, JNC could be accomplished as follows.

```
jc shut             ;we really want to jump if CF = 0
jmp up              ;CF must be 0, or we would not be here
```

JNE   Jump on Not Equal
JNZ   Jump on Not Zero

Mnemonic	Displacement	Range	Operation	Flags
JNZ	SHORT-LABEL	Short	$(IP) \leftarrow LABEL$ IF $ZF = 0$	None
JNE	SHORT-LABEL			

Examples:

```
jnz notyet          ;jump to label "notyet" if quantity not zero
jne notsame         ;jump to label "notsame" if two things not equal
```

Operation:

JNZ (or JNE) will replace the Instruction Pointer with the address of the short label if the Zero flag is cleared to 0. These jumps are the reverse of the JZ/JE jumps, which take the jump if the Zero flag is set to 1.

As was the case for JZ and JE, two mnemonics are used for the same op-code to enable the programmer to signal the sense of the program decision that is to be made. If a data operand is not zero as the result of some opera-tion, and this is the decision that is to be made, then JNZ will take the jump for the condition that the quantity is nonzero. If two data operands have been compared (using a CMP or SUB for instance) and they are not equal, then JNE jumps on the fact that the data are not equal.

Remember, the ZF is set to a 1 if the result of some opcode is all zeroes. If the opcode result is 0, then it is true that the result was 0, and ZF is set to 1 to indicate a true condition.

As was the case for the JNC jump opcode, we could accomplish a JNZ/JNE using JZ/JE followed by an unconditional jump. If the Zero flag is 0, then the JZ/JE will not take the conditional jump and will "fall through" to the uncon-ditional jump and jump on the fact that the Zero flag is equal to 0.

UNCONDITIONAL JUMPS

JMP   Jump

Mnemonic	Displacement	Range	Operation	Flags
JMP	SHORT-LABEL	Short	(IP) ← SHORT-LABEL	None
JMP	NEAR-LABEL	Near	(IP) ← NEAR-LABEL	
JMP	FAR-LABEL	Far	(CS:IP) ← FAR-LABEL	

Examples:

jmp here	;JMP to label "here"; "here" could be short or near
jmp short here	;force assembler to use a short-range jump to "here"
jmp far here	;JMP to another segment

Operation:

JMP is an unconditional jump and is always taken. This is the *only* jump that may be taken over all three ranges, depending on how the label name is specified to the assembler. The words short, near, and far tell an assembler what type of address number (how many bytes are needed) for the label in the assembled code for the JMP address. The A86 assembler, however, does *not* require the short or near qualifiers for intrasegment jumps.

5.2 Flow Charts

Now that we have a small bag of instructions to work with, we shall develop a way to diagram small programs. Diagrams of programs are called *flow charts* because they visually show the way a program operates or "flows" as it runs. These charts are not unique to computer programming and are used in many other fields such as business and construction planning.

Figure 5.2 ■ Flow chart elements.

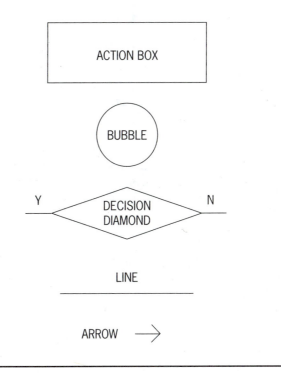

Flow charts can be a valuable aid in visualizing programs. Let us concentrate on concepts before becoming engrossed in opcode details. As mentioned in Chapter 4, flow charts have been supplanted by modular programming as a technique for organizing and documenting programs. Flow charts are, however, very useful learning tools for the beginning programmer.

Flow Chart Elements

For our purposes, we shall use five elements, or parts, to draw a flow chart. These are: the box, the bubble, the diamond, the line, and the arrow. The five flow chart elements are shown in Figure 5.2.

ACTIONS

The *action box* is used to contain statements of actions that the computer program will execute. The box is entered from the top and exited from the bottom. The action statements may be data moves, math operations, or any other instructions that denote a program action. Figure 5.3 shows several boxes that contain action instructions for the 8086 microprocessor.

DECISIONS

Decisions are shown on a flow chart by using the *decision diamond* element. The diamond is entered from the top corner and exited through any

Figure 5.3 ■ Action box elements.

one of the other three points of the diamond: the left, right, or bottom corner. Decisions are indicated by a flag or condition statement. The usual answer to a decision is yes (Y) or no (N). Normally only one decision per diamond is shown. Figure 5.4 shows several diamonds that indicate 8086 decision instructions.

FLOW

Program flow is indicated by connecting the boxes and diamonds of a program together with *lines* and *arrows*. Each line begins at a box, or diamond, and has an arrow at the termination end that shows the direction the program is to go. Traditionally, the flow of a program begins at the top of the page and flows down.

BUBBLES

Bubbles, or small circles, are used to begin and end programs. They are also used as markers to indicate where a line leaves an area and where it reenters an area. Using bubbles as markers tends to make the diagram more readable when there are many lines of flow, or when the program leaves one page and resumes on another. Figure 5.5 shows a complete flow chart for a simple program.

We should again note here that large programs are not usually put into flow chart form. The advantage of flow charting is to show, hopefully on a single page, a single simple program. Flow charts are also useful for diagramming concepts within a larger program. The trend in programming today is to write programs in modular form and to copiously document and comment those modules. Chapter 6 discusses some features of modular programs, which are also called *procedures*.

Figure 5.4 ■ Diamond decision elements.

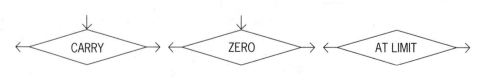

Figure 5.5 ■ A flow chart.

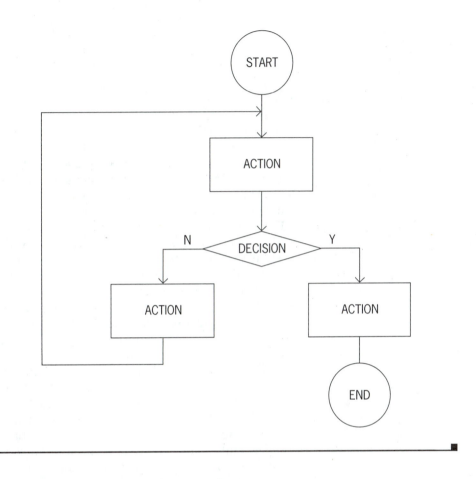

5.3 Sequential Programs

Suppose that we wish to write a program that will perform multi-byte addition. That is, we wish to construct data in data memory so that a single number is contained in 3 bytes of data. A 3-byte number means our numbers may be as small as 000000h or as large as FFFFFFh. Such numbers are called *unsigned* integers because they are considered to be whole (integer) numbers, and no provisions have been made to assign a positive or negative sign to the numbers. Because there are 3 bytes per number, we must have some way to distinguish the highest order byte from the lowest order byte. Typically, for a multi-byte number, the high order byte is called the *most significant byte* or MSB and the low order byte is the *least significant byte* or LSB. The middle byte(s) gets no name at all.

Let us assume that all of the 3-byte numbers are kept in memory, say from location 60000h to location 60011h, which means there are six, 3-byte numbers in memory. The first number is found at addresses 60000h, 60001h, and 60002h, and the last number is found at addresses 6000Fh, 60010h, and 60011h.

Now, we have to decide in what order the numbers are to be stored in memory—that is, should we put the LSB or the MSB at the lower address in memory? To find this answer we should inspect how the 8086 goes about storing data words in memory. A review of Chapter 3 will show that the 8086 stores all data words in ascending order in memory. As one goes from a low address to a higher address, the LSB is stored first, followed by the MSB. There is no obvious reason why this should be so, but it is the way the 8086 designers (and the 8080 designers before them) decided to store words of data in memory. It seems reasonable, therefore, to use the same low-high scheme for our data. So we shall put the LSB of the first number in location 60000h, the next byte in address 60001h, and the MSB in location 60002h. We continue this low-to-high storage scheme until all six numbers are stored.

The program we will write will add together any two of the six numbers and put the resulting sum in location 60020h (LSB) to location 60023h (MSB). The alert reader will notice that the sum is stored in 4 byte locations, whereas the numbers used to form the sum are 3 bytes long. A little thought will reveal that when two 3-byte numbers are added together, it is possible that the result will take 4 bytes to hold the sum. For instance, FFFFFFh plus FFFFFFh will yield a sum of 01FFFFFEh, which is a 4-byte answer.

If you are a beginning programmer, and somewhat intimidated by the details that must be considered in this simple program, then the example has worked its intended result. *Every* detail must be examined and defined by the programmer. The programmer must decide, for instance, how large the numbers are to be, where in memory to store the numbers, and how to arrange the numbers in memory. The programmer must also define whether the numbers are signed, unsigned, packed BCD, or unpacked BCD.

Those who program in a high-level language do not need to worry about such mundane things as exactly how the data is physically stored in memory. But assembly language programming is inexorably linked with the physical structure of the computer, and we must be aware of the physical results of our program.

A flow chart of the addition program is shown in Figure 5.6. The program must do three things in order to add two numbers and store the result. The three actions are as follows.

1. Address the two data numbers
2. Add the numbers together
3. Store the results

The way any individual programmer decides to perform each of these tasks is completely arbitrary. By *arbitrary* I mean that there is no such thing as the "right" way to write this program. The only real criterion for any program is that it work! A decree for a working program is followed closely by the demand that the program be finished *on time*. Beyond that we can decide that the *shortest* program (in code bytes) that works is the "best" one out of a number of right programs.

For the sake of example, we shall examine four programs that work equally well. Each example will add the 3-byte number that is stored at

Figure 5.6 ■ Addition program flow chart.

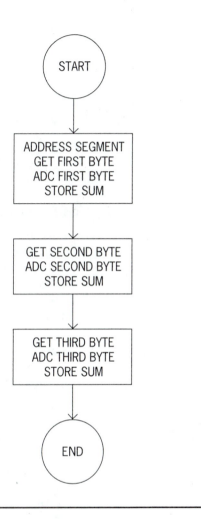

address 60003h to the 3-byte number that is stored at address 6000Ch and store the double-word result at addresses 60020h to 60023h. All of these programs are *sequential*, because they run from start to finish, with no jumps backward.

Example Program 1: Packed Number Addition

This program will use direct addressing to get each number and add them 1 byte at a time. The LSB of each word will be simply added together, and the result stored. All higher-order bytes will have any carry out of a lower-order byte added to them before storing the resulting sum.

```
code segment              ;                                    See Note
starthere:
        mov ax,6000h      ;get the DS segment to point to 60000h    1
        mov ds,ax         ;the only Intel way to get a number to DS
```

```
mov al,[03h]        ;get the LSB of number one                 2
add al,[0ch]        ;add the LSB of number two
mov [20h],al        ;store the result of adding the LSBs
mov al,[04h]        ;get the middle byte of number one
adc al,[0dh]        ;add, with any carry, middle byte of number two
mov [21h],al        ;store the result
mov al,[05h]        ;get the MSB of number one
adc al,[0eh]        ;add, with any carry, MSB of number two
mov [22h],al        ;store the result
mov al,00h          ;prepare to add any final carry from MSB
adc al,00h          ;if there was a carry it is now in AL
mov [23h],al        ;store MSB of result
jmp starthere       ;go again, if desired
code ends
```

NOTES ON PROGRAM 1

1. Why did the programmer use AX to load 6000h to DS, when any 16-bit register would do as well? AX was chosen at random. There is no "correct" 16-bit register used to load a segment register with an immediate number. But many programmers use one working register as a "scratch pad" register and never use it to hold any permanent numbers.

2. AL is used to do the addition by arbitrary choice. Any 8-bit working register could work as well. (A *very fine point:* Instructions involving the A registers often use one opcode byte to specify A; other registers use two opcode bytes to encode.)

Also note that the program could easily add 3-byte numbers found anywhere in data memory offsets 0003h and 000Ch just by changing the contents of the DS data segment register.

DEBUGGING EXAMPLE PROGRAM 1

As a practice exercise, type Example Program 1 into a text file and assemble it using A86. Then, call up D86 and debug Example 1.

Your debugging session should proceed in the following steps.

a. Make sure you have really loaded Example Program 1. If D86 comes up as nothing but NOP instructions, you did not load the correct .COM program.

b. Place trial data at locations DS:0003h–0005h and DS:000Ch–000Eh by using data moves such as mov b[0003h],trialdata. Move the data to memory *after* DS is set to 6000h by the program. Note that DOS has loaded your program into a segment far from segment 6000h.

c. Display memory on line 1, using 1,b,0000h, to have D86 show you what is happening in segment 6000h, offset 0000h. Use Ctrl-N to display the next set of bytes in data memory.

d. Step your program, using F1, and watch what happens. *Take your time,* and make *sure* the program works.

e. Repeat the test using different data until you are satisfied that the program works for data ranging from 000000h to FFFFFFh.

There is no substitute for adequately testing your program, and testing it again.

The next example addresses memory directly also but makes use of the fact that addition can be done on words as well as bytes. The next program depends on having the data stored as words in memory. The arrangement of data used for Program 1 is called *packed* data because we have made the most efficient use of memory to store the data. Program 1 could really have stored the bytes in any order, such as storing the LSB higher than the MSB in memory. But word-oriented programs must have the data words organized correctly (LSB first) or the data moves will not work as planned.

Example Program 2: Word Addition

Data will be moved as words (except for the MSB of each number) and added by word.

```
code segment           ;                                      See Note
starthere:
        mov ax,6000h    ;get the DS to point to 60000h
        mov ds,ax       ;DS now set to 6000h
        mov ax,[03h]    ;get LSB and next byte of number one
        add ax,[0ch]    ;add LSB and next byte of number two
        mov [20h],ax    ;store result at 60020,21h
        mov ah,00h      ;prepare to add MSBs                    1
        mov al,[05h]    ;get MSB of number one
        adc al,[0eh]    ;add, with carry, MSB of number two     2
        adc ah,00h      ;add any carry to AH
        mov [22h],ax    ;store result at 60022,23h
        jmp starthere   ;go again, if desired
code ends
```

NOTES ON PROGRAM 2

1. AH was set to 00h so that any carry from the MSB addition could be added to AH. Any carry from AL is added to AH. It is only necessary to make sure AH is 00h before starting the high-byte addition.

2. The carry added to the MSBs is a carry from the previous 16-bit addition. The carry is added to an 8-bit MSB addition process. It does not matter how the carry was generated; it can be added to word or byte operands.

Program 2 is more efficient because the first 2 bytes are added as words and stored as a word.

It is very tempting to store the 3-byte numbers using 4 bytes instead of 3, even though a byte of memory would be wasted (set to 00) for each 3-byte number in memory. Wasting memory space is not a real problem unless we are storing hundreds of thousands of 3-byte numbers in data memory.

If we use 2 words of data memory to store each 3-byte number, we will have arranged our data in a manner that is convenient for our computer. The 8086 is a 16-bit CPU and it makes good sense to handle data in word lengths whenever possible.

Example Program 3: Unpacked Number Addition

Let us assume that we have word-aligned six, 3-byte numbers on 2-word boundaries. The first number takes up byte addresses 60000h to 60003h, and the last of our six numbers occupies addresses 60014h to 60017h. Our two numbers that used to start at 60003h and 6000Ch now begin at 60004h and 60010h. The high order byte of each high order word is 00 because we still are using numbers made up of 3 bytes. The sum of two numbers will continue to require 4 bytes of storage at locations 60020h–23h.

When the six numbers are originally stored in memory, one assumes by a previous program action, the previous program action *must* ensure that each high order byte of the high order word is zero.

```
code segment           ;                                       See Note
starthere:
        mov ax,6000h    ;point to segment 60000h with DS
        mov ds,ax       ;DS now set
        mov ax,[04h]    ;get first word of number one at 04h         1
        add ax,[10h]    ;add first word of number two at 10h
        mov [20h],ax    ;store first word results at 20,21h
        mov ax,[06h]    ;get second word of number one
        adc ax,[12h]    ;add second word, with carry
        mov [22h],ax    ;store final result at 22,23h
        jmp starthere   ;go again, if desired
code ends
```

NOTE ON PROGRAM 3

1. How does the assembler "know" that the data is word-length, instead of byte-length? Because the register used in the opcode, AX, is word-length.

Program 3 is very efficient because the data is stored in a convenient manner for word-length math operations.

Example Program 4: Indirect Addressing

For our fourth program let's use an indirect addressing mode for our operands. This addressing mode is not really the best way to access our data but becomes very desirable if we wished to add many numbers together. Index registers DI and SI will be used to hold the addresses of the two numbers that are to be added together, and register BX will hold the address of the storage location for the result. We will assume that the data is word-aligned, as in Example 3.

```
code segment           ;                                       See Note
starthere:
        mov ax,6000h    ;point to the data segment with DS
        mov ds,ax       ;DS initialized
        mov si,0004h    ;get SI to point to number one              1
        mov di,0010h    ;get DI to point to number two
        mov bx,0020h    ;get BX to point to the storage address     2
        mov ax,[si]     ;get first word of number one
        add ax,[di]     ;add first word of number two
        mov [bx],ax     ;store first word of result
```

```
        inc si              ;increment si to point to second word      3
        inc si              ;of number one
        inc di              ;increment DI to point to second word
        inc di              ;of number two
        inc bx              ;increment BX to point to second word
        inc bx              ;of the result
        mov ax,[si]         ;get second word of number one
        adc ax,[di]         ;add, with carry, number two              4
        mov [bx],ax         ;store the high order result
        jmp starthere       ;go again, if desired
code ends
```

NOTES ON PROGRAM 4

1. There are three, general-purpose pointing registers that use DS as a default segment register: BX, SI, and DI. The choice of which to use to point to any location in memory is arbitrary.

2. The choice of BX to hold the address of the sum is not arbitrary, once SI and DI are assigned. BP is the last remaining general-purpose pointing register, and it uses SS, not DS, as a default segment register. The program could use BP by overriding the default SS with optional DS.

3. Registers increment by 1 byte. A word takes up 2 bytes, so a double increment is needed to point to the next word. Adding an immediate number of 2 (example: add di,02h) to the pointing registers would work, as well.

4. INC instructions do not affect the Carry flag, so it is as we left it before the string of INCs in the program.

Example 4 is a very long program because of the necessity of having to load the indirect address registers and incrementing them to point to the next word of each number. But, what if we wished to add together *all* six numbers stored from location 60000h to 60017h? Then the indirect address approach lets the computer take care of getting each set of numbers and adding them together. We shall explore using indirect memory addressing in the next set of programs.

5.4 Looping Programs

Our challenge now is to write a program that adds together all of the numbers stored as double words from location 60000h to 60017h. As was the case for the simple addition problems we have just finished for Examples 3 and 4, the numbers are stored as double words, but the high order byte of the high order word is 00.

We must use register indirect addressing for the next example, because only indirect addressing allows the CPU to change an address contained in a register. The program *loops* because it repeats a program similar to Example 4 until all six numbers have been added together. Loops are formed in a program when the program jumps backward in code, executing the same code sequence over and over. The program must have some way to "get out" of the loop, usually by counting loops until enough have been done.

Figure 5.7 ■ Looping addition program.

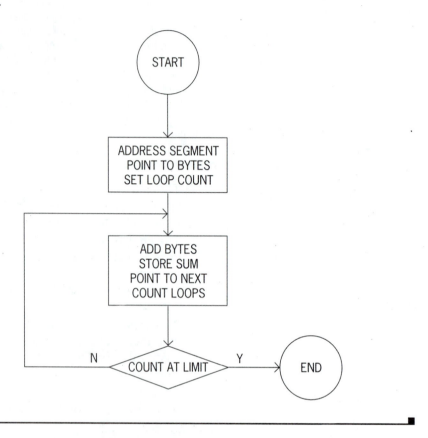

The largest possible result obtainable, for our six-number example, will be 05FFFFFAh, if all six numbers are 00FFFFFFh. The largest possible sum will still fit into 2 words. The flow chart for this program is shown in Figure 5.7. This flow chart includes a diamond decision element that decides if the program has looped the required number of times.

Any program that loops through a set of instructions repetitively must have some means to terminate the loop, or the program will run forever in the loop. We will use register CX to count how many times we have added the numbers, and we will leave the loop when all six numbers have been added together. How we use CX to count the loops is, again, up to the programmer. Two examples will be shown for this program, each using CX in a different way to count the loops.

Example Program 5: Loop Counts Up

In Example 5 we will count CX *up* from 0000h to the value that signals that the program is finished.

```
code segment              ;                              See Note
starthere;
       mov ax,6000h       ;point to the data segment with DS
       mov ds,ax          ;DS initialized
```

```
        mov w[20h],0000    ;set low order word of sum to 0000          1
        mov w[22h],0000    ;set high order word of sum to 0000
        mov si,0000h       ;get SI to point to number one
        mov cx,0000h       ;get CX to count the loops                  2
lp:
        mov bx,0020h       ;BX points to sum low order word
        mov ax,[bx]        ;add old sum LSW to number LSW
        add ax,[si]        ;add first word of the number
        mov [bx],ax        ;store current sum LSW
        inc si             ;SI points to MSW of number
        inc si             ;increment SI twice for MSW of number
        inc bx             ;increment BX to high word of sum
        inc bx             ;increment BX twice for sum MSW
        mov ax,[bx]        ;get second word of sum
        adc ax,[si]        ;add second word of number, with carry
        mov [bx],ax        ;store high order word of sum
        inc cx             ;stop looping after 5 loops
        inc si             ;point to first byte of next number
        inc si             ;increment SI twice
        cmp cx,0006h       ;use a CMP to set the flags for equal
        jne lp             ;if CX is not 0006h then loop again         3
        jmp starthere      ;go again, if desired
code ends
```

NOTES ON PROGRAM 5

1. Our loop adds the *current sum* to each set of numbers. We have no way of knowing what numbers are in the sum memory locations when the program starts, so we "make sure" it is zero. Note that we must use the w (word) assembler directive, or the assembler will move 00h to the *byte* at offset address 0020h.

2. Choosing CX to count the number of additions in the loop is arbitrary, but from a limited choice of registers. AX, BX, and SI have been chosen for other duties, leaving CX, DX, BP, SP, and DI. Pointing registers are normally not used for counting. CX was chosen because it is used as a counter in other, more advanced, instructions.

3. Try to place the jump instruction as close as possible to the opcode that sets the jump flags, in this case the CMP instruction.

Note that SI and BX have to be incremented twice to point to the first byte of the next word.

The use of indirect addressing lets the computer generate the operand addresses *as the program runs*. This is advantageous only when the data to be used is organized so that it is grouped into consecutive locations in memory. When data is organized into consecutive memory addresses, it is said to occupy a *block* of memory. The program written here could add quite a few word-aligned 3-byte numbers together simply by changing the CMP operand of 0006h to a larger number. A hundred numbers could be added simply by changing the CMP operand to 0064h, and changing the storage address of the sum to a location above the hundred numbers.

Note that the JNE form of the JNE/JNZ opcode is used to signify that CX was not equal to 0006h, so the jump is taken. When CX reaches 0006h, CMP will set ZF to 1, meaning that the result of the CMP instructions was 0, thus CX and 0006h are equal. Opcode JNZ could have been used instead of JNE, but the reader might assume that the programmer was going to loop until CX is 0. Testing CX for a zero condition is exactly the reverse of the CX action that will terminate the loop.

Example Program 6: Loop Counts Down

This time we shall use CX to count the number of loops by counting CX *down* instead of up. One advantage of this method is that the CMP instruction will not be needed. Loop counters, such as CX, are often counted down to avoid using a CMP instruction.

```
code segment           ;                                   See Note
starthere:
        mov ax,6000h    ;point to the data segment with DS
        mov ds,ax       ;DS initialized
        mov w[20h],0000 ;set low order word of sum to 0000
        mov w[22h],0000 ;set high order word of sum to 0000
        mov si,0000h    ;get SI to point to number one
        mov cx,0006h    ;get CX to count the loops
lp:
        mov bx,0020h    ;BX points to sum low order word
        mov ax,[bx]     ;add old sum LSW to number LSW
        add ax,[si]     ;add first word of the number
        mov [bx],ax     ;store current sum LSW
        inc si          ;SI points to MSW of number
        inc si          ;increment SI twice for MSW of number
        inc bx          ;increment BX to high word of sum
        inc bx          ;increment BX twice for sum MSW
        mov ax,[bx]     ;get second word of sum
        adc ax,[si]     ;add second word of number, with carry
        mov [bx],ax     ;store high order word of sum
        inc si          ;point to first byte of next number
        inc si          ;increment SI twice
        dec cx          ;stop looping after 6 (6-0) loops        1
        jnz lp          ;if CX is not 0000h then loop again
        jmp starthere   ;go again, if desired
code ends
```

Note on Program 6

1. CX is decremented *after* the addition operation, so the sixth addition operation was done when CX = 1. After performing the sixth addition, CX decrements to 0000h, ending the loop. If CX had been decremented *before* the addition process, a starting number of 0007h would be needed in CX to do six loops.

Note that the negative form of the conditional jump is used. As long as the Zero flag is not set to 1 (ZF = 1 means CX is 0), the jump is taken, and the looping continues. Also, note that the JNZ opcode immediately follows the DEC CX instruction so that the Zero flag is tested immediately, before any intervening instruction can change it.

Using a loop counter, such as CX, is a standard programming technique for determining when to quit looping the program. Another technique is to inspect the number in the register that is used to address the data, and stop the loop when a register has pointed beyond the last data address that is of interest. For the problem we have been doing, the last data address of interest is at location 60017h, which is the high order word of the sixth number. Thus, when the indirect register that addresses data (SI in this case) reaches 0018h, all of the numbers have been added together.

Example Program 7: End Loop at Limit

A CMP instruction shall be used to detect that register SI has reached offset address 0018h in memory, thus terminating program looping.

```
code segment          ;                                        See Note
starthere:
        mov ax,6000h      ;point to the data segment with DS
        mov ds,ax         ;DS initialized
        mov w[20h],0000   ;set low order word of sum to 0000
        mov w[22h],0000   ;set high order word of sum to 0000
        mov si,0000h      ;get SI to point to number one
lp:
        mov bx,0020h      ;BX points to sum low order word
        mov ax,[bx]       ;add old sum LSW to number LSW
        add ax,[si]       ;add first word of the number
        mov [bx],ax       ;store current sum LSW
        inc si            ;SI points to MSW of number
        inc si            ;increment SI twice for MSW of number
        inc bx            ;increment BX to high word of sum
        inc bx            ;increment BX twice for sum MSW
        mov ax,[bx]       ;get second word of sum
        adc ax,[si]       ;add second word of number, with carry
        mov [bx],ax       ;store high order word of sum
        inc cx            ;stop looping after 5 loops
        inc si            ;point to first byte of next number
        inc si            ;increment SI twice
        cmp si,0018h      ;use a CMP to set the flags for equal      1
        jne lp            ;if SI is not 0018h then loop again
        jmp starthere     ;go again, if desired
code ends
```

Note on Program 7

1. SI points to the *next* number to be added to the current sum, at the end of the addition process. Six 2-word numbers are stored from location in memory as follows:

Number	Offset Memory Addresses (Hex)
1	0000–0003
2	0004–0007
3	0008–000B
4	000C–000F
5	0010–0013
6	0014–0017

A seventh number could be stored starting at offset address 0018h, where the program stops getting numbers.

If SI is initialized at numbers other than the number 0000h used in this example, then the terminating address, 0018h, must also be changed.

Example Program 8: Names for Constants

Our previous seven examples have used numbers to identify all the data locations in the program. Numbers help us to think about exactly where, in memory, everything is placed, but numbers tend to make the program hard to change. For instance, if we decide to add numbers in absolute address locations 60100h to 60117h, and store the results at 60240h, we must make sure we load all of the pointing registers with the new values. We might forget to change a value, buried deep inside the program, and the program would not work.

Numbers also make the program difficult to read, because numbers cannot imply how the data is used or what its purpose is. Names, however, are very good ways to make the program changeable, readable, and less confusing to the reader.

Not surprisingly, most assemblers have a directive that gives a name to a *fixed* number. For A86, the directive is *equate* (EQU). EQU is used by writing name equ number.

The EQU assembler directive lets the programmer assign a *name* to an *absolute constant number*. Example 8 is the same as Example 6, but uses the EQU directive to assign names to all of the address numbers.

```
                                     ;                                       See Note
        code segment                 ;give the data numbers names
        starthere:
                dat_seg equ 6000h    ;data segment is at 6000h
                num_one equ 0000h    ;first number at offset 0000h
                sum_at equ 0020h     ;the sum is placed at offset 0020h
                count_it equ 0006h   ;count six loops in CX
        ;
        ;every data address now has a, hopefully, meaningful name
        ;
                mov ax,dat_seg          ;point to the data segment with DS
                mov ds,ax               ;DS initialized
                mov w[sum_at],0000h     ;set low order word of sum to 0000
                mov w[sum_at+2h],0000h  ;set high order word of sum to 0000    1
                mov si,num_one          ;get SI to point to number one
                mov cx,count_it         ;get CX to count the loops
        lp:
                mov bx,sum_at           ;BX points to sum low order word
                mov ax,[bx]             ;add old sum LSW to number LSW
                add ax,[si]             ;add first word of the number
                mov [bx],ax             ;store current sum LSW
                add si,0002h            ;increment SI twice for MSW of number  2
                add bx,0002h            ;increment BX to high word of sum
                mov ax,[bx]             ;get second word of sum
                adc ax,[si]             ;add second word of number, with carry
                mov [bx],ax             ;store high order word of sum
                add si,0002h            ;point to first byte of next number
```

```
        dec cx              ;stop looping after 6 (0–5) loops
        jnz lp              ;if CX is not 0000h then loop again
        jmp starthere       ;go again, if desired
code ends
```

NOTES ON PROGRAM 8

1. The assembler can form absolute numbers from constant operands using the ordinary math operators +, −, / (division), and * (multiplication). Note that the plus (+) operator, when used as an assembler directive, is *not* the same as the + used in an instruction, to add a displacement to a pointing register. Assembler operands are generated, by the assembler, at *assembly time* from known absolute numbers. Addresses, formed by adding a displacement to a pointing register, are formed by the CPU, at *run time*. Assembly time is when the assembler forms numbers from known, absolute constants. Run time numbers are calculated, by the CPU, at the instant the opcode is executed.

2. Adding 2 to the pointing registers points them to the next word.

Example 8 makes use of the EQU assembler directive to assign names to program numbers. The names given to the address numbers are purely up to the programmer, as long as reserved names, such as the mnemonics mov or jnz, are not used as names. Some programmers like to use underlines (_) as part of the name, to differentiate an operand name from a code label name.

Should we wish to use different numbers in our program, we only have to change the numbers after the EQU assembler directives and assemble the program. The assembler will change any line of code that uses an EQU name to the name's new number. The EQU statements are normally placed at the beginning of the code segment so the programmer does not have to hunt all through the program to find names for operands.

Readability of the program improves when names are used for numbers in a program. By clever selection of names, the programmer can help make the program more understandable.

Example Program 9: Variables and Data Segments

Example programs 1 through 8 use pre-assigned addresses (6000h:0000h–0023h) to store data numbers to be added and the location of the sum. Many times the programmer does not care where the data is placed, as long as the data and the code spaces do not interfere with each other.

Data can be stored in a data segment, and *names assigned to the data addresses*, just as we assigned names to constants using the EQU directive in Example Program 8. Data names are called *variables* because the variable name address can hold *any* number, and the program can change the *contents* of the variable address as the program runs.

We shall repeat our first problem, in which only two packed 3-byte numbers are added together.

```
                            ;                                    See Note
data segment                ;declare a data segment                 1
org 1000h                                                           2
```

```
            num1lsb db ?        ;name all the data addresses              3
            num1mb db ?         ;variable names can be any name
            num1msb db ?        ;the ? reserves space in data memory
            num2lsb db ?        ;"db" reserves a byte of space
            num2mb db ?
            num2msb db ?
            num3lsb db ?
            num3mb db ?
            num3msb db ?
            num4lsb db ?
            num4mb db ?
            num4msb db ?
            num5lsb db ?
            num5mb db ?
            num5msb db ?
            num6lsb db ?
            num6mb db ?
            num6msb db ?
            sumlsb db ?
            sum2b db ?
            sum3b db ?
            summsb db ?
    data ends                                                            4
    code segment
    starthere:
            mov al,num2lsb      ;get the LSB of number one               5
            add al,num5lsb      ;add the LSB of number two
            mov sumlsb,al       ;store the result of adding the LSBs
            mov al,num2mb       ;get the next byte of number one
            adc al,num5mb       ;add, with any carry, number two byte
            mov sum2b,al        ;store the result
            mov al,num2msb      ;get the MSB of number one
            adc al,num5msb      ;add, with any carry, number two MSB
            mov sum3b,al        ;store the result
            mov al,00h          ;prepare to add any final carry from MSB
            adc al,00h          ;if there was a carry it is now in AL
            mov summsb,al       ;store MSB of result
            jmp starthere       ;go again, if desired
    code ends
```

NOTES ON PROGRAM 9

1. The assembler directive data segment signals to the assembler that a data segment, not a code segment, is defined.

2. The org directive is used to place data memory space in memory, above code memory. Originate, or org, directs the assembler to reserve address space in data memory for each byte, word, and the like that is defined. Using org 1000h tells A86 to leave space for data beginning at offset address 1000h. The number 1000h is a guess. Program 9 is very short, so it probably does not extend beyond 1000h in memory. Should the guess be wrong, and code and data memory overlap, A86 will *not* issue a warning. Should you wish to know how much code memory is used, assemble the program. Use D86 to step to the end of the program and note the last address used

in the program. Data *can* begin above the last address used for code, but you would be wise to leave some room for more code in case you need to expand your program.

3. Variable addresses are named simply by typing a variable name in front of the db (or dd, dw, etc.) define directive. Program 9 adds numbers of byte size, therefore every byte of data has a variable name. Variable names, like all names used in a program, are chosen by the programmer.

4. Assembler directive data ends ends the data segment.

5. The variable name num2lsb is used to get the contents of the variable into register AL. No "contents of" brackets ([ ]) are necessary.

Variables make the program more readable and informative. Variable names also help during debugging, because D86 will use the variable names, not variable address numbers.

You may have a question at this point: "Why define the data variables first, before starting the code?" It is possible, though *not* recommended, to place the data variable definitions after the code. The reason the data variables are named before the code is written has to do with A86 and a phenomenon called *forward referencing*. When A86 encounters a name in a program, it must "guess" what type of name it is. The name could be a label, a data variable, or a data constant. Labels are noted by colons (:), but other names *could* be data variables or constants. Usually, A86 guesses that any undefined name found in a program is a constant (label or number).

If you define the name first, by using equ for constants and db and the like for variables, A86 *knows* the nature of the name. If you use the name in a code line, and then define it later in the program ("forward reference the name"), then A86 will guess the first time it sees the name and *not correct that guess* when it finally comes to your forward-referenced name. So, define your data names *before* writing the code.

5.5 Program Numbers: Constants, Labels, and Variables

Assembly language instructions are made up of action mnemonics and operands. The assembler program encodes the mnemonics and the operands into binary code. Thus, every mnemonic and every operand become real, fixed, binary numbers in code memory. There are no names in code memory, just numbers. Assemblers, however, let the programmer use *symbols* (names) in place of *code numbers*.

Instructions are encoded by the assembler depending on the mnemonic used and the operands associated with the mnemonic. Operands, no matter what the addressing mode, must be numbers, symbols, or the name of a register.

Register names are assigned by the CPU designers. The programmer, then, has no choice but to use the register names (AX, DI, BP, etc.) chosen by the manufacturer. Numbers in a program, however, are chosen by the programmer as *absolute constant* numbers or *symbolic* numbers.

Absolute constant numbers are numbers that are entered as part of the program and do not change. Absolute constant values can be seen when the program is read. For instance, the following lines all use an absolute constant as one of the operands.

```
mov ax,1000h            ;immediate data constant of 1000h
mov w[1000h],bx         ;constant memory address 1000h
mov w[2000h],1234h      ;move the constant 1234h to constant memory
                        ;location address 2000h
```

As can be seen in the preceding example, absolute constant numbers are just plain numbers that do *not* change each time the program is assembled.

Symbolic constants, however, are numbers that are assigned names *by the programmer*, and become numbers when the program is *assembled*. There are *three types* of symbolic names available to the programmer: constants, labels, and variables.

Constant names are assigned by the programmer using the EQU directive. There is very little difference between an absolute constant and a symbolic constant, except for the convenience of using meaningful names in place of obscure numbers. For instance, the following constant name may make the resulting program easier to understand.

```
code segment
        beernum equ 99      ;99 bottles of beer
        fall equ 1          ;1 less bottle of beer
        mov cx,beernum      ;count the bottles of beer
one_less:
        sub cx,fall         ;if one happens to fall
        jnz one_less        ;continue until no more
code ends
```

The names used with the EQU directives are not variable names (although they do not end in a colon), they are the names of constants. A86 assumes that any undefined name in a program (except jump operands) is the name of a constant.

Labels are names assigned to address location numbers in code or data memory. The assembler knows a name is a label because it ends in a colon (:). The absolute number that A86 assigns to a label is determined when the program is assembled into code. For example, the label one_less: in the preceding example is assigned to memory location 0103h by A86 when the program is assembled. The label symbol becomes a fixed number only during assembly. If the program is rewritten, and a new line of code inserted before the label, then the number assigned to one_less: will change during assembly. For instance, if a 1-byte instruction, such as NOP (no operation), is inserted in the program before the label, the assembler will assign the number 0104h to the symbol one_less:. Thus, as the program is rewritten, and lines are added or deleted, numbers assigned to labels may change each time the program is assembled.

Variables are names assigned to address location numbers in data memory. The assembler knows a name is *not* a label because it *does not* end in a colon. Whenever A86 encounters a variable name in a program, it assumes that the programmer wants to *use the contents* of the variable location, *not the address number* associated with the variable name. Variable names are assigned by the programmer using the various *define* directives such as DB,

DW, and so on. For example, the following program assigns a variable name to a data memory location in the program.

```
data segment
      org 1000h              ;begin assigning data locations starting at 1000h
      can_vary dw ?          ;reserve a word space at 1000h in data memory
data ends
code segment
      mov can_vary, 1234h    ;move constant to data memory location 1000h
      mov ax,can_vary        ;move contents of variable to AX
      mov ax,w[1000h]        ;does exactly the same thing
```

The assembler uses the correct opcode for "contents of" whenever it encounters a variable name as one of the instruction operands. The address number assigned to a variable name may change as we alter the program. Should we modify the last program, and ORG the data segment at some other location, say 2000h, then the assembler will assign the address number 2000h to the variable name can_vary when the program is assembled.

The programmer can *override* A86, however, and force it to use a label as a variable, or a variable as a label, in an instruction. The override directive that makes A86 use the variable address number is the *offset operator*. The override directive that makes A86 treat a label number as the "contents of" is the bracket [] operator. For instance, in the next program we shall define a variable and a label and use them interchangeably.

```
data segment
      org 1000h              ;start data memory at 1000h
      label_name: dw ?       ;A86 assigns the number 1000h to the label
      vary_name dw ?         ;A86 assigns the number 1002h to the variable
data ends
code segment                 ;COM programs are assembled beginning at 0100h
start_here:                  ;the label "start_here" = 0100h
      mov ax,label_name      ;the constant number 1000h is moved to AX
      mov vary_name,ax       ;the 1000h in AX is moved to contents of 1002h
      mov ax,offset vary_name ;AX = 1002h (number assigned to variable)
      mov w[label_name],ax   ;move 1002h to contents of 1000h
      mov bx,codeptr         ;BX = 0117h
      mov cx,[bx]            ;CX = 4567h
      mov cx,w[codeptr]      ;CX = 4567h—brackets force contents of
      jmp start_here         ;jump to location 0100h in program code
codeptr: dw 4567h            ;codeptr: is a label = 0117h
code ends
```

5.6 Summary

Eighteen instructions are chosen as representative of the most often used of the 97 distinct 8086 instructions. The instructions covered are the following:

ADD	JC/JB/JNAE	NOT
ADC	JE/JZ	OR
AND	JMP	RCL
CMP	JNC/JAE/JNB	SBB

```
DEC      JNE/JNZ      SUB
INC      MOV          XOR
```

Instructions that have two mnemonics, such as JE/JZ, have two names for the same operational code. The name chosen by the programmer should best represent the sense of the program at that point.

Flow charts of programs are composed of five elements that may be used to diagram simple programs. The elements and their uses are listed here.

Element	Use
Box	For program actions
Diamond	For program decisions
Line	To connect boxes and diamonds
Arrow	To show direction of program flow
Bubble	To connect parts of flow chart

Nine programming examples demonstrate how flow charts and some of the 18 instruction mnemonics can be used to solve a problem.

5.7 Problems

Write programs that solve each problem using only the instructions included in this chapter. Use any addressing mode that works unless specifically requested to do otherwise.

Place comments at the end of each line of code. Include a short explanation of how the program was tested.

All numbers are unsigned unless specifically designated as signed.

Problems that search memory begin the search at the *lower* memory address.

If a program refers to the contents of a register or memory operand, do NOT program a fixed number as the operand. Use the D86 debugger to load trial numbers into unspecified operand locations.

Exercise Problems

1. Take a word placed in BP and perform *ONLY logical* operations on that word so that it equals F0F0h, no matter what the starting value in BP was.

2. Make Bit 3 of any word placed in AX a 0 without changing any other bit in the word.

3. Make Bit 12 of any word placed in AX a 1 without changing any other bit in the word.

4. Change Bit 13 of any word placed in AX to its complement without changing any other bit in the word.

5. Add 1234h to 5678h and place the sum in DX.

6. Move the data in register AX into memory location 60000h using indirect addressing.

7. Move the data in memory location 60FFFh into register SI using direct addressing.

8. Move the data in register BX into memory location 61234h using indirect addressing.

9. Rotate the carry bit into Bit 4 of DX. *Note:* The carry bit may be set to 1 by the instruction sequence mov al,0ffh/add al,al.

10. Add the byte in AL to the byte in AH and put the result in BP.

Beginning Problems

1. Add together the numbers F123h, F456h, and ABCDh. Store the result in registers AX (least significant word) and BX (most significant word).

2. Add together the bytes F2h, E4h, C6h, 78h, and 9Ah using registers AL and BL to hold the sum. AL is to hold the least significant byte of the sum and BL the most significant byte of the sum.

3. Move the words stored in memory between DS:1000h and DS:1100h to memory space DS:1200h to DS:1300h.

4. Compare the word in AX with the one in CX. Jump to label big: if AX is larger than CX, to the label small: if AX is smaller than CX, and to the label same: if they are equal.

5. Compare the byte in AL to the byte in memory location DS:1234h. Jump to label big: if the contents of AL are equal to or larger than the contents of memory location DS:1234h. Jump to label small: if the contents of AL are smaller than the contents of DS:1234h.

6. Increment register CX from a count of 0000h until it equals the number placed in register DX.

7. Decrement register DX from a count of FFFFh until it equals the number found in memory location DS:1000h.

8. Make any word placed in AX equal to 0000h without using any MOV or AND instructions.

9. Take whatever random number is placed in register CX and move all the even numbered bits (bit 0, 2, 4 . . .) into the same bit position of register DX. The odd bits of DX (bit 1, 3, 5 . . .) are to be 0.

10. Subtract the number in register DX from the number in register BX. If no borrow is generated, store the result in register CX. If a borrow occurs, store the result in register AX.

11. Rotate registers AX (least significant word) and DX (most significant word) 1 bit to the left as if they were one 32-bit register. The most significant bit of DX is rotated to the least significant bit of AX, and the most significant bit of AX is rotated to the least significant bit of DX.

12. Repeat Problem 11, but rotate the registers to the right.

13. Take the smaller of 2 words stored in memory locations DS:1234h and DS:5678h and store it in register AX.

14. Find the smallest byte in register AX and store it in register CL.

15. Add together the *nibbles* of register AL and put the result in register BH.

16. Take whatever random number is placed in CX and compute a number such that when it is added to the number in CX, the result will be 0000h. Place the computed number in DX.

17. Swap the bytes in register DI, such that the least significant byte of DI becomes the most significant, and the reverse.

18. Compare the byte in AL against the bytes in memory from DS:0400h to DS:0420h. If a match is found, jump to label equal:. If no match is found during the search, jump to label nomatch:.

19. Compare the byte in AL against the bytes in memory from DS:0400h to DS:0420h. Count the number of matches found in register DX.

20. Count the number of bytes located in memory from DS:1000h to DS:1200h that are greater than 77h. Put the count in register CX.

Intermediate Problems

Note: When a range of addresses in memory is given, the range is inclusive. Address A to B means use everything *including* A and B.

1. Count the number of bits, in the double word that starts at memory location DS:1234h, that are a 1. Place the count in register AL.

2. Count the number of bytes, between location 60000h and 60100h, that are an even number. Place the count in register BH.

3. Move the 100h bytes located from 60000h to 600FFh to locations 60200h to 602FFh, in reverse order. The byte located at 60000h is stored at location 602FFh and so on until the byte located at 600FFh is stored at location 60200h.

4. Move every byte of data from memory locations 60000h to 600FFh to memory addresses 601C0h to 602BFh. As the data is moved, count the number of bytes that are greater in magnitude than 7Fh, using register DX to hold the count.

5. Repeat Problem 4 for data of word length. Count the number of words that exceed 7FFFh in magnitude, using register DL to hold the count. State your assumptions as to how the bytes of the word are stored in memory.

6. Count the number of bytes, stored from memory address 60000h to 60200h, that are smaller than or equal to the number 4Bh. Use register CX to hold the count.

7. Add together all of the 2-word numbers that are stored from address 60000h to 60FFFh in memory. Store the sum starting at location 61020h. Each 2-word number may range in size from 00000000h to FFFFFFFFh. State your assumptions as to how the words for each number are stored in memory and how large the final sum can be.

8. Repeat Problem 7 using the same address range of 60000h to 60FFFh for numbers that are 4 words in length. A number may vary in size from 0000000000000000h to FFFFFFFFFFFFFFFFh.

9. Two 31-byte numbers are stored beginning at locations 60000h and 60020h in memory. Add the two numbers together and place the sum of the two numbers in location 60040h and higher addresses. State all of your assumptions as to how the numbers are stored in memory and how the sum is stored.

10. Numbers, which are 2 words in length, are stored in two separate areas of memory. One set of numbers is stored from location 60000h to location 6003Fh; the other set is stored from location 60080h to 600BFh. Add each corresponding pair of numbers together and store the results beginning at memory location 61000h. State all of your assumptions as to how the numbers are stored in memory and how the sum of each pair is stored.

11. Repeat Problem 10 for numbers that are 4 words in length.

12. Three sets of 1-byte numbers are stored in three separate areas of memory. Each set consists of 32 bytes. One set is stored from address 60000h to 6001Fh, the second set from address 60030h to 6004Fh, and the third set from address 600A0h to 600BFh. Add together the 3 bytes, 1 from each set, that are stored in the same order in memory. Place the sum of each group of 3 bytes in memory beginning at location 600D0h. State your plan for arranging the sum of each trio in memory and the total memory address range required to store the sums.

13. Repeat Problem 12 for numbers that occupy the same memory spaces but are word-, not byte-, length.

14. Arrange the bytes of registers AX and DX in numeric order such that the largest byte is in AL and the smallest byte is in DH. For instance, if AX is F43Ch and DX is C623h before the program is run, AX should be C6F4h and DX should be 233Ch after the program is run.

15. Put a random word in offset location DS:1000h and another random word in location DS:1010h. Increment the word at offset 1000h, and then decrement the word at offset 1010h, until the two words are equal in magnitude. *Warning!* Do not let the words "pass" each other by forgetting to check for equality before incrementing or decrementing.

16. The number A3h is somewhere in memory between DS:1000h and DS:1400h. Find the first offset address that contains A3h and put the offset address in register BX.

17. Place any binary number in register DX and count the number of binary 1s in the number in DX in register AL. For instance, the binary number 3F62h contains nine binary 1s.

Advanced Problems

1. Somewhere in memory, between locations 60000h and 60100h, lurks the message "Hello." Find the offset address where Hello begins and store the address in register BX. If you find more than one Hello, use the first one you find. *Beware!* There may be false Hellos such as hello, hElLo, or Helo. Find the first, real, Hello.

2. Construct the message "Wild Crazy U" from characters found in memory locations 60000h to 60100h. Store the offset address of each character you

find beginning at memory location 60200h. Use the first offset address you find for each character.

3. Search memory between addresses 60000h and 60100h for the smallest and largest bytes. Store the offset address of the smallest number in location 60102h and the offset address of the largest number at location 60104h. In case of duplicate numbers (ties for large or small), store the address of the *last* number found.

4. Find the largest double-word number in memory, between locations 60000h and 60100h, and place the number in CX(MSW) and DX(LSW).

5. Test memory locations 60000h to 60100h by writing 13h to each byte address and then reading it back to ensure that 13h was correctly written. If a write/read operation fails, then store the offset address at which up to four failures occurred in memory locations 60200h (the first failure found starting at offset 0000h), 60202h, 60204h, and 60206h.

6. There is one, and only one, secret message stored between locations 60000h and 60100h in memory. You can tell it is a message because it begins and ends with an "!". You do not know how long it is. Find the beginning of the message and put the offset address in CX. Then find the end of the message and put the offset address in DX. Put the length of the message, in bytes, in register BX.

7. You are writing a program for a production-monitoring computer. Every bit of the byte at memory location 60000h represents the output of a production machine so that eight machines are monitored. Every time any bit of the byte at 60000h changes from a 1 to a 0, that machine has produced one item. Each machine runs so slowly that the program will not miss a change of 1 to 0 in a bit position.

   Write the program so that each machine's production count, for an 8-hour shift, is stored beginning at location 60010h in memory. It is possible for a machine to produce up to 10,000d units during an 8-hour period. Indicate in your documentation how you store the data for each machine.

8. The security system used to guard a computer room is activated at 7 P.M. every evening and deactivated at 7 A.M. every morning. After activation, the person entering the room has 30 seconds to enter the correct code, which is the number sequence 1, 5, 6, 2.

   The computer that controls the security system has the following memory assignments:

Location	Function	Format
60000h	Time of day, minutes from midnight	Word, 0000 to 1440
60010h	Time since door opened, seconds	Byte, 0 to 30
60020h	Sequence entered	Four ASCII bytes

   Write a program that will place an alarm number of FFh in location 60030h if the room is entered any time between 7 P.M. and 7 A.M. and 30 seconds expires before a correct sequence is entered.

9. Write a program that will arrange the words placed in registers AX, BX, CX, DX, SI, and DI in ascending order with the largest word in AX and the smallest word in DI.

10. Take any 6 bytes found from memory address DS:1000h to DS:1005h and arrange them in descending order starting at address DS:1000h. The largest byte should be in address DS:1000h and the smallest byte in address DS:1005h.

11. Arrange the nibbles of any word found in register AX so that the smallest nibble is in AL (least significant nibble), the next smallest nibble is in AL (most significant nibble), the largest nibble in AH (most significant nibble), and the next-largest nibble in AH (least significant nibble). For instance, if AX = 1234h at the start of the program, then AX should be 4321h at the end of the program.

12. Somewhere in memory, between addresses 60000h and 60100h, is at least one sequence of *five* ASCII numbers. A sequence is defined as any set of consecutive decimal numbers. For instance, 34567 is a five-number sequence, as are 01234 and 56789. Find the address of the beginning of the first five-number sequence found starting from 60000h and put that address in register DX. *Beware!* There are many sequences that are not five characters long or are over five characters long.

6

PROCEDURES

Chapter Objectives

On successful completion of this chapter you will be able to:

- Use push and pop stack instructions.
- Call procedures.
- Organize your programs so that they are easier to write and debug.
- Choose between assembly language programming and high-level language programming.
- Pass data between the main program and procedures using registers, memory, and the stack.
- Use immediate, direct, and indirect addressing modes to call procedures.
- Describe how procedures may be called from high-level language programs.

6.0 Introduction

Assembly code programming was the original programming discipline at the beginning of the computer age in 1943. Over the past 50 years, as in all technical disciplines, assembly coding practices have changed and evolved. One major step was to write assembler programs that convert programs written in assembly language instructions into assembled code. Other innovations centered around methods that speed up the programming process.

One contemporary programming practice is to divide long, monolithic programs into smaller pieces. The trend toward writing many compact programs, in place of one long program, has come about for two reasons. First, the division of a single large program into separate smaller programs allows *many* people to work simultaneously on the overall project. Second, certain small program pieces become so generic that they need be written only once. The generic program can then be *used* many times without further effort.

A second trend in contemporary programming practice is the growing use of high-level programming languages in areas once reserved for assembly language code. High-level language compilers have now been written, for most microprocessors and microcontrollers, that generate *relatively* memory-efficient code. The appearance of memory-efficient compilers has also coincided with the manufacture of ever-cheaper DRAM computer memory chips. Cheaper memory facilitates the use of high-level programs because the system designer can afford to add the extra memory required by memory-hungry high-level programs.

With the growing use of efficient high-level languages, assembly language programming has become more and more a discipline of writing smaller, highly specialized programs. Assembly language programs can then be used by high-level language programs for interfacing to specific equipment and for those instances in which speed of program execution demands the most compact assembly code possible. Today's system programmer needs to be proficient in both high-level *and* assembly code programming.

6.1 Procedures

Assembly language programs, no matter what their length, have always simply been called *programs*. The names for smaller parts of assembly language programs, however, have been given several different names. Large parts of a program became known as *routines*, and smaller parts were then identified as *subroutines*.[1]

High-level languages, such as Pascal and C, did not follow the nomenclature established for their ancestors. High-level languages are designed to be modular in form and consist, mainly, of a collection of smaller parts called *procedures*, or *functions*. The term *procedure* has now been borrowed by 8086 assembly language programmers and has replaced the venerable *subroutine* in the lexicon of today.

---

[1]*Computer Dictionary* (Redmond, WA: Microsoft Press, 1991), p. 332.

A procedure is a small program that is part of a larger *main* program. Procedures are usually designed to perform one overall action (a *task*) in the context of the larger program. Procedures, in turn, may use other procedures in their operation. Procedures must be "well-behaved," that is, they must not cause the main program (or any other procedures that use them) any problems.

Problems that could be caused by procedures include altering the contents of any of the CPU registers that are used by the main program, except those it is "allowed" to change. The main program, for instance, might assume that register AX is always used for temporary storage and allow any procedure to change the contents of AX randomly. Or, the main program may assume that all the CPU working registers are changed by a procedure. The important point is that the main program must ensure that any register content changes by procedures do not adversely affect main program operation. Conversely, procedures that are *not* allowed to permanently change any CPU register must take care to *save* and *restore* to their original state the CPU registers it uses.

Let us now inspect a hardware feature of the 8086 that enables the program to access procedures.

6.2 The Stack

In Chapter 3, we discussed an area of RAM called the *stack*. A stack is nothing more than a group of RAM data locations that are indirectly addressed by the Stack Pointer (SP) register working in conjunction with the Stack Segment (SS) register. Figure 6.1 shows how a stack is implemented using the SS and SP registers.

Most programs maintain a stack by loading SP with a number that can

Figure 6.1 ■ Stack addressing.

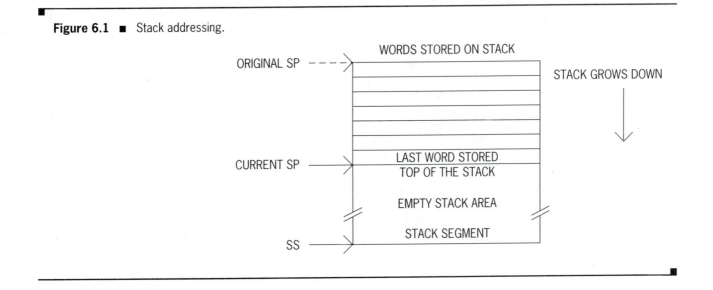

safely go *down* as the stack *grows*. A program can establish new stacks simply by loading SP with a new number that points to a new area of stack RAM. For our purposes, a single stack will normally meet program needs.

It has become commonplace to refer to the SP and SS registers, which *hold the address of a location* in the stack, and the area of RAM that is addressed by SP and SS as the *stack*. We should be careful to note, however, that the *stack* is part of RAM memory and the indirect addressing registers that point to the stack are the *stack pointers*.

Very few 8086 instructions are designed to use the stack area of RAM. In keeping with our 80/20 rule of programming, we shall study six of the most common stack instructions.

PUSH and POP the Stack

The Stack Pointer register indirectly addresses, or points to, an offset address in RAM. The segment register used with SP is SS, and *no* segment override is possible. The address *currently* pointed to by the SS:SP combination is called the *stack*. The address pointed to represents a location in RAM that is the *last* address used for data storage. Only *words* of data are handled by stack operations, and no B or W data-length assembler directives are needed by stack instructions.

The top of the stack is the address at which the stack begins to grow downward in memory, from higher addresses in RAM to lower addresses in RAM. The stack "grows down" as it is used to store data. The mental image of a stack, however, remains that of a growing mound of stored words.

The way the stack grows down is set by the action of the following data storage instruction, which has the mnemonic PUSH.

PUSH   Push Source on Stack

Mnemonic	Operand	Size	Operation	Flags
PUSH	SOURCE	W	$(SP) \leftarrow (SP) - 2$ $W[SP] \leftarrow SRC$	None

Addressing Modes
(Source)

register
direct
indirect

Examples:

```
push ax          ;push register AX on the stack
push cs          ;push register CS on the stack
push [bx]        ;push word pointed to by BX on the stack
push [1000h]     ;push word in location 1000h on the stack
```

Operation:

PUSH causes the SP register to be immediately decremented by a count of 2. The source data *word* is then placed in RAM at the address pointed to by SP. The data word is stored low byte first at the address pointed to by SP, and the high byte is stored at the address pointed to by SP + 1. Register combination SS:SP holds the indirect address of the *last* address used to store data on the stack. Only *words* of data may be pushed onto the stack.

Notice that the destination for the data is properly [SS:SP], but this is not stated explicitly in the PUSH mnemonic. We might call such an omission "implied indirect," because the destination address is the address held in the SS, SP register combination. Note also that PUSH, along with MOV, is one of the few instructions that can access the segment registers.

PUSH is commonly used for temporary CPU register storage in the stack area of RAM. Programmers often have need of a place to put the contents of some CPU registers while they detour from the main program to another, temporary, program. Before leaving the main program, the programmer stores the CPU register contents on the stack. Once the register contents are safely stored on the stack, the programmer may use the same CPU registers for the temporary task. After the temporary task is completed, the programmer retrieves the original register data from the stack, and the main program resumes using the original register data. The only precaution the programmer must take is to *ensure* that the SP is not pointing to a new stack area.

STACK PUSH PRECAUTIONS

There is one overriding caution that the programmer must keep in mind when using the stack for any reason: *The stack is not infinite in depth.* It is entirely possible to have the stack grow down until SP becomes 0000h. On the way down, the stack may overwrite program data and, finally, program code. Normally, the stack area is placed at a high address in the stack memory segment. However, if code, data, and stack are in the same segment as they are for .COM programs, the stack *can* "grow down" and erase some code bytes. Disaster usually follows uncontrolled stack growth. The most common symptom is that the program will run for a few milliseconds, and then crash as the stack quickly overwrites vital data or code addresses.

Another common mistake, usually made by beginning programmers, is to *forget* to set SS and SP to safe stack addresses. When power is first applied to the 8086, SS and SP can conceivably contain any number. If SS and SP are not set to an area of RAM that physically exists then the first PUSH will not save anything. Be sure you *set* SP and SS, for non-DOS applications, in the first lines of your program.

Users who are running programs under DOS will have the luxury of having *DOS set the SS:SP register pair for them.* It is *good* practice however, to set SP for all programs that use the stack.

Now that we know how to temporarily store data on the stack, let us find out how to get data back from the stack using a POP instruction, which is examined next.

POP   Pop from Stack to Destination

Mnemonic	Operand	Size	Operation	Flags
POP	DESTINATION	W	(DST) ← W[SP] (SP) ← (SP) + 2	None

Addressing Modes
(Destination)

register
direct
indirect

Examples:

```
pop cx          ;pop word on stack to register CX
pop ds          ;pop word on stack to register DS
pop [bp]        ;pop word on stack to memory location in BP
pop [1000h]     ;pop word on stack to memory location 1000h
```

Operation:

POP moves the data *word* found at the address pointed to by SP to the destination. The byte addressed by SP is copied to the low order byte of the destination operand. The data byte at address SP + 1 is then copied to the high order byte of the destination operand. The SP register is then increased by 2.

CAUTION

- It is possible to repeatedly execute POP operations until the SP register overflows, from FFFFh to 0001h.
- POP CS is undefined. Attempting to pop anything to the CS register would cause an unconditional jump in the program.

Note that POP, PUSH, and MOV are the only instructions that can directly access the segment registers.

POP is the exact reverse of PUSH. Data is copied from the stack to the destination, and then the SP register increments by 2 so that SP now addresses the next *word* of data on the stack.

The operations of push and pop are often referred to as "pushing the stack" and "popping the stack," or to "push the stack" and "pop the stack." The stack area of RAM soon becomes thought of as an endless supply of scratch paper, where you can quickly store and retrieve any amount of data simply by using the stack. Once the programmer has set the stack to a *safe* area in memory, it can be used with little (or no) thought as to exactly where SS:SP happens to point at the moment.

STACK POP PRECAUTIONS

The warnings that apply to uncontrolled growth of the stack by using excessive push operations also apply to pop operations. Every pop increases SP by 2. Eventually, SP can overflow, say from FFFEh to 0000h (FFFEh + 2), and invade another memory area by accident.

In general, the programmer goes to great pains to *keep the stack "balanced."* That is, for every PUSH opcode in the program there is a matching POP opcode that exactly undoes the effect of the PUSH. The stack is balanced if, at the end of the program, it returns to where it was set at the start of the program. A balanced stack cannot grow uncontrollably even if the program is executed many times.

Any deviation from balancing the stack must be accounted for by the programmer. One way to balance the stack is to reset the SP to a number that represents any PUSH and POP discrepancy. For example, if there have been 20 pushes and 14 pops, then adding 12 (6 words × 2 bytes/word) to the SP will restore the SP to its original count before any pushes were done.

A final note: Remember that a PUSH decrements the Stack Pointer so that the *stack grows toward address 0000h.* A POP increments the Stack Pointer and the *stack shrinks toward address FFFFh.*

Stacks operate on a *last-in first-out* or LIFO principle. The *last word pushed* on the stack is the *first word popped* from the stack. The first word pushed on the stack will be the last word popped off the stack. Figure 6.2 shows the operation of the stack for a series of pushes and pops.

Flags and the Stack

PUSH and POP can save and retrieve any register or memory location in the computer except one: the Flag register. We shall need two very special push and pop instructions that deal only with the flags, PUSHF and POPF, which are shown next.

Figure 6.2 ■ Push and pop operations.

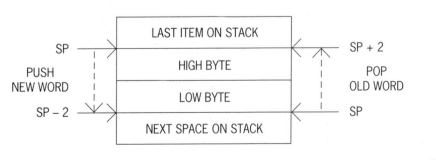

PUSHF   Push Flags on Stack
POPF   Pop Flags from Stack

Mnemonic	Size	Operation	Flags
PUSHF	W	$(SP) \leftarrow (SP) - 2$ $W[SP] \leftarrow Flags$	None
POPF	W	$Flags \leftarrow W[SP]$ $(SP) \leftarrow (SP) + 2$	None

Addressing Modes

register

Examples:

```
pushf        ;push the Flag register on the stack
popf         ;pop the word on the stack to the Flag register
```

Operation:

PUSHF and POPF push and pop the Flag register onto and off of the stack. PUSHF and POPF are the *only* two instructions that can access the Flag register as an *entire* 16-bit register.

Saving the CPU flags is essential if the complete "state of the machine" is to be preserved at any instant. The state of the machine can be defined as the contents of *all* the CPU registers at any given time in the program. PUSH can be used to save all of the working and segment registers, and PUSHF can then be used to save the Flag register. Pushing saves the current state of the CPU working registers and all of the program flags contained in the Flag register.

PUSH and POP Examples

We will discuss several examples of using stack operation codes using the two programs that follow next. The first program is meant to be used under DOS but does set SP anyway.

In the first program, the stack pointer is set to 1000h. Choosing 1000h as the address in RAM to start the stack is purely *arbitrary* and could be any number that sets the stack comfortably *above* program data and code memory areas. Setting SP is not necessary for programs loaded by DOS. DOS will allocate a stack address when the program is loaded (SP = FFFEh) for .COM type programs.

Should the program be written for a computer with no operating system, such as a board-level computer, then SS and SP registers must *both* be set by the program. Board-level computers typically have RAM in only a few physical addresses, and the SS:SP pair must point to actual RAM addresses. Program 2 demonstrates the type of program that might be used with a board-level computer.

STACK EXAMPLE PROGRAM 1

```
code segment
main:
        mov sp,1000h        ;initialize SP to point to stack
        mov ax,1234h        ;AX = 1234
        mov bx,5678h        ;BX = 5678
        mov cx,9abch        ;CX = 9ABC
        push ax             ;SP = 0FFE
        push bx             ;SP = 0FFC
        push cx             ;SP = 0FFA
        pop ax              ;AX = 9ABC SP = 0FFC
        pop bx              ;BX = 5678 SP = 0FFE
        pop cx              ;CX = 1234 SP = 1000
        mov bx,0200h        ;BX = 0200
        mov w[bx],1234h     ;address 0200 holds 1234
        push [0200h]        ;push 1234 SP = 0FFE
        push [bx]           ;push 1234.SP = 0FFC
        mov bx,0210h        ;BX = 0210
        pop [bx]            ;address 0210 = 1234
        pop [0212h]         ;address 0212 = 1234
        jmp main            ;demo again
code ends
```

Example Program 1 establishes the stack at 1000h and puts some random numbers in registers AX, BX, and CX. AX is pushed on the stack that stores it at SS:0FFFh(MSB) and SS:0FFEh(LSB). BX and CX are stored in a similar manner with an attendant decrement of SP by 2 for each push. The contents of the stack are then popped from the stack, in reverse order.

The contents of CX were the *last* to be pushed onto the stack, thus they are the *first* to be popped off of the stack. For illustration purposes, we have popped the stack in reverse order. AX is loaded with the data that was pushed from CX. The contents of BX do not change, because it is in the middle. CX receives the original contents of AX. SP is incremented by 2 for each pop operation. After three pops, SP returns to the value it had (1000h) before the program was run. Pushes and pops using direct and indirect memory address modes are also demonstrated.

INSPECTING THE STACK IN EXAMPLE PROGRAM 1

D86 can be used to demonstrate how the stack operates as Example Program 1 is exercised. The bottom memory line (the SP line) of D86 shows the contents of the stack as the program executes. When SP is set to 1000h, a full stack is displayed. (You may wish to use Ctrl-t to reset the stack display, after changing the contents of SP.) After each push, D86 will show the last data word pushed on the stack, at the far left of the SP line. As more data is pushed on the stack, old stack data moves rightward, and the new data appears on the left. As data is popped from the stack, the old data moves toward the left of the SP line, and the last data popped disappears from the line.

Remember, D86 *displays* words left to right as we humans are accustomed to: high byte then low byte. To see how the stack physically stores data, use memory line 1, in byte mode, to display the stack area of memory.

To see the stack as it is *actually* stored in RAM, type the command 1,b,SS,0ffah to watch the stack grow down from 1000h to 0FFAh.

The direct and indirect memory pops may be viewed by setting memory display line 1 as 1,w,bx.

STACK EXAMPLE PROGRAM 2

Example Program 2 could be used with a single-board computer. The single-board computer used in this program has a small 8K RAM mapped at addresses 20000h–21FFFh. The only significant difference between Example Programs 1 and 2 is the need to *immediately* set a valid Stack Segment register in Program 2.

```
code segment
        mov ax,2000h        ;set SS register via AX
        mov ss,ax           ;SS now points to real RAM
        mov sp,0200h        ;set SP at 0200
        mov si,1234h        ;SI = 1234
        mov di,5678h        ;DI = 5678
        push es             ;SP = 01FE
        push ds             ;SP = 01FC
        pop es              ;SP = 01FE
        pop ds              ;SP = 0200
code ends
```

This program must ensure that the Stack Segment register points to an area of RAM on the board that is physically present. The computer of this example uses an 8K RAM located at physical address 20000h to 21FFFh. Within the stack segment, the Stack Pointer is set to begin the stack at 0200h. The stack is addressed at the absolute address 20200h. Registers ES and DS are swapped using the stack for temporary storage. ES is pushed first, followed by DS. Then, ES is popped from the stack, receiving the contents of DS. DS is popped last, receiving the contents of ES.

PUSH (PUSHF) and POP (POPF) instructions can save the contents of all of the CPU working registers except one: the Instruction Pointer. In the next section we shall discuss how, and why, the IP is saved on the stack.

6.3 Calling a Procedure

A *procedure* is a small program that is used by another program, for example, the main program. To be able to use procedures requires that some *automatic* mechanism is built into the CPU that lets the IP of the main program be *saved on the stack*. Using a procedure means that the IP of the procedure replaces the IP of the main program. The main IP register is saved because we wish to be able to *restore* the main IP to its original contents (place in the main program) when the procedure program is finished.

Procedures may be used many times in a main program. Each time a procedure is used it must be able to "return" back to the proper place in the main program. Saving the main IP on the stack lets the CPU return to the proper place in the main program.

We have seen in Chapter 5 how it is possible to modify the IP using jump instructions. Jump instructions do *not* save the original program IP contents. The branch taken by the program, when a jump is taken, is permanent and the original program IP contents are *lost*. But, if we wish to use procedures, we must be able to resume our place in the original program *after* the procedure has been used. Jumps are not suitable instructions for calling and returning from procedures.

As we have seen, a procedure is a small program that can be used by another program. The mechanism by which one program uses another program is to have the using program *call* the program to be used. The using program is titled the *calling* program, and the program to be used is the *called* program.

In practice, the calling program executes a CALL instruction, followed by an address in code memory for the called program. *All* of the 8086 addressing modes may be used to specify the address of the called program.

When a CALL opcode is executed, the *called* program's address in program code is placed in the CPU *after* the *calling* program's address has been *saved* on the stack. When the called program is finished, the calling program's address is *restored* to the CPU from the stack, and the calling program resumes operation. The calling program address saved on the stack is termed the *return address* of the calling program.

We saw, in Chapter 3, that as the CPU executes code, the IP register contains the address of the *next* instruction to be fetched. If instruction A is in the BIU queue and instruction B is waiting to be fetched, the IP contains the address of instruction B. If instruction A is a CALL to a procedure, the IP address of instruction B (the instruction that follows the CALL instruction) is *automatically* saved on the stack by the CPU in response to the CALL instruction.

The IP address of instruction B is thus saved as a return address on the stack. The procedure will *return* the IP to the address of instruction B that was saved on the stack, when it has finished executing, by using a return instruction.

A CALL instruction functions as shown next.

CALL   Call a Procedure

Mnemonic	Operand	Size	Operation	Flags
CALL	NEAR_PROCEDURE	NA	(SP) ← (SP) − 2 W[SP] ← (IP) (IP) ← NEWIP	None
CALL	FAR_PROCEDURE	NA	(SP) ← (SP) − 2 W[SP] ← (CS) (CS) ← NEWSEG (SP) ← (SP) − 2 W[SP] ← IP (IP) ← NEWIP	None

Addressing Modes

immediate near
register near
direct near
indirect near
immediate far
direct far
indirect far

Examples:

```
call addit          ;addit IP is a displacement from current IP
call cx             ;procedure IP in CX
call w[1000h]       ;procedure IP contents of location 1000h
call w[bx]          ;procedure IP contents of address pointed to by BX
call faraddit       ;faraddit CS:IP in far code segment
call d[2000h]       ;double word location 2000h holds CS:IP of far call
call d[bx]          ;double word location pointed to by BX contains
                    ;CS:IP
```

Operation:

CALL causes the return address of the calling program to be saved on the stack. The nature of the return address (push IP or push CS:IP) is determined by the address (far or near) of the procedure that is called. After saving the return address (IP or CS:IP) on the stack, the CPU replaces the IP (or CS:IP) with the address of the procedure that is called.

CALL is used to access a procedure located at *any* address in code memory. Procedures may be located in the *same* segment as the current Code Segment register. Procedures in the same code segment as the calling program are considered to be *intrasegment* or *near* procedures. Procedures located in a code segment that is *different* from the code segment of the calling program are *intersegment* or *far* procedures.

Near Calls

Near calls replace the old IP with a new IP (NEWIP). NEWIP can be found two ways. If the near call uses immediate addressing, then a *displacement* number is added to the current IP so that the new IP (old IP + displacement) points to the address of the procedure. If register, direct, or indirect addressing is used for the near call, then the new IP is found in a CPU register or a memory location. Near calls cause the contents of the old IP (the next instruction after the CALL instruction) to be pushed on the stack. A new IP (NEWIP) is found, using the call instruction addressing mode, and replaces the old IP.

Near calls let the procedure be located anywhere in the same 64K segment as the calling program. The new IP can point to any code address in the code segment, from 0000h to FFFFh. The programmer bears the responsibility for *ensuring* that the correct address for the procedure is used.

We shall use near immediate calls for most of the example programs.

NEAR IMMEDIATE CALLS

Immediate calls contain a fixed number (a displacement number) that is added to the old IP to form a new IP address. The new IP points to a called procedure, which then executes.

Procedures that are in the same 64K code segment as the call instruction can be accessed by a near call. In immediate addressing, a *label* is used to specify the address of the procedure. The assembler calculates the displacement between the return address and the called procedure. For example, if the opcode call checkit is used, and the label checkit: is 10h bytes beyond the call, A86 will calculate the displacement as 0010h bytes and code the opcode accordingly.

NEAR REGISTER CALLS

For register addressed near calls, the new IP is the contents of any 16-bit working register—AX, BX, CX, DX, SI, DI, BP, or SP. Note that SP is a *poor choice* for the register, because loading SP with the address of the procedure also means that a *new* stack is defined.

Placing the correct procedure address in a register must be handled by the programmer. The A86 assembler has no way to force any register to contain a certain IP number.

NEAR DIRECT AND INDIRECT CALLS

A new IP for a near call can also be directly addressed as the contents of a word in memory. The new IP may also be indirectly addressed in memory by one of the base, index, or base-index pointing registers. Indirectly addressed IP values are pointed to by BX, BP, SI, or DI. The new IP is the contents of the memory address pointed to by one of the pointing registers.

Far Calls

Far calls use immediate, direct, and indirect addressing in order to replace the old CS:IP *pair* with a new CS:IP pair. Register mode addressing is *not* used for far calls because no single 8086 register can hold a double word CS:IP pair. Far calls involve replacing *both* the old CS and IP register contents with totally new CS (NEWSEG) and IP addresses found immediately or in memory. Far calls may address procedures located anywhere in the 1M memory space.

The use of near and far distances is similar to near and far jumps as discussed in Chapter 5. The terms *near* and *far* are used solely to alert the assembler as to how the various call instructions are to be translated to hex. If no declaration of the *range* of the call is made by the programmer, then the A86 assembler will assume that the procedure is *near* and assemble the call opcode accordingly.

Formal declaration of a procedure means that the procedure is uniquely identified using the keyword PROC. The entire line used to define a procedure is: name PROC near (or far). The programmer must declare the procedure to be near or far if a procedure is formally defined in a program.

FAR IMMEDIATE CALLS

Immediate far calls use immediate code segment and instruction pointer words found in the call opcode as sources for NEWIP and NEWSEG. Far calls that use immediate addressing will have their segment and displacement address components specified when the far procedure is given a label or is declared as a far procedure in a formal procedure declaration. Memory addresses directly specified in the instruction hold a CS:IP procedure address as a fixed double word that follows the CALL opcode in *code* memory.

FAR DIRECT AND INDIRECT CALLS

Direct and indirect far calls may only be done using double words that are located in memory. Far procedure addresses are pointed to by indirect address registers or are at the direct memory address given in the instruction. Memory must be used for indirect far calls because no register in the 8086 CPU can hold the 4 bytes required to specify NEWIP and NEWSEG. The intersegment address is arranged in memory with NEWIP as the low word and NEWSEG as the high word. Each word is arranged in the standard low byte in the lower address, high byte in the higher address format.

Base, index, and base-index addressing using registers BX, BP, SI, and DI may be used for far indirect calls. Indirect and memory far calls depend on the *programmer* to have correctly placed the proper code segment and instruction pointer numbers in the memory addresses directly specified or pointed to by the indirect addressing register.

Returning From a Call

We have seen that pushing data on the stack requires a balancing pop to retrieve the data. In a similar manner, a call causes interruption of the original program sequence and pushes a return address for the calling program on the stack. Figure 6.3 shows the contents of the stack after a near call and after a far call are executed.

The called program must have a way to pop the return address of the calling program. Instructions that enable the main program to resume operation at the instruction that follows a CALL mnemonic are called *returns* and are discussed next.

Figure 6.3 ■ Stack contents after a call.

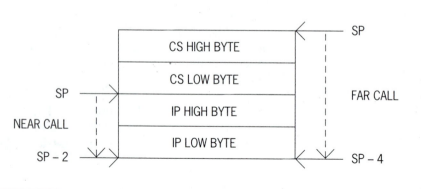

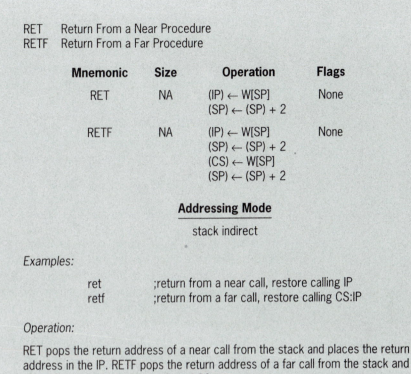

RET   Return From a Near Procedure
RETF   Return From a Far Procedure

Mnemonic	Size	Operation	Flags
RET	NA	(IP) ← W[SP] (SP) ← (SP) + 2	None
RETF	NA	(IP) ← W[SP] (SP) ← (SP) + 2 (CS) ← W[SP] (SP) ← (SP) + 2	None

Addressing Mode

stack indirect

Examples:

```
ret        ;return from a near call, restore calling IP
retf       ;return from a far call, restore calling CS:IP
```

Operation:

RET pops the return address of a near call from the stack and places the return address in the IP. RETF pops the return address of a far call from the stack and places the return address in IP and CS.

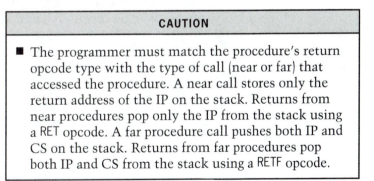

CAUTION
■ The programmer must match the procedure's return opcode type with the type of call (near or far) that accessed the procedure. A near call stores only the return address of the IP on the stack. Returns from near procedures pop only the IP from the stack using a RET opcode. A far procedure call pushes both IP and CS on the stack. Returns from far procedures pop both IP and CS from the stack using a RETF opcode.

It may come as a surprise to discover that the opcode RETF does not *formally* exist as an original Intel 8086 opcode. When the 8086 mnemonics were first published, all returns were given the mnemonic RET. The assembler was told how to properly assemble a near or far return by using *formal* procedure declarations. Over time, however, programmers (such as Eric Isaacson) found the formal procedure declarations to be so cumbersome that they invented the RETF mnemonic to mean a far return opcode. A86 generates the proper hex code for RET and RETF mnemonics, therefore no formal call declarations are necessary. The programmer, of course, must *remember* to use the correct return mnemonic to match the call range used.

We shall not use the procedure declaration practice for our programs, because all of our procedures are near, and near is the default call type for A86. To aid in understanding programs that do use procedure declarations however, the procedure syntax is discussed next.

Formal Procedure Declaration

A procedure is identified using the keywords PROC and NEAR or FAR as follows:

```
First Keyword line:    NAME          PROC NEAR (FAR)
                       Body of procedure program
Last Keyword line:     ENDP
```

NAME is any legal name such as might be used for a program label. No colon (:) is necessary at the end of the name. Keyword PROC tells the assembler that everything that follows is part of a procedure. Keywords NEAR or FAR instruct the assembler as to the nature of the call and return code for the procedure. Near procedures push and pop only the IP as a return address; far procedures push and pop both CS and IP return addresses. ENDP is a keyword that informs the assembler that the procedure is finished.

6.4 Using Procedures

Modular Design

Programs that are designed to make use of procedures extensively are said to be written using a *modular* design approach. Modular programming design involves defining the overall program objective and then dividing the overall program into smaller tasks, or *modules*.[2] The degree of "modularity" may vary with the wishes of the program designer. In this book, a module is defined as follows:

> *Module:* A procedure that performs *one* clearly definable program task, has *one* entry point, has *one* exit point, and has *no* jumps outside of itself.

Modules are called by the main program and may, in turn, call other modules. *How* the overall problem is divided into modules gives rise to organizational approaches called *top-down design* and *bottom-up programming*.

Top-down program design involves starting with the overall problem and dividing it into a series of overall tasks. Each task is then further divided into modules. Bottom-up programming involves developing (programming and testing) each module and then combining the modules so that the overall tasks are accomplished.

One goal of modular programming is to attempt to make each module as *generic* as possible. For example, suppose that we wished to write a procedure

---

[2]*Computer Dictionary* (Redmond, WA: Microsoft Press, 1991), p. 231.

that takes two numbers, multiplies them together, and returns the product. The numbers could vary from 2 to 6 bytes in length. A set of *five* modules—one each for numbers of length 2, 3, 4, 5, and 6 bytes—could be written. The main program could call the proper multiplying procedure based on the lengths of the numbers at hand. A more generic approach is to write *one* module that can multiply numbers of any length, from 2 to 6 bytes. To use the generic multiply module, the calling program informs the module how many bytes to multiply and the location of the bytes. Calling programs often must "communicate" with the procedures that they call. The calling program may have to supply the module with data or the addresses of data, and the module often returns some data values to the calling program.

Interprogram Communication

To effectively use generic procedures, there must be communication between the calling program and the called program. Communication between programs is commonly done by having the calling program pre-load a set of CPU registers, or an area of data memory, with the numbers that are the raw material for the called program. The procedure is then called and operates on the raw data that has been stored for it by the calling program. The procedure then loads the results of its action into a set of CPU registers, or an area of data memory, and returns to the calling program.

The practice of interchanging data in certain CPU registers, or in data memory, is termed *passing* the data. The calling program passes raw data to the procedure that is called, and the procedure passes finished data back to the calling program.

Communication between programs requires very careful *planning* so that the calling program and the called program both know what to pass and where to pass it. Planning becomes even more important when the calling and called programs are written by *different* people. Generally, one individual is assigned the job of defining all of the protocols that are to be used by the calling and called programs. Often, a uniform approach is tried wherein all data passing between programs is done using a standard set of registers, or a standard data area.

Because it is used extensively for temporary data, the AX register is often used for passing 1 or 2 bytes of data between programs. Whenever larger amounts of data must be passed, say 4 to 20 bytes, the rest of the CPU *registers* can be used. Data may also be passed on the *stack* by having one program push data on the stack and the second program pop data from the stack. If large amounts of data are passed between modules, then an area of data can be set aside in *memory*, and the calling program can pass a *pointer* (say SI or DI) to the data area to the called program.

COMMUNICATION VIA REGISTERS

Register passing has the advantage of simplicity, speed, and precision. The calling program loads raw data into a set of registers and calls a procedure. The procedure processes the data passed in the registers and returns the results in a similar set of registers. Registers involved in the transactions are well defined, and the data is available to both programs immediately. Although only 7 words of data may be passed using CPU registers (using SP is not recommended), modular programs that interchange more than 7 words of data may probably be further divided into *simpler* pieces.

COMMUNICATION VIA THE STACK

Using the stack to pass data has the advantage of extending the number of data words passed between programs beyond the available CPU registers. Data may be interchanged by pushing and popping data on and off the stack. Or a stack data area may be defined by setting SP to point to a stack address *below* its current location. The stack area between the old SP and new SP location is now available for passing data. Once the data has been passed, the SP can be restored to its former place in the stack.

COMMUNICATION VIA MEMORY

If very large amounts of data must be passed, say hundreds to thousands of bytes, then a data *array* is established in data memory, and the *address* of the array is all that is passed between modules. An array is any area of data in memory that has a single name; for example, an array of numbers might be named nmbers. Each number in the array has an address within the array. For instance, the 18th number could be addressed as [nmbers+11h]. The address of the array can be passed in a register (SI or DI are recommended) or the array address may be stored in a memory variable.

MAKING COMMUNICATION SAFE

One important precaution must be taken by the modular procedure programmer. The procedure must not permanently alter any of the CPU registers, unless *allowed* to do so. The calling program programmer and the procedure programmer may agree that some registers, such as AX, may be used by the procedure at will. All other non-allowed registers used in the procedure must have their contents saved before they are used by the procedure. The calling program may assume that the state of the CPU is absolutely unaltered by the procedure, *including the flags*. In such a situation, procedures that alter the flags should first save the old Flag register contents and restore the old flags before returning.

If the procedure changes any register or flag contents, *other than those allowed*, the changes can cause unpredictable behavior in the calling program. These unwanted changes, dubbed *pollution*, are particularly vexatious because they may be random in nature and occur in a sporadic manner. Random procedure pollution is very, very difficult to debug. To make communication safe between programs, all *disallowed* CPU registers used by the procedure are stored on the stack by the procedure before any other actions are taken. The saved registers are restored by the procedure just before returning to the calling program.

An alternative to having the procedure save CPU registers is to have the calling program save CPU registers before calling a procedure. The procedure is then free to alter any CPU register before returning. The calling program then pops the saved CPU registers. Occasionally the calling programmer will lose count and push more registers than are popped, or the reverse. The stack then slowly grows or shrinks, every time the procedure is called, until the program crashes. Depending on the conditions that call the procedure to be called, the crash may take hours or days to occur. Then the fun begins.

Naming Procedures

Assembly code programmers, over the years, have evolved personal preferences regarding label and procedure names. Over time, these preferences have become de facto "standards" as they were found to make sense for a majority of assembly code programmers. One common procedure name standard is to use the underline (_) as part of the procedure name. Using an underline lets the programmer break a long, confusing name into something readable. For instance, a procedure that adds numbers could be named addthenumbers. Using underlines makes the procedure name add_the_numbers, a much clearer way to say the same thing. In general, names used for procedures and labels should "make sense" to the program and help document the program so that it is understandable. *Understandable* programs are easier to debug and change.

6.5 Examples of Procedures

We shall first examine several procedures that involve passing data using registers. The procedures involved in all of the examples perform the rather simple task of adding several data numbers and returning the sum. Although the procedures are simple, documentation of how data is passed and returned between calling and called programs must be clear and thorough. We shall use only procedures that are near. All the examples assume that any register or flag used in a procedure, *other than those allowed*, is saved by the procedure.

Passing Data in AX

Our first procedure adds the byte of data in AL to the byte of data in AH, returning the sum of the 2 bytes in AX. The calling program loads the bytes to be added into registers AH and AL and calls the procedure. The procedure adds the bytes found in registers AH and AL and returns the sum in register AX. The Flag register and register BX are saved by the procedure because they are the only CPU registers affected by the procedure routine.

```
example_one:              ;a near procedure
    pushf                 ;save the original flags
    push bx               ;save original BX contents
    mov bl,ah             ;align bytes for addition
    mov ah,00h            ;clear AH
    mov bh,00h            ;clear BH
    add ax,bx             ;add the bytes, get a word result
    pop bx                ;restore old BX
    popf                  ;restore the old flags
    ret                   ;return to the calling program
```

Our first example saves the Flag register and BX because the original contents of these registers are altered by the procedure.

Note the order in which registers FR and BX are pushed and popped. The stack is a last in, first out structure. Register BX is the last register to be pushed onto the stack and must be the first to be popped from the stack in order to retrieve the data in the order in which it was saved. Also, note that the number of pushes and pops are equal to each other.

Procedure example_one is called by program example_two, which we examine next.

```
code segment   ;passing data in a working register
      byte1    equ 0ffh              ;example data
      byte2    equ 3ah               ;example data
example_two:   mov sp,2000h          ;set the stack
               mov al,byte1          ;AX = raw data
               mov ah,byte2
               call example_one
               jmp example_two       ;loop back to beginning
```

Program example_two loads register AX with the numbers to be added and then calls procedure example_one. Note that there are no keywords used in the call, simply the *label* of the procedure, example_one. Example_one is *assumed* by A86 to be a near procedure when it is written, and the assembler will always generate a near call code whenever example_one is called. The assembler will always generate a near return in example_one because the RET opcode was used in example_one.

The complete program consists of main module example_two and procedure example_one. The complete program appears as follows.

```
code segment   ;passing data in a working register                    See Note
      byte1    equ 0ffh              ;example data
      byte2    equ 3ah               ;example data
example_two:   mov sp,2000h          ;set the stack                      1
               mov al,byte1          ;AX holds raw data
               mov ah,byte2
               call example_one
               jmp example_two       ;loop                               2
;
;*********** do not allow main program to run into procedure *******
;
example_one:                         ;a near procedure                   3
               pushf                 ;save the original flags
               push bx               ;save original BX contents
               mov bl,ah             ;align bytes for addition
               mov ah,00h            ;clear AH
               mov bh,00h            ;clear BH
               add ax,bx             ;add the bytes, get a word result
               pop bx                ;restore old BX
               popf                  ;restore the old flags
               ret                   ;return to the calling program
code ends
```

NOTES ON THE PROGRAM

1. Even though DOS will set the SP, set the stack to keep in practice. The initial stack number, at 2000h, is arbitrary. Set the stack above the code and data parts of the program.

2. Procedures must be called, *not run into*. If the jmp opcode were not present, the main program would fetch the first opcode in

example_one, and execute it. The rest of example_one would then be executed, including the RET instruction. There is no valid return address on the stack, and the procedure returns to whatever address happens to be on the stack at the time.

3. Place procedures at the end of the main program.

DEBUGGING EXAMPLE_TWO

Using D86 to debug example_two is very instructive. The following procedure is recommended.

   a. Single step down to the call. Note that DOS has set the stack when the program is loaded.

   b. Watch the stack memory line, at the bottom of the screen, as the call is executed. Observe that the return address is saved on the stack and that procedure example_one is executed. (You may use Ctrl-t to "clean up" the new stack when it is set to 2000h.)

   c. Step through the procedure and observe the Flag register and BX saved on the stack, then popped from the stack.

   d. Step through the RET instruction, and note the procedure returns to the instruction after the CALL.

Our examples will use near procedures for simplicity, but keep in mind that it is incumbent on the programmer to remember to call a program correctly (as near or far) and to ensure that the procedure ends in the proper type of return, RET or RETF.

Large assembly code programs, that may exceed 64K of code, may have a number of near and far procedures. One method of keeping track of which procedures are near and which are far involves having *only* far procedures. All procedures are put in a unique set of segments reserved for procedures only. The main program will always call far procedures, and the bother of remembering who is near and who is far is avoided. Making all procedures far does have the distinct disadvantage of using large amounts of memory and is not practical for small programs that run from ROM instead of under DOS.

Passing Data in Registers

We shall now expand our procedure technique by adding together 8 *bytes* of data passed in registers AX, BX, CX, and DX by the calling program. The result is returned in AX, which is the *only* register allowed to be changed by the procedure. The procedure, example_three, is listed next.

```
example_three:          ;must be called as near
        pushf           ;save the original flags
        add al,ah        ;add AL and AH—result in AL
        mov ah,00h       ;clear AH for high byte of result
        adc ah,00h       ;add any caries to AH
        add al,bl        ;add BL
        adc ah,00h
        add al,bh        ;add BH
        adc ah,00h
        add al,cl        ;add CL
```

```
        adc ah,00h
        add al,ch              ;add CH
        adc ah,00h
        add al,dl              ;add DL
        adc ah,00h
        add al,dh              ;add DH
        adc ah,00h             ;done
        popf                   ;get old flags
        ret                    ;return to the calling program
```

The calling program, example_four, loads the passing registers and then calls the procedure, as is shown in the next program.

```
code segment
        byte1 equ 00h          ;example data
        byte2 equ 12h
        byte3 equ 34h
        byte4 equ 56h
        byte5 equ 78h
        byte6 equ 9ah
        byte7 equ 0bch
        byte8 equ 0deh
example_four:
        mov sp,0000h           ;stack at top of segment
        mov al,byte1           ;load raw data into registers
        mov ah,byte2
        mov bl,byte3
        mov bh,byte4
        mov cl,byte5
        mov ch,byte6
        mov dl,byte7
        mov dh,byte8
        call example_three     ;call addition procedure as near
        jmp example_four       ;continue on with program
```

Example_four uses most of the working registers to pass data to procedure example_three. The complete program appears as shown next.

```
code segment                   ;passing data in registers        See Note
        byte1 equ 00h          ;example data
        byte2 equ 12h
        byte3 equ 34h
        byte4 equ 56h
        byte5 equ 78h
        byte6 equ 9ah
        byte7 equ 0bch
        byte8 equ 0deh
example_four:
        mov sp,0000h           ;stack at top of segment          1
        mov al,byte1           ;load raw data into registers
        mov ah,byte2
        mov bl,byte3
        mov bh,byte4
        mov cl,byte5
        mov ch,byte6
        mov dl,byte7
```

```
            mov dh,byte8
            call example_three      ;call addition procedure as near
            jmp example_four        ;continue on with program
example_three:                      ;must be called as near
            pushf                   ;save the original flags
            add al,ah               ;add AL and AH—result in AL
            mov ah,00h              ;clear AH for high byte of result
            adc ah,00h              ;add any caries to AH
            add al,bl               ;add BL
            adc ah,00h
            add al,bh               ;add BH
            adc ah,00h
            add al,cl               ;add CL
            adc ah,00h
            add al,ch               ;add CH
            adc ah,00h
            add al,dl               ;add DL
            adc ah,00h
            add al,dh               ;add DH
            adc ah,00h              ;done
            popf                    ;get old flags                    2
            ret                     ;return to the calling program
    code ends
```

NOTES ON THE PROGRAM

1. SP can be set to 0000h because the first push will set the stack to FFFEh and store the first word pushed at FFFEh and FFFFh. (Note that D86 will not display the stack at FFFEh–FFFFh on the SP memory line, but the return address is stored.)

2. No register is altered in the procedure except AX and Flag. AX returns the sum, therefore it is not necessary to save AX.

Example_four uses up all of the byte-addressable registers. If we wished to add together more bytes, we could pass the bytes in registers BP, SI, and DI. The procedure could move the bytes from the word-addressable registers to the byte-addressable registers for byte-sized addition. For adding more bytes than can be held in all of the working registers, we turn to the stack.

Passing Data on the Stack

In this example, we shall pass 12 bytes to the addition procedure on the stack by having the calling program (example_six) push the data on the stack. The called procedure (example_five) will pop the data, add it, and return the sum on the stack. The calling program (example_six) then pops the sum from the stack. Note that the stack can only store *words* of data. If bytes are passed on the stack, they must be grouped as words. It is assumed that the called procedure *does not* preserve any CPU working registers or flags. The complete program appears as follows.

```
;program example_six calls procedure example_five
code segment                                        See Note
        word1 equ 1234h       ;data for example_six
        word2 equ 5678h
```

```
              word3 equ 9abch
              word4 equ 0def0h
              word5 equ 4321h
              word6 equ 8765h
example_six:
      push word1              ;store twelve bytes for addition           1
      push word2              ;word1 to word6 are data variables
      push word3
      push word4
      push word5
      push word6
      call example_five       ;call addition procedure
      jmp example_six         ;continue with program
example_five:                 ;get data from the stack
      pop di                  ;get the return address                    2
      mov ax,0000h            ;clear AX for addition
      pop bx                  ;get first bytes and add them
      mov al,bl               ;add byte one
      add al,bh               ;add byte two
      adc ah,00h              ;add any carry after each add
      pop bx                  ;get bytes three and four
      add al,bl               ;add byte three
      adc ah,00h
      add al,bh               ;add byte four
      adc ah,00h
      pop bx                  ;get bytes five and six
      add al,bl               ;add byte five
      adc ah,00h
      add al,bh               ;add byte six
      adc ah,00h
      pop bx                  ;get bytes seven and eight
      add al,bl               ;add byte seven
      adc ah,00h
      add al,bh               ;add byte eight
      adc ah,00h
      pop bx                  ;get bytes nine and ten
      add al,bl               ;add byte nine
      adc ah,00h
      add al,bh               ;add byte ten
      adc ah,00h
      pop bx                  ;get bytes eleven and twelve
      add al,bl               ;add byte eleven
      adc ah,00h
      add al,bh               ;add byte twelve
      adc ah,00h
      push di                 ;restore the return address                3
      ret
code ends
```

NOTES ON THE PROGRAM

1. Immediate data push.

2. Remember, the last thing pushed on the stack is the *return address* of the calling program. Popping the return address to DI *saves* it for later use.

3. The *return* address of the calling program is *pushed* back onto the stack. The RET instruction will then use it to return to the calling program. Saving the return address is the most foolproof way to ensure that it is not lost.

Note: D86 (and other PC functions) uses the stack area JUST BELOW THE USER STACK to store temporary data. You MAY NOT assume that "old" stack data (data that has since been popped from the stack) is still available.

The next program, example_seven, calls the addition routine, example_eight, after moving SP below its original setting, thus opening up a free data area on the stack. Data is moved to the free stack area by example_seven, using BP, and example_eight is then called. Example_eight uses BP to get the data by moving it from the stack area. After example_eight returns, example_seven restores SP to its original value. The calling program assumes that all working registers, flags, and BP are altered by the procedure.

```
code segment                                                  See Note
        word1 equ 0ffffh      ;data for example_seven
        word2 equ 1111h
        word3 equ 2222h
        word4 equ 3333h
example_seven:
        sub sp,08h            ;free up eight bytes on the stack        1
        mov bp,sp             ;BP is now a base pointer in the stack    2
        mov w[bp], word1      ;store the data on the stack, from the base
        mov w[bp+2],word2
        mov w[bp+4],word3
        mov w[bp+6],word4
        call example_eight
        add sp,08h            ;restore the stack                        3
        jmp example_seven     ;loop to exercise again
example_eight:
        pop di                ;save return address
        mov ax,0000h          ;clear AX for addition
        mov bx,[bp]           ;get first bytes and add them             4
        mov al,bl             ;add byte one
        add al,bh             ;add byte two
        adc ah,00h            ;add any carry after each add
        mov bx,[bp+2]         ;get bytes three and four
        add al,bl             ;add byte three
        adc ah,00h
        add al,bh             ;add byte four
        adc ah,00h
        mov bx,[bp+4]         ;get bytes five and six
        add al,bl             ;add byte five
        adc ah,00h
        add al,bh             ;add byte six
        adc ah,00h
        mov bx,[bp+6]         ;get bytes seven and eight
        add al,bl             ;add byte seven
        adc ah,00h
        add al,bh             ;add byte eight
        adc ah,00h
```

```
        push di              ;restore return address
        ret
code ends
```

NOTES ON THE PROGRAM

1. Subtracting a number from SP, in effect, causes the stack to grow down by the number of bytes *subtracted*. Because no data was pushed into the area between the old SP and the new SP, the stack growth area is *free* to store data.

2. Any pointing register may be used to store data in the freed-up stack area, but BP uses SS as the default segment register, thus BP is a logical choice for the pointing register.

3. SP is restored to its original value by adding 8 back to it. Failure to restore SP would result in the stack growing, in this case by 8 bytes, every time procedure example_eight is called. Over time, say a few seconds, hours, or days, the stack will eat its way into places where it can overwrite data or code. Stack growth problems are *very* difficult to find.

4. BP was passed by the calling program, to point to the *base* of the stack area where the data is stored.

Passing Data in Memory

The next program, example_nine, is a high-to-low sorting program that uses a small procedure named sort_it. Data is passed to the procedure via an array of data in memory. The data array sortthis is set to hold 100h bytes for example_nine.

Example_nine acts to sort data bytes by size. The largest numbers are stored in low memory, and the smallest numbers are stored in high memory.

The sort used in example_nine compares each byte with the next byte above it in memory. If the next byte is greater, the two are swapped in memory, with the smaller byte placed higher in memory. If the next byte is not larger, then nothing is done.

The 2-byte swap is done for all 100h bytes. At the end of the first loop, all byte pairs have the larger byte lower in memory. The program then loops back to the beginning of memory and begins to compare and swap again. The program loops through memory 100h times, each time moving the largest byte of a pair lower in memory, until the largest byte(s) are now at the low end of the memory space.

Program example_nine demonstrates how an entire program can be easily changed if one of its subroutines is changed. Should the jump in procedure swap_it be changed from jnc equalorless to jc equalorless, for instance, the entire program will sort bytes with the smallest numbers in low memory and the largest numbers in high memory.

Sorting programs are very common in certain computer applications involving arranging data in a numerical order. Sort programs are also popular for use as teaching examples. Program example_nine is a simple sort that sorts numbers in memory such that the smallest numbers are placed in high memory. Chapter 8 contains more sorting problems.

```
;example_nine, a program that sorts 100h numbers high to low. Larger
;numbers are placed in low memory and smaller numbers in high memory.
;The calling routine assumes that no working register or flag is
;preserved by the called procedure.

data segment                            ;                              See Note
                org 1000h               ;place array at 1000h
sortthis        dw 100h dup ?           ;define a data space of 100h bytes      1
data ends
code segment
                size equ 100h           ;size of the test array
example_nine:
                mov sp,0a000h           ;set the stack
                mov cx,size             ;pass array size to sort_it in CX
                mov si,offset sortthis  ;pass location of array in SI
                call sort_it
                jmp example_nine        ;loop to demonstrate this again
;
sort_it:
                mov dx,si               ;calculate limit on data array
                add dx,cx               ;and store the limit in DX
                dec dx                  ;100h bytes = 00h to FFh                2
                mov di,si               ;save array starting point
start:          mov si,di               ;restore SI to base of array
nextpair:                               ;compare this byte with the next byte
                call high_low           ;high_low sorts larger byte downward    3
                inc si                  ;point to next byte
                cmp si,dx               ;we done yet? (with one loop of the sort)
                jne nextpair            ;if not, swap next pair
                dec cx                  ;done the complete sort?
                jz sorted               ;if completely done, return
                jmp start               ;else swap array again
sorted:         ret
;
high_low:
                mov ax,[si]
                cmp al,ah               ;compare this byte (AL) with next byte (AH)
                jnc equalorless         ;if CF = 0 then next = or < this byte
                mov [si+1],al           ;if CF = 1 then swap the bytes
                mov [si],ah
equalorless:    ret                     ;or just return if next byte is smaller
                                        ;or the same size
;
;the data array is initialized FOR EXAMPLE PURPOSES ONLY
                org 1000h               ;data for example_nine                  4
                dq 20h dup 0123456789abcdefh                                    ;5
code ends
```

NOTES ON THE PROGRAM

1. 100h bytes of data space are defined with the dup keyword. The *contents* of the array variable name is sortthis. The *address* of the variable array is offset sortthis.

2. The array is 100h bytes long and *starts* at address 1000h, thus it extends from 1000h to *10FFh*.

3. Procedure sort_it calls another procedure, high_low. Both procedures are modular in form.

4. The assembler is used to define 100h bytes of test data. The data is assembled as part of the code segment, because the data is now a *permanent* part of the program and will not change as the program is run and rerun. DOS loads the program, and the test data, into DRAM where the test data may be rearranged in DRAM. Should the program have been used in system where code is in ROM, the test data could not be altered, because of the read-only nature of ROM.

5. The dq directive defines a quad-word of data. The 20h dup directive repeats the quad-word 20h times, for a total of 100h bytes.

6.6 Register, Direct, and Indirect Procedure Calls

Our previous examples have used immediate addressing (a label *name*) to call the procedures. Immediate addressing means that the name of the procedure is used as part of the call mnemonic; the assembler can assemble the numeric equivalent of the procedure name in the hex code. All of our previous calling examples assigned a label to the procedure, such as example_one, example_two, and so on. The assembler converts these label names to addresses and computes the offset number to be added to the IP when a call opcode is found in the program.

For example, say a call opcode is located at offset address 0200h in the assembled code and the next instruction is at address 0203h. Assume the procedure to be immediately called is located at 0300h in the program. The assembler will compute the number of bytes from 0203h to 0300h (00FDh) and code that number after the hex code for the near call. The resultant code is E8 (near call) FD 00 (offset) to call a procedure at address 0300h from a call mnemonic located at address 0200h.

A far immediate call, to a procedure located in code segment 4000h at offset 1234h would be coded FF (far call) 34 12 00 40 (address of procedure). The procedure address numbers are known to the assembler because the programmer has assigned a label name to the far call and formally declared it to be a far call.

Immediate, Register, and Indirect Near Procedure Calls

Register addressed near calls are made by using the *contents* of any 16-bit working register to hold the *address* (IP) of a near call. Only near calls may be made via register addressing, because far calls require 2 words to specify new CS and IP numbers.

Direct near procedure calls use a procedure address found in a direct memory *word* address.

Indirect near calls employ a pointing register to hold the address of a word in memory where the procedure address is found.

Register, direct, and indirect near calls have one thing in common: the programmer, not the assembler, must ensure that the correct procedure IP address is in place.

Immediate, Direct, and Indirect Far Procedure Calls

A far call requires *2 words* of data to specify both the offset into the code segment and the code segment itself. The working registers can hold only 1 word, so indirect calls to far procedures can only be done using 2 words in memory.

Immediate far procedure addresses can be included in the call opcode by using the actual address of the far procedure, a label in the far segment, or a PROC FAR assembler directive.

Direct memory addressing specifies the first word of a double-word CS:IP pair that has been placed in the direct memory location.

Indirect addressing involves using BP, BX, SI, or DI pointing registers to point to a double-word CS:IP procedure address in memory.

As was the case for near calls, the programmer must make sure that the indirect address fetched from memory does indeed point to a procedure in the far segment and that the far procedure ends in an RETF opcode to ensure that both the IP and CS registers of the calling program are restored.

Program example_ten, shown next, *demonstrates* near and far calls using immediate, register, direct, and indirect addressing modes. Program example_ten should be run using D86 in order to observe how the various calls are handled by the CPU. Most of the problems in this text will not need to use any other type of call except immediate near types. Far calls will find employment, however, in programs involving interfacing to the ISA bus in Chapter 11 and the single-board computer of Chapter 12.

An immediate far call is shown *with a warning*, because we cannot know in advance in which code segment DOS will place the program. The particular computer and version of DOS that was used to debug programs for this book loads code in segment 2C7Fh. Other computers, and DOS versions, *may load programs to be debugged in a totally different segment.*

```
example_ten:                        ;an example that calls a procedure from near or far
                                    ;using all of the call addressing modes
;***************DEMONSTRATION PURPOSES ONLY ********************
                                    ;                        See Note
code segment                        ;start with near calls
        mov sp,3000h                ;set the stack
        jmp over                    ;jump over procedure _local        1
_local:                             ;_local is low in code memory
        ret
over:
        call _local                 ;call immediate (backward displacement)
        call _forward               ;call immediate (forward displacement)
        mov cx,offset _forward      ;CX contains code address of _forward
        call cx                     ;call register mode addressing
        mov [1000h],cx              ;now store CX (address) in memory
        mov bx, 1000h               ;point to address of _forward using BX
        call w[bx]                  ;call indirect addressing via BX
        call w[1000h]               ;call direct addressing to memory
```

```
        mov w[bx],offset _far      ;get offset of simulated far procedure   2
        mov w[bx+2],cs             ;store current code segment
        call d[bx]                 ;call_far indirect addressing            3
        call d[1000h]              ;call_far direct addressing
        call 2c7fh:0135h           ;ONLY FOR DOS ON THIS COMPUTER           4
        jmp over                   ;loop back to beginning
_forward:
        ret                        ;a very simple near procedure
_far:
        retf                       ;a very simple far procedure
code ends
```

NOTES ON THE PROGRAM

1. If the program does not jump over _local, it will go into it and execute the ret instruction, _local will not have been called, and no valid return address will be on the stack.

2. Far call addresses are stored as IP in low memory, and CS in high memory. The directive offset _far gets the IP of _far.

3. The use of the double word (d) qualifier tells the assembler to code a double word fetch from memory. A (d) qualifier is not necessary for a near call, and a single word is fetched if (d) is not specified.

4. The immediate far call uses whatever CS that DOS chooses to load the example program. This may not work for your computer and your DOS version. Place a semicolon (;) in front of this line before debugging the program. Use D86 to show you the CS for your computer, and substitute your CS for 2C7Fh.

6.7 Calling Procedures From High-Level Language Programs

"Most *standard* programming should be done using high-level languages." A rather odd statement in a book that is concerned mainly with assembly-level programming, but valid nevertheless. High-level languages have supplanted assembly code in many *mundane* applications mainly because of two factors: programmer time and memory cost.

The history of personal computers began with 8-bit microprocessors, 4K of NMOS dynamic RAM, tape recorders, and boot-up panel switches. Memory space, in early microprocessor-based computers, was very small by contemporary standards. Memory was also very expensive. When memory is *extremely* limited, programs are normally written in assembly code. In the beginning, microcomputer programmers, by the very nature of their limited memory computing "environment," became proficient in assembly language. Every programmer of early personal computers soon became familiar with an 8080, 6800, 6502, or Z80 8-bit instruction set. Assemblers for 8-bit machines were very inexpensive, often supplied at no cost by the microprocessor manufacturer. Assembly code programming, for microprocessors, became popular subjects at schools of engineering and technology using board-level computers.

Programs written in high-level languages, however, at the beginning of the personal computer age, were suitable *only* for use on minicomputers and mainframes. High-level languages required enormous resources of solid-state memory and, in some cases, mass disk memory. High-level language compilers were expensive and consumed additional computer system assets. High-level languages, and the teaching of same, were and are popular subjects at schools of computer science.

Technology moves forward. Memory costs have plummeted. Microprocessor CPU designs have evolved to handle high-level languages, and personal computers have more power, at far less cost, than older minicomputers. Compilers for popular high-level languages are inexpensive and available from a number of vendors. Personal computers can now economically be programmed in high-level languages. Most schools of engineering and technology now require that their students take *at least* one high-level programming course. Also, programmer time costs more money, and everyone wants their program developed *now*.

Assembly Language Versus High-Level Language

Clearly, we are at a point where using computers as a tool for solving industrial and commercial problems involves more than the choice of computer model. A programming *language strategy* should be developed that uses the best features of high-level *and* assembly code programming. Chapter 3 lists some reasons why assembly language programming is advantageous. This section seeks to re-explore some of the issues involved in the trade-off for each type of language.

ADVANTAGES OF ASSEMBLY LANGUAGE PROGRAMMING

There are many advantages to be gained when using assembly code:

- Programs written in assembly code are as *compact* as possible.
- Programs written in assembly code can be *fully customized* to use exactly the hardware that is present in the system.
- Because of their compactness, programs written in assembly code execute as *quickly* as possible.
- Programs written in assembly code are *relatively* easy to debug because of the fact that each line of code produces an observable result.

High-level languages, by their very nature, *hide* details of computer operation from the programmer. When debugging a high-level language program, however, it is often helpful to have a good *knowledge* of assembly code programming. High-level language compilers have been known to introduce exotic bugs during compilation that can only be discovered by inspecting the resultant assembly code.

DISADVANTAGES OF ASSEMBLY LANGUAGE PROGRAMMING

There are many disadvantages that hinder the programmer who uses assembly language:

- The programmer must spend a great deal of time learning the *exact* details of a particular CPU architecture, mnemonics, memory maps, and I/O hardware.
- Programs written in assembly language for one microprocessor cannot be used for a *different* microprocessor.
- Programs written in assembly language sometimes require *large* numbers of program lines to perform common mathematical operations.

ADVANTAGES OF HIGH-LEVEL LANGUAGE PROGRAMMING

High-level languages began to appear in the world of personal computers at the same time that large memories and floppy disks became affordable, and common. First BASIC appeared, followed by FORTRAN, then Pascal, and now C.

High-level languages offer many attractive alternatives to assembly language programming. Some advantages of high-level language programming include:

- High-level languages are *transportable*, that is, a program written in Pascal for one computer CPU type can run on a computer with a completely different CPU type.
- High-level languages do *mathematical operations* with programming ease.
- High-level languages are *easier to learn* because they use program statements that are similar to standard English (for instance) terms and standard mathematical expressions.
- High-level languages require the user to have absolutely *no knowledge* of computer hardware. Children can (and often do, to their elders' amazement) write a BASIC program for a computer they have seen for the first time.
- High-level languages are *efficient* (from a programmer's *time* standpoint) to write. One line of high-level code might require tens of lines of assembly code.

DISADVANTAGES OF HIGH-LEVEL LANGUAGE PROGRAMMING

There is a downside to writing programs in high-level languages, and most of the disadvantages involve the enormous hardware resources required to support a high-level language:

- *More software* is required to use a high-level language: operating systems, compilers, linkers, locators, and, lately, debuggers.
- More and *more memory* is required, solid-state and disk—megabytes of RAM and tens of megabytes of hard disk space. The high-level programmer seems to have no sense of memory limitations.
- *Debugging* high-level language programs can often take on some aspects of a murder mystery: You have the clues, but the suspects are many. Many subtle bugs occasionally creep into high-level language programs, and they can remain latent for years until just the right set of circumstances set them free.

The very thing that makes high-level languages appealing also limits their use in certain applications. Many computer applications are run using very limited memory space, such as in small controllers that are embedded in machinery or appliances. High-level languages may take up too much code memory space.

Computers that are interfaced to very high speed (microsecond duration) events cannot run high-level programs fast enough to capture the event because of the greater amount of high-level code that must be executed. Computers that are interfaced to *nonstandard* peripherals cannot use *standard* high-level peripheral routines.

High-level Language Calls to Assembly-Language Programs

As demonstrated in the lists of advantages and disadvantages, neither assembly code or high-level languages alone can efficiently solve many computer application problems.

Most computer application problems, however, can be solved by using a high-level language wherever possible in order to reduce programmer time and using assembly code where *compactness, speed, and customizing* are essential. The marriage of high-level and assembly code languages is particularly convenient when using a computer to control an industrial manufacturing process.

Industrial programs, and others, can often be thought of as having two parts: a "number crunching–report generating" part, and a high-speed part. High-level languages are perfect for extensive data manipulation and report generation. Assembly code can be used to get high-speed data and store it for later use by the high-level program. High-level languages and assembly languages are *partners* in the computer application world, not opponents.

Programs written in high-level languages call assembly coded subroutines using whatever calling rules the high-level language obeys. This section will investigate how three high-level languages—BASIC, Pascal, and C—can call assembly coded subroutines.

The examples of calling assembly coded subroutines from high-level languages that are presented in the following sections are not intended to be a complete tutorial on the subject of high-level languages. Some problems with generality arise because many software manufacturers who produce a high-level language compiler *also* produce assembly code assemblers. These manufacturers make it *somewhat* difficult for the high-level language programmer who uses their compiler to use any other assembly code assembler but theirs.

High-level languages often involve *linking* several .OBJ type files together to form a single .EXE file. Linking two or more programs involves an intermediate step between assembling (compiling) each program into an .OBJ form and production of the final .EXE code. Program A, for instance, may call a procedure in program B. When program A is assembled (or compiled) the address of the procedure in program B is unknown. Linking the programs resolves the unknown address in program B when both assembled programs are put together.

Linking methods vary among high-level languages, and among various companies' versions of the same high-level language. Entire books are devoted to each of our example languages, BASIC, Pascal, and C. Linking high-level languages to assembly code involves knowing the particulars of the link programs involved for each language and of each manufacturer's version of the language.

We look here at an *outline* of some of the methods used to link assembly code to high-level languages. The really hard parts, such as passing and returning variables from the high-level language to the assembly code subroutine, are left to textbooks that cover the particular high-level language in detail. Or, you may have to read the manufacturer's manuals that cover their particular version of the high-level language and its associated assembler and linker.

Including Assembly Code in BASIC Programs

John Kemeny and Thomas Kurtz of Dartmouth College developed BASIC in the mid-1960s as a teaching tool. BASIC (which is an acronym for Beginner's All-purpose Symbolic Instruction Code) attracted the attention of minicomputer-, and then microprocessor-, based computer programmers.

Despite annual predictions of its demise, BASIC still remains the language of choice for many "nonprogrammer" programmers. BASIC has the attraction of ease of use and very good human input-output features. In addition, there seems to be a BASIC statement that can handle just about any situation.

BASIC is produced in several dialects, notably Microsoft's GWBASIC, and IBM's BASIC and BASICA. There are also several ANSI standard BASICs. Most versions of BASIC perform the same functions. Very limited versions of BASIC can be found incorporated into the ROM of microcontrollers, enabling "quick and dirty" solutions to embedded control problems by persons with no particular programming skills.

BASIC is an interpretive language. That is, the BASIC interpreter inspects each line of a BASIC program and converts it into CPU code, on the spot. Interpretive BASIC runs very slowly, because each line of a BASIC program must first be interpreted and then executed. BASIC compilers are also available from companies such as Borland and Microsoft and should be used if speed (relatively speaking) is important.

We shall use a version of Microsoft's GWBASIC interpretive BASIC to call an assembly-language program. The assembly-language program, program.com, prints an *A* character to the system printer.

CALLING ASSEMBLY CODE FROM BASIC

BASIC has a CALL xxxx instruction that can be used to call a machine-language program that starts at address xxxx in BASIC's data area. The trick is to know exactly where xxxx is. The BASIC function VARPTR(variable) must be used to "find" where the subroutine has been loaded in memory.

Remember, DOS will load the BASIC interpreter program into RAM and then the BASIC program will load your program written in BASIC into memory and run it. Your assembly code must be included as part of your BASIC program, or it will not be loaded and run by the BASIC interpreter.

There are, generally, three methods available to the BASIC programmer for including assembly code into the BASIC program.

INCLUDING ASSEMBLY CODE IN DATA STATEMENTS

The first method used to include assembly code in BASIC programs is to use the BASIC DATA statement to define an array of bytes in the BASIC data area. The contents of the array is the assembled hexadecimal machine code of the assembly coded subroutine of interest.

To use the DATA method you must take the hex code listing output of the assembler and type each byte of hex code on a DATA line of the BASIC program. Having to type each byte of assembly code on a DATA line is impractical except for the shortest (say less than 50 bytes) of assembly code subroutines. One error of omission or commission in transcribing bytes from the assembly listing to the DATA line will cause the assembly code to crash. One solution to the typing problem is to write a small BASIC program that will get an assembled .COM file and convert it into a hex file that can be included in the BASIC program.

The assembly code contained in the DATA lines is READ into the BASIC data area as an array of bytes. A CALL to the subroutine can be issued after the address of the array (the address of the subroutine now in data memory) is assigned a name using VARPTR.

The sequence of a BASIC program that relies on using DATA statements (with the BASIC statements capitalized) is as follows:

DIMension an ARRAY I to hold the assembly code bytes

Write the assembly code bytes in DATA lines

READ the assembly code DATA into data ARRAY I

Name the subroutine as: name = VARPTR(ARRAY I)

CALL the subroutine name

POKE ASSEMBLY CODE

A second method for calling assembly code routines from a BASIC program uses a POKE command to place the DATA items in the data array in lieu of a READ loop. The work involved in typing each byte of assembly code on a DATA line is not lessened by using POKEs. The possibility of errors remains just as likely.

POKE involves the same steps as listed for DATA and READ statements. First, an assembled machine code version of the assembly code procedure must be obtained from the assembler. The following steps are then done in the BASIC program:

DIMension an ARRAY I to hold the assembly code bytes

Write the assembly code bytes in DATA lines

POKE the assembly code DATA into data ARRAY I

Name the subroutine as: name = VARPTR(ARRAY I)

CALL the subroutine name

BLOAD ASSEMBLY CODE

BASIC provides a third, much less labor-intensive, method for including assembly code as part of a BASIC program. Means exist for loading and running assembly language programs with the BLOAD filename command, which stands for "Byte Load a disk file identified in filename." BASIC will load an assembly code .COM from disk into an area of BASIC memory and execute it when the subroutine is called.

The only inconvenience involved with BLOAD is that a unique set of bytes must precede the assembly code in the file. BLOAD normally uses filenames saved with a BSAVE (Byte Save) command, so we must make our assembly code file *appear* as if it was saved by an earlier BSAVE command in the BASIC program.

BSAVE creates a group of 7 bytes when a binary file is saved. The order of the 7 BSAVE-created bytes is as follows:

Byte Number	Contents	Comments
1	FDh	BSAVE identifier
2–5	00h	Fill characters
6–7	0000h–FFFFh	Size of subroutine in bytes

The BSAVE header can be included in the assembly code source file (.asm file) using DB or DW directives to define the header bytes.

Warning! The BASIC interpreter program loses control over the computer while the subroutine is running. One mistake in the subroutine, and BASIC will never be seen again (at least not until you reboot the computer). Such dire warning of crashing is common to all assembly-language routines called from a high-level language. Be sure the assembly-language routine is well debugged *before* calling it from a high-level program.

EXAMPLE OF USING BLOAD TO CALL AN ASSEMBLY CODE PROCEDURE

One nice thing about BASIC is that it lets you write all sorts of diagnostic messages as it runs. You can track the performance of your program by printing messages to the screen so you will know what has happened. The BASIC program, INT.BAS, follows.

```
1  'INT.BAS, a program in BASIC that calls assembly language program
2  'print.com
3  '
100  DEFINT S                'define variable names starting with S as integer
110  DIM SUB(100)            'sub is a 202 byte (101 integer) array
120  PRINT "here goes"       'display before calling subroutine
130  SUBRT = VARPTR(SUB(0))  'location of first code byte in array
140  BLOAD "b:print.com",SUBRT  'get assembly code routine from disk
                             'B: and load it into memory at SUBRT
150  PRINT "loaded subroutine"  'so far so good
160  CALL SUBRT             'call the subroutine at address SUBRT
150  PRINT "I'm back!!"      'we hope
160  END
```

INT.BAS uses the following key statements:

DEFINT   We wish our array to store numbers as 16-bit integers

DIM SUB(100)   101 integers (202 bytes) is more than enough room for our assembly code routine

BLOAD "filespec",SUBRT   Load the subroutine in the array beginning at the address of the first element, SUB(0)

```
                         CAUTION

■ Variable memory in BASIC is assigned in a dynamic
  manner, that is, it is increased every time the BASIC
  interpreter finds a line of code with a new variable in
  it. Arrays are placed in memory above simple vari-
  ables. If your BASIC program gets the address of the
  first byte in the assembly code program array (using
  SUBRT = VARPTR), and you then declare another vari-
  able (such as A = 6) before CALLing SUBRT, you will find
  that the array (your assembly code program) has
  moved upward in memory. Your CALL will not be to
  the correct address in memory.
```

The assembly code routine, print.com, is loaded into BASIC's data area starting at the first element of the SUB array. Two program bytes are loaded into each array element.

BLOAD *cannot* be used in a ROM-based application, such as might be found in a single-board computer. The READ or POKE approach is used whenever BASIC does not have access to a disk and operating system.

The assembly code program print.com will print the character *A* to the printer, as follows.

```
;print.com, a program that prints "A" to a printer             See Note
code segment
              db 0fdh,00h,00h,00h,00h     ;the BSAVE header          1
              dw codesize          ;the number of bytes in the program  2
start:                             ;start = initial IP program value    3
              mov dl,'A'           ;use DOS to print an A               4
              mov ah,05h
              int 21h
              mov ah,05h
              mov dl,0ah           ;follow A by a LF                    5
              int 21h
              mov dl,0dh           ;follow LF by CR
              mov ah,05h
              int 21h              ;force any buffered printer to print
              retf                 ;FAR return                          6
codesize equ $ - start            ;codesize = number of program bytes   7
code ends
```

NOTES ON THE PROGRAM

1. BSAVE will not load the file unless it is "fooled" into thinking the file was created by a BLOAD command. The file is a .COM assembled file.

2. A86 can compute the number of bytes in the program by subtracting the beginning IP from the ending IP.

3. start is a label for the beginning value of the program counter. A86 keeps track of IP and assigns label start the beginning IP number.

4. DOS has procedures that can be called to print characters on the printer, among other things. See Chapter 7 for more details on DOS procedures.

5. Many printers (those equipped with internal RAM data buffers) will not print anything until the end of a line (CR followed by LF) is detected.

6. A far return is used because BASIC treats all calls as calls to far procedures.

7. $ is used to determine the IP value at the end of the program. $ means (to A86) the current program IP value.

Compiled BASIC is a BASIC that has been converted into machine code hexadecimal bytes, rather than the shorthand ASCII notation of standard BASIC. Compiled BASIC is much easier to use when assembly code subroutines are included and behaves much the same as our next language, Pascal.

Including Assembly Code in Pascal Programs

Pascal is a compiled language, which means that the compiler can link to assembly code during compilation. Compiling the code fixes all code in known addresses, which removes the problem of BASIC's dynamic memory allocation.

Niklaus Wirth, of Switzerland, headed the team that developed Pascal in the late 1960s. Pascal, like BASIC, was designed to be a teaching language for students learning about structured (high-level modular) programming.

As was the case with BASIC, Pascal was soon adopted by programmers outside of academe. Borland Corporation's release, in the early 1980s, of an eminently affordable version known as *Turbo Pascal* greatly spread the popularity of Pascal. The name *Pascal* is not an acronym; it is the last name of Blaise Pascal, a 17th-century French mathematician.

Pascal, as is true for all structured languages, takes more time and effort to learn than BASIC. Structured languages, although they have very few program statements, use very rigid rules for syntax and punctuation. Pascal programs are a set of user-defined modules (procedures or functions) that may be used in the main program once they have been defined.

Pascal also features the passing of variables from the calling routine to the called subroutine, and the reverse. Care must be taken to ensure that the data passed (the type of data—byte, word, and so on) is the type expected by the receiving program. Many structured-language program problems come about from subtle errors in passing data from one procedure to another.

We shall use the assembly code routine from our BASIC example in our Pascal program, int.pas, which is written in Borland's Turbo Pascal. We will need to modify our assembly code program from the BASIC version to one that is compatible with Turbo Pascal. For instance, we shall not need the BSAVE header, and our assembly code program must be assembled as an .OBJ type, so that it may be linked to the compiled Pascal program.

INCLUDING ASSEMBLY CODE USING INLINE ASSEMBLY CODE

Turbo Pascal has two mechanisms for including assembly code programs as part of the overall Pascal program.

One method uses an *inline* structure that lets the user include assembly code directly in the Pascal program. The inline method is very similar to using BASIC's DATA line. Each byte of assembly code must be entered manually in the inline portion of a Pascal program. Writing inline code, as was the case for BASIC DATA code, is very laborious and prone to error. Inline code should only be used for very short assembly code programs.

LINKING TO ASSEMBLY CODE

The second method used to include assembly code as part of a Pascal program is to link the assembly code object program as part of the Pascal program when the Pascal program is compiled. Linking enables the programmer to include huge assembly code programs into a Pascal program with very little effort. Linking assembly code programs is the preferred method for all but the shortest assembly programs.

The following Turbo Pascal program enables the Pascal program to call the assembly-coded print program. The Pascal program, named int.pas, is a procedure that links the assembly code into the Pascal program. The Pascal program is as follows:

```
{ int.pas, a program that links assembly code into a Pascal program }
     procedure print; external;   {declare print as an external}
     {$L b:print}               {instruct the compiler to link print}
     var
          A: integer          {A is a numeric key on the keyboard}
  begin
     A := 1;   {set A to 1 to enable first loop}
     while A = 1 do          {loop until a 1 is not entered}
       begin
          print;             {call the print assembly-code routine}
          writeln('hit a 1 key to go again');
          readln(A)          {get key from keyboard}
     end
end.
```

Int.pas keywords:

$L   Directs the Pascal compiler to load the file

print   from drive B: and link it into the Pascal program

while-do   sets up loops in Pascal

print   Pascal does not use a call instruction; just naming the procedure or function will call it

writeln   A Pascal instruction to write the message in parentheses to the CRT screen

readln   A Pascal instruction to read a key from the keyboard and use it as variable *A*

Our Pascal program, int.pas, will print an *A* on the printer until we press some key other than a 1 key on the keyboard.

The assembly code program, print.asm, must also be changed to interface to Pascal. The most important change is to assemble print.asm into print.obj, not print.com. As is discussed in Chapter 4, .OBJ files are assembled so that they may be fed into a linking program (or a compiler, like Pascal's). The linker (compiler) adjusts any memory references necessary to include the assembly code program into the Pascal-compiled total assembly code.

Assembling our program print.asm into print.obj requires only that the name of the output program be print.obj instead of print.com, when invoking A86. The A86 command line for producing .OBJ files is as follows:

```
A86 print.asm print.obj
```

A second change to print.asm is to declare the code segment to be public. The Pascal compiler will only link external routines that are declared to be public (available to all Pascal programs) in nature.

A third change to print.asm is the inclusion of the label print: on the first line of the program. Pascal's compiler will look for a beginning label in the assembly code program that matches the name declared as an external procedure in the Pascal program.

The last change to print.asm is to change the return from retf (a far return done for BASIC) to ret (a near return for Pascal).

The resultant print.asm program is as follows.

```
;print.obj, a program that will link to a Pascal program
        code segment public    ;declare this code to be public
print:      mov dl,'A'         ;use DOS to print an A
            mov ah,05h
            int 21h
            mov ah,05h
            mov dl,0ah          ;follow A by a LF
            int 21h
            mov dl,0dh          ;follow LF by CR
            mov ah,05h
            int 21h             ;force any buffered printer to print
            ret                 ;NEAR return
        code ends
```

Note that the BSAVE header and the code necessary to calculate the size of the program for BASIC have been removed.

Pascal, in recent years, has been joined by the programming language *C*, which is another structured programming language. I conclude this section with a short example of linking assembly code to C.

Including Assembly Code in C Programs

About the same time that BASIC and Pascal were making their appearances on the programming stage, Dennis Ritchie of Bell Labs (Murray Hill, New Jersey) was developing the *C* programming language for the DEC PDP-11 minicomputer. (One wonders, at times, why the 1960s saw the birth of three of the most popular high-level programming languages at present.)

C was meant to be an internal language for Bell Telephone. The UNIX operating system was also written in C at Bell, and UNIX includes a C compiler and several C libraries. Bell, in a generous and also wise gesture, made UNIX available, at little or no cost, to many computer science departments at colleges and universities around the United States. Needless to say, budding computer science students were soon writing programs in C. On graduation, the newly hatched C programmers carried C into the outside world.

C has now been *ported* to DOS-based computers and many versions of C have appeared, including an ANSI standard C, Borland's Turbo C, and Microsoft's C. C is a structured language, much like Pascal, and the mechanisms C uses to get assembly-coded .OBJ programs incorporated into C programs are similar. C names subroutines *functions*, as opposed to the procedures of Pascal. For both C and Pascal, just naming the subroutine calls it.

INLINE ASSEMBLY CODE

Some C compilers allow the programmer to use the keyword asm to indicate that everything contained within the curly brackets ({}) that follow asm are assembly code mnemonics. The structure of a typical C program that includes inline assembly code appears as follows.

```
main()
{
        asm
         {
             ...
             ;body of assembly-language routine
             ...
        }
}
```

On encountering the asm keyword, the C compiler will call a *default* assembler, usually one sold by the same vendor that produced the C compiler.

Using asm directives in C is superior to including assembled inline hex code as is done when programming in Basic or Pascal, because no lengthy machine hex code needs to be entered in the C program.

One disadvantage of using asm is that the assembler called by the C compiler must be *very* compatible with the C compiler, or the assembly process will not work. For instance, Borland's C++ compiler requires Borland's Turbo assembler in order to assemble inline assembly code. Other C compilers may *assume* the user is employing Microsoft's MASM assembler and will not work with another assembler.

Because this book features the A86 assembler, we shall assemble our print program separately and call it as a function.

LINKING ASSEMBLY CODE TO C PROGRAMS

A C program, int.c, which duplicates the previous results of our Pascal program by printing a character until a key other than 1 is pressed, is written as follows.

```
#include <conio.h>              /* include getch and printf*/
main()
{
    char more_print;            /* print until other than a 1 key pressed */
    more_print = getch();
    while(more_print == 49)     /*49d = 31h = ASCII number "1" */
        {
            print();            /* call our assembly-code routine */
            more_print = getch();
        }
    return 0;
}
```

Int.c keywords:

conio.h   A standard set of Input/Output C routines

getch()   A C instruction for getting a keyboard key

while   Sets up a loop in C

==   The C symbol for "equals"

Every C program must have a main function, and our C program main function serves solely to hold the while loop that gets a 1 from the keyboard and calls our assembly code function, print(). The assembly code routine, print.obj, is almost identical to the one used for linking in the previous Pascal program. The assembly code program, print.obj, that is linked to the C program is shown next.

```
;print.obj, a program that will link to a C program
        code segment public      ;declare this code to be public
_print:     mov dl,'A'           ;use DOS to print an A
            mov ah,05h
            int 21h
            mov ah,05h
            mov dl,0ah            ;follow A by a LF
            int 21h
            mov dl,0dh            ;follow LF by CR
            mov ah,05h
            int 21h               ;force any buffered printer to print
            ret                   ;NEAR return
        code ends
```

The only change made in the assembly code file print.obj from our Pascal example is that the Pascal version's label print: has been changed to _print:.

6.8 Summary

· Modular program design involves dividing long, monolithic programs into smaller, more specialized parts. The parts are called modules or subroutines or procedures.

Program procedures are integrated into a complete main program using the CALL mechanism. Calling allows individual procedures to be run as needed by the main program. The main routine calls a procedure, which then executes. When the procedure is finished, program execution resumes at the next instruction that follows the call instruction in the main program.

A call causes a return address to be stored on the stack, and the address of the procedure is placed in the IP (near calls in the same code segment) or IP and CS (far calls in another code segment).

A stack is an area of RAM pointed to by the SS:SP register pair. Data is placed on the stack with a PUSH opcode or retrieved from the stack with a POP opcode. A CALL opcode causes an automatic push of the return address to the main program. Procedures execute an RET (for near procedures) or an RETF (for far procedures) to return to the main program. A return opcode pops the return address from the stack to the IP or CS:IP registers.

Calls may be done using immediate, direct, register, or register indirect addressing for procedures.

Assembly-language programs may be called from high-level language programs—such as BASIC, Pascal or C—by linking the assembly code to the high-level programs or by loading the called program from disk. Each high-level programming language has its own methods of linking to assembly-coded programs.

6.9 Problems

Write programs that solve the problems using only mnemonics that have been previously defined in the text. All stack addresses used in each problem are offsets into the data segment where DOS loads your program. All procedures are to be NEAR procedures unless specifically directed to be FAR procedures.

1. Create a stack at 1000h and push all of the CPU working registers, the Flag register, and all of the segment registers on the stack. Inspect the stack and make a list of each stack address and the contents of that address. Load unique data into as many registers as possible in order to test the program.

2. Create a stack area at 1000h. Push 10h bytes, located at data memory addresses 60000h to 600FFh, on the stack using indirect addressing. Then pop the stack to memory addresses 62000h to 620FFh using indirect addressing. Load unique data into the 10h bytes in order to test the program.

3. Create a stack at 1000h. Use the stack to interchange the contents of all of the working registers. Exchange AX with DX, BX with CX, and DI with SI. Use unique data for each register to test the program.

4. Use the stack to move the Flag register to the AX register.

5. Use the stack to get the IP register to the AX register.

6. Create a stack at 1000h. Use the stack to interchange the contents of memory addresses 60000h to 60010h with memory addresses 60030h to 60040h. Interchange 60000h with 60040h, 60001h with 6003Fh and so on until 60010h is swapped with 60030h.

7. Call a simple procedure that restores the stack pointer to its original contents (prior to the call) and then jumps to the label noreturn in the main program.

8. Call a procedure that calls a second procedure that, in turn, calls a third procedure. Have the last procedure return to the original main program with the stack pointer restored to its original (pre-call) number.

9. Write a small procedure that adds 2 words together and stores the results in CX (high word) and DX (low word). Call the procedure from a program that loads the words to be added together in registers AX and BX.

10. Write a small procedure that adds 2 words together and stores the results in CX (high word) and DX (low word). Call the procedure from a program that stores the words to be added together on the existing stack.

11. Write a small procedure that adds 2 words together and stores the results in CX (high word) and DX (low word). Call the procedure from a program that creates a new stack area below the existing stack area and passes the 2 words that are to be added together on the new stack. Have the calling program pass the new stack pointer to the procedure on the existing stack.

12. Write a simple near procedure (such as a return) and call it using direct addressing.

13. Write a simple near procedure (such as a return) and call it using register addressing.

14. Write a simple near procedure (such as a return) and call it using indirect addressing.

15. Write a simple far procedure (such as a return) and call it using memory direct addressing.

16. Write a simple far procedure (such as a return) and call it using register indirect addressing.

17. Write a simple far procedure (such as a return) and call it using immediate addressing.

18. Write a simple program that does nothing more than return to the calling program and link it to a BASIC program using BLOAD.

7

OPERATING

SYSTEMS

AND

SYSTEM

INTERRUPT

SERVICES

Chapter Objectives

On successful completion of this chapter you will be able to:

- Describe the function of operating system software.
- Describe how operating system interrupts work.
- Describe how operating system interrupts are used.
- Describe CRT monitor types and capabilities.
- Describe the coding scheme used for the keyboard.
- Use BIOS interrupts to display characters on the monitor screen.
- Use BIOS interrupts to get key codes from the keyboard.
- Use BIOS interrupts to read the computer real-time clock.
- Use DOS interrupts to display characters on the monitor screen.
- Use DOS interrupts to get key codes from the keyboard.
- Write your own interrupts.

7.0 Introduction

A computer, to the average person watching old movies, is a huge box covered with blinking lights and surrounded by whirring tape drives. Large computer systems are popularly thought to reside in special, raised floor and air-conditioned rooms, attended by white-smocked impersonal humans.

Computers, of course, range in size from postage stamp sized single-chip microcontrollers to true supercomputers that occupy good-sized rooms along with their air-conditioning units.

The physical size of a computer tells us very little about its capabilities. Most personal computers take up slightly more space than a television set, yet personal computers have more raw computing power than room-sized computers of just 15 years ago. The inexorable trend of putting more and more computing power in smaller and smaller spaces, at ever-decreasing costs, means we should expect desk-sized supercomputers by the end of this century.

Classification of Computers

Classifying computers by size tells us very little about their computing abilities. Computers are actually classified by several indicators, which are, in order of importance: cost, speed, CPU type, and memory address range. Inspection of advertisements for personal computers shows that price, CPU type, and operating frequency are emphasized, followed by memory size. Details of internal CPU construction are rarely mentioned.

Computers can be classified by application, that is, by price and computing power. There are five general classes of computers: dedicated controllers, personal computers, minicomputers, mainframe computers, and supercomputers. Examples of models from each of these classes are included in the following list.

Dedicated controllers   The Intel 8051, Motorola 65HC11, Zilog Z-8

Personal computers   Apple Macintosh, Commodore Amiga, IBM PS/1

Minicomputers   DEC VAX 6000, Hewlett Packard 9000, NCR 3000, IBM AS/400

Mainframes   CDC Cyber 972, DEC VAX 9000, Hitachi EX 420, IBM ES/9000, NCR System 3600, Unisys A19

Supercomputers   Convex C3800, Cray Research C90, Cray Computer Cray-3

There are gray areas between each classification. Dedicated controllers also may include single-board computers based on personal computer CPUs. Some people may wish to add a sixth class of computers, called *workstations*, to the list. (Workstations are personal computers that have the power of low-end minicomputers.) The Sun SPARC family, HP Apollo, and IBM RISC 6000 are examples of workstation models.

Within each classification there is a range of capability. For instance, is a 66MHz, 486DX-based unit a personal computer, a workstation, or a minicomputer? Perhaps price alone is the determining factor.

Dedicated controllers, also dubbed *embedded controllers,* differ from the other classes of computers in one fundamental aspect. Embedded controllers generally have *no* operating system program. Embedded controllers are computers that often execute a single program, and that program is tailored exactly to the controller hardware. Examples of embedded controllers include automobile controls (fuel injection, brakes, dashboard displays), robot arms, appliance controls (TVs, VCRs, toys, kitchen appliances, etc.), and the myriad of other "smart" gadgets that surround us in our daily lives.

All of the rest of the computer family, from personal computers to supercomputers, are equipped with operating system programs. Sometimes, as in the case of PC personal computers, the operating system, DOS, is considered to be an *inseparable* part of the computer.

Operating Systems

An operating system is a program that *manages the assets* of the computer system. Common computer assets are the *keyboard, CRT monitor, printer, disk drives, and internal RAM memory.* The operating system is a program that enables us to use the computer.

In the following section, we investigate the operating system that is most commonly used for PC computers, the *Disk Operating System* or DOS. For more information on DOS, please refer to Appendix D.

7.1 The Structure of a Computer System

A computer can be thought of as a building, such as the one shown in Figure 7.1.

Figure 7.1 ■ Building a computer.

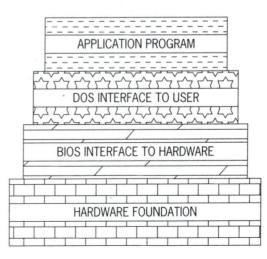

APPLICATION PROGRAM

DOS INTERFACE TO USER

BIOS INTERFACE TO HARDWARE

HARDWARE FOUNDATION

Hardware Foundations

Digital and analog integrated circuits (ICs), resistors, capacitors, and connectors, all soldered together on a printed-circuit board (PCB), make up the *hardware foundation* of a computer. Computer electronic and mechanical parts are called the *hardware* of the computer, to differentiate the physical part of the computer from the program or *software* part.

It generally requires several years of study to learn the electronic details that must be known in order to interconnect all of the hardware that makes up a computer (one that works reliably and that can be made at a profit). Chapter 12 discusses some aspects of constructing a minimal 8088 computer, starting with a "clean sheet of paper." The major issues involved in building a computer include logic design, memory timing restraints, and detailed knowledge of CPU logical and *electrical* behavior. At high CPU operating frequencies (say in excess of 20MHz), a knowledge of transmission line theory and practice is also required. In most cases, however, most users do not have to know how to build a computer; but they do need to know how to use a computer.

BIOS

Resting on the computer hardware foundation, we find the first layer of computer software programs stored in ROM. When the computer is booted up, the very first program instruction will be fetched from the ROM-based program. Clearly, the memory that holds the first computer instructions must be in ROM, because power may not have been applied previously to the computer. On the application of power, the CPU is designed to fetch its first instruction at address FFFF0h from the nonvolatile ROM program area of memory, called the *basic input output system* or BIOS.

BIOS contains a *collection* of programs that have been tailored, exactly, to *interact* with the particular *hardware* used to build the computer. Should some detail of the computer hardware change, then the BIOS program(s) must be changed also, to be able to interface to the new hardware configuration. The BIOS program is normally provided by the computer manufacturer as an inseparable part of the computer.

On power-up or reset, BIOS programs direct the initialization of all of the computer parts. Initialization includes programming some hardware parts or testing other parts, such as memory. Once the hardware is ready, BIOS loads the operating system program from the mass storage memory (normally a magnetic disk drive) into RAM. CPU control is then switched from ROM BIOS to the operating system program in RAM.

Disk Operating System

The second level of the computer building houses the operating system. The operating system program completes the initialization of the computer hardware begun by BIOS. The operating system also contains programs that interface with computer peripherals, although generally at a higher level than do BIOS programs. In particular, the operating system has many programs that deal with disk storage, hence the name *disk operating system* or DOS.

DOS may be supplied by the computer manufacturer or bought from DOS manufacturers such as Digital Research, IBM, or Microsoft.

Whenever the computer is not executing a specific application program, all computer I/O operations are under the control of the operating system program. For instance, the operating system lets the user type commands on the keyboard and see results on the monitor screen. We type *commands* to the operating system (DIR, COPY, TYPE, etc.), load and run programs (such as EDLIN), and, in general, make use of the computer. Each command is really the name of a small program that is loaded and run by the operating system.

DOS manages the interface between the computer *user* and the computer, translating our commands into computer action. DOS uses BIOS to get our keystrokes as we type them on the keyboard, interprets the meaning of the keystrokes, and directs BIOS to carry out the user's command. DOS is the building manager, BIOS the janitor, and the building occupants are application programs.

DOS and BIOS are made up, in large part, of a number of procedures that have been *standardized* by the industry. Standardized procedures serve as *interpreters* between computer hardware details and user application programs. BIOS and DOS procedures, or *service routines*, can be used by application programs, particularly application programs written in assembly code.

Applications

The last floor of the computer structure is made up of the programs we, the users, supply to the computer. User programs are called *application programs* because they perform a unique function, or application, at the command of the user. Typical application programs include word processors, spreadsheets, CAD programs, assemblers, debuggers, and communication programs. We supply the programs to the computer by inserting program diskettes into one of the computer disk drives.

When we wish to use the application program we type its *path* (drive:\directory\filename) after the DOS screen prompt (>). DOS loads our application program from the disk into RAM and directs the CPU to run it.

An application program can make *use* of DOS and BIOS service routines to operate computer hardware input and output devices. Using DOS and BIOS service routines relieves the application programmer from having to know *exact* hardware details. As long as DOS and BIOS routines are *standardized*, then the application programmer can use the service routines without any *concern* for hardware details.

The history of DOS is told in the versions issued by DOS manufacturers; that history is summarized in Appendix D. Currently, there are three primary DOS suppliers: Microsoft (MS-DOS), IBM (DOS), and Digital Research (DR DOS). You are not forced to use DOS on your PC. Other operating systems—such as Digital Research's CP/M-86 or multitasking Concurrent CP/M-86, Pick, or UNIX—will work just as well as (some say better than) DOS.

Microsoft is currently the dominant manufacturer of DOS, because IBM is pushing the use of OS/2. Initially, however, IBM and Microsoft were both vigorous developers of DOS. Refer to Appendix D for a short history of DOS, an explanation of how DOS works, and a listing of DOS service routines.

7.2 Introduction to BIOS and DOS Interrupt Services

Interrupts

There are three methods available to the programmer that will alter the flow of a computer program. We have discussed two methods in previous chapters: jumps and calls.

Program flow changes when a jump from one location in the program to another location is made. Jumps force the computer to an entirely new address in the program by replacing the IP, or the CS:IP combination with new program addresses. Elementary jumps are studied in Chapter 5.

A second method that will alter program flow is to call a procedure from the main or calling program and then return to the main program after the procedure is finished. Procedures force temporary suspension of the calling program and are examined in Chapter 6.

The third method used to alter program flow is termed an *interrupt*. Interrupts are much like calls in that a return address to the main program is saved on the stack and program flow continues in the main program once the interrupt procedure is finished. The major differences between calls and interrupts are the mechanism by which the interrupt procedure address is found and the fact that interrupts *can* be caused by hardware events *external* to the CPU. External events can force an interrupt via hardware actuation, without any direct software instructions. Interrupts can also be caused by using *software* interrupt instructions in the program.

Hardware interrupts are discussed in Chapters 11 and 12. This chapter is concerned with software-generated interrupts.

It is perhaps unfortunate that the word *interrupt* has become associated with both software and hardware means of changing program flow. A software interrupt is actually a convenient way to perform a *memory indirect far* call (see Chapter 6) using a predefined area of memory to hold the addresses of the called procedures. Software interrupts are under complete control of the programmer and never occur in a random fashion. A hardware interrupt, in contrast, generally involves external signals that are applied to the CPU at unknown, random times. Hardware interrupts are not under the control of the programmer, except that the programmer may block them by clearing the Interrupt flag.

In this chapter, we shall use the term *interrupt service routine* or the interrupt instruction mnemonic INT when *software* interrupts are discussed. The generic term *interrupt* will be used to describe the use of software and hardware interrupts.

Interrupt Memory Organization

When a computer program jumps from one place in the program to another, no stack operations take place. A jump is a one-way trip, and there is no need to save a return address on the stack in order for the program to find its way back to the jumping off place.

A call instruction, however, causes the CPU to automatically save the address of the instruction following the call on the stack. At the end of the procedure that was called, a return instruction in the procedure program pops the return address off of the stack. The program resumes at the address that was originally saved on the stack.

Calls have one common feature: The address of the procedure is part of the call instruction. Procedure addresses may be immediate addresses, as in call add_it, where add_it is the label for the call. Call addresses can be in a register, for example call ax, or in memory, as in call [0400h]. The programmer can call procedures as long as the address of the program that is called is known.

Interrupts are calls with *no* specified procedure address. Interrupts are always memory *indirect* and use a special area of memory, named the *interrupt vector table*, to hold the *address* of the procedure that is called by the main program. The term *vector* is used to describe a memory location that contains the *address* of a procedure, not the procedure itself.

A vector is also another name for the CS:IP double-word memory address of a procedure. The term *vector* originated in mathematics and engineering to define a quantity that has both magnitude and direction. Programmers appropriated the term for its directional abilities. A vector "points to" the procedure. A vector *table* is a collection of CS:IP word pairs.

To understand how the CPU uses an interrupt instruction to find a procedure, we must review the *low memory* map for the 8086 and draw some hardware inferences about how the memory is physically constructed.

PHYSICAL MEMORY

An 8086 CPU may address up to 1M of memory, from address 0000 to FFFFFh. The way this memory can be physically constructed is not as arbitrary as it may appear. By physical construction, I mean the mix of ROM and RAM chips that makes up the memory space.

As suggested earlier in this chapter, the 8086 CPU resets CS:IP to a memory location in the BIOS (address FFFF:0000h) whenever the computer is reset. There must be a ROM-type memory at that location because power has just been applied to the system and volatile RAM memory could not hold a permanent boot program. There will always be some amount of ROM in any system to enable the CPU to initialize itself and to enable the CPU to load the operating system program from disk.

If the system is not disk-based, as in a board-level system, then the majority of the memory space will be ROM. Board-level systems have all code stored in ROM and small amounts of RAM as may be required by the application. Disk-based systems will have small amounts of ROM in the memory space, because almost all programs will be loaded into RAM from the disk. Disk-based systems are generally characterized by small amounts of ROM and large amounts of RAM.

Board-level or disk-based systems may have one memory area in common, however. Memory from address 00000 to 003FFh will be available in both types of systems if interrupts are used. Typically, the disk-based system will use RAM to hold the vector table, and the board-level system may use RAM or ROM.

The 8086 interrupt *mechanism* uses memory addresses from 00000 to 003FFh to hold the CS code segment and IP offset *address* of interrupt procedures. Interrupt service procedure addresses must be stored in *low memory*, the site of the interrupt address vector table.

Figure 7.2 ■ Interrupt vector memory.

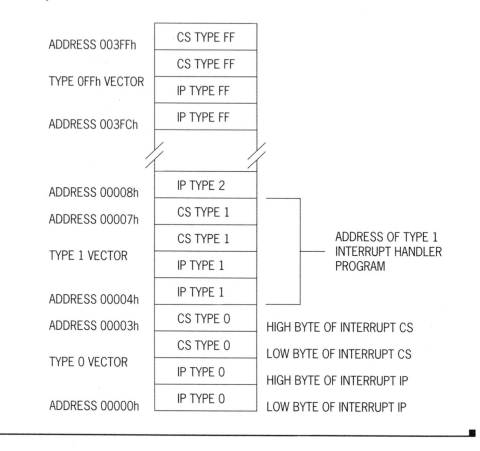

ADDRESS 003FFh — CS TYPE FF

CS TYPE FF

TYPE 0FFh VECTOR — IP TYPE FF

ADDRESS 003FCh — IP TYPE FF

ADDRESS 00008h — IP TYPE 2

ADDRESS 00007h — CS TYPE 1

CS TYPE 1

TYPE 1 VECTOR — IP TYPE 1

ADDRESS 00004h — IP TYPE 1

ADDRESS 00003h — CS TYPE 0

CS TYPE 0

TYPE 0 VECTOR — IP TYPE 0

ADDRESS 00000h — IP TYPE 0

ADDRESS OF TYPE 1 INTERRUPT HANDLER PROGRAM

HIGH BYTE OF INTERRUPT CS
LOW BYTE OF INTERRUPT CS
HIGH BYTE OF INTERRUPT IP
LOW BYTE OF INTERRUPT IP

INTERRUPT VECTORS AND INTERRUPT TYPES

Figure 7.2 shows the first 400h bytes of low memory space, from addresses 00000 to 003FFh. Memory contents between addresses 00000 and 003FFh are divided into 100h 4-byte pieces. The first 2 bytes of each piece hold an IP offset number, and the second 2 bytes hold a CS code segment number. Taken together, the CS and IP bytes *specify* a CS:IP *address* of an interrupt procedure that may be located *anywhere* in code memory. The address words are stored in the usual low-byte, high-byte format as one ascends from low to high addresses in memory.

Procedures that are accessed by interrupts are called *interrupt service routines,* or *interrupt handlers.* An interrupt handler is written to deal with the interrupt that called it, thus to "handle" the interrupt.

As stated, there are 256, 4-byte address groups in the 1,024 byte addresses that are found from address 00000 to 003FFh. Each 4-byte group is assigned a *type* number. The first type is *type 0* for the group that begins at address 00000 and extends to address 00003. The last type number is *type FFh* for the group that begins at address 003FCh and extends to address 003FFh. Table 7.1

Address	Type	Address	Type	Address	Type	Address	Type
00000	00	00100	40	00200	80	00300	C0
00004	01	00104	41	00204	81	00304	C1
00008	02	00108	42	00208	82	00308	C2
0000C	03	0010C	43	0020C	83	0030C	C3
00010	04	00110	44	00210	84	00310	C4
00014	05	00114	45	00214	85	00314	C5
00018	06	00118	46	00218	86	00318	C6
0001C	07	0011C	47	0021C	87	0031C	C7
00020	08	00120	48	00220	88	00320	C8

Table 7.1 ■ Sample Interrupt Vector Type Addresses in Low RAM

shows the starting address (where the IP offset to the handler is stored) for several representative interrupt type groups. Refer to Appendix C for the complete interrupt-type low-memory address listing. All addresses and type numbers in Table 7.1 are in hex.

Note that multiplying the *type number by 4* will yield the *address* in low memory for that type number. For instance, interrupt type number 7Dh multiplied by 4 is 1F4h. Address 001F4h is the low memory address "pointed to" by interrupt type 7Dh. The 2-word contents of the address pointed to by an interrupt type is the CS:IP address of a procedure used for that interrupt type. Put another way, the contents of the address pointed to by an interrupt type is the vector to the handler procedure. Vectors contain all of the information necessary (the handler address) to find the address of the handling procedure *anywhere* in memory.

There are 256 interrupt types, but not *all* of the interrupt vector types should be used by beginning DOS programmers. Interrupt types 00, 01, 02, and 04, for instance are "dedicated" interrupts on 8086-based personal computers. Dedicated means the CPU manufacturer has designed the CPU so that it uses certain interrupt vectors to handle program problems. Vectors 00 and 04 are for mathematic opcode interrupts, and their use is discussed in Chapter 8. Vector 01 is a debugging aid, used for instance by D86, and Vector 02 is a hardware-actuated vector.

Interrupt types above type 04h are not dedicated to the CPU and may be used by the operating system or an application program. DOS and BIOS are made up, in large part, by software interrupt procedures that have been standardized by the industry. The standard DOS/BIOS interrupt procedures may be used by application programs to gain access to computer assets.

The following sections discuss some popular interrupt types that may be used to write programs featuring keyboard input and CRT screen output. Chapter 10 discusses interrupt service routines that enable the user to access the printer and serial data ports. There are hundreds of DOS/BIOS interrupt service routines. Refer to Appendices C and D for a listing of additional DOS and BIOS interrupts.

7.3 Assembly Code Interrupt Instructions

INT Type

The 8086 instruction set includes one general-purpose interrupt instruction, which has the mnemonic INT Type. I will not use our normal shorthand format to explain the operation of an INT instruction. The sequence of events that occurs during an interrupt is sufficiently complex to require a narrative explanation of how INT works.

INT Type directs the CPU to cause a software interrupt sequence to be performed. Type can be any of the interrupt type numbers shown partly in Table 7.1 and fully in Appendix C, that is, Type is any number from 00 to FFh. On execution of an INT instruction, the following sequence of automatic CPU operations takes place:

1. The CPU stores the Flag register on the stack.
2. The CPU stores the CS:IP address of the instruction that follows the INT mnemonic on the stack. The stored address is a return address, exactly the same type of return address as is stored for a CALL opcode.
3. The CPU takes the type number specified by the programmer (from 00 to FFh) and multiplies it by 4, thus forming an address to a vector located between 00000 and 003FCh in low memory.
4. The CPU replaces the current contents of the IP and CS registers with the contents of the 4-byte CS:IP group (the vector) found at low memory address (= type number × 4).
5. Program execution resumes at the handler procedure found at the new CS:IP location in code memory.
6. A return instruction, IRET, normally placed at the end of the handler, pops the CS:IP return address and the Flag register off of the stack. Program execution resumes at the instruction following the INT opcode.

Examples of INT instructions include:

```
INT 7fh
INT 30h
INT 21h
INT 0ffh
```

Remember to use *h* for hex, because the interrupts discussed in this book, and in most references, generally use the hex form of the interrupt type number.

We see that the operation of an interrupt instruction is very much like that of a call sequence. The major differences are that the *Flag register* is stored and retrieved during an interrupt sequence and all interrupts behave like far calls, because CS is always saved. Saving the Flag register is called "saving the state of the machine" and is very important for the proper operation of external hardware interrupts.

As must be done for a call, the programmer must have written a handler procedure that takes care of the particular type of interrupt generated. The programmer (or the operating system) is also responsible for making sure that the vector address that has been placed in low memory points to the desired interrupt service procedure.

Writing and Using Interrupts

To use interrupts, the programmer must perform the following steps:

1. Write a handler procedure that will be used by an interrupt type.
2. Decide where, in the 1M memory space, to put the interrupt handler procedure.
3. Decide what interrupt type number to assign to the interrupt handler procedure.
4. Place the address of the handler in a low memory location equal to the interrupt type number multiplied by 4.
5. Use the handler in a program by including INT Type in the program.

Clearly, there is room for error in these steps. The most probable source of mistakes arises in making sure the right vector is at the correct vector address in low memory.

A quick scan of the details a programmer must follow to use an interrupt gives rise to the obvious question, "Why use interrupts, if they are so much trouble?" Because the final result of an interrupt is to call a procedure, it would seem to be a lot less fuss just to call the procedure in the first place and avoid all the bother of the trip through the interrupt vector table. One answer to the question will be covered later (in Chapters 11 and 12) when external hardware interrupts are discussed. Another answer to the question has to do with personal computers, operating systems, and the issue of *program compatibility*.

7.4 Interrupts to BIOS Service Routines

BIOS, as we have noted in this chapter, stands for "basic input output system." BIOS can be used as a software interface between assembly language programs and computer system hardware.

Compatibility Issues

A complete computer system is equipped with many peripherals. A keyboard, a CRT monitor, various disk drives, a printer port, and a serial port are all features of even the least expensive personal computer. Each of these peripherals may be controlled by integrated circuits (ICs) made by a dozen different manufacturers. Each IC may handle the interface to the peripherals in a slightly different manner. Moreover, most interface ICs are *intelligent*, that is, they must be programmed on power-up or system reset.

Intelligent ICs are designed to operate under many different configurations. Each configuration is chosen by the hardware designer to interface with the particular type of I/O peripheral included as part of the PC system. For instance, there are over 20 manufacturers of disk drives, and each manufacturer offers various models. Assume that one intelligent IC, which controls disk drives, can be programmed to operate any of the many disk drive models.

Thus, PC model X may require one IC program for drive type A, and a totally different program for drive type B.

The use of intelligent interface ICs from various IC manufacturers in order to control a galaxy of different peripherals made by a number of peripheral manufacturers creates a nightmare for programmers attempting to write standard programs that will run on *any* PC-compatible computer. If every computer can have different peripheral chips and each chip must be programmed to control any number of different I/O devices, then (theoretically) any "standard" programs could only run on *one* manufacturer's model. And then only on models that have the same peripheral IC configuration.

THE NEED FOR STANDARD BIOS

Clearly, having to write numerous versions of a program to match numerous PC hardware configurations is intolerable to application program developers. Not only application programmers but computer buyers also would balk at being restricted to only one version of software for their particular PC. No mass market can develop unless products are *standardized*, and the personal computer is no exception.

The problem of interfacing between standardized application software and nonstandard PC hardware is solved by a group of programs that interface between application programs and computer hardware using *standard* interrupt types: the BIOS. The question "Why use interrupts?" is answered: "So standard means of accessing nonstandard computer peripherals can exist."

BIOS software interfaces save the programmer the considerable trouble of having to discover exactly what make and model of interface IC is present in the PC and writing unique code for those particular interface ICs.

BIOS input/output programs are normally placed in ROM, so that they remain valid when power is removed from the system. The whole assemblage of input/output programs stored in ROM is collectively named *ROM-BIOS*.

CONFIGURATION FILES

Input/output programs may also be loaded from the disk into RAM, if the input/output needs of the computer change. I/O programs loaded from disk into RAM are called *system drivers* and usually have the suffix *.SYS*. DOS can be instructed, via a CONFIG.SYS file that contains the file names of the system drivers, to load system driver files into RAM. System driver files are used by devices that may not have existed when the first PCs were designed. Pointing devices, such as mice, are good examples of I/O devices that normally use system files instead of BIOS for their control.

Every manufacturer of personal computers, by using the BIOS and system driver concepts, is theoretically free to build its computer system with few restraints on I/O IC device driver selection. Each manufacturer simply equips its PC model with a set of BIOS routines that serves to translate between application programs and the particular hardware configuration used in that model.

The term *PC compatible* was born along with the concept of BIOS. Early in the development of the PC marketplace, some BIOS proved to be more compatible than others, and there was the usual legal challenges over BIOS copyrights. Happily, the maturing of the industry has seen a corresponding improvement in BIOS compatibility, and a decrease in lawsuits.

Including BIOS Interrupts in Assembly Code Programs

Application programs can use BIOS interrupts to take care of such details as getting key codes from the keyboard every time the operator strikes a key. BIOS can place the character for the key press on the screen and print output to the printer.

BIOS performs all of its services by using small programs that have been specifically designed for each task involved in getting keys, showing characters, and printing text. BIOS is written for application programmers who write programs and do not want to be bothered with the details of getting keystrokes, showing characters on the screen, loading data from the disk, or printing output. We have much more important things to do, such as learning how to put a "hello" on the screen.

There is one law that must be followed by the manufacturer's BIOS programmers: All BIOSs, for all different models, must get the same interrupt service routine, at the same vector address, in precisely the same way. BIOSs must be *standardized*. Happily, and after a lot of thought and work by the industry, PC-compatible BIOS has *standard* service routines. Each BIOS service routine is accessed, by the program, using *standard interrupt type numbers*.

Appendix C lists standard BIOS service routines and how the programmer can access them. We shall discuss, in this section, a few of the most common BIOS service routines, ones that a beginning programmer might find handy. Later sections cover several useful DOS routines for keyboard and monitor I/O. Later chapters will identify other BIOS services, as they are needed.

BIOS has the programs for many mundane I/O tasks written, so it is only natural for us to want to use them. Interrupts are the mechanism by which we can "get at" the BIOS service routines. Using BIOS service routines also frees us from worrying about exactly the make (and model) of our computer. If the BIOS is matched by the manufacturer of a particular model of personal computers to BIOS standards, then the programmer can use BIOS service routines with no concern as to the make, model, or manufacturer of the PC in use.

BIOS INTERRUPT FUNCTIONS

BIOS interrupt types are designated by the hexadecimal number standardized by BIOS programmers. Numerous *functions*, or options, within each interrupt type are also possible. A function is any of a number of unique actions, within a given interrupt type, that is identified in the BIOS service interrupt documentation. Appendix C lists the BIOS service interrupts and the functions that are available within each interrupt type. Some interrupt types have numerous functions defined; others have only one function defined.

Choosing a function usually consists of *loading register AH* with the function *number* desired, and some other working register (normally AL) with any data the function might require. An INT Type opcode is then placed next in the program, and BIOS performs the desired function. For most functions, data is sent from the program to the BIOS function. Occasionally data is returned by the BIOS function to the program.

DOS Service Routines

Many types of I/O services supplied by the BIOS are also available using interrupts to DOS. DOS services tend to be more general than BIOS services, which means that BIOS gives the application program more control over the hardware interface than does DOS. DOS, in fact, uses many BIOS services to carry out most standard DOS operations. DOS interrupts do not control system hardware to the degree that BIOS does, except in the areas of disk access. Disk file services are one reason DOS exists, and interrupts to DOS service routines often involve disk file operations. Application programs run on a computer are loaded by DOS from disk into RAM and then executed. Our assembler (A86), debugger (D86), and all application programs pass through DOS.

DOS disk file service routines are covered in Chapter 9. Common DOS I/O services are covered in this chapter.

A Warning: Interrupt Service Routines and Program Flags

BIOS and DOS interrupt service routines come with a small price tag attached. *Most software interrupts can change any working register they wish, including, occasionally, the Flag register.* Do not *assume* that your program flags, or any other working register, are the same before and after using a software interrupt. If you must save any register before invoking an interrupt then do so and pop it back after the interrupt returns. This warning is particularly important as regards the Flag register. If you are counting on some flags being unchanged before and after an INT, then save the flags before using the INT service routine.

7.5 BIOS CRT Output Services

The CRT video *monitor* is the most ubiquitous output peripheral available to the program. Every personal computer must have a visible display of some type, whether LCD displays or CRT screens. CRT monitors are popular output devices because of their low cost.

CRTs are called *monitors* because they permit easy viewing (monitoring) of program output results. As the PC industry has progressed, monitors have evolved from grainy broadcast quality black-and-white TV sets to VGA and super VGA color monitors with near-photographic picture quality.

In order to understand monitor interrupts that can display output on the CRT, we need to investigate the world of PC monitors.

Monitors, Screens, and Pages

Any CRT monitor used with a computer is viewed by BIOS as a group of *screens*. A screen is one complete 25-line CRT display. Only one screen at a time may be viewed by the user, but several screens can be defined, filled with characters, and stored in RAM memory for display at any time. We shall begin our study of displaying characters on a screen by investigating the nature of PC displays.

There are two common types of displays used with PCs: monochrome and color. Monochrome monitors are available in two configurations, *direct digital drive* and *composite analog video*. Direct monochrome monitors are the standard, amber or green, single-color display that is the least expensive type of PC terminal. Direct monochrome monitors produce very high resolution text and graphics.

Color displays are generally divided into two types: *RGB* (direct Red-Green-Blue) and *VGA* (video). RGB monitors are supplied direct digital red, green, and blue primary color signals. VGA monitors are supplied analog video type signals for extremely good picture detail.

VGA color monitors are compatible with programs written for monochrome and RGB monitors. So-called "page white" VGA monitors, which may display many shades of one color (gray) on a white background, are also a popular way to get VGA resolution at low cost.

Monitor types are also known by the type of display card that is plugged into the computer motherboard in order to drive the display. Monitor plug-in cards are known as *adapters* because they adapt the PC to display on the desired monitor type.

The term *display adapter* is now becoming obsolete because most contemporary PCs have all display functions integrated on the computer motherboard. We still refer to adapter when discussing monitors, such as "monochrome display adapter," "color graphics adapter," "Hercules graphics adapter," and so on. As with all things technical and multisyllabic, monitor adapter types have been given the following acronyms:

CGA	Color Graphics
EGA	Extended Graphics
HMGA	Hercules Monochrome Graphics
MCGA	Multicolor Graphics
MDA	Monochrome
VGA	Video Graphics
SVGA	"Super" VGA

Monochrome and color adapters are supported by BIOS service routines that enable the programmer to completely control all aspects of a screen display. Older versions of BIOS will support those adapter types that were available when the BIOS was written; modern versions of BIOS will generally support all adapter types.

MONITOR ADAPTERS

Adapters use a block of reserved DRAM memory, addresses A0000h to BFFFFh, to hold the characters that are to be displayed on the monitor. Every displayed character requires 2 bytes of memory for definition, 1 byte for the ASCII code for the character and a second byte for character *attributes*. A complete block of characters to be displayed makes up one screen.

A screen can display 25, 80-character lines, or 2,000 characters per screen. A screen of characters thus requires 4,000 total *bytes* to define all of the characters; 2,000 bytes for the characters and 2,000 bytes for each character's attribute. A screen is also referred to as a *page*. Pages of text seem a more natural expression than screens, and we shall use the term *page* when referring

Adapter	Memory Address	Bytes	Pages	Monitor Type
MDA	B0000h–B7FFFh	32K	1	Direct Monochrome
CGA	B8000h–BFFFFh	32K	4	RGB, Composite Monochrome
EGA\VGA	A0000h–BFFFFh	128K	8	VGA and EGA emulate MDA, CGA

Table 7.2 ■ Display Adapter Memory Buffer Addresses and Monitors

to display screens. Table 7.2 lists the page capabilities of three common display adapters, as well as the monitor types each adapter will drive.

Note, from Table 7.2, that MDA and CGA share the reserved memory buffer space, whereas EGA\VGA takes all the memory buffer space. An adapter may not use all of the buffer memory reserved for it. For instance, MDA uses only 4K (one page) of the 32K reserved for it.

It is possible to write text data directly to the memory segments that hold text pages *without* using BIOS. Direct write to display memory is not recommended for some adapters, particularly CGA. Video displays are controlled by various types of CRT chips that access the page display area of memory with little regard as to what the CPU is doing. When the program is writing to page memory at the same instant the CRT chip is reading the same page memory, *snow* may result. Snow is any interference that results from uncoordinated access to page memory by the program and the CRT chip at the same instant. Using a BIOS service interrupt for CRT text display will avoid snow, and other blizzards.

DISPLAY MODES

Display *modes* refer to how the video screen is configured. BIOS display interrupts can set the CRT screen to one of several modes by supplying a *mode number* to the mode service routine. For instance, the programmer can display text or graphics on a screen by specifying the desired screen mode in the program. As an example, text can be displayed on a CGA monitor in mode 0 as 25 lines of 40 characters per line or in mode 02 as 25 lines of 80 characters per line.

Graphic displays can choose screen resolutions from 320×200 to 1024×768 *pixels*, and from 2 to 256 colors per pixel. A pixel is one dot on the screen. Most screens are rectangular and have more horizontal than vertical pixels. The term 320×200 refers to a screen that can display 320 horizontal and 200 vertical pixels.

We shall not investigate graphic displays in this book. For our purposes, we shall display text pages of 25 lines of 80 characters per line. The mode numbers that *should* be used in all text display programs for this book are *mode 02*, which is supported by CGA, EGA, and VGA adapters, and *mode 07*, which is supported by MGA adapters.

HOW TEXT IS DISPLAYED

As mentioned previously, each character to be displayed on a page requires 2 bytes of page memory. One byte, the character itself, is stored in ASCII code at even byte addresses. The other byte, stored at odd byte addresses, contains the *attributes* for the ASCII character stored at the even address.

Monochrome Display Attribute Desired for Display	Attribute Byte Bit Settings
	Bits 7 6 5 4 3 2 1 0
Normal character	0 x x x x x x x
Blinking character	1 x x x x x x x
Dark background	x 0 0 0 x x x x
Light background (Inverse)	x 1 1 1 x x x x
Normal intensity	x x x x 0 x x x
High intensity	x x x x 1 x x x
Light foreground	x x x x 0 0 0
Underlined light foreground	x x x x 0 0 1
Dark foreground	x x x x 1 1 1

Table 7.3 ■ Monochrome Character Attribute Byte Codes

An attribute specifies the exact *way* the character is to be displayed. A character, such as the letter *A*, is actually made up of a *foreground* A surrounded by a *background* box. The foreground character color, intensity, underline, and blink characteristics and the background box color may be controlled by individual bits in the attribute byte.

The attribute byte to be used depends directly on the type of monitor that is used with the computer, that is, monochrome or color. Table 7.3 lists the monochrome character display byte attributes (Table 7.4 lists the attribute byte choices for color monitors).

The attribute byte code shown in Table 7.3 shows bit settings (1 or 0) for various character display features, which are explained next.

> Blinking characters toggle between foreground and background color.
>
> Foreground is the character color, light or dark.
>
> Background refers to the color, light or dark, that surrounds a character.
>
> Underlined characters place a dark line under the character.

To use the table, decide what attributes you want the character to possess and fill in the appropriate 1 or 0. Suppose, for instance, you wish to use a "normal" character, which is considered to be a light character against a dark background. The attribute code would then be 00000000b or 00h. If you want the reverse, a dark character against a light background, the attribute code is 01110111b or 77h. A blinking normal character is coded as 80h, and a high-intensity reverse character is coded as 7Fh.

It is very important that you use the *correct* attribute for your monitor type. For instance, attribute byte 00h, which works well for a monochrome monitor, will result in an invisible character (character and box the same color) on a color monitor. For other suggestions on displaying characters, refer to the D86 debugger instructions in Appendix B.

Color monitors use the memory attribute byte as shown in Table 7.4.

The *b* in the code stands for background color around the character; the *f* stands for the character foreground color. Sharp-eyed readers will note that we may choose any one of *16* foreground colors, because there are four *f*-bits in a color attribute byte. Background bits are limited, however, to three in

Color Display Attribute Desired for Display		Attribute Byte Bit Settings
		Bits 7 6 5 4 3 2 1 0
Normal character		0 x x x x x x x
Blinking character		1 x x x x x x x
Background color		x b b b x x x x
Foreground color		x x x x f f f f
		Color
Background color	bbb	
Foreground color	f f f f	
	0000	Black
	0001	Blue
	0010	Green
	0011	Cyan
	0100	Red
	0101	Magenta
	0110	Brown
	0111	White
	1000	Gray
	1001	Light blue
	1010	Light green
	1011	Light cyan
	1100	Light red
	1101	Light magenta
	1110	Yellow
	1111	High-intensity White

Table 7.4 ■ Color Character Attribute Byte Codes

number, or 8 colors. Background colors may be values 000 to 111, or black to white. Foreground colors may be any of the 16 listed in Table 7.4.

If you wish to display a red character against a green background, for instance, the attribute code 00100100b (24h) would be used. Reversing the character, to give a green character against a red background, would require an attribute code of 42h. A light magenta character against a white background would be coded 7Dh in the attribute byte for that character.

Unreadable characters may be defined in both monochrome and color attribute bytes. You may choose to display dark characters against a dark background for monochrome, or red characters against a red background for color. Nothing will be seen using similar foreground-background colors, but a crude form of graphics (using character-sized pixels) could be developed.

VIDEO SCREEN TEXT ORGANIZATION

A standard CRT screen is organized as 25 horizontal lines (rows) of text characters, each line containing up to 80 characters (columns), for a total of 2,000 characters.

The line of text at the very *top* of the screen is row number 0, and the *leftmost* character position is at column number 0. Text rows (lines) are numbered from row 0 (top of the screen) to row 24 (bottom of the screen).

Characters may be displayed from column 0 (fully left on the screen) to column 79 (fully right on the screen).

Characters are generally divided into two types: *printable* characters (those imprinted on the keyboard keys), and *control* characters. Control characters are those ASCII codes meant to control some aspect of a printer or CRT screen. (Refer to Appendix E for a discussion of ASCII codes.) The control characters with which we are most familiar are the LF (Line Feed) and CR (Carriage Return) control characters. Every time we hit the Return key on our keyboards, a LF,CR control sequence is sent to the screen to move the cursor down one line, and over to the left side of the screen.

A cursor is a visual indicator of where, on the screen, the next text character will be placed by BIOS. BIOS allows the program to define the shape of the cursor (within reason) for the application.

BIOS Screen Service INT Type 10h

BIOS uses INT type 10h to control the *display* of text characters on the CRT monitor. We shall investigate the most common type 10h screen control functions in this chapter. Refer to Appendix C for a listing of other type 10h functions.

Each function, within an overall INT type, is identified to BIOS by the contents of a register, usually AH. To get BIOS to perform the function, the program loads AH with the desired function number and then issues the associated INT Type opcode. Other CPU registers may also be used to pass data to the BIOS INT. Up to 256 (00–FFh) functions are possible, per INT type, if AH is used to pass the function number to BIOS.

I shall document each function by listing the function number in AH, any other register used to *pass* data to the function, and a description of what the function does. If data is returned from BIOS by the function, it will be identified as *Returned* register contents, with explanation. If no data is returned from the interrupt, then *Nothing* will be listed as the returned data.

Function 00   Set Display Mode

Register	Contents
AH	00
AL	Mode number
Returned	Nothing

Operation:

Function 00 of INT 10h sets the adapter display mode using one of the following modes:

Mode	Display Format	Adapter Type	Colors
00	Text, 25 rows of 40 characters	CGA, EGA, VGA	16
01h	Text, 25 rows of 40 characters	CGA, EGA, VGA	16
02h	Text, 25 rows of 80 characters	CGA, EGA, VGA	16
03h	Text, 25 rows of 80 characters	CGA, EGA, VGA	16

04h	Graphics, 320 by 200 pixels	CGA, EGA, VGA	4
07h	Text, 25 rows of 80 characters	MDA, EGA, VGA	Mono
08h	Graphics, 160 by 200 pixels	PCjr	16
09h	Graphics, 320 by 200 pixels	PCjr	16
0Ah	Graphics, 640 by 200 pixels	PCjr	4
0Bh	Reserved for Future		
0Ch	Reserved for Future		
0Dh	Graphics, 320 by 200 pixels	EGA, VGA	16
0Eh	Graphics, 640 by 200 pixels	EGA, VGA	16
0Fh	Graphics, 640 by 350 pixels	EGA, VGA	Mono
10h	Graphics, 640 by 350 pixels	EGA, VGA	16
11h	Graphics, 640 by 480 pixels	MCGA, VGA	2
12h	Graphics, 640 by 480 pixels	VGA	16
13h	Graphics, 320 by 200 pixels	MCGA, VGA	256

Note:

Every time the display mode is set for the computer, *all* video pages are blanked, and a thin (two-line) cursor is defined at location row number 0, column number 0 on each page. The display mode should be set only once in a program and not re-used, to avoid erasing each display page. If you *wish* to quickly erase a page, however, setting the display mode is effective.

We shall use display mode 02h, for CGA, EGA, or VGA monitors, and mode 07h for MDA monitors. Both modes 02h and 07h will give us a 25 line by 80 character display format to work with.

A program code fragment that can be used to set the display adapter to mode 02h appears as follows.

```
mov ax,0002h        ;select function 00, mode 02h
int 10h             ;BIOS INT type 10h sets mode number 2
```

BIOS will set mode 2, and return to the instruction that follows INT 10h.

Once the adapter mode is set, the cursor shape can be set by the program using INT 10h, function 01h, as is discussed next.

Function 01h   Set Cursor Shape

Register	**Contents**
AH	01h
CH	Starting line for cursor (00 to 0Ch)
CL	Ending line for cursor (00 to 0Ch)
Returned	Nothing

Operation:

Function 01h determines how many pixel lines will be used to make up the cursor. A cursor is formed by displaying some number of lines from those available in a character box. Register CX is used to hold the starting line number (CH) and ending line number (CL) of the cursor. The cursor shape is *global* for all pages. Global means that only one cursor shape is defined for all pages, and all cursors will be shaped as determined by the last use of function 01h.

The CRT screen cursor is made up of 8 lines (CGA) or 13 lines (MDA, EGA, VGA) of dots. The top line of the cursor is line 00, and the bottom line is 07h (CGA) or 0Ch (MDA, EGA, VGA). Fat cursors use all the available lines, and thin cursors use 1 or 2.

Function 01 allows you to customize the cursor by specifying the beginning and ending lines you wish to use for the cursor shape. To set the cursor shape, load AH with 01h (for function 01h). Load the low nibble of CH with the starting line number (00 to 0Ch) and the low nibble of CL with the ending line number (00 to 07h or 0DH).

To make the fattest possible cursor using CGA, CX would be loaded with 07h. For MDA, EGA, and VGA adapters, CX would be set to 0Ch. If the program does not set the cursor shape, the default values for the cursor are 06–07h for CGA and 0Bh–0Ch for EGA, MDA, and VGA. The selected cursor shape appears on all screen pages in the shape last selected by the program.

Once the cursor shape is determined, the cursor must be *positioned* on the screen. The cursor determines where, on the screen, the next text character will appear. Function 02h, which is discussed next, sets the cursor position.

Function 02h   Set Cursor Position

Register	Contents
AH	02h
BH	Page number, 00 to 07h
DH	Row number, 00 to 18h
DL	Column number, 00 to 4Fh
Returned	Nothing

Operation:

Function 02h places the cursor at a row,column location of a selected page. Pages are complete CRT screens that may be stored in RAM buffer memory as listed in Table 7.4. MDA has one page (00), CGA four pages (00h–03h), and EGA and VGA eight pages (00h–07h). Default cursor positions are set at location 0,0 for all pages when the display mode is set using function 00.

You must *move* the cursor in order to place text characters on a page. The very upper lefthand corner of a page screen has the coordinates row 0, column 0. The very lower righthand corner of the screen has coordinates of row

24,column 79. Cursor coordinates are written in row,column notation. For instance, the upper left corner has coordinates 0,0 and the lower right corner has coordinates 24,79. Characters sent to the screen will be positioned at the place you have set the cursor. The cursor does *not* move by itself as characters are displayed. The program must move the cursor to display a line of text.

To position the cursor on a page, load register AH with function number 02h. Load register BH with the page number you are filling up. Then load register DL with the column number (00 to 4Fh), and register DH with the row number (00 to 18h). Use an INT 10h next, and the cursor is positioned. A program that will place the cursor in the center of the screen could be written as follows:

```
mov ah,02h        ;position the cursor
mov bh,00h        ;page 0
mov dh,0ch        ;row 12
mov dl,27h        ;column 39
int 10h           ;position the cursor
```

If you specify a row or column number that does not physically exist, such as row 50d, column 200d, the cursor will *disappear*.

Having to position the cursor means that your program must know where the cursor is on the screen. One way for the program to fix the cursor location is to have it store the current cursor location as a memory variable. Another way to fix the cursor location is to have Function 03h, discussed next, *return* it to the program.

Function 03h   Get Cursor Position and Shape

Register	Contents
AH	03h
BH	Page number, 00 to 07h
Returned	
DH	Row number, 00 to 18h
DL	Column number, 00 to 4Fh
CH	Cursor starting line
CL	Cursor ending line
BH	Page number

Operation:

Function 03h returns to the program the row and column position of the cursor in DX, the cursor shape in CX, and the page number of the cursor in BH.

Using function 03h relieves the programmer of having to keep track of where the cursor happens to be on the screen. Function 03h is also useful when the cursor has been *automatically* moved by a display string interrupt, such as DOS interrupt 21h, function 09h.

Positioning the cursor does not mean that it will appear on the screen page currently displayed on the CRT. The program must select the page to be displayed on the screen using function 05h, discussed next.

Function 05h   Select Displayed Page

Register	Contents
AH	05h
AL	Page number, 00 to 07h
Returned	Nothing

Operation:

Function 05h determines *which* screen page in the RAM is to be displayed on the CRT screen. Function 05h is invalid for MDA adapters, which only have one page.

The ability to store multiple pages enables the programmer to put new screens on the CRT at a rapid pace. Programs that write to a screen, while it is visible to the user, appear to be sluggish. It may take up to several seconds to erase and write a new screen character by character. Pages that have been stored, and then suddenly displayed by choosing that page with display function 05h, seem to instantly appear on the face of the CRT.

The screen to be viewed is selected by loading register AH with function number 05h and loading register AL with the page number to be displayed (from page 00 to page 07). The following sample program will display page number 6 for viewing, for EGA and VGA adapters.

```
mov ax,0506h        ;choose page 6 for viewing
int 10h             ;display it
```

Functions 00, 01h, 02h, and 05h let us set the display mode, the cursor shape, the cursor position, and the page to be viewed. We have yet to put any text on the selected page. Function 09h allows the program to write text characters to a page at the page cursor position. If the cursor is positioned on the currently selected screen page, then the text characters will appear on the CRT as they are written.

Function 09h   Write a Text Character at the Cursor Position

Register	Contents
AH	09h
AL	ASCII code for character to display
BH	Page number, 00 to 07h
BL	Attribute for monitor type, 00 to FFh
CX	Number of characters to be displayed
Returned	Nothing

Operation:

Function 09h writes an ASCII character, contained in register AL, to the display page number contained in register BH. The character will have the attributes of the byte in register BL. Attributes have been summarized previously in Tables 7.3 and 7.4. CX contains the number of times the character in AL is to be written to the page in BH.

7.6 BIOS CRT Output Examples

We are ready, finally, to actually write a character to the CRT screen! Before any characters can be displayed, the following display functions must be programmed:

Function 00: Set display mode (02h or 07h) and *erase* all pages.

Function 01h: Define the cursor shape (optional).

Function 02h: Position the cursor where the character is to be written.

Function 05h: Select the page for screen display.

Once we begin to display characters, however, not all of the preparatory functions (00, 01h, 05h) have to be repeated.

A character is displayed by loading register AH with 09h. Register AL is loaded with the ASCII character to be displayed, BL with the attributes of the character loaded into register AL, and BH with the page number to which the character is written. Register CX is loaded with the number of identical characters that will be written starting at the current cursor position.

It is important to note that the first character will be written at the current cursor position, *not* to the right of the cursor, and that the cursor does *not* move as a character is written over it.

Normally CX will be 1, unless we wish to write a number of repeated characters, such as blanks (ASCII 20h) to the screen.

Example Program 1: Linear Hello

A small program, hello1.asm, that will write the message "hello" in the center of a color screen follows next. The characters will have the attribute of normal, blue foreground, and a red background. The program is *linear*, that is, no calls to procedures are made in the program.

```
;"hello1.asm" is a very simple routine for writing the message "hello" to
;page 3 of a color screen.

code segment                                                      See Note
        mov sp,8000h        ;SET THE STACK
        mov ax,0002h        ;select mode 02h display for CGA,
                            ;EGA, or VGA color monitor
        int 10h             ;BIOS sets mode via int 10h              1
        mov ah,01h          ;set a half-sized cursor
        mov cx,0006h        ;assume EGA or VGA for half size
        int 10h             ;BIOS sets cursor size                   2
```

```
hello1:
        mov ah,02h              ;position cursor on page 3
        mov bh,03h              ;cursor on page 3                          3
        mov dx,0c25h            ;position on row 12, column 37
        int 10h                 ;cursor now positioned
        mov ax,0503h            ;now, display page 3 on CRT
        int 10h                 ;CRT screen now shows page 3              4
        mov ah,09h              ;display character function
        mov al,'h'              ; h
        mov bh,03h              ;page 3, for this example
        mov bl,41h              ;blue character, red background
        mov cx,0001h            ;one h
        int 10h                 ;display it
        mov ah,02h              ;move the cursor over one space           5
        mov bh,03h              ;on page 3
        mov dx,0c26h            ;next column
        int 10h
        mov ah,09h              ;display character function
        mov al,'e'              ; e
        mov bx,0341h            ;page 3, normal character attributes
        mov cx,0001h            ;one e
        int 10h                 ;display it
        mov ah,02h              ;move cursor over one space
        mov bh,03h              ;on page 3
        mov dx,0c27h            ;next column
        int 10h
        mov ah,09h              ;display character function
        mov al,'l'              ; l
        mov bx,0341h            ;page 3, normal character attributes
        mov cx,0002h            ;two l
        int 10h                 ;display it
        mov ah,02h              ;move cursor over two spaces
        mov bh,03h              ;on page 3
        mov dx,0c29h            ;two spaces
        int 10h
        mov ah,09h              ;display character function
        mov al,'o'              ; o
        mov bx,0341h            ;page 3, normal character attributes
        mov cx,0001h            ;one o
        int 10h                 ;display it
        mov ah,02h              ;move cursor one more time
        mov bh,03h              ;on page 3
        mov dx,0c2ah            ;move to next column
        int 10h                 ; "hello" in blue and red
        mov ax,0500h            ;get back to D86                          6
        int 10h
        jmp hello1              ;loop
code ends
```

NOTES ON EXAMPLE PROGRAM 1

These notes apply when D86 is used to debug the program (highly recommended).

    1. The screen will blink, and a small cursor will appear at 0,0 on the screen. D86 uses page 0 to display the program under test. D86, you

may have noticed, has a cursor at the bottom of the screen. D86 comes up in screen 0, and setting the display mode using function 00 places a default cursor size at default cursor location 0,0. *Remember!* All pages are erased (set to blanks) by function 00. D86 is not blanked because D86 constantly rewrites page 0.

2. A blinking FAT cursor appears at screen location 0,0. Every page has its own cursor, and the cursor setting affects all cursor sizes on every page that has a cursor.

3. A blinking fat cursor is now on page 3, but we cannot see it yet, because we are not currently displaying page 3.

4. The CRT now shows an empty page 3. Every time the display mode is set (function 00), all pages are blanked. D86 rewrites the screen on every step, so page 0 appears to blink when the display mode is set by function 00. Note blinking, fat cursor, at the center of the screen. D86 is still running, but we can no longer see the D86 screen, page 0. Commands can still be blindly typed to D86.

5. Move it, or lose it.

6. Recover the D86 page 0 screen. As an experiment, at this line in the program, type the command mov ax,0503h followed by the command int 10h. The "hello" message should appear, in a blink of the eye.

Example Program 1 seems a bit complicated to just be able to display a simple "hello." Example Program 1 does point out some things to look for, however. For instance, when using function 09h, it would seem unnecessary to have to reload CX, BH, and AH for every character. Why not load them once and just change the display character in AL? The answer is that you cannot be *sure* what will be in registers AH, BH, and CX when BIOS *returns* from performing the function you chose. As the old saying goes, "Trust your friends—and cut the cards." A modern version of that chestnut is "Trust your BIOS—and reload your registers."

We can shorten the "hello" program by using the full power of our CPU and assembler. A programming scheme is discussed next that will let us display messages of arbitrary length using very little code.

Example Program 2: Displaying Strings of Text

A *string* is a sequence of characters, usually printable ASCII, stored in memory. We tend to think of strings as messages that are displayed on a screen, or printed. The "hello" message displayed in the previous program is a string. This chapter of text is a very long string. The only requirement that a string must meet is that the characters occupy sequential addresses in memory.

A program that will display strings of arbitrary lengths will now be developed. The program must know two things: where the string begins, and where it ends. The beginning address of the string is passed to the program as a label that identifies the location in memory where the first string character may be found. The end of the string is denoted by using a special character. The character chosen to mark the end of the string is the ~, a character that rarely appears in any English text. Other candidates could be the ^, |, or #. If you decide not to use a printable character as a terminator, then a control character, such as EOT (ASCII 04), could be included in the message string using a DB 04 directive.

Example Program 2, named hello2.asm, is shown next.

```
;example program two, named "hello2.asm." A pointer to a string is
;passed to the program which displays characters until a terminating
;character, ~, is encountered. A CGA, EGA, or VGA adapter is
;assumed. Mono adapter users should change all appropriate functions
;from page 1 to page 0, and choose a monochrome attribute byte.

        data segment                                            See Note
                org 1000h           ;define variable address        1
                cursnum dw ?        ;name the variable "cursnum"
        data ends
        code segment
                ;EQUATES            equate names to constant numbers
                color equ 0002h     ;color mode
                plump equ 0007h     ;fat cursor
                middle equ 0c26h    ;row and column for screen center
                page1 equ 01h       ;page one of screen display
                charcolor equ 9fh   ;display characters as blinking white
                                    ;on a blue background

                ;END EQUATES
                mov sp,0a000h       ;SET THE STACK                   2
                mov ax,color        ;chose color text mode 02 (25 × 80)
                int 10h             ;BIOS INT to set mode
                mov ah,01h          ;define a plump cursor
                mov cx,plump        ;use 8 lines
                int 10h             ;BIOS INT to set cursor
                mov cursnum,middle  ;place the cursor on screen
                call curses         ;routine curses moves cursor
                mov ah,05           ;display on page 1
                mov al,page1        ;choose page 1 for message
                int 10h             ;BIOS INT for page
                mov di,msg1         ;DI points to message address    3
        showit: mov ah,09h          ;write character at [DI]
                mov bh,01           ;page 1 character attributes
                mov bl,charcolor    ;blinking white on blue
                mov cx,0001h        ;write one character
             cs mov al,[di]         ;get character FROM CODE SEG     4
                cmp al,'~'          ;quit when at end
                je dun              ;if not at end, go on
                int 10h             ;display character
                inc di              ;point to next character
                call curses         ;increment cursor
                jmp showit          ;loop until done
        dun:
                mov ax,4c00h        ;return to DOS (see section      5
                int 21h             ;on DOS interrupts)
;the subroutine "curses" gets the last cursor location from RAM
;location cursnum, increments it by one, and places the cursor
;at the new location. The new cursor location is stored at cursnum
;before returning. All affected registers are saved.
```

```
curses: push dx              ;save all registers used
        push ax
        push bx
        mov dx,cursnum       ;get last cursor position
        inc dx               ;increment it
        mov ah,02h           ;position cursor
        mov bh,01h           ;on page 1
        int 10h              ;BIOS INT places cursor
        mov cursnum, dx      ;save new cursor location
        pop bx               ;restore all old registers
        pop ax
        pop dx
        ret
msg1:   db 'hello~'          ;end messages in ~                    6
        code ends
```

Example Program 2 may be run, full speed, by just typing hello2 opposite the DOS > prompt. This program includes an INT to DOS (see Note 5) that allows DOS to regain control of the computer at the end of the program. Unfortunately, DOS returns using the last page defined in Example Program 2: page 1. Use the DOS mode co80 command to reset a color adapter and return to page 0. Monochrome users will not have this problem but will not be able to use D86 to debug the program, because D86 uses page 0 also and overwrites the "hello" message. Monochrome users should change the initial cursor number from 0C26h to 1826h (line 24, at the bottom of the D86 screen) to see the hello message on the D86 screen.

NOTES ON EXAMPLE PROGRAM 2

1. The choice of 1000h for the data variable address is arbitrary. Equates are used here to demonstrate how a program might be made more readable by using names in place of numbers.

2. The choice of where to set the stack is above the data.

3. Remember, msg1 is a label, so the address in code where the hello message is stored is known to the assembler.

4. Note the use of the cs to override default DS choice.

5. DOS interrupts work much as do BIOS interrupts. DOS INT type 21, function 4C00h returns the computer to DOS control.

6. Forgetting the terminating character, ~, will cause Example Program 2 to run through all of memory, displaying everything it finds. Very spectacular.

The display program has one flaw: If the programmer forgets the ~ character, the routine will display *everything* from msg1 to the end of memory. Other schemes for displaying messages include making all messages the same length and passing the address of the message, and the number of message characters, to the showit procedure. Both of these approaches will be addressed by problems at the end of this chapter.

Another problem with using any character, such as ~, as a message terminator, is that we cannot print any messages with a ~ embedded inside. A control character (assuming that the system will not respond to the control character) can be added (and checked for) to the end of the string. For instance, using the control character 00h, (NUL) lets the string line of code from Example Program 2 be written as:

```
msg1: db 'hello~',00h          ;end messages in NUL
                               ;which allows the ~ to be displayed.
```

Example Program 3: Writing to Display Memory

If you wish to try writing directly to the display memory for displaying text, despite warnings to the contrary, you may try the following program. The next program, "direct.asm," accesses color text memory (starting at B800h) and will display a very rapid sequence of screen images.

```
;a program, "direct.asm," that accesses display memory directly.
code segment
        mov sp,0C000h          ;set the stack
        mov ax,0b800h          ;set DS at display memory
        mov ds,ax              ;CGA page 0
        mov dx,0064h           ;100 screens and back to DOS
begin:
        mov bx,0fffeh          ;BX points to character space
char:
        inc bx                 ;point to next character
        inc bx                 ;and attribute byte
        mov b[bx],'O'          ;display a light "O"
        mov b[bx+1],07h        ;dark background
        cmp bx,0f9fh           ;write 2000 characters
        jc char                ;loop until screen full
lpy:
        dec cx                 ;delay to be able to see it
        jnz lpy                ;delay loop
        mov bx,0fffeh          ;next screen
chr:
        inc bx                 ;point to next character pair
        inc bx
        mov b[bx],'x'          ;display an "x"
        mov b[bx+1],07h        ;LCD display, dark character
        cmp bx,0f9fh           ;stop after a full screen
        jc chr
lp:
        dec cx                 ;delay to see this screen
        jnz lp
        dec dx                 ;if done then exit to DOS
        jz back
        jmp begin
back:
        mov ax,4c00h
        int 21h                ;back to DOS
```

The program will run 100 screens and return to DOS. Depending on the speed of your computer, you will see an alternating series of screens full of "O"s and "x"s that change faster than can be seen by the eye.

7.7 BIOS Keyboard Input Services

A keyboard is the generic *input* device for personal computers. Keyboards may range from the simple hex keypad, used on board-level computers, to 83-, 84-, or 101-key keyboards used with personal computers. Not surprisingly, BIOS has an INT, type 16h, that deals with the PC keyboard.

PC keyboards are well known to PC users. We know that we may press character keys, shift keys, function keys, and so on. The character of the key pressed appears on the screen, or some control function is performed. Keyboards are not simple, however. A keyboard is a serial device that transmits a code for any depressed key to the main computer. The code byte, or bytes, transmitted from the keyboard to the PC is called a *scan code*. Scan codes for the PC keyboard are shown in Appendix E. Keyboard scan codes are *not* ASCII codes.

Keyboard keys fall into two general types: text and control. Text keys, such as alphabetic or numeric, are those keys that will be displayed on the monitor when they are pressed. Control keys, such as the Return key, generate so-called *nonprintable* characters, that are not normally displayed on the monitor by DOS or most word processor programs. The control keys can be displayed, however, on the monitor if direct screen display BIOS interrupts are used. Control characters are displayed on the monitor as graphic characters such as "happy faces," playing-card pips, arrows, musical notes, and other small icons.

Almost all PC keyboards contain a microcontroller that handles the chores of deciding what key has been pressed, converting the key switch closure to a scan code, serializing the code, and transmitting the key code down the coiled wire to the PC. Once the serialized character scan code reaches the PC, it is converted back to parallel format and converted to ASCII, if the key pressed is assigned an ASCII code.

The PC reacts to each incoming character by automatically using BIOS interrupt 09h to process the key code. Interrupt 09h stores the incoming scan and ASCII codes for the key in a small memory called the keyboard *buffer*. Buffers are normally used to temporarily save small amounts of data, such as a sequence of keystrokes, for later access by the program. If the program is occupied with other tasks, the keyboard buffer saves any user keys until the program can deal with the keystrokes. The standard keyboard buffer is limited to a capacity of 15 keys. We are all familiar with the *beep* from the PC when too many keys are pressed and the buffer fills up while the computer is busy with other tasks. BIOS is responsible for the beep.

Application programs can use BIOS INT type 16h to detect that characters have been placed in the keyboard buffer by interrupt type 09h and to get characters. Only two INT type 16h keyboard functions, 00 and 01h, will be needed for our purposes in this chapter. Refer to Appendix C for a discussion of other type 16h BIOS keyboard service functions.

BIOS Keyboard INT Type 16h

INT type 16h has been assigned to BIOS keyboard functions. Type 16h INT functions return data to the program, as opposed to monitor interrupts, which get data from the program. Two type 16h functions, numbers 00 and 01h, can be used to efficiently get keys from the keyboard.

Function 00   Get Key from Keyboard Buffer

Register	Contents
AH	00
Returned	
AL	ASCII code for key
AH	Scan code for key

Operation:

Function 00 is invoked by placing 00 in register AH and using INT 16h. Function 00 will *wait* until a key is pressed and then return the ASCII code for the key in register AL and the key scan code in register AH. A warning. Two special key combinations, Ctrl-Alt-Del and Shift-PrtSc, will not be returned by BIOS. Ctrl-Alt-Del will cause the execution of an immediate system reset by BIOS. Shift-PrtSc will cause the contents of the current screen to be printed.

Function 00 may be used to get key inputs for your program, but it does have the distinct disadvantage of stopping the computer program from performing any other action. The INT will wait until a key is hit before returning. The next function, 01h, can be used to determine if a key is waiting; if not, the program could go on its way.

Functions such as function 00 are described as "well-behaved." A well-behaved function will work as a reasonable person might expect and get the key from the keyboard buffer. The buffer is emptied by one or two (for extended key codes) key characters and any keys that follow will be the next to be fetched from the buffer. The next function, 01h, is not well-behaved and does not empty the buffer. Function 01h is useful, however, as a signal to the program that a key is available in the buffer.

Function 01h   Read Keyboard Status

Register	Contents
AH	01h
Returned	
AH	Scan code for key
AL	ASCII code for key
Flags	No key waiting, zero flag = 1
	Key waiting, zero flag = 0

> *Operation:*
>
> Function 01h uses the Zero flag to indicate that a key has been pressed and the key code is available. If the Zero flag is *set* to 1 after the INT returns, *no* key is waiting. A 0 in the Zero flag indicates that a new key is waiting to be read by the program.

Software interrupts, such as function 01h, can be used by the programmer in a *polling* mode. Polling means that the program repeatedly inquires as to the status of a particular device and takes action when the device is active. In the case of keyboards, polling will reveal if a key is available. If no key is available, the program can go on to other things or wait in a loop until a key has been pressed.

Function 01h does, in fact, return the scan code of the key (if a key has been pressed) in register AH and the ASCII code for the key in register AL. Function 01h should not be used, however, to get the key code. Function 01h does *not* empty out the keyboard buffer and will *continue* to return the first key detected, forever. Its only use is for polling to see if a key is present. If no key is waiting, then the program could proceed on to other things and return to poll the keyboard a few milliseconds later. If a key is waiting, then function 00 can get it and clear out the keyboard buffer for a fresh key.

A short example program, which will get the ASCII code for a key by polling the Zero flag returned by function 01h, is written next.

```
poll:
        mov ah,01h          ;check for a key
        int 16h             ;ask BIOS
        jz no_key           ;go do something else
        mov ah,00h          ;get the key codes
        int 16h
        mov bl,al           ;store for later use
no_key:
        nop                 ;rest of program
        jmp poll            ;check for key
```

Note that function 01h *destroys* the original contents of the Flag register. Because function 01 uses the Zero flag as a signal to the program, it does not return with an IRET (which would restore the original flags); rather it returns with a RET opcode.

Function 01h does flush the original stack, so the stack remains balanced, but the original flags are lost. If you need to save the contents of the Flag register, push them on the stack immediately before issuing the INT 16h instructions. Pop the Flag register from the stack after your program has examined the flags returned by function 01h. Our previous example, used to demonstrate polling, could be altered to save the flags, as is shown next.

```
poll:
        pushf               ;save original flags
        mov ah,01h          ;check for a key
        int 16h             ;ask BIOS
```

```
          jz no_key              ;keep polling until a key
          popf                   ;get original flags back
          mov ah,00h             ;get the key codes
          int 16h
          mov bl,al              ;store for later use
          jmp poll               ;go again?
no_key:
          popf                   ;get original flags back
          nop                    ;rest of the program
          jmp poll               ;check for a key
```

We can use BIOS INT services to write simple text display programs. Several text I/O programs are discussed in the next section.

7.8 Input/Output Example

Display Keys

The I/O example, a program named showkey.asm, will display any printable key hit on the screen. Control keys will display an empty box. The program returns to DOS when the Esc key is pressed.

```
;I/O example, the program showkey.asm, that displays any printable
;keyboard character on the screen, until Esc is hit.
code segment                                                    See Note
          mov sp,7000h           ;set the stack
          mov ax,0002h           ;select mode 02h display for CGA,
                                 ;EGA, or VGA color monitor
          int 10h                ;BIOS sets mode via int 10h
poll:
          mov ah,01h             ;check for a key
          int 16h                ;ask BIOS
          jz poll                ;keep polling until a key            1
          mov ah,02h             ;position cursor on page 0
          mov bh,00h             ;cursor on page 0
          mov dx,1801h           ;position on row 24,column 1 for D86  2
          int 10h                ;cursor now positioned
          mov ah,00              ;get the key ASCII code in AL
          int 16h
          cmp ah,01h             ;stop if Esc key hit                  3
          jne screen             ;if not then display key
          mov ax,4c00h           ;if so, back to DOS
          int 21h
screen:
          mov ah,09h             ;display character function
          mov bh,00h             ;page 0, for this example
          mov bl,1fh             ;white character, blue background
          mov cx,0001h           ;one character
          int 10h                ;display it
          mov ah,02h             ;move cursor so we can see character
          mov bh,00h
          mov dx,1802h           ;position at 24,2
          int 10h
          jmp poll               ;next character
code ends
```

Notes on the Example

1. Using D86 to debug a program using function 01h is awkward, because D86 assumes any key you type in is a command and tries to interpret the key. Step over the JZ opcode and go on with the program. Once at function 00, D86 will wait for a key to be hit once INT 16h has been single-stepped.

2. Line 24 of the D86 screen is the only part of the D86 screen that is not constantly overwritten by D86.

3. The CMP opcode is looking for the scan code for the Esc key in register AH. Scan codes are useful for detecting control keys that have no ASCII counterparts, such as the Insert key or Home key. Also, be aware that the Ctrl, Shift, Caps Lock, Num Lock, Scroll Lock, and Alt keys are not returned by INT type 16h functions 00 or 01h. Function 02h can be used to determine the status of the special keys.

7.9 BIOS Real-Time Clock Services

The PC is equipped with a system timer that generates an INT 08h, by *hardware* means, every 54.925493 milliseconds. (See Chapter 11 for more details on hardware interrupts.) One function INT 08h performs is to update the time-of-day clock *tick* counting registers. The tick counters start at 0000, 0000 at midnight, and count each tick arriving at 0018h,00b0h (1,573,040) by the end of a 24-hour period. User programs can access the tick counters using INT 1Ah and read the contents to generate accurate timing delays. We shall examine function 00 of INT 1Ah, which reads the tick registers in this section. Refer to Appendix C for a discussion of other INT 1Ah functions.

Function 00   Read Clock Tick Counters

Registers	Contents
AH	00h
Returned	
AL	01h if past midnight, 00 if not
CX	High order counter
DX	Low order counter

Operation:

Function 00 returns the total ticks since midnight in registers CX,DX. Register DX may be used for delays ranging from approximately 54 milliseconds to 1 hour.

An example program that places a message on the screen every 10 seconds, using INT 1Ah, is shown next.

```
;"timedelay.asm" uses INT 1Ah to get the current time for a real-
;time delay


code segment
timedelay:                                                      See Note
        mov sp,1000h        ;set stack above program
        mov di,10xd         ;do delay and message 10 times
timout:                     ;return here every 10 seconds
        mov ah,00h          ;get time of day
        int 1ah             ;read DX for number of ticks
        mov bx,dx           ;save current (old) time in BX
timer:
        mov ah,00h
        int 1ah             ;get new time
        sub dx,bx           ;calculate delay since last inquiry    1
        cmp dx,182xd        ;CF will be 1 until delay = 182 ticks   2
        jc timer
        mov ah,09h          ;display message
     cs mov dx,msg
        int 21h

        dec di              ;count DI for 10 messages
        jnz timout
        mov ax,4c00h        ;return to DOS
        int 21h
msg:
        db 'Ten more seconds have passed $'
code ends
```

NOTES ON THE PROGRAM

1. The correct difference is calculated even if DX has rolled over at FFFFh to 0000. For instance, assume the "old" time is FF00h, and DX has rolled over to 0000—0000 minus FF00h yields 0100h (and a Carry flag), which is the correct result.

2. There are 18.20648193 interrupts per second, or about 182 in a 10-second period. Note that in a 24-hour period there are exactly 1,573,040.039 ticks, so that the PC clock could be off by as much as .039 ticks, or .0002 seconds, a day.

DOS also has INT service routines that can be accessed by assembly code programs. Interrupts to DOS are discussed in the next section.

7.10 DOS Service Routines

BIOS interrupts go to a hardware level in the computer that is just above the I/O chips themselves. Interrupts to BIOS let the assembly language programmer use computer system resources, such as the monitor and keyboard, without having to directly program the chips that perform the physical interface to system I/O devices.

DOS is also equipped with a number of interrupt service routines for system I/O devices. DOS is loaded from disk, into RAM, and serves to operate system resources for use by application programs.

Interrupts to DOS suffer from two disadvantages. First, DOS interrupts tend to run more *slowly* than do BIOS interrupts. Second, DOS does not usually have the *detail of control* that BIOS does.

DOS is fundamentally concerned with interpreting commands from the keyboard, such as COPY, TYPE, DIR/W, and the like, and operating the disk drives found in the computer system. DOS does, however, have interrupts that deal with the other system I/O devices, including the CRT monitor and the keyboard.

Chapter 9 discusses BIOS and DOS disk operations. Here, we shall investigate the common DOS monitor and keyboard interrupts. Please refer to Appendix D for a listing of DOS INT services.

The most common DOS INT is type 21h, which offers more than 100 functions and subfunctions. DOS INT type 21h is the *primary* DOS INT type. Most of the DOS type 21h functions deal with disk access, but INT functions for other computer assets are included.

DOS Keyboard Input and CRT Output Services

DOS interrupts for keyboard access do not provide as much control of the keyboard as do the BIOS INT functions. Loss of precise control is generally true for all DOS interrupts, which operate at a more general level than do BIOS interrupts of the same type.

Function 01h   Keyboard Input with Echo

Register	Contents
AH	01h
Returned	
AL	ASCII character code

Operation:

DOS INT 21h, function 01h is a dual keyboard-monitor service routine. The term *echo* means that any key pressed on the keyboard is also displayed on the screen. INT function 01h waits until a key is pressed, then displays the character on the monitor and returns.

The action of function 01h is desirable if you want all of the keyboard characters placed on the monitor. However, if you are using some characters to control program operation and not for display, monitor display may not be desirable. Use DOS function 07h or one of the BIOS keyboard INT 16h functions to gain full control of every key entered from the keyboard.

DOS INT 21h, function 01h is accessed from the program by loading AH with 01h, and then using an INT type 21h. The INT waits for a key to be pressed, displays it, and then returns.

On return from the INT, AL will contain the 8-bit ASCII code for the character displayed. If AL is zero on INT return, then another call must be done to find the scan code for the key, because the key has an *extended* ASCII code. Extended ASCII codes are generated for special keys, such as the function and cursor control keys. Extended code sequences consist of a 00 followed by a second ASCII code, as shown in Appendix E.

A warning. The key sequences Ctrl-Break and Ctrl-C will cause *immediate* suspension of any program using function 01h with a subsequent return to DOS. DOS screens all keys entering the program via function 01h and *traps* the Ctrl-C or Ctrl-Break keys. (Trap means to intercept something before it gets to a normal destination.) Ctrl-C and Ctrl-Break are universal (for DOS) keys a user employs to stop the execution of an application program.

DOS function 01h will display the keyboard character on the last page number selected by BIOS INT 10h function 05h. DOS, when it *boots up*, will select page *0* for display.

Function 02h   Display Output

Register	Contents
AH	02h
DL	ASCII code for character
Returned	Nothing

Operation:

Function 02h displays the ASCII character on the current page and at the current cursor position.

Function 02h of INT type 21h is a very handy way to display characters. However, you will find that the output character is placed where DOS "left" the cursor when your program is executed—on the line following your command. If you wish to place the cursor on the screen at a place of your choosing, and use a screen other than page 0, use the BIOS type 10h interrupts to arrange the display.

A character may be displayed, using function 02h by loading AH with 02h and DL with the desired ASCII code for the character. INT type 21h is then placed next in the program, and the character in DL will be displayed at the current cursor location. Function 02h returns no data on completion.

Function 07h   Read Keyboard Character

Register	**Contents**
AH	07h
Returned	
AL	ASCII code for key

Operation:

Function 07h returns the ASCII key code for keyboard keys in register AL. Extended code keys (a sequence of two key codes) require a second INT to retrieve the second code byte. Function 07h waits for a key to be pressed on the keyboard if a key is not ready in the keyboard buffer.

Function 07h is well-behaved and truly removes the key code from the buffer. Any additional key codes that are stored in the buffer will move up to fill the space left by the key code just read by function 07h. Function 07h is accessed by loading register AH with 07h and using INT 21h to actuate DOS. No special escape key codes, such as Ctrl-Break or Ctrl-C, are trapped by DOS function 07h.

Because function 07h waits for a key to be pressed, no other computer action will take place *until* a key is present. DOS function 0Bh (see below) can be used to check the keyboard to determine if a key code is to be read. The program can then get the key with Function 07h or continue on if no key is waiting.

Function 09h   Display String

Register	**Contents**
AH	09h
DS	Segment address of string
DX	Offset address of string
Returned	Nothing

Operation:

Function 09h displays character strings pointed to by DS:DX. The string must end in a $ character for function 09h to work properly. The $ is not displayed, but serves to tell function 09h when to stop displaying characters in the string.

Clearly, because a $ marks the end of the string, no $ character may be displayed using function 09h. Function 09h could be used to display almost all strings, and function 02h could be used to display the $ character when necessary.

To use function 09h, register AH is loaded with 09h, and register DX is loaded with the address in memory of the first character in the string. Register DS must hold the data segment address of the segment that contains the string.

Function 0Bh Get Keyboard Status

Register	Contents
AH	0Bh
Returned	
AL	FFh if character present

Operation:

Function 0Bh will return an FFh in register AL if a character is available in the keyboard buffer. If no key is available, some other value (other than FFh) is returned. Function 0Bh is accessed by loading register AH with 0Bh and using an INT 21h.

Function 0Bh does *not* return an ASCII code for the key, but alerts the program if a key is waiting to be read. Function 0Bh is useful as a polling scheme for the keyboard. The polling program can get the key, if available (using functions 01h or 07h), or continue on to other actions if no key is waiting.

Function 4Ch Return to DOS

Register	Contents
AX	4C00h
Returned	Nothing

Operation:

Function 4Ch halts execution of an application program and returns control of the computer to DOS.

Previous examples have made use of function 4Ch. Function 4Ch terminates our program and returns control of the computer back to the DOS command interpreter program, COMMAND.COM.

We *invoke* (cause DOS to load and run) our programs by typing the name of the program after the DOS command (>) prompt. DOS loads the program and turns control of the computer over to our program. The program finishes and returns control of the computer back to DOS using INT type 21h, in conjunction with function number 4Ch in register AH. Register AL may be used to pass *return codes* to DOS. For our purposes, AL should always return a 00 code.

Other return functions exist (INT type 20h; and INT type 21h, function 00), but function 4Ch is currently the desired standard exit function from an application program back to DOS.

7.11 DOS Input/Output Examples

Example Program 1: Display Keys

A short example, showkey.asm, of using DOS INT type 21h, function 01h follows next. Assume that the program has already set the monitor to text mode.

```
;"showkey.asm" displays any printable key pressed until a q, Ctrl-C
;or Ctrl-Brk key is pressed.
code segment                                                   See Note
        mov sp,7000h        ;set the stack
_loop:
        mov ah,01h          ;get key and echo                  1
        int 21h             ;DOS does it                       2
        cmp al,'q'          ;exit from program if quit key hit
        jnz _loop
        mov ax,4c00h        ;return to DOS
        int 21h             ;return gracefully
code ends
```

Run this program and find out where the key echoes begin on the screen. You can verify all of the ASCII key codes, including the control codes, by using this program.

NOTES ON EXAMPLE PROGRAM 1

1. The jump label is _loop, not the reserved assembler mnemonic loop.
2. D86 boots up using screen 0, with the cursor set to line 24 on the screen. Any printable keys you hit will be displayed at the bottom of the D86 screen.

Example Program 2: Display a Message

A sample program, hellodos.asm, that displays the familiar "hello" message is written next.

```
;the program "hellodos.asm" uses dos INT 21h, function 02h to
;display a message

code segment                                                   See Note
        mov sp, 0d000h        ;set the stack
        mov bx,offset msg1    ;point to the message with BX      1
        mov cx,0005h          ;five characters in "hello"
```

```
more:
        mov ah,02h          ;display function
        mov dl,[bx]         ;get character
        int 21h             ;INT to DOS
        inc bx              ;point to next character
        dec cx              ;quit after five
        jnz more            ;go until done
        mov ax,4c00h        ;return to DOS
        int 21h
msg1    db 'hello'          ;assemble message here
code ends
```

FIRST NOTE ON EXAMPLE PROGRAM 2

1. The variable name msg1 is converted to a constant address number by the offset directive. If msg1 were written as msg1: (a constant), then the offset directive would be redundant.

The hellodos.asm program uses CX to count any five-character message. A better way to display messages of any length would be to load CX with the message length, in bytes, or terminate each message with a special character. The assembler can be used, in Example Program 2, to calculate the byte length of an arbitrary message, as is shown next.

```
;the program "hellodos.asm" uses dos INT 21h, function 02h to
;display a message. A86 calculates the length of the message.
code segment                                              See Note
        mov sp, 0d000h        ;set the stack
        mov bx,offset msg1    ;point to the message with BX
        mov cx,mlength        ;message length
more:
        mov ah,02h            ;display function
        mov dl,[bx]           ;get character
        int 21h               ;INT to DOS
        inc bx                ;point to next character
        dec cx                ;quit after five
        jnz more              ;go until done
        mov ax,4c00h          ;return to DOS
        int 21h
beginm equ $                  ;IP = start address
msg1 db 'hello'
endm equ $                    ;IP = end address
mlength equ endm-beginm       ;length of message          1
code ends
```

SECOND NOTE ON EXAMPLE PROGRAM 2

1. Remember, A86 cannot make *calculations*, such as subtraction, using forward-referenced symbols for constants. Trying to use mov cx,endm-beginm will get an error message from A86, because the constant values for each name are not yet known. But, when the symbol mlength is calculated, the values of beginm and endm are known.

Example Program 3: BIOS and DOS Display a Message

Example Program 3, dosbios.asm, uses BIOS interrupts to set up the desired page and cursor location, then displays "hello" on screen 1 using function 02h. The program then polls the keyboard and waits for a q(quit) key to be pressed using function 07h, before returning to DOS. DOS sets the monitor to text mode.

```
;use bios and dos calls in "dosbios.asm" to display a message
code segment
        mov sp,8000h        ;set the stack
        mov ax,0501h        ;choose display page 1
        int 10h
        mov ah,02h          ;move cursor of page 1
        mov bh,01
        mov dx,0c30h        ;row 12,column 48
        int 10h
        mov bx,offset msg1  ;point to first character
        mov cx,0005h        ;five characters in "hello"
more:
        mov ah,02h          ;display function
        mov dl,[bx]         ;get character
        int 21h             ;INT to DOS
        inc bx              ;point to next character
        dec cx              ;quit after five
        jnz more            ;go until done
dun:
        mov ah,07h
        int 21h             ;get a key
        cmp al,'q'
        jne dun
        mov ax,0500h        ;back to display page 0
        int 10h
        mov ax,4c00h        ;return to DOS
        int 21h
msg1    db 'hello'          ;assemble message here
code ends
```

Example Program 4: Display a String

Example Program 4, hellostr.asm, displays our favorite message, "hello," as a string.

```
;Use DOS function 09h to display a string ending in a $
code segment                                            See Note
        mov sp,6500h        ;set the stack
        mov ax,0501h        ;choose display page 1
        int 10h
        mov ah,02h          ;move cursor of page 1
        mov bh,01
        mov dx,0c30h        ;row 12,column 48
        int 10h
        mov dx,offset msg1  ;point to first character
        mov ah,09h          ;display function          1
        int 21h             ;INT to DOS
```

```
dun:
        mov ah,07h
        int 21h                 ;get a key
        cmp al,'q'
        jne dun
        mov ax,0500h            ;back to display page 0
        int 10h
        mov ax,4c00h            ;return to DOS
        int 21h
msg1    db 'hello$'             ;assemble message here
code ends
```

2

Notes on Example Program 4

1. If assembled as a .COM program, then DS and CS are the same segment. If DS and CS are not the same number, then load DS with the contents of CS.

2. Do not forget to end all strings with a $, or the program will display the contents of memory. Using a $ to terminate a message suffers from the same problem encountered by our BIOS examples that terminate in a ~. Messages with $ in them can not be displayed with this INT.

Example Program 5: Drawing

Example Program 5, drawit.asm uses INT 21h, function 09h to draw two simple rectangles on the screen. The larger rectangle is drawn using several strings for each part of the rectangle. The smaller rectangle is drawn using a very long, single string. The smaller rectangle embeds carriage return and line feed characters into the string to position the cursor to the next line of the figure.

Note that the assembler uses the single quote mark (') to convert keyboard characters to ASCII. No quotes are needed for the characters of line feed and carriage return as the actual ASCII bytes for LF (0Ah) and CR (0Dh) are defined by the program. The single line approach is more tedious to write and debug than the multiple line approach. Multiple lines let you "draw" the box using a text editor.

```
;"drawit.asm" uses DOS function 09h to draw with character strings

        cr equ 0dh              ;equate label cr to ASCII 0Dh
        lf equ 0ah              ;equate label lf to ASCII 0Ah
code segment
        mov sp,0fffeh           ;set the SP
        mov ax,cs               ;make sure DS point to CS
        mov ds,ax               ;no problem for .COM files
        mov dx,0313h            ;draw large rectangle at row 3
        call curses
        mov dx,line1            ;point to first line of outer
        call draw               ;draw second line of rectangle
        mov dx,0413h            ;move cursor to next row
        call curses
```

```
                    mov dx,line2          ;point to next line
                    call draw
                    mov dx,0513h          ;draw third line
                    call curses
                    mov dx,line2          ;repeat line 2 four times
                    call draw
                    mov dx,0613h          ;draw fourth line
                    call curses
                    mov dx,line2          ;repeat line 2 four times
                    call draw
                    mov dx,0713h          ;draw fourth line
                    call curses
                    mov dx,line2          ;repeat line 2 four times
                    call draw
                    mov dx,0813h
                    call curses
                    mov dx,line1          ;last line at bottom
                    call draw
                    mov dx,0a16h          ;now draw the next box
                    call curses
                    mov dx,line3          ;draw as continuous string
                    call draw
                    mov ax,4c00h          ;return to DOS
                    int 21h
draw:
                    mov ah,09h
                    int 21h               ;draw first line
                    ret
curses:
                    push ax               ;save all registers used
                    push bx
                    mov ah,02h
                    mov bh,00h
                    int 10h               ;BIOS INT places cursor
                    pop bx                ;restore all old registers
                    pop ax
                    ret
line1:   db 'XXXXXXXXXXXXXXXXXXXXXXXXXXXXXXXXXXXXXXXX$' ;40X
line2:   db 'X                                      X$'
;single string box follows next. Note you must "db" each line.
line3:   db 'XXXXXXXXXXXXXXXXXXXXXXXXXXXXXXXXXXXXXX',cr,lf
         db '                                    X'
         db '                                          X',cr,lf
         db '                              ,
         db 'XXXXXXXXXXXXXXXXXXXXXXXXXXXXXXXXXXXXXX$'
```

Example Program 6: Fill the Screen

Example Program 6 fillitup.asm, fills the screen with any key character pressed. Function 0Bh is used to poll the keyboard, function 07h gets the key, and function 02h displays the key on the screen.

```
;press any printable key and "fillitup.asm" will display an entire
;screen of that character
```

```
code segment
        mov sp, 5000h         ;set the stack
pollit:
        mov ah,0bh            ;check keyboard status
        int 21h              ;INT to DOS
        cmp al,0ffh          ;key = FFh?
        jne pollit           ;loop until key ready
        mov ah,07h           ;get the key
        int 21h              ;DOS
        mov bl,al            ;save character
        mov cx,07d0h         ;write 2,000 characters
nxt:
        mov ah,02h           ;display character
        mov dl,bl            ;retrieve character
        int 21h              ;DOS
        dec cx               ;go 2,000 times
        jnz nxt              ;loop until done
        cmp bl,'q'           ;time to quit?
        jne pollit           ;if not, get next screen
        mov ax,4c00h         ;if so, exit nicely
        int 21h              ;DOS
code ends
```

Example Program 6 will slowly fill screen number 0 with any printable character entered from the keyboard. The routine is so slow that you may type ahead a few characters. The program will empty the keyboard buffer, as the screen fills, and display all of the keys selected (until the keyboard buffer fills up).

For those who keep track of such things, notice how slowly DOS writes to the screen compared with direct character writing to the display memory area in RAM in a previous BIOS example.

Example Program 7: Clock Display

This example, daytime, will display a series of digits on the screen that show the time as hours, minutes, and seconds, with each separated by a colon. The program draws characters on the screen about 1 inch high and .5 inches wide.

```
;a time-of-day program that uses DOS INT 21h, function 09h to display
;user-defined characters on the screen. Also featured are BIOS
;interrupts for cursor location (INT 10h, function 03h) and clock
;ticks (INT 1Ah, function 00h).
;                                                          See Note
;
data segment                    ;set aside variables in data memory
        org 8000h
        hr10 dw ?               ;pointer to current hours tens digit
        hr1 dw ?                ;pointer to current hours units digit
        cln1 dw ?               ;pointer to colon between hours and minutes
        min10 dw ?             ;pointer to current minutes tens digit
        min1 dw ?              ;pointer to current minutes units digit
        cln2 dw ?             ;pointer to colon between minutes and seconds
        sec10 dw ?            ;pointer to current seconds tens digit
        sec1 dw ?             ;pointer to current seconds units digit
```

```
            ante dw ?              ;pointer to A or P of AM / PM
            meridian dw ?          ;pointer to M of AM or PM
            last db ?              ;set to FFh to signal end of clock display
    data ends
    code segment
    ;EQUATES
            tenhours equ 0c06h     ;begin display at row 12,column 6
            pagenum equ 06h        ;use page 6 for this example
    ;END EQUATES
            mov sp,0fffeh          ;set the SP
            mov ax,cs              ;make sure DS points to CS for table lookup
            mov ds,ax              ;no problem for .COM files
            mov ax,0002h           ;set text mode and clear all screens
            int 10h
            mov ah,05h             ;use page defined by "pagenum"
            mov al,pagenum
            int 10h
            mov hr10,zero          ;move the "zero" character pointer into digits
            mov hr1,zero           ;so that 00:00:00 is displayed
            mov cln1,colon         ;move the "colon" character pointer as needed
            mov min10,zero
            mov min1,zero
            mov cln2,colon
            mov sec10,zero
            mov sec1,zero
            mov ante,early         ;point to A of AM
            mov meridian,mtime     ;point to M of AM or PM
            mov last,0ffh          ;"last" is a flag to the display program to
                                   ;stop digit update
    tick:
            call _showtime         ;"showtime" gets memory pointers & displays
            call _time             ;"time" delays about 1 second (1% error)
            add sec1,25xd          ;point to next seconds unit number (0–9)      1
            cmp sec1,colon         ;if next pointer is to colon, reset to "zero"
            jb tick                ;else point to next seconds unit number
            mov sec1,zero          ;start again pointing to "zero" character
            add sec10,25xd         ;point to next seconds tens number (0–5)       1
            cmp sec10,six          ;reset when seconds tens = 6
            jb tick                ;if not 6 then point to next number
            mov sec10,zero         ;else reset to point to "zero" character
            add min1,25xd          ;point to next units minute number (0–9)       1
            cmp min1,colon         ;reset when pointing to colon character
            jb tick                ;else point to next number
            mov min1,zero          ;reset units minute to point to "zero"
            add min10,25xd         ;point to next tens minutes number (0–5)       1
            cmp min10,six          ;reset to point to "zero" if pointing to "six"
            jb tick                ;else increment to point to next number
            mov min10,zero         ;reset to point to "zero"
            add hr1,25xd           ;point to next units hour (0–9)                1
            cmp hr10,one           ;if the tens hour is one, then roll 12 to
            je rollover            ;1 o'clock
            cmp hr1,colon          ;reset units hour when next character is colon
            jb tick                ;else point to next number
            mov hr1,zero           ;point to "zero" for units hour
            add hr10,25xd          ;point to tens hour character                  1
```

```
rollover:
        cmp hr1,three          ;check to see if it is 13 o'clock
        jb tick                ;if not , go on
        mov hr1,one            ;else, roll over 13 to 01, point to "one" in hr1
        mov hr10,zero          ;point to "zero" for tens hour
        cmp ante,early         ;change A to P, or P to A as appropriate
        jne change             ;change contents of "ante" to the other choice
        mov ante,late          ;change A to P
        jmp tick               ;keeps on tickin'
change:
        mov ante,early         ;change P to A
        jmp tick               ;keep timing until "q" is pressed
dun:
        mov ax,0500h           ;return to page 0
        int 10h
        mov ax,4c00h           ;return to DOS
        int 21h
;END PROGRAM
;BEGIN PROCEDURES
_curses:
        mov ah,02h             ;passed cursor position in DX
        mov bh,pagenum         ;position cursor on current page
        int 10h
        ret                    ;returns nothing
_draw:                         ;this procedure will draw a character defined as
                               ;five bytes on five consecutive rows. Byte 5 = $
        mov cx,05h             ;The character is drawn any place the cursor
                               ;happens to be when called

nextline:
        mov ah,09h             ;The calling program passes the address of the
        int 21h                ;strings that define the character in DX.
        push dx                ;save DX character pointer
        push cx                ;save count
        mov ah,03h             ;get current cursor position from INT 10h (F.03)
        mov bh,pagenum         ;returns position in DX, cursor shape in CX
        int 10h

        pop cx                 ;retrieve count
        sub dl,04h             ;back cursor up by four characters (byte 5 = $)
        inc dh                 ;cursor to next row
        call _curses           ;position cursor for next line
        pop dx                 ;get pointer to character strings and update
        add dx,0005h           ;point to next line of character (each line = 5 bytes)
        dec cx                 ;stop after 5 passes
        jnz nextline
        ret
_showtime:                     ;"_showtime" displays whatever characters are
        mov si,0000h           ;pointed to by the contents of the clock digit memory
                               ;locations "hr10," "hr1', etc.
        mov dx,tenhours        ;DX is set to point to the first digit position
nextdigit:                     ;SI is used to increment through data memory
        call _curses           ;position cursor at upper left corner of tens hour
        push dx                ;save cursor position
        mov dx,w[hr10+si]      ;load DX with contents of hr10, hr1, etc.
        call _draw             ;DX is now a pointer to a character to be drawn
```

```
            pop dx                  ;retrieve cursor position
            add dl,07h              ;point to next clock digit to be displayed
            add si,0002h            ;update SI to point to next word in memory
            cmp b[hr10+si],0ffh     ;see if at end of digit pointer memory
            jb nextdigit            ;if not at end, display next digit
            ret
_time:                              ;"time" returns after 18 ticks (.989 seconds)    2
            mov ah,01h              ;plus the time taken to run the procedure
            int 16h                 ;the procedure also checks the "q" key
            jz nokey                ;and returns to DOS if hit by the user
            mov ah,00h              ;get the key and empty the buffer
            int 16h
            cmp al,'q'              ;check for quit command
            je dun
nokey:
            mov ah,00h              ;if no key has been hit, read clock counter
            int 1ah
            mov bx,dx               ;save old time in BX
tock:
            mov ah,00h              ;get new time
            int 1ah
            sub dx,bx
            cmp dx,18xd             ;wait until 18 clock ticks have gone by
            jb tock
            ret                     ;return to calling program
;END PROCEDURES
;BEGIN DEFINED MESSAGES
zero:
            db ' 00  $'             ;each character will draw ANYWHERE on the screen
            db '0   0$'             ;because the "draw" procedure will always draw
            db '0   0$'             ;the character as a series of 5 bytes on 5 rows.
            db '0   0$'             ;Note each line of the character ends in a $ for
            db ' 00  $'             ;the 5th byte. Four columns and 5
                                    ;rows/character

one:
            db ' 11  $'             ;The "draw" procedure will draw the first row, back
            db '111  $'             ;the cursor up by 4 bytes, and position the
            db ' 11  $'             ;cursor at the next row. The next row is drawn and
            db ' 11  $'             ;the procedure repeated until all 5 rows are
            db '1111$'              ;drawn.

two:
            db ' 222$'
            db '2   2$'
            db '   2 $'
            db ' 2   $'
            db '2222$'

three:
            db '3333$'
            db '    3$'
            db '3333$'
            db '    3$'
            db '3333$'
```

```
four:
        db '4   4$'
        db '4   4$'
        db '4444$'
        db '   4$'
        db '   4$'

five:
        db '5555$'
        db '5   $'
        db '5555$'
        db '   5$'
        db '5555$'

six:
        db '6   $'
        db '6   $'
        db '6666$'
        db '6  6$'
        db '6666$'

seven:
        db '7777$'
        db '   7$'
        db '  7 $'
        db ' 7  $'
        db '7   $'

eight:
        db '8888$'
        db '8  8$'
        db '8888$'
        db '8  8$'
        db '8888$'

nine:
        db '9999$'
        db '9  9$'
        db '9999$'
        db '   9$'
        db '   9$'

colon:
        db ' :: $'
        db ' :: $'
        db '    $'
        db ' :: $'
        db ' :: $'

early:
        db ' AA $'
        db 'A  A$'
        db 'AAAA$'
        db 'A  A$'
        db 'A  A$'
```

```
late:
        db 'PPPP$'
        db 'P   P$'
        db 'PPPP$'
        db 'P    $'
        db 'P    $'

mtime:
        db 'M   M$'
        db 'MMMM$'
        db 'MMMM$'
        db 'M   M$'
        db 'M   M$'
code ends
```

NOTES ON EXAMPLE PROGRAM 7

1. To speed the program up, add 250xd to all of the units digits and 150xd to the tens digits.

2. To really speed the program up, simply have _time return immediately.

7.12 Writing Custom Interrupts

We are not *restricted* to using only the software INT interrupts supplied by BIOS and DOS. We are perfectly free, within certain restrictions, to write our own INT routines. Customized hardware interrupts prove useful when unique hardware peripherals are installed in the computer. Chapter 11 discusses custom interrupts as applied to expansion cards plugged into the computer expansion bus. Custom interrupts become absolutely necessary for non-DOS computers, as is discussed in Chapter 12. To prepare for Chapters 11 and 12, a short example of setting up a custom software INT is discussed now.

Low memory locations 00000 to 003FFh are reserved by the CPU for the interrupt vector table. Inspection of Appendixes C and D reveals that there is no standard BIOS or DOS INT type above 70h, which is used by BIOS for real-time clock interrupts. INT type 70h will be converted to table addresses 01C0h to 01C4h when INT type 70 is multiplied by 4. Thus, the only restriction on our INT subroutines is that we can use any INT type above 70h.

To use our INT, we choose a type number, write the INT subroutine, and place the address of our INT subroutine in the INT vector table at the address of our type number multiplied by 4. A custom INT program example, myint.asm, is shown next. The custom INT is chosen, at random, to be type 80h.

```
;a custom INT, "myint.asm," type 80h
code segment                                                   See Note
        mov sp,1000h         ;set SP above program
        push ds              ;save current DS
        mov ax,0000h         ;set DS to low memory segment
        mov ds,ax            ;DS points to low RAM
        mov w[200h],here     ;IP address of custom routine        1
        mov [202h],cs        ;CS address of custom routine        2
```

```
          pop ds                ;restore original DS
          int 80h               ;execute our custom INT
          mov ax,4c00h          ;return to DOS
          int 21h               ;back to >
here:     ;the custom interrupt handler is located at this address
          mov ah,09h            ;our custom INT
       cs mov dx,offset msg     ;pass address of string           3
          int 21h               ;DOS displays message
          iret                  ;return from our INT              4
msg       db 'hello$'           ;hello with $ terminator
code ends
```

NOTES ON THE PROGRAM

1. 80h times 4 is 200h. IP is stored first, at the vector address. The label here is the IP of our custom INT in whatever CS DOS has loaded program myint.asm.

2. CS is stored as the high-order word of the INT vector, at 202h in low RAM. Whatever CS DOS has loaded the program myint.asm in, is now at location 0202h in low RAM.

3. For this short .COM example, CS and DS are the same, so that the override for DS is not necessary. If the handler were in another code segment, however, the CS override is necessary.

4. IRET must be used to return from interrupts, in order to pop the flags from the stack.

When we write out type 80h INT, we do not know the absolute address of our INT subroutine labeled here. DOS will load the program into RAM at some convenient address.

The initial part of the INT program serves to load our INT vectors (IP and CS) in the vector table in low memory. As we have arbitrarily chosen INT type 80h, we must place the address of our INT in locations 0200h (IP) and 0202h (CS).

Address 0200h is 80h multiplied by 4, and holds the IP offset of the CS:IP vector to our INT routine. The code segment value is stored at vector address 0202h. DS is restored to its original value, and our custom INT, type 80h, is used by the program. DOS function 4Ch, type 21h then returns our program to DOS.

Our custom INT uses DOS INT 21h, type 09h, to display a "hello" message. Displaying "hello" lets us test our concept of interrupts; any INT routine could begin at label address here.

The INT routine returns to the main program by executing an IRET to pop the return address and Flag register from the DOS stack.

7.13 Summary

DOS stands for disk operating system. The DOS program uses permanent ROM routines called the BIOS (basic input output services) and RAM-resident DOS programs loaded from disk. Six major versions of DOS have been released over the years 1981 to 1993.

On system reset, or power-up, BIOS routines load DOS into low RAM and transfer control to the DOS programs. DOS operates the system peripherals and responds to user inputs to the computer. Many DOS and BIOS peripheral service routines are available to the user's assembly code program via the INT instruction mechanism. Interrupt action may be initiated by hardware signals or software INT instructions. Interrupts are indirect calls that use low memory addresses 00000 to 003FFh to hold the address of INT procedures.

Instruction INT Type causes the Flag register and return address to the main program to be stored on the stack. The INT type number is then multiplied by 4, and the address of the INT procedure is found at the address in low memory equal to the type number multiplied by 4.

The CS:IP address located at address type number times 4 is loaded into the CS and IP registers respectively, and the INT routine is executed. An IRET opcode, which is the last mnemonic in the INT procedure, pops the main program return address and Flag register. Execution of the main program resumes at the instruction that follows the INT Type instruction in the main program.

Personal computers are equipped with INT routines, collectively called the BIOS. BIOS interrupts enable the programmer to use system I/O devices such as keyboards and CRT terminals and to use system resources with no detailed knowledge of exact circuit operation.

Interrupt type service routines are subdivided into various functions, which are identified by a function number placed in one of the CPU registers before the INT is executed. Refer to Appendix C for a discussion of BIOS INT types.

Common BIOS interrupts include the following:

INT 10h, function 00:	Set Video Mode
INT 10h, function 01h:	Define Cursor
INT 10h, function 02h:	Set Cursor Position
INT 10h, function 03h:	Get Cursor Position
INT 10h, function 05h:	Select Display Page
INT 10h, function 09h:	Write Character and Attribute
INT 16h, function 00:	Read Keyboard Character
INT 16h, function 01h:	Read Keyboard Status
INT 1Ah, function 00:	Read Clock Tick Counters

DOS service INT routines may also be performed. DOS service routines tend to operate at a higher level and are primarily concerned with disk operations. DOS INT 21h is the main DOS INT type and is made up of over 100 functions. Common DOS I/O INT type 21h functions include the following:

Function 01h:	Echo Keyboard Character
Function 02h:	Display Output
Function 07h:	Get Keyboard Character
Function 09h:	Display Character String
Function 0Bh:	Get Keyboard Status
Function 4Ch:	Return Control to DOS

The programmer is free to write custom INT handlers using any type number greater than 70h. Before the custom INT may be used, the programmer must load the address vector of the custom INT address in the low memory vector table.

7.14 Problems

Write a program that solves the problem posed. Comment each line of code, where appropriate, and attempt to use as few lines of code as possible. Use direct memory access to the screen RAM at your own risk.

Exercise Problems

1. Display all of the printable ASCII codes, from 20h to 7Fh, on the screen starting at line 1,column 1.
2. Display all of the unprintable ASCII codes, from 00 to 1Fh, on the screen starting at line 1, column 1.
3. Display all of the codes from 80h to FFh on the screen starting at line 1,column 1.
4. Display the character *A* using all possible attributes. Use the attributes (monochrome or color) suitable for your type of computer monitor.
5. Write a program that displays the message "key" every time any key is pressed.
6. Write a program that does nothing but wait until the q key is pressed on the keyboard. When the q key is pressed, the program is to return to DOS. Only the q key is acted on by the program; all others are ignored.
7. Have the program display any printable ASCII key pressed.
8. Have the program display a screen full of the first printable ASCII key pressed.
9. Have the program display a cursor and back it up each time the Backspace key is pressed.
10. Have the program display a cursor and move it to the right each time the > key is pressed.
11. Make the colon character (:) blink on and off on the screen at a rate of 1 second on and 1 second off.

Intermediate Problems

1. Have your program determine if two numeric keys (0–9) are pressed in sequence. A sequence could be 1–2, or 8–9, or 4–5. The program displays the message "sequence" any time two sequential keys are pressed, and the message "not sequential" whenever two keys that are not sequential are pressed.
2. Write a program that will display the hex key code for any key, or key combination, that is pressed on the keyboard. Display the key code in the center of the screen. For instance, if the key number 0 is pressed, display

a 30 on the center of the screen. If the function key F1 is pressed, then a 00 appears on the screen.

3. If your computer is equipped with a CGA adapter, or above, write the message "this is page 1" on page 1, the message "this is page 2" on page 2, and so on until the message "this is page 7" is written to page 7. Center the messages on the pages, in exactly the same place. Have the program then alternate the messages and determine if you can detect the fact that each message is rewritten. Have the program terminate and return to DOS when the q key is pressed.

4. Draw on the screen a series of four concentric rectangles made of the character o. The outer rectangle is to extend from row 0,column 0 to row 24, column 79. The remaining rectangles must be separated from each other by at least 1 blank space in all directions. Have the program terminate and return to DOS when the q key is pressed.

5. Draw on the screen a "happy face" made up of the character O. The face must be close to a circle and still be as large as possible. Remember: A happy face has no nose. Have the program terminate and return to DOS when the q key is pressed.

6. Draw on the screen a hatchmark pattern made up of the character x. Hatchmarks are made by drawing diagonal lines on the screen from upper right down to lower left, and from lower left to upper right. Separate each diagonal line by at least 4 blank lines. Have the program terminate and return to DOS when the q key is pressed.

7. Draw on the screen a checkerboard pattern that fills the screen. For monochrome monitors, use the character w for the white squares and blanks for the black squares. For color monitors, make the squares red and black.

A checkerboard consists of 8 rows and 8 columns of alternating squares of black and white (red). The lower left square is white (red). Have the program terminate and return to DOS when the q key is pressed.

8. Draw on the screen three concentric squares made up of the character Q. The outer square is to be as large as possible, and the remaining squares separated from each other by 1 blank space in all directions. Center the squares in the middle of the screen. Remember that these are squares, not rectangles. Have the program terminate and return to DOS when the q key is pressed.

9. Draw on the screen a set of three equilateral triangles. Make the outer triangle as large as possible, and separate the remaining triangles by 2 spaces in all directions. Have the program terminate and return to DOS when the q key is pressed.

10. Draw on the screen a giant "WCU" sign in block letters at least 4 inches high and 2 inches wide. Make the letters out of any character you wish. Make the "WCU" blink, one character at a time. Have the program terminate and return to DOS when the q key is pressed.

11. Make an approximately 3-inch-square rectangle appear to move across the screen, from left to right across the screen and top to bottom down the screen.

Start the rectangle at the upper lefthand corner of the screen, and have it move across the screen to the right. On reaching the right side, the rectangle is to move immediately back to the left side and down on the screen by 1 rectangle height. The rectangle continues to move until it is at the lower right corner of the screen, at which time it repeats its path from the upper lefthand corner.

Have the program terminate and return to DOS when the q key is pressed.

12. Create a vertical rainbow on the screen consisting of rows of solid rectangles colored, from bottom to top, blue, green, white, brown, and red.

Make the colors rotate, that is, the blue rectangle slowly moves up the screen, row by row, followed by the other colored rectangles, while the blue rows that disappear on the top reappear at the bottom and follow the red rectangles upward.

Have the program terminate and return to DOS when the q key is pressed.

13. Repeat Problem 12 using a horizontal rainbow and have the colors rotate from right to left instead of from bottom to top.

14. Write a stopwatch program that will display elapsed time in letters approximately 2 inches high and 1 inch wide. Display minutes, seconds, and tenths of seconds. Have the elapsed time reset to 0.00.00 when the r key is pressed, and then begin to time when the t key is first pressed. Stop timing, and freeze the display, when the t key is pressed the second time.

Advanced Problems

1. Draw a Tic-Tac-Toe game on the screen, as shown below, with each square numbered as shown.

```
         |     |
     1   |  2  |  3
    -------------------------
     4   |  5  |  6
    -------------------------
     7   |  8  |  9
         |     |
```

The player can place and X or an O by first pressing the number of the square where the character is to go, and then an X or an O. The program then places the character in the indicated square, replacing the number. The players *must* alternate X and O characters on the screen.

A player can *change* the *placement* of the last X or O by typing the c key *right after* having placed an X or an O on the screen. The square number reappears in place of the last X or O, and the player can try again.

The program is to ignore any illegal numbers or characters and will not overwrite an X or O already placed in a square. Once every square is filled, the program is to display the message "GAME OVER" under the game.

Restart the program *whenever* the s key is pressed.

Have the program terminate and return to DOS when the q key is pressed.

2. Write a password program that operates as follows:

   1. Display the message "enter password."

   2. The user enters the password, which is "agent 007" followed by the Return key. Do not display what is actually typed in as the password. Display asterisks (*) as the user types in the password.

   3. If the password is correct then display "access granted."

   4. If the password is incorrect, then display "sound the alarm."

   An incorrect password is anything that is not *exactly* "agent 007," including the space between "agent" and "007." The program must allow the user to type in a mistake, and backspace over the error(s) and retype the message. A backspace works by backing up the cursor to the left and erasing any character found there.

   Do not allow the user to backspace beyond the left side of the screen or type a message beyond the right side of the screen.

   When the Return key is pressed, the program is to display one of the two messages and then start over again.

   Return to DOS when the q key is pressed.

3. Write a 1-page text editor that will position the cursor and type user text on the screen.

   The user can position the cursor 1 space each time an arrow key (up, down, left, right) is pressed or use the Return key to move the cursor to the far left column and down 1 row. The Home key moves the cursor to the top left side of the screen; the End key moves the cursor 1 space beyond the *last* character typed on the page.

   Have the program display whatever character the user types at the cursor location, and then move the cursor 1 space to the right. If the Backspace key is used, back the cursor up 1 space to the left and erase any character found there.

   Do not allow the user to move beyond any screen border using any key. Display only keys that have an ASCII code.

   Return to DOS when the Esc(ape) key is pressed.

4. Write a simple drawing program that will draw lines on the screen in response to certain keys. All other keys are to be ignored by the program.

   Use the underline character (_) for the right or left arrow key and the pipe character (|) whenever the up or down arrow key is pressed. Draw the characters right or left, or up or down, depending on which arrow key is pressed.

   Move the cursor, but draw no character, if the Backspace key is pressed *before* the arrow key. Draw a foreslash (/) if the Tab key is pressed *before* an arrow key, and draw a backslash (\) if the Esc key is pressed *before* an arrow key. Erase the character if the Space bar key is pressed *before* an arrow key.

Do not allow the cursor to go beyond the boundaries of the screen.

Start the program with the cursor in the center of the screen. Have the program terminate and return to DOS when the q key is pressed.

5. Write a program that will draw a rectangle, made up of asterisks (*), as directed by the user.

   The user may move a visible cursor to any point on the screen, using the arrow keys, and press the l key. The program puts a period (.) at the cursor location. The user then may move the cursor to any other point on the screen and press the r key. The program is to draw a rectangle that has its lower lefthand corner at the place the l key was pressed, and its upper righthand corner where the r key was pressed. The rectangle drawing sequence may be repeated as desired by the user. The program is not to draw the rectangle if the right corner is to the *left* of the left corner.

   The program should ignore all other keys and only draw rectangles that begin with the l key, and end with the r key.

   Return to DOS if the Esc key is pressed.

6. Write a simple "paint" program.

   Move the cursor in response to the user up, down, left, and right arrow keys. Do not place a colored square at the cursor location unless directed to do so.

   Make a colored square (white for monochrome displays) at the current cursor position as directed by the user typing the following keys:

   > r = red; b = blue; g = green; w = white; c = cyan

   If the Backspace key is pressed, then *erase* any colored square found *at* the cursor location.

   If a number key (2–9) is pressed *before* the color key, then make as many colored squares as indicated by the number key, in the *last* direction indicated by the arrow key.

   Do not allow the cursor to go beyond the screen boundaries.

   Have the program terminate and return to DOS when the q key is pressed.

7. Write a program that matches each of your classmates' last names to one of the secret code names listed below.

   Display the message "Last Name?" on the screen and await an entry from the user by positioning the cursor on the screen. Display the name as it is typed in and move the cursor over to the right to indicate where the next entry will go.

   The user can type in a student name and then press the Return key. Allow the user to use the Backspace key to erase name characters, if desired, before pressing the Return key. Using the Backspace key erases the character to the left of the cursor and moves the cursor to the left.

   Print, after the name is typed in, the message "code name xxx" where xxx is the code name you have assigned to the person. If an illegal name is typed in by the user, have the program ignore it and display the message "no code name for that student: please re-enter." Illegal names are any that do not *exactly* match a student name.

Have the program begin again when the r key is pressed.

Do not allow the user to type beyond the limits of the screen when entering a name, or when backspacing.

The *allowable* code names are: swift, osprey, cardinal, wren, jay, junco, heron, swan, shrike, hawk, owl, eagle, phoebe, sparrow, plover, puffin, penguin, gull, pelican, finch, and martin.

Have the program terminate and return to DOS when the Esc key is pressed.

8. Write a simple Ping-Pong game program. Create two "paddles" on each side of the screen as vertical stripes. Place the paddles about 8 inches apart.

   Play is as follows:

   Have the left side begin the game by pressing the l key. The "ball," which is shown as an asterisk (*), is to move from the left-side paddle toward the right-side paddle. Move the ball at about 1 inch per second.

   The right-side player must "hit" the ball by pressing the r key when the ball is *within an inch, or less,* of the right-side paddle.

   If a hit is successful, then the ball moves toward the other side of the screen, where the opposing player must "hit" it when it is within 1 inch of the paddle by pressing the l key.

   Points are scored if a player "misses" the ball by hitting *before* the ball is within 1 inch of the paddle, or the ball *passes* the paddle. The victor serves after the opposite player misses.

   Keep score by displaying missed balls (0–11) below each paddle. Stop the game when one player misses 11 balls, and begin play again when the p key is pressed. Return to DOS by pressing the q key.

9. Create a "pitch-the-penny" game program. The object of the game is to see how close a player can let the asterisk (*) get to a line, without going past the line.

   Have the program create a vertical line at the rightmost column of the screen and two areas labeled "Player 1" and "Player 2."

   Play begins by Player 1 pressing the 1 key, or Player 2 pressing the 2 key. After the player key is pressed, an asterisk moves from the leftmost column of the screen to the right at a rapid rate, say approximately 3 inches per second. The player then stops the moving asterisk by pressing the s key. The program displays the distance (in columns) from the line, in the appropriate player area, as a score. The next player then attempts to beat the other by stopping the asterisk closer to the line than the opponent. A player automatically loses if the asterisk hits the line before being stopped by a player.

   After a round is played, flash the winning player's score. The game may be repeated by pressing the r key, or return to DOS by pressing the q key.

10. Write a program that matches your classmates' first names and middle initials to their last names.

    The program begins by placing the line "Please enter first name and middle initial" on the screen. A visible cursor then appears on the screen to indicate where the name is to go. Letters are displayed on the screen as they are typed in, and the cursor moves to the right 1 space.

Allow the user to enter a first name (up to 10 letters), a space, and a middle initial, followed by the Return key. If the Backspace key is pressed, erase the last character entered and move the cursor 1 space to the left. Do not allow the user to type beyond the area allowed for a name or backspace further to the left than the original starting point.

After the Return key is pressed, display the person's last name 1 space beyond the middle initial typed on the screen. If the name and middle initial are not spelled correctly, then display a question mark (?).

The user can repeat the program by pressing an r key, or exit to DOS by pressing the q key.

11. Write a program that places an analog alarm watch face on the screen. The face is to consist of a second hand, an alarm hand, and numerals from 00 to 59 placed around the center. The numerals do not have to be in a circle, but may be arranged around the screen as desired.

Set the alarm hand to point at the seconds entered by the user. The user enters an alarm time by pressing the a key followed by a two-digit number, from 01 to 60. Set the second hand to 00 when the r key is pressed. Have the second hand begin to sweep around the clock face when the t key is pressed. The second hand must move after every second passes. Have the second hand stop when the s key is pressed. If the second hand reaches the alarm hand, stop the second hand and display the message "TIME IS UP."

Make the hands using any character you wish.

Restart the program if the r key is pressed, or return to DOS if the q key is pressed.

8

CONVENIENT AND
SPECIALIZED
INSTRUCTIONS

Chapter Objectives

On successful completion of this chapter you will be able to:

- Multiply and divide binary numbers.
- Adjust ASCII, unpacked BCD, and packed BCD sums and differences.
- Adjust results of multiplication and division to unpacked BCD.
- Convert two's complement bytes to words, and words to double words.
- Let the CPU compute effective addresses during run time.
- Repeat operations.
- Move data strings.
- Search data strings.
- Use lookup tables in a program.
- Construct modular programs.
- Use structured programming techniques.
- Input and output data with ports.

8.0 Introduction

Chapters 5, 6, and 7 discuss a small set of instructions that can be used for writing most 8086 assembly language programs. As a programmer matures in the programming art, however, the need is felt for additional instructions.

The 8086 is equipped by its designers with more than 80 instructions beyond those discussed in preceding chapters. Most of the additional instructions aid in the timely creation of programs that take less time to write and execute. Some of the added instructions are highly specialized, such as those used for multiprocessor systems. Other instructions are very convenient and applicable to most problems. In this chapter, we shall investigate additional instructions that the programmer should find useful for most programming challenges.

8.1 Convenient Arithmetic Instructions

Mathematics, when done using assembly code, becomes a very tedious business indeed. Fortunately, for today's assembly language programmer, high-level languages can be used for any serious "number crunching" in a program. Assembly code mathematics can be reserved for those instances when speed or unique device-dependent interface code is needed.

If needed, however, assembly-code algorithms for performing mathematical operations are well developed and include fixed and floating point math routines. Should the system designer need faster mathematical execution, math coprocessors can be added to augment CPU capabilities. Also, many libraries of math subroutines are available for most popular microprocessors. But even with the availability of high-level languages, coprocessors, and libraries of math subroutines, there may be instances when short math routines must be written in assembly code.

The 8086 family is equipped with a full complement of binary arithmetic instructions that can add, subtract, multiply, and divide signed and unsigned binary numbers.

Multiplication and Division

Multiplication and division are commonly done in microprocessors more primitive than the 8086 by software subroutines. One feature of modern CPUs, such as the 8086, is the ability to perform multiplication and division in hardware. Hardware execution greatly speeds up program running time. The 8086 has instructions for signed and unsigned binary number multiplication and division. In keeping with our 80/20 rule, only the unsigned multiply and divide operations are presented next.

DIV   Divide Destination by Source

Mnemonic	Operands	Size	Operation	Flags
DIV	SOURCE	B	(AX)/(SRC)	All Undefined
DIV	SOURCE	W	(DX:AX)/(SRC)	All Undefined

Addressing Modes
(Destination, Source)

register, register
register, memory

Examples:

div bx	;DIV DX:AX by the word contents of register BX
div cl	;DIV AX by the byte contents of register CL
div w[1234h]	;DIV DX:AX by the word stored at address 1234h
div b[di]	;DIV AX by the byte at address [DI]

Operation:

DIV results in dividing the unsigned contents of the AX register, or the DX:AX register pair, by the unsigned contents of the source operand. If the source data is *byte*-sized (dividend is in AX), then the quotient is placed in AL and the remainder is placed in AH. If the source data is *word*-sized (dividend is in DX:AX), then the quotient is placed in AX and the remainder placed in DX.

CAUTION
■ The quotient *must* fit within a single byte, or single word. If the quotient is larger than FFh for byte dividends, or FFFFh for word dividends, then an INT type 0 will occur.
■ No segment register may be used as a source operand.
■ The state of the flags listed are undefined.

MUL   Multiply Destination by Source

Mnemonic	Operands	Size	Operation	Flags
MUL	SOURCE	B	DX:AX ← (AL) × (SRC)	CF, OF
MUL	SOURCE	W	DX:AX ← (AX) × (SRC)	CF, OF
				AF, PF, SF, ZF are Undefined

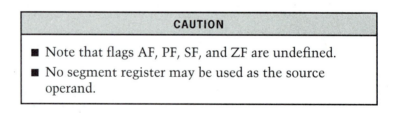

Addressing Modes
(Destination, Source)

register, register
register, memory

Examples:

mul bx	;MUL AX by the contents of register BX
mul cl	;MUL AL by the contents of register CL
mul b[1234h]	;MUL AL by the byte stored at address 1234h
mul w[di]	;MUL AX by the word stored at address [DI]

Operation:

MUL multiplies the unsigned contents of the source operand by the unsigned contents of AX or AL. If the source data is *byte*-sized, then the source data is multiplied by the byte of data in AL. The result is placed in AH (most significant byte) and AL (least significant byte). If the source data is *word*-sized, then the source data is multiplied by the word in AX. The result is returned in DX (most significant word) and AX (least significant word). If the most significant part of the result is greater than zero, then CF and OF are both set to 1. If the most significant part of the result is zero, then both CF and OF are cleared to 0. Setting CF and OF serves to signal the program that the upper half of the result is nonzero and is available for use by the program.

CAUTION
■ Note that flags AF, PF, SF, and ZF are undefined.
■ No segment register may be used as the source operand.

ASCII and BCD Numbers

Binary arithmetic operations are the foundation of digital mathematics. There are, however, instances when the programmer may wish to use binary coded numbers in special circumstances. The most common binary coded numbers in general use are ASCII-coded decimal numbers and binary coded decimal numbers (BCD numbers). The 8086 family is equipped with specialized instructions that enable the programmer to perform arithmetic with ASCII and BCD numbers.

ASCII-coded numbers are binary codes for numeric characters. ASCII 30h to 39h are the codes for the characters 0 to 9.

BCD numbers are decimal-equivalent binary numbers that vary from 0000b to 1001b, or 0 to 9 decimal. The binary numbers from 1010b to 1111b are not used in BCD because there are no single-digit decimal equivalents to Ah (10d) through Fh (15d). BCD is inefficient, from a purely storage standpoint, because the largest BCD number that can be stored in a byte is 99h, whereas pure binary allows storage of a number as large as 255 (FFh). Wasting six numbers above 9 can be readily excused, though, if a program interfaces with human operators.

Humans have steadfastly refused to use hexadecimal numbers in everyday life, despite the obvious advantages of hex over decimal. Programs that interface with humans, therefore, commonly receive numbers from decimal keypads and display data for human viewing in decimal. If the input data to a program is in decimal, and the output is in decimal, it often makes good sense to conduct internal arithmetic operations in decimal also.

BCD numbers can exist in two forms, *packed* (pbcd) and *unpacked* (ubcd). Packed BCD numbers each occupy one nibble, so that a packed BCD byte is made up of two nibbles, each of which ranges from 0 to 9h (00 to 99 BCD). Unpacked BCD numbers take up an entire byte, with the leading nibble set to 0. Unpacked BCD byte numbers range from 00 to 09h (0 to 9 BCD.)

ASCII numbers are binary codes for decimal numbers that range from 30h to 39h, or 0 to 9 decimal. ASCII can be converted to unpacked BCD by the simple expedient of making the leading 3 a 0. For instance, ASCII 35h becomes 05h, or unpacked BCD number 5. Unpacked BCD numbers can be converted to ASCII simply by making the leading nibble a 3 instead of a 0. Clearly, ASCII numbers and unpacked BCD numbers are closely related to each other, as each is separated from the other by a binary 3.

ASCII-based arithmetic is advantageous, again, when dealing with humans. All PC keyboards and displays use ASCII to generate and display characters, including numbers. Programs that receive numeric data from a keyboard, and display data for human interpretation, can use ASCII-based arithmetic to eliminate the need for converting from ASCII to binary for input data, and binary to ASCII for output data.

There remains, nevertheless, one problem. *All* arithmetic is done by the CPU in pure binary. The CPU is a digital machine and takes no notice of how any binary numbers may be coded. Thus, BCD numbers added to BCD numbers become hexadecimal sums. For instance, ASCII 30h (0d) added to ASCII 31h (1d) becomes 61h (a), not ASCII 31h (1d).

Rather than have unique BCD and ASCII arithmetic CPU logics, the 8086 family makes use of conversion instructions that restore BCD and ASCII mathematical operations to their correct format. Conversion instructions are said to *adjust* the pure binary result of a BCD or ASCII operation back to its correct format. All of the BCD and ASCII instructions use the word *adjust* as part of the mnemonic.

There is, importantly, one restriction on the use of adjusting instructions that the programmer must remember. To get meaningful results from an adjusting opcode, the numbers to be adjusted *must* have been generated from BCD (or ASCII) numbers originally. For instance, adding hex numbers to hex numbers and then adjusting the sum will not yield an equivalent BCD result. BCD numbers must be added to BCD numbers, and the resulting hexadecimal sum then adjusted *back* to BCD. All of the example programs used to illustrate adjusting instructions will include one ERRONEOUS example to demonstrate the old adage of "garbage in, garbage out," or, "hex in, errors out."

ASCII and Unpacked BCD Instructions

There are four ASCII-based instructions—AAA, AAS, AAD, and AAM, which stand for ASCII Adjust for Addition, Subtraction, Division, and Multiplication. The following notation is used to indicate the type of binary number involved under the "Operation" headings.

hex = Hexadecimal number

pbcd = Packed (00 to 99h) BCD number

ubcd = Unpacked (00 to 09h) BCD number

AAA   ASCII Adjust for Addition

Mnemonic	Size	Operation	Flags
AAA	W	(AX)ubcd ← (AL)hex	If (AL) AND 0Fh > 9 or AC = 1 then: CF = 1 AC = 1 OF, PF, SF, ZF are Undefined

Addressing Modes
(Destination, Source)

register, register

Operation:

AAA takes the binary sum of ASCII, or BCD, or a mix of ASCII and BCD, numbers in register AL and converts the sum to equivalent unpacked BCD numbers in AX. AH must be zero before the numbers are added, for proper operation. The AC and C flags are set if the adjusted BCD number is greater than 9d, or AC was 1 before the AAA instruction. The OF, PF, SF, and ZF flags are undefined.

AN AAA EXAMPLE

An example of the proper, and improper, use of AAA follows. Several ASCII, BCD, and a mix of ASCII and BCD numbers are added to demonstrate the workings of AAA. The last set of numbers that is added results in ERRONEOUS results, because they were not ASCII or BCD *before* addition.

```
;AAA example program
code segment
        mov ah,00h        ;AH must be zero
        mov al,35h        ;AL contains an ASCII 5
        add al,39h        ;add ASCII 9, the sum will be 6Eh
        aaa               ;adjust the result to AX=0104h
                          ;(unpacked 14 BCD)
        mov ah,00h        ;AH must be zero
```

```
        mov al,05h          ;AL is unpacked BCD 5
        add al,09h          ;add an unpacked BCD 9 for a sum of 0Eh
        aaa                 ;adjust answer to AX = 0104h
        mov ah,00h          ;AH must be zero
        mov al,35h          ;AL is set to ASCII 5
        add al,09h          ;add a BCD 9 for a sum of 3Eh
        aaa                 ;adjust to AX = 0104h
        mov al,3Ah          ;AL contains a non-ASCII number
        add al,27h          ;add a non-ASCII number to it
        aaa                 ;the result is >>>>>ERRONEOUS<<<<<
code ends
```

Note that AAA is as equally adept at adjusting unpacked BCD sums as it is for ASCII sums. The result of AAA will be an unpacked BCD number in AH (00 or 01) and AL. ORing the result in AX with 3030h will result in an ASCII number in AX. AAA keys on the AC flag and the C flag in order to make the adjustment correctly. Make sure that no instructions that might alter the AC flag, or the Carry flag, intervene between the addition operation and AAA.

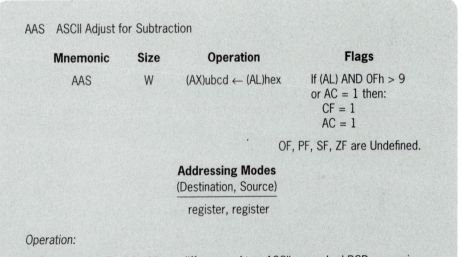

AAS   ASCII Adjust for Subtraction

Mnemonic	Size	Operation	Flags
AAS	W	(AX)ubcd ← (AL)hex	If (AL) AND 0Fh > 9 or AC = 1 then: CF = 1 AC = 1 OF, PF, SF, ZF are Undefined.

Addressing Modes
(Destination, Source)

register, register

Operation:

AAS serves to adjust the binary difference of two ASCII, unpacked BCD, or a mix of ASCII and unpacked BCD numbers in AL to an equivalent unpacked BCD number in AX. AH must be zero before using AAS, and the difference in AL must have been formed from a mix of two ASCII or unpacked BCD numbers. The states of the OF, PF, SF, and ZF flags are undefined.

An AAS Example

An example of the proper, and improper, use of AAS follows. Several ASCII, unpacked BCD, and a mix of ASCII and unpacked BCD numbers are subtracted to demonstrate the workings of AAS. The last set of numbers that is subtracted results in ERRONEOUS results, because they were not ASCII or unpacked BCD *numbers before* subtraction.

```
;AAS example
code segment
        mov ah,00h              ;AH must be zero
        mov al,35h              ;AL contains an ASCII 5
        sub al,39h              ;subtract ASCII 9, result is FCh&borrow
        aas                     ;adjust the result to AX=FF06h
                                ;(15 BCD – 9 BCD = –6 BCD)
        mov ah,00h              ;AH must be zero
        mov al,05h              ;AL is unpacked BCD 5
        sub al,09h              ;subtract BCD 09, result is FCH&borrow
        aas                     ;adjust answer to AX = FF06h
        mov ah,00h              ;AH must be zero
        mov al,35h              ;AL is set to ASCII 5
        sub al,09h              ;subtract a BCD 9 for a 2Ch&borrow
        aas                     ;adjust to AX = FF06h
        mov al,3Ah              ;AL contains a non-ASCII number
        sub al,27h              ;subtract a non-ASCII number from it
        aas                     ;the result is >>>>>ERRONEOUS<<<<<
code ends
```

AAS converts any borrow during the previous subtraction operation into an FFh (–1 in two's complement math) in register AH and an unpacked BCD difference in register AL. AAS treats a borrow *as if* the upper nibble of AL is a BCD 10; thus BCD 05 – BCD 09 *acts* as if one had BCD 15 – BCD 09 = BCD 6. AAS keys on the AC flag and the C flag in order to make the adjustment of the upper nibble to FFh (–) correctly. Make sure that no instructions that might alter the AC flag, or the Carry flag, intervene between the subtraction operations and AAS. AAS works equally well with a difference formed by ASCII, unpacked BCD, or a combination of ASCII and unpacked BCD numbers.

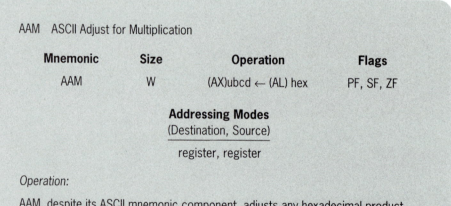

AAM   ASCII Adjust for Multiplication

Mnemonic	Size	Operation	Flags
AAM	W	(AX)ubcd ← (AL) hex	PF, SF, ZF

Addressing Modes
(Destination, Source)

register, register

Operation:

AAM, despite its ASCII mnemonic component, adjusts any hexadecimal product found in AL that ranges from 00 to 63h into two unpacked BCD numbers in AH and AL. How the product is formed is of no interest whatsoever, and, in fact, multiplication need not have formed the number in AL. The unpacked BCD number formed by AAM may be ORed with 3030h to convert it to ASCII in AX. The Sign flag is set to bit 7 of AX (the most significant bit of AL), and the Zero flag is set on the state of AL. The Parity flag is set to reflect the parity of AX. Hex numbers in AL that exceed 99d (63h) will result in an ERRONEOUS result.

AAM EXAMPLES

The first AAM example that follows takes whatever number is placed in AL and converts it into an unpacked BCD number in AX. The last number converted by AAM in the example exceeds 99d (63h) and results in an ERROR after AAM is executed.

```
;AAM example
code segment
        mov al,63h          ;set AL to 99d
        aam                 ;convert AL to AX = 0909h (99 BCD)
        mov al,4dh          ;set AL to 61d
        aam                 ;convert AL to AX = 0601h (61 BCD)
        mov al,71h          ;set AL to 113d
        aam                 ;result is AX = 0B03h >>ERROR<<
code ends
```

AAM operates by dividing the number in AL by 10d and placing the quotient in AH and the remainder in AL. The erroneous result of placing 71h in AH, as shown in the last example, and using AAM results in non-BCD numbers in AX. The contents of AH do not affect the AAM operation and are overwritten by AAM.

The next AAM example shows how AAM is used to correct the product of two unpacked BCD numbers from hex to unpacked BCD. To work correctly, the numbers to be multiplied must be unpacked BCD (not ASCII) before multiplication takes place. Using AAM to correct the product of two ASCII numbers results in an error in AH. ASCII numbers should be converted to unpacked BCD *before* multiplication.

```
;AAM multiplication example
code segment
        mov ax,0009h        ;set AL = 09 unpacked BCD
        mov bl,05h          ;set BL = 05 unpacked BCD
        mul bl              ;multiply, result is 002Dh (45d)
        aam                 ;convert hex to unpacked BCD 0405h
        mov al,09h          ;largest numbers possible
        mov bl,09h
        mul bl              ;result = 0051h (81d)
        aam                 ;result = 0801h unpacked BCD
        mov ax,0039h        ;same example using ASCII
        mov bl,35h          ;set BL = 35h
        mul bl              ;result is 0BCDh
        aam                 ;convert to unpacked BCD 1405h << ERROR >>
code ends
```

Instructions AAA, AAS, and AAM are meant to be used *after* an addition, subtraction, or multiplication operation in order to return the previous ASCII/unpacked BCD numbers to ASCII/unpacked BCD results.

The next opcode, AAD, is used *before* division, in order to get the ASCII/ unpacked BCD numerator to the proper form for division by an ASCII/ unpacked BCD denominator.

AAD   ASCII Adjust for Division

Mnemonic	**Size**	**Operation**	**Flags**
AAD	W	(AL)hex ← (AX)ubcd	PF, SF, ZF

Addressing Modes
(Destination, Source)

register, register

Operation:

AAD takes the unpacked BCD number in AX and converts it into a single hexadecimal equivalent number in AL. AH is zeroed after the operation. The number in each byte of AX must be between 00 and 09h for AAD to work properly. The Z flag is set on the state of AL, and the Sign flag is set to the state of bit 7 of AX (the most significant bit of AL). Numbers in AH or AL that exceed 09h will cause AAD to return an erroneous result in AL.

AAD Examples

AAD is the mirror image of AAM. Unpacked BCD numbers in AX are converted into equivalent hex numbers, between 00 and 63h, in AL. The following example shows that the unpacked BCD number to be converted into hex does not have to be the result of a previous division operation. The last number used in the example is illegal, because it is not unpacked BCD, and results in an ERRONEOUS result in AL.

```
;AAD example
code segment
        mov ax,0909h            ;set AX to unpacked 99 BCD
        aad                     ;convert unpacked 99 BCD to 63h
        mov ax,0502h            ;set AX to unpacked 52 BCD
        aad                     ;convert it to 34h
        mov ax,0abcdh           ;NOT unpacked BCD
        aad                     ;results in >>ERRONEOUS<< number in AL
code ends
```

AAD takes the number in AH, multiplies it by 10d, and adds the result to AL. The multiplication is done modulo 8, so that any result larger than FFh is discarded before adding the contents of AL. Thus, in our example, 0909h becomes 90d + 9d = 63h, whereas an error occurs when ABCDh becomes 1710d − 1536d = AEh + CDh = 7Bh. Or, ABh (171d) × 10d = 1710d = 6AEh = AEh + CDh = 7Bh.

The next example shows how AAD might be used to perform BCD division. As was the case for AAM, AAD works only if the number in AX is unpacked BCD, not ASCII. Convert ASCII numbers to unpacked BCD before using AAD.

```
;AAD division example
code segment
        mov ax, 0808h        ;AX = unpacked 88 BCD
        mov bl, 02h          ;BL = unpacked 2 BCD
        aad                  ;convert AX to 0058h
        div bl               ;divide AX by BL, result = 002Ch
        aam                  ;result = unpacked BCD 0404h (88/2 = 44)
code ends
```

DIV produces remainders in AH. AAM will overwrite any remainders in AH, so the program must deal with any remainders if high accuracy is desired.

Packed BCD Arithmetic Instructions

Packed BCD refers to one BCD digit per nibble, or packed BCD bytes that range in size from 00 to 99h. No nibble may exceed binary 9, or the packed BCD instructions will not function correctly. There are two packed BCD instructions, DAA and DAS.

DAA   Decimal Adjust for Addition

Mnemonic	Size	Operation	Flags
DAA	B	(AL)pbcd ← (AL)hex	AC, CF, PF, SF, ZF

Addressing Modes
(Destination, Source)

register, register

Operation:

DAA takes a hexadecimal number in AL, which is the result of the previous addition of two packed BCD numbers, and adjusts the answer to packed BCD. The Carry flag is set if a carry from the BCD sum exists, the Sign flag is set to bit 7 of AX (the most significant bit of AL). The Parity flag is set based on the parity of AL, and the Zero flag is set based on the state of AL. The Overflow flag is undefined.

DAA works only on the packed BCD number in AL. AH is completely unaffected by DAA, and the contents of AH do not change after DAA executes. The hexadecimal number in AL must be the sum of the previous addition of two packed BCD numbers, or the resulting number in AL, after DAA is executed, will be erroneous. DAA operates based on the state of the AC flag, so no instructions that alter the AC flag should intervene between the addition opcode and DAA. DAA should immediately follow any addition opcode that uses packed BCD numbers.

A DAA Example

The next example program shows DAA adjusting the sums of several sets of packed BCD numbers. The last addition in the example will use non-packed BCD numbers to show that the failure to use packed BCD numbers to form the sum will result in an ERROR when DAA is executed.

```
;DAA example
code segment
        mov al,99h          ;set AL to packed BCD 99
        add al,99h          ;add packed BCD 99, sum is 32h and CF set
        daa                 ;adjust the sum to packed BCD 98 and carry
        mov al,42h          ;set AL to packed BCD 42
        add al,68h          ;add packed BCD 68 for a sum of AAh no CF
        daa                 ;adjust to packed BCD 10 and carry
        mov al,0abh         ;add non-BCD numbers together
        add al,0cdh         ;the sum is 78h and CF
        daa                 ;the result is DEh >>ERRONEOUS<<
code ends
```

To adjust the hexadecimal number in AL correctly, the sum in AL must have been formed from two packed BCD numbers. DAA will not convert a random hexadecimal number to packed BCD.

DAS   Decimal Adjust after Subtraction

Mnemonic	Size	Operation	Flags
DAS	B	(AL)pbcd ← (AL)hex	CF, PF, SF, ZF

Addressing Modes
(Destination, Source)

register, register

Operation:

DAS adjusts the hexadecimal number in AL, which is the result of a subtraction operation involving two packed BCD numbers, to an equivalent packed BCD number. All affected flags are set based on the state of the number in AL—the Carry flag is set if a borrow exists as a result of the subtraction; the Parity, Sign, and Zero flags are set based on the results in AL. AH is not affected by DAS and remains the same before and after DAS is executed.

A DAS Example

The example program that follows next demonstrates the proper and improper use of DAS. Several legal packed BCD subtraction and adjustments are demonstrated, followed by a final example using numbers that are not packed BCD. The result of DAS on a difference based on nonpacked BCD numbers is ERRONEOUS.

```
code segment
        mov al,33h              ;set AL to 33 packed BCD
        sub al,99h              ;subtract packed BCD 99,
                                ;for 9Ah&borrow result
        das                     ;adjust the result to 34&borrow
        mov al,76h              ;set AL to 76 packed BCD
        sub al,29h              ;subtract packed BCD 29 and get 4Dh
                                ;with no borrow.
        das                     ;adjust to packed BCD 47, no borrow
        mov al,0dch             ;set AL to a non-BCD number
        sub al,0cch             ;subtract a non-BCD number = 10h (16BCD)
        das                     ;yielding an >>ERRONEOUS<< result (10BCD)
code ends
```

DAS does not affect, and is not affected by, the byte in AH. DAS will not adjust a random hex number in AL to packed BCD.

Two's Complement Byte Adjustments

As discussed in Chapter 2, complementary math requires that negative numbers be expressed in two's complement form. A negative number has the most significant bit set to 1, and the magnitude of the number is in two's complement form. Ideally, the programmer knows ahead of time the largest numbers needed in the program and makes all complements based on the known size of the largest possible number in the program. For instance, if all numbers used in the program are between +127d and −128d (7Fh to 80h), then a byte can hold all possible signed numbers. If, however, the magnitude of numbers in the program needs to be larger, then word-sized signed numbers that can range in size from +32767d to −32768d (7FFFH to 8000h) may be used. Larger signed numbers can be formed by using additional bytes or words to hold the numbers.

Problems arise, however, when a byte-size signed number must be combined with a word-size signed number, or a word with a double word, and so on. Such problems are easily solved by *extending* the sign of the smaller size number to match the length of the larger size number. As an example, −1h when converted to a byte-size signed number is formed by subtracting 1 from 100h, yielding FFh for the two's complement of −1h (byte-size). The word-size equivalent to −1h is formed by subtracting 1h from 10000h, yielding FFFFh for the two's complement of −1h (word-size).

Clearly, the only thing needed to be done to a byte-size −1h to convert it to a word size −1h is to *extend* the sign bit of the byte (bit 7, which is a 1) to the rest of the word above bit 7. Converting −1h byte-size to −1h word-size means the byte is converted to a word with FFh in the most significant byte. Positive numbers will extend a 0 into the upper byte of the word. For instance, positive 7Fh becomes positive 007Fh, when extending bit 7 from byte-to word-size. The same arguments apply when extending words to double words, and so on, from any size signed number to a larger size signed number.

Two 8086 instructions extend byte-to-word and word-to-double-word signed numbers, CBW and CWD. CBW converts the byte in AL to a word in AX by extending the most significant bit of AL (bit 7) up through AH. CWD converts the word in AX to a double word in DX and AX by extending the high bit of AX through DX.

CBW   Convert (Signed) Byte to (Signed) Word
CWD   Convert (Signed) Word to (Signed) Double Word

Mnemonic	Size	Operation	Flags
CBW	W	W ← B	None
CWD	D	D ← W	None

Addressing Modes
(Destination, Source)

register, register

Examples:

```
code segment
        mov al,80h      ;set AL to -128d
        cbw             ;AX = FF80h
        cwd             ;DX = FFFFh, AX = FF80h
        mov al,7fh      ;set AL to +127d
        cbw             ;AX = 007Fh
        cwd             ;DX = 0000h, AX = 007Fh
code ends
```

Operation:

CBW and CWD extends the high order bit of AL (CBW) or AX (CWD) to the next byte (AH) or word (DX) involved in the conversion. If the high order bit is a 1, then FFs will appear in the next byte or word. If the high order bit is a 0, then 00s appear in the next byte or word.

8.2 Run-Time Effective Addresses

The effective address of program locations (code or data) is usually known to the programmer as the program is written. The programmer typically assigns symbols (names) to program locations by the following methods:

■ Variable names to data address contents using name DB ? or name DW ? in a data segment.

■ Constant names to numbers using name EQU 1234 or name EQU 3Ch statements in a code segment.

■ Labels to code addresses and fixed data by the use of names followed by a colon, as in name:.

Symbols are used for numeric constants and numeric code and data addresses to make the program more readable and to enable the programmer to spend less time on numeric details.

A symbolic assembler, such as A86, is of value to the programmer because it keeps track of each symbol and the symbol's corresponding numeric value. Symbols are numbers that are known to the assembler at what is known as *assembly time*. Assembly time is the moment when the source code is assembled, and all symbols in the program are assigned their actual numeric values.

For example, one use for a label is to load pointing registers with the base address of some area in memory that is to be processed by the program. For instance, a lookup table can be given the label lookup by the programmer. The instruction

```
mov bx,lookup
```

will move the offset address number of the table into register BX. During assembly time the assembler, knowing the numeric value of the label lookup, will assemble the code for the mov instruction using the number that the label represents.

The numeric equivalent of the label lookup will be placed into BX during *run time*. Run time is the time when the CPU is actually operating and executing the program code. During run time for our example above, the number represented by the label lookup ends up in BX. Whenever you run D86 to debug a program, you see, for the first time perhaps, the actual numeric equivalents of the symbols you have used in your program.

Normally the programmer can assign labels, constant names, and variable names to all items of interest to the program. By using symbols, most address and constant numbers are known by the assembler at assembly time. There are occasions, however, when the addresses of program data cannot be known, by the programmer or the assembler, at assembly time. These cases of unknown addresses arise when data is stored in an unpredictable manner by the program during run time. For example, suppose that a program is making a table of names followed by age. A person's name can be of any length, so that the place (address) in memory where the age begins, after a name, is variable and unpredictable. One solution to the "unknown-name-length dilemma" is to set aside a fixed-length area for each name. By making the fixed-length space longer than any reasonable name, the ages can begin at the end of the data area reserved for each name. Setting aside a fixed data area for each name is very wasteful of memory, because most names will not need the area set aside for the worst-case long name.

A second solution to the name and age storage dilemma is to have the CPU calculate the offset address of the age data during run time. The offset address cannot be known at assembly time, because it is an unknown number, usually contained in one of the pointing registers. To enable the CPU to compute effective addresses during run time will require a new instruction, one that can load the effective address of data into a pointing register during run time. For the 8086 the opcode is LEA, Load Effective Address.

LEA Load Effective Address

Mnemonic	Operands	Size	Operation	Flags
LEA	DESTINATION,SOURCE	W	(REG) ← Address	None

> **Addressing Modes**
> (Destination, Source)
>
> register, memory
>
> *Examples:*
>
> ```
> lea ax,[bx+table]      ;load AX with the offset number BX + table
> lea bp,[si+bx]         ;load BP with the offset number SI + BX
> lea cx,[bp+1000h]      ;load CX with the offset number BP + 1000h
> ```
>
> *Operation:*
>
> LEA has the CPU calculate an offset address from the contents of memory point-ing registers and fixed numbers. The resulting offset address must be placed in a register. The offset address must be calculated from legal memory operands, such as pointing registers and fixed constants.

Note the difference between the instructions mov ax,[bx+table] and lea ax,[bx+table]. The mov instruction loads the contents of the address number formed from [bx+table] into AX, whereas the lea opcode loads the address number [bx+table] into AX. Note also that the same result (as an LEA) is obtained by moving BX into AX, and then adding the constant table to AX.

The program will execute faster if MOV instructions are used wherever possible in place of LEA instructions. It is better to have the assembler calcu-late addresses during assembly time than to have the CPU calculate addresses during run time.

LEA EXAMPLES

The next program, lea_legal.asm, demonstrates how LEA can determine off-set addresses using memory pointers and fixed numbers. Running the pro-gram in D86 permits you to change the values of the pointing registers and view the results after LEA is run.

```
;the program lea_legal, which tries some legal lea commands
    code segment
    mov si,1234h          ;load some pointing registers
    mov di,5678h
    mov bx,2222h
    mov bp,3333h
    lea dx,lable          ;a "mov dx,lable" works as well
    inc bp                ;change bp
    lea ax,[bp+1234h]     ;compute at run time
    ;
    mov ax,bp             ;could also be done this way
    add ax,1234h
    ;
    lea dx,[bx+lable]     ;compute and load offset addresses
    lea dx,[si+lable]
    lea dx,[di+lable]
    lea dx,[bp+lable]
```

```
                lea dx,[si+bx+lable]
                lea dx,[di+bp+lable]
lable:    nop
code ends
```

A second program, name_age.asm, shown next, demonstrates how LEA is used for finding offset addresses at run time.

```
;name_age.asm is a program that calculates and stores the offset
;address of a person's age found in a name-age table named "names."
;The offset of the age is a number that specifies how far from the
;beginning of the "names" table the age for that person is stored.
;The first part of the program finds each age in the "names" table,
;and stores the offset in a data table called "effective." Remember
;that the contents of "effective" are the offsets of each person's
;age in the "names" table, not the absolute address of the age in the
;code segment. The last part of the program finds an age for one of
;the names, fred.
```

```
data segment                                                            See Note
        org 1000h           ;base of effective address table for ages
effective dw 10h dup ?      ;set aside 16 words for age offset table       1
data ends
code segment
        mable equ 0000h     ;the list of names equated to numbers
        fred equ 0001h      ;find an age by entering the person's
        henry equ 0002h     ;number. Note the different lengths.
        ken equ 0003h
        anne equ 0004h
        heather equ 0005h
        freidreich equ 0006h
        mo equ 0007h
        stanislaus equ 0008h
        henrietta equ 0009h
        bubu equ 000ah
        ron equ 000bh
        chad equ 000ch
        brian equ 000dh
        elijah equ 000eh
        barbara equ 000fh
        lea si,effective    ;point to offset address table                 2
        mov dx,10h          ;make a 16-address offset table
        lea bx,names-1      ;point to pre-defined name/age table
        mov al,' '          ;look for the space name-age separator
search: inc bx              ;point to next character in name/age table
    cs  cmp al,[bx]         ;find offset address of age for name            3
        jne search          ;continue until the blank (20h) found
        inc bx              ;point to age offset address
        mov cx,bx           ;form offset address of age in name-age table
        sub cx,names        ;subtract "names" address from age address
    cs  mov [si],cx         ;save offset of age in effective table
        add si,0002h        ;point to next word in effective table
        inc bx              ;skip over age of person
        dec dx              ;store 16 age offsets into "names" table
        jnz search          ;continue until the offset table is made
```

```
;try an example,—find the age of fred
nxt:        lea si,effective+ 2*fred   ;Example: find age of fred
            ;get offset of fred's age from effective table
    cs      mov si,w[si]                ;offset of age in SI, now find age address
            lea bx,[names+si]           ;BX points to fred's age in "names" table
    cs      mov ah,b[bx]                ;get fred's age in AX (as ASCII numbers)
    cs      mov al,b[bx+1]              ;stored as bytes, high byte FIRST
            jmp nxt                     ;want to try the other names?
;a sample table of names and ages
names:      db 'mable 23fred 21henry 34ken 52anne 51heather 20'      ; 4
            db 'freidreich 43mo 17stanislaus 35henrietta 75'
            db 'bubu 43ron 25chad 37brian 22elijah 57barbara 64'
code ends
```

NOTES ON SECOND LEA EXAMPLE

1. Set aside 16d word-length variables in memory, the first named effective, the second effective+1, and so on. Each variable holds the offset, from the start of the names table, of the age for the person at that variable number. For instance, mo is person number 0007, and variable effective+14 holds the offset of mo's age (003Eh) in the names table of the code section.

2. The code mov si, offset effective would do the same thing, but lea si,effective is clearer. (*Note to this note:* A86 will code this LEA as a MOV because it results in shorter machine code. Eric Isaacson has found that LEA may often be replaced with more efficient MOV opcodes and has A86 behave accordingly.)

3. Use the segment override cs, or the CMP opcode will use DS as the segment register associated with BX. For .COM programs this is generally not needed, because DS and CS are the same.

4. The names-age table has to be carefully constructed so that each name is separated by a space (ASCII 20h) from its age, and there are no other spaces in the table.

Another useful function that LEA performs is to help in documenting the program. The instructions found in Chapter 5 can perform the LEA function, but the mnemonic LEA leaves no doubt in the reader's mind as to the intent of the program.

8.3 String Operations

A *string* is any *sequential* group of data items. Normally, strings are considered to be a sequence of ASCII-coded bytes representing a printable message. Strings may be, however, sequences of data items of any kind and size.

String processing is pervasive in certain computer applications, such as word processors and data communication programs. String operations are sufficiently common that the X86 family includes several mnemonics that are adapted specifically for handling the most common string processing problems. As is the case with many specialized instructions, the same results may be obtained using more general purpose mnemonics. String operations are

Index Register	Default Segment Register	Optional Segment Register
Source (SI)	Data segment (DS)	CS, ES, SS
Destination (DI)	Extra segment (ES)	None

Table 8.1 ■ String Operation Segment Registers

much more efficient, however, from a coding standpoint than are standard mnemonics. String operations provided by the X86 family are string moves and string compares, which are discussed next.

The Role of DI and SI in String Operations

Index registers DI (Destinations Index) and SI (Source Index) play an important part in string operations. Most string operations involve moving strings from one memory location to another, or searching a string for a certain data item. Generally, SI is used to point to an area of memory (the source of the string) containing an existing string, and DI is used to point to an area of memory (the destination for the string) where the existing string is to be placed.

SEGMENT REGISTERS USED IN STRING OPERATIONS

Moving data with a MOV instruction involves using the DS segment register as one of the default segment registers. DS can be overridden by specifying one of the other segment registers—CS, ES, or SS—as a prefix to the MOV mnemonic. String moves use *both* the DS and the ES segment registers to form the final physical address of the string source and the string destination addresses. The default segment registers used in string operations are shown in Table 8.1.

We can see from Table 8.1 that the *source* of data for a string operation, that is, any string operation that uses SI, can be in a segment addressed by DS, CS, ES, or SS. The *destination* of a string operation, that is, all string operations that use DI, *must* be in the segment addressed by segment register ES.

The Role of the Direction Flag in String Operations

Pointing registers SI and DI can point to the items of a string in one of two ways. SI and DI can begin their operations by pointing to low memory addresses and then increment up toward higher memory addresses. Or, DI and SI can start at high memory addresses and decrement down toward lower memory addresses.

The direction DI and SI take in memory (low to high, or high to low) is specified by the Direction flag in the Flag register. If the Direction flag is set to 1, then SI and DI will decrement from high to low in memory. If the Direction flag is cleared to 0, then SI and DI will increment from low to high in memory. The applicable instructions that select the SI and DI direction in memory are as follows.

CLD   Clear Direction Flag

Mnemonic	Operand	Size	Operation	Flags
CLD	DF	bit	DF ← 0	DF

Operation:

CLD clears the Direction flag (DF) in the Flag register to 0. DI and SI will increment when used in string operations.

STD   Set Direction Flag

Mnemonic	Operand	Size	Operation	Flags
STD	DF	bit	DF ← 1	DF

Operation:

STD sets the Direction flag (DF) in the Flag register to 1. DI and SI will decrement when used in string operations.

The decision to process memory from low to high or from high to low may seem to be arbitrary, and often it is. There are instances, however, when memory must be processed in only one direction. For example, when the source and destination addresses overlap in the same segment. Say a 20h-byte string located at addresses 1000h to 101Fh is to be moved to addresses 1010h to 102Fh in the same segment. The source and destination addresses *overlap* from addresses 1010h to 101Fh. Should the strings be moved from low to high, then the data at addresses 1000h–100Fh will *overwrite* the data at addresses 1010h–101Fh. Moving the strings from high to low will have moved the data from addresses 1010h–101Fh to addresses 1020h–102Fh, so that the data from 1000h–100Fh can safely overwrite addresses 1010h–101Fh.

The assembler, using define (DB, DW, etc.) directives, assembles defined data from low to high in memory. Because of assembler action, we usually think of strings beginning at low memory and extending upward into high memory. Strings do not have to begin or end in any consistent memory direction, but "read" better, at least to Westerners, in the debugger memory windows when arranged from left to right (low to high) in memory.

Moving Strings

String operations consist of initializing DI and SI so that they point to the desired areas of memory, and setting DF to establish the direction of the move process in memory. Once the location and direction are set, string operations are done.

MOVSB   Move a String of Bytes
MOVSW   Move a String of Words

Mnemonic	Size	Operation	Flags
MOVSB	B	(DI) ← (SI)	None
MOVSW	W	(DI) ← (SI)	None

Addressing Modes
(Destination, Source)

memory indirect, memory indirect

Operation:

MOVSB and MOVSW move a byte (MOVSB) or a word (MOVSW) from the location pointed to by SI to the location pointed to by DI. SI and DI are both changed to point to the next data item. SI and DI increase if DF is 0 and decrease if DF is 1. MOVSB causes DI and SI to change by a count of 1; MOVSW causes DI and SI to change by a count of 2. DI and SI change *after* the move. DI is always paired with ES to specify the destination address, whereas SI defaults to DS.

A short example program, which uses both forms of the MOVS instruction, follows.

```
;a MOVS program which demonstrates single byte and single word
;string moves.

code segment
        source equ 1000h    ;address of source string in memory
        dest equ 2000h      ;destination in memory for string
        mov si,source       ;set SI register to string source address
        mov di,dest         ;set DI register to string destination address
        cld                 ;process string from low to high
repeat:                     ;loop if desired
        movsb               ;move byte at (SI) to (DI)-increment SI and DI by 1
        movsw               ;move word at (SI) to (DI)-increment SI and DI by 2
        jmp repeat
org 1000h
        db '123456789'
code ends
```

Observe this program using D86. Set memory line 1 to text display starting at address 2000h. You should see that byte moves will move a single number, and word moves will move a double number. Note also that, for .COM programs, the default data segment and extra segment are the same.

STOSB   Store AL in a Byte String
STOSW   Store AX in a Word String

Mnemonic	Size	Operation	Flags
STOSB	B	(DI) ← AL	None
STOSW	W	(DI) ← AX	None

Addressing Modes
(Destination, Source)

memory indirect, register

Operation:

STOSB and STOSW move the contents of register AL (STOSB) or AX (STOSW) to the memory address pointed to by ES:DI. DI is changed to point to the next address location, *after* the move. DI increases if DF is 0 and decreases if DF is 1. STOSB causes DI to change by a count of 1; STOSW causes DI to change by a count of 2. ES is the segment register used with DI.

An example of using a string store is listed as follows.

```
mov di,1000h        ;DI points to ES:1000h
mov ax,1234h        ;AX = 1234h
stosb               ;offset location 1000h = 34h
stosw               ;offset location 1001h,1002h = 1234h
```

STOS, if used repetitively, may be used to initialize an area of memory with the repetitive data. Note that SI plays no part in the operation.

LODSB   Load String Byte to AL
LODSW   Load String Word to AX

Mnemonic	Size	Operation	Flags
LODSB	B	AL ← (SI)	None
LODSW	W	AX ← (SI)	None

Addressing Modes
(Destination, Source)

register, memory indirect

Operation:

LODSB and LODSW move the contents of memory pointed to by SI to AL (LODSB) or AX (LODSW). LODS is the reverse of STOS and moves data from memory to the AL or AX register. SI will increment or decrement by 1 for byte moves or 2 for word moves, as determined by DF, *after* the move takes place. DS is the default segment register for SI. Note that DI plays no part in the operation.

A short example program that uses LODS follows.

```
mov si,1000h      ;set SI to point to offset 1000h
lodsb             ;move the contents of 1000h to AL
lodsw             ;move the word at 1001h to AX
```

Repeated String Move Operations

String moves MOVS and STOS have very little appeal if only one byte or word is moved. The overhead involved in setting up SI, DI, and DF makes moving one item tedious. If moving is repeated many times, however, then the overhead effort is negligible compared to the many repeat moves accomplished.

The programmer can set up program loops to move long strings or make use of specialized, string-operation *repeat* instructions. Repeat instructions will repeat string operations until certain *conditions* are met.

Repeat instructions are called *prefix* instructions because they may be typed by the programmer in front of the string operation instruction that is to be repeated. A prefix instruction will cause the CPU to perform a repeated string operation. We use a similar idea when we type a prefix segment override in front of a data move instruction. Segment override causes the assembler to assemble the opcode using the desired segment.

Repeat instructions are useful *only* when used with string instructions. And, only *certain* repeat instructions should be used with *specific* string operations.

REP   Repeat (MOVS or STOS) Until CX = 0

Mnemonic	Size	Operation	Flags
REP	NA	Repeat MOVS,STOS until CX = 0	None

Operation:

REP is a *prefix* instruction for MOVS and STOS instructions. The string operation that follows REP is done until CX = 0. CX decrements *after* the string operation. Note that CX decrements by 1 for the repeated *operation*, not the number of bytes affected. Thus CX will decrement by 1 for a byte or a word move.

REP may also be used with LODS instructions, but its use would rarely make any practical sense because AX\AL could only contain the last item in the string. Using REP with a LODS instruction would result in the last data quantity pointed to by SI copied to AX\AL. All earlier moves from memory pointed to by SI to AX\AL would be overwritten by the last move of the repeated operation.

A short example of using a repeat prefix is listed next. Source address here and destination address there are assumed to have been previously defined using equates. If source and destination overlap, then the *Destination flag* should be set to process data from high to low in memory.

```
        mov si,here       ;point to string source
        mov di,there      ;point to string destination
        mov cx,0100h      ;set CX to count 100h bytes
    rep movsb             ;repeat MOVSB 100h times
        mov di,there      ;point to string destination
        mov cx,0200h      ;set CX to count 200h words
        mov ax,2020       ;set AX to ASCII "space space"
    rep stosw             ;store 400h spaces at there
```

If you exercise the repeat example using D86 you should see that the repeat operations are done once if F1 is pressed, and run to completion if F2 is pressed.

Searching Strings

Searching for a unique data item contained in a string of data items is another standard string operation. For example, many word processors make use of search procedures to "find and replace" one text word with another. Strings may be compared using the instructions that are discussed next.

CMPSB   Compare String Bytes
CMPSW   Compare String Words

Mnemonic	Size	Operation	Flags
CMPSB	B	(SI) – (DI)	AF, CF, OF
CMPSW	W		PF, SF, ZF

Addressing Modes
(Destination, Source)

memory indirect, memory indirect

Operation:

CMPSB and CMPSW cause the destination byte (CMPSB) or word (CMPSW) pointed to by ES:DI to be subtracted from a similar data type pointed to by SI. The comparison flags are set, but no changes are made to the data pointed to by SI or DI. DI and SI are changed *after* the comparison to point to the next data item. If DF is 0, then DI and SI count up by 1 (CMPSB) or by 2 (CMPSW). If DF is 1, then DI and SI are decremented by 1 (CMPSB) or 2 (CMPSW). DS is the default segment register for SI.

CMPS instructions allow two strings to be compared until a match is found. A short example of comparing two strings follows next.

```
mov di,onestr         ;set DI to point to one string
mov si,anthrstr       ;set SI to point to another string
cld                   ;set DF to move up in memory
cmpsb                 ;compare byte in each string
cmpsw                 ;compare word in each string
```

Strings are searched for an item using the next instruction.

SCASB   Scan a String Byte (and Compare with AL)
SCASW   Scan a String Word (and Compare with AX)

Mnemonic	Size	Operation	Flags
SCASB	B	(AL) – (DI)	AF, CF, OF
SCASW	W	(AX) – (DI)	PF, SF, ZF

Addressing Modes
(Destination, Source)

register, memory indirect

Operation:

SCASB and SCASW cause the destination byte (SCASB) or word (SCASW) pointed to by ES:DI to be subtracted from AL (SCASB) or AX (SCASW). The comparison flags are set, but no changes are made to the data pointed to by DI or the contents of AX. DI is changed *after* the comparison to point to the next data item. If DF is 0, then DI counts up by 1 (SCASB) or by 2 (SCASW). If DF is 1, then DI is decremented by 1 (SCASB) or 2 (SCASW). ES is the default segment register for DI.

SCAS searches for one item in a string that matches an equivalent item in AX or AL. SCAS should be used to look for a single item in a string. An example of using SCAS in a program is listed next.

```
mov di,lookhere       ;set DI to point to string to be scanned
cld                   ;search from low to high in memory
mov al,'A'            ;search for an "A"
scasb                 ;scan byte in string
mov ax,'AB'           ;search for an "AB"
scasw                 ;scan word in string
```

REPEATED STRING SEARCHES

Just as the REP prefix instruction may be used to repeat string move operations, the REPZ (REPE) and REPNZ (REPNE) prefixes may be used to repeat string search instructions.

Mnemonic	Size	Operation	Flags

REPZ   Repeat String Compare (While ZF = 1 AND CX <> 0)
REPE

Mnemonic	Size	Operation	Flags
REPZ (REPE)	NA	Repeat CMPS or SCAS until CX = 0 or ZF = 0	None

REPNZ   Repeat String Compare (While ZF = 0 AND CX <> 0)
REPNE

Mnemonic	Size	Operation	Flags
REPNZ (REPNE)	NA	Repeat CMPS or SCAS until CX = 0 or ZF = 1	None

Operation:

REPZ (REPE) repeats the string compare operation until CX reaches 0, *or* the Zero flag (ZF) is set to 0. REPE is usually the best form of the mnemonic and can be thought of as "Repeat (*While*) Equal."

REPNZ (REPNE) repeats the string compare operation until CX reaches 0, *or* the Zero flag (ZF) is set to 1. REPNE is usually the best form of the mnemonic and can be thought of as "Repeat (While *Not*) Equal."

Set the Direction flag to search from low to high or high to low as seems appropriate for the search.

REPE operations are *looking for something* in the destination string that does *not* match the source data. REPE expects to *normally* find a match. So long as the items that are compared are equal (ZF = 1), the search continues. As soon as an exception is found (when ZF = 0, signifying not equal), the repeat operation is terminated. Counting register CX is decremented *after* each compare operation. The loop will terminate when CX = 0, if no exception to equality is found.

REPNZ operations are looking for something in the destination string that *does* match the source data. REPNE expects to normally *not* find a match in the string that is under search. So long as the items that are compared are not equal (ZF = 0), the search continues. As soon as a match is found (when ZF = 1, signifying equality), the repeat operation is terminated. Counting register CX is decremented by 1 *after* the compare operation is done. The loop will terminate when CX = 0, if no match is found.

Note that the condition CX = 0000h does *not* affect the Zero flag. The program must inspect the Zero flag to find out how the repeat operation turned out as follows:

■ When a REPE is exited, and the Zero flag is 0, then a *mismatch* has been found. If the Zero flag is 1 after a REPE operation, then a match was found for each compare instruction.

■ When a REPNE is exited, and the Zero flag is 1, then a *match* was found. If the Zero flag is 0 after an REPNE operation, then no matches were found in the string compare operations.

An example program that uses repeat prefixes with string compares is shown as follows.

```
        mov di, lookhere    ;set DI to point to string to be scanned
        cld                 ;search from low to high in memory
        mov cx,0100h        ;set CX for a 100h byte search
        mov al,'A'          ;search for an "A"
repnz scasb                 ;scan 100h bytes in the string for the first "A"
```

String Instructions Example

The program movit.asm, listed next, demonstrates many string manipulation instruction features.

```
;the program movit.asm explores the possibilities of moving
;and comparing enormous strings of data in memory
code segment                                                            See Note
        putithere equ 0a000h    ;destination of data
        getithere equ 2000h     ;source of data
        amount equ 1000h        ;number of data items
movit:
        cld                     ;move from low to high addresses
        mov di,getithere        ;destination of data in DI          1
        mov cx,amount           ;move this many items               2
        mov ax, 'ab'            ;move ascii characters a and b
rep     stosb                   ;fill memory with many "b" bytes    3
str1:
        mov di,getithere        ;destination of data in DI          4
        mov cx,amount           ;now overwrite memory
rep     stosw                   ;memory contains many "ab" words    5
str2:
        mov cx,amount           ;move string to new memory location
        mov di,putithere        ;destination of data in DI
        mov si,getithere        ;source of data in SI
rep     movsw                   ;move many words from here to there
str3:
        mov cx,amount           ;now compare the strings
        mov di,putithere        ;destination of data in DI
        mov si,getithere        ;source of data in SI
repe    cmpsw                   ;the strings should compare exactly
str4:
        jnz error               ;if not then the computer is broken  6
        mov [putithere+1feh],'ac'  ;insert an "error" word 200h bytes
                                ;from start at offset location A1FEh
        mov cx,amount           ;compare the strings again
        mov di,putithere        ;destination of data in DI
        mov si,getithere        ;source of data in SI
repe    cmpsw                   ;compare should fail at CX = 0F00h
str5:
        jz error                ;if not then the computer is broken  7
        mov cx,amount           ;look for the 'ac'
        mov ax,'ac'             ;set AX for the search word
        mov di,putithere+2*(amount-1)    ;search from high end
        std                     ;set direction down in memory
```

```
            repne  scasw                ;the error should be at CX = OFFh
                                        ;offset = CX x 2 = 1FEh from start
    str6:
                    jnz error           ;if not then the computer is broken        8
                    jmp movit           ;debug again?
    error:          nop                 ;problems if we end up here
    code ends
```

NOTES ON STRING EXAMPLE

1. Remember DI must use ES as a companion segment register. As long as all program segments are the same, as for a .COM program, then using ES with DI will cause no problems. If destination data is in some other segment, however, then be sure to change ES in the program to point to the destination segment.

2. If CX is set to 0000, *no* repeat operation will take place.

3. 1000h bytes will be stored. Use D86 memory windows to see it happen.

4. Labels—str1:, str2:, and so on—are added to the program for debugging convenience. D86 will *single step*, using key F1, the repeated string operation until the end condition (CX = 0, or ZF) occurs. Labels after repeat prefixes will enable you to set them as breakpoints and go to the label quickly. Should you not wish to see a few repeat operations, then key *F2* will run the repeat operation to completion.

5. 1000h *words* are stored. CX counts down 1 for each operation, not 1 for each byte.

6. The program must test the Zero flag after every repeat compare operation to find out *how* the operation terminated. We know that every word in the two identical strings should match, so we expect ZF to be set to 1 when the loop terminates, and CX to be 0000.

7. We expect the loop to find a mismatch, and ZF should be 0.

8. We expect to find a match, and ZF should be 1.

Loading Memory Pointers for String Operations

String operations necessarily need up to four registers that point to the memory areas involved in the string operation. MOVS, for instance, uses SI and (usually) DS as the pointing registers for the source of the string. Registers DI and ES are used for the pointing registers for the destination of a MOVS operation.

Registers DS, ES, SI, and DI can all be loaded using MOV instructions, but loading the MOVS pointing registers could take as many as six data moves (two for SI and DI, and four for DS and ES). Two convenient single opcode instructions are available for loading the string operation pointing registers, as is discussed next.

LDS   Load DS (and pointing register)

Mnemonic	Operands	Size	Operation	Flags
LDS	Register,Source	DW	Register ← (source) DS ← (source+2)	None

LES   Load ES (and pointing register)

| LES | Register,Source | DW | Register ← (source)
ES ← (source+2) | None |

Addressing Modes
(Destination, Source)

register, memory

Operation:

LDS and LES load the specified 16-bit pointing register with the word contents of the memory address specified by the source memory operand. Segment register DS (LDS) or ES (LES) is loaded from the next word in memory after the word addressed by the source memory operand.

An LDS/LES Example

LDS and LES are very handy ways to load complete sets of pointer addresses into the segment and pointing registers involved in string operations. *Any of the 16-bit general-purpose registers may be specified to receive the first word addressed by the source operand, but the segment register is fixed by the opcode used.* Our next program, load_ptrs.asm, shows how the string pointer can be quickly loaded before a string move is done.

```
;Initialize the string pointers; program load_ptrs.asm
code segment                                                          See Note
        cs   lds si, pointer1     ;point to source of data
        cs   les di, pointer2     ;point to destination of data           1
             mov cx,0005h         ;move five bytes
        rep  movsb                ;move string
dun:         nop                  ;finished
pointer1     dw 0300h,5000h       ;source of bytes in 5000h:0300h          2
pointer2     dw 0500h,6000h       ;destination of bytes in 6000h:0500h
code ends
```

Notes on LDS/LES Example

1. We *must* use a CS override for the LES opcode. The reason for this is because our first opcode, LDS, changes DS to 5000h from the common .COM segment values assigned by DOS when the program is loaded. The memory operands pointer1 and pointer2 always use DS as

a default register. The CS override for LES insures that pointer2 is loaded from our code segment, where it is defined.

2. Note that the double-word contents of memory address pointer1 and pointer2 are loaded, not the numeric addresses pointer1 and pointer2. Symbols pointer1 and pointer2 are variable names, not address labels.

Our pointers to data could easily have been defined as variables in a data segment, permitting the source and destination of our data move to vary as the program executes.

8.4 Tables and Arrays

Throughout the development of the programming art, many standard techniques have evolved to solve common problems. Certain programming "tricks" were long ago given names by programmers. Recently, academics have gotten into the act, so that a list of quasi-standard names has been established for some of the more common programming practices. Two common programming tricks are the use of tables and arrays.

Lookup Tables

One of the more common programming techniques is the *lookup table*. A lookup table conjures up mental images of a single entry row, multiple column arrangement of data, such as in a table of trigonometric functions.

A list of trig functions in a human lookup table is entered by specifying an angle. The trig table is organized as rows of angles, and columns of trig function numbers. The table columns list many functions such as the sine, cosine, tangent, cotangent, secant, and cosecant for the table rows of angles. To use a trig table, it is only necessary to enter at the desired angle row, and trace across the row while reading the various trigonometric values from the function columns.

You do not have to use a trig table to arrive at a trigonometric function of an angle. You *could* calculate all the trig functions for an angle by using appropriate series formulas. (Or use a calculator, which then simply refers the trig problem to the calculator programmer.) Calculating the angle functions, although possible, is very time-consuming compared to simply looking it up in a table.

The major features of a table, such as our trig table, is one entry point and one or more data items associated with the entry point. Other tables we use constantly are telephone directories (entry by name, multiple items of telephone number and address) and appointment calendars (entry by date, multiple items of things to do). Lookup tables are time-consuming to make up, but easy and *fast* to use.

Arrays

An *array* is any group of items that are all of the same type. A class role is an array of student *names*. Opposite each name is an array of student *grades*. All of the sines in a trig table form an array of *sines*. An array can be thought of as a single-item table; one entry point and one data item. Conversely, a table can be thought of as made up of a number of arrays, or a table of arrays. We shall simplify the fine line between tables of arrays and arrays by calling everything that is organized into *related* data a *table*.

Table Construction

Suppose a number is to be squared. There are two ways to find the square of a number: multiply the number by itself, or look the answer up in a book of the squares of numbers. Someone had to do the multiplications in order to compile the book, but, once done, the answer can be found by looking it up.

The only complication involved in using the squares table is the means used to find the answer in the book. One way to use the table that seems obvious is to list each number and its square in numerical sequence. The number to be squared is called the *entry* into the table. Entering the table allows one to find the desired data at a unique place in the table.

The appendices of this book have many lookup tables: one is the ASCII codes for keyboard characters, arranged by ASCII code number; another is the 8086 mnemonics, arranged alphabetically; and another lists the BIOS and DOS interrupt routines arranged by interrupt number. All of the book tables employ one concept: use the known quantity to find the unknown answer. For example, use the number to find (look up) its square, use the ASCII code to find its character, use the interrupt type and function number to find a description of the interrupt service.

A computer can use the concept of looking up answers very effectively by means of pointers to memory. The known quantity (entry into the table) helps form a unique memory address. The contents of memory at that address contains data related to the known address quantity. The known number is said to "point to" the desired answer.

The address of the first entry in the table is called the *base* of the table. The table can be addressed by a variety of schemes, all involving the use of one register to hold the base address of the table, and another register to hold the entry address (related to the known number) into the table.

Given the segmented nature of memory in the X86 family, it is very tempting to use a segment register to hold the base address of the table, and one of the indirect addressing registers (BX, BP, SI, or DI) to hold the offset. To use a segment register to point to the base of a table requires that the table base address begin on physical memory addresses that end in *0h*. (Remember, the segment register is shifted left 1 nibble before the offset address is added to it to form the final address in memory.)

A second approach, used to point to table data, is to set a segment register (usually DS) to the segment where the table is located. An index register (SI or DI) is set to the offset address of the table in the segment, and then a base register (usually BX) is set to the value of the known entry number. Based indexed indirect addressing can then be used to form the final table entry address by adding the index register to the base register to find the entry offset address into the code segment table.

Figure 8.1 shows how a unique table entry point can be found using a single or a double register to form the offset address into the table.

No matter how we point to the table, the trick is to construct data in the

Figure 8.1 ■ Lookup table entry.

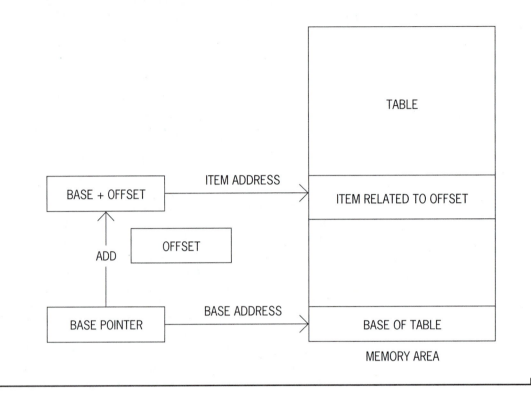

table so that the right data (the answer) just "happens" to be where the known number will point. The table is purposefully constructed so that the address formed by the known number will be the address of whatever data is associated with the known number.

Lookup tables are normally assembled in a *code* segment using define data (DB, DW, etc.) assembler directives to "build" the table. The programmer typically has to build the table manually, or use other predefined tables. Once built, however, the table can be re-used easily.

Lookup Table Examples

To use the square of a number as an example, assume that a table of the squares of all integers from 00 to FFh is to be made. Once this table is constructed, the square of any number from 00 to FFh can be found by using the known number to address the square of that known number from offset address 0000 (00 squared) to offset address 01FEh (FFh squared). The table can be placed in code memory, either on a zero nibble boundary for segment base addressing or anywhere in memory if two addressing registers are used.

Lookup tables work best (can be made up easily) if the known numbers are sequential. If the known numbers are sequential, then the data associated with the known numbers can be arranged in the table sequentially also. For our example of finding the square of a number from 00 to FFh, the construction of the table is very straightforward. The answer to 00 squared is found at offset word address 0000; the answer to 01h squared is found at word address 0002h; and so on until the answer to FFh squared is found at address 01FEh. The lookup table for the square of any number is defined in code space using DW directives, because the square can be as large as a word.

Lookup tables are used for two reasons: speed of program execution, or because there is no other way. An example of speed of execution is finding the trigonometric function of some angle. This can be done using a semi-infinite series to calculate the function (a lengthy process of multiplication and division) or having the angle form the address of an entry in a table of functions. The program can find the trig function very quickly (a few instructions) in a lookup table (although it requires a lot of programmer time to make up the table).

Lookup tables *must* be used if there is no mathematical way to link a quantity and a result based on that quantity. One example of "no other way" is when coding a secret message. Each ASCII character becomes some other character on a random basis, and only a table can make the translation.

We shall use several methods of table addressing to look up the squared numbers, beginning with using a segment register as the pointer to the base of the table. The DS register is the most natural choice to point to the base of the table, as move operations use DS as the default segment register.

Example 1: Segment Register Addressing

```
;a program named sqrit.asm that finds the square of any number
;from 00 to FFh placed in AL, and places the answer (0000 to FE01h)
; in AX. DS is used to point to the base of the lookup table.
       code segment                                                See Note
sqrit:    mov bx,cs          ;start at the table CS value              1
          add bx,20h         ;add segment address of table            2
          mov ds,bx          ;DS now points to base of table
          mov bx,0000h       ;clear BX to receive the number
          mov bl,al          ;the number to be squared is in AL
          add bx,bx          ;double it to point to the word answer
          mov ax,[bx]        ;get answer in AX. DS is default for BX
          jmp sqrit          ;loop for more demonstration
;
;begin table of squared words at address "table". NUMBERS IN DECIMAL
          org 200h           ;place table on an address that ends in 0h
table:    dw 0000xd          ;00 squared is 0000                      3
          dw 0001xd          ;01 squared is 0001
          dw 0004xd          ;02 squared is 0004
          dw 0009xd          ;03 squared is 0009
          dw 0016xd          ;04 squared is 0016
          dw 0025xd          ;05 squared is 0025
;and so on until, at location 01FEh in the table segment:
          org 03feh          ;last entry for number = FFh
          dw 65025xd         ;255 squared is 65025                    4
       code ends
```

NOTES ON EXAMPLE 1

1. DOS has put this program into some code segment; we cannot know which one. We do know, however, that our table is built at CS + 200h because of our org 200h for the square table.

2. 20h, when shifted to the left 1 nibble, becomes 200h, the offset of the square table in the code segment.

3. Using xd makes sure A86 does not assemble the number as 0000d hexadecimal.

4. In the absence of a terminating *h* or *xd*, A86 assumes *decimal* if the number begins with a decimal numeral greater than 0 (1–9). A86 assumes *hex* if the number begins with a 0. Use xd for decimal numbers, just as good documentation.

Example 2: Multiple Lookup Tables

The lookup table does not have to be constrained to a single table of data. The base register that points to one table can be changed to point to the base address of other tables. The offset register can point to the same offset in many different tables. Many tables can be accessed by the same entry number to yield answers that can be extended to whatever accuracy desired by the programmer.

Example 1, above, to find the square of a 1-byte number that yields a 2-byte answer, could have been done by using two tables. One table holds the least significant byte of the answer, and the second holds the most significant byte of the answer. The number in AL points to both bytes in each separate table. DS is changed to point to the high and low order byte of the answer. The original squaring program, sqrit.asm, can be rewritten to use two tables as sqritII.asm, which follows next.

```
;a program named sqritll.asm that squares any number from 00 to FFh
;placed in AL, and places the answer (0000 to FE01h) in AX.
;DS is used to point to the base of two tables, the first for the
;high order byte, and the second for the low order byte.
code segment                                                    See Note
sqritll:  mov bx,cs        ;start at the table CS value
          add bx,20h       ;add segment address of high table
          mov ds,bx        ;DS now points to base of high table
          mov bx,0000h     ;clear BX to receive the number
          mov bl,al        ;the number to be squared is in AL        1
          mov ah,[bx]      ;get high byte to AH; DS is default for BX
          mov bx,cs        ;start at the table CS value
          add bx,40h       ;add segment address of low table
          mov ds,bx        ;DS now points to base of low table
          mov bx,0000h     ;clear BX to receive the number
          mov bl,al        ;the number to be squared is in AL
          mov al,[bx]      ;the low byte of the answer is in AL
          jmp sqritll      ;loop for more demonstration
;
;begin high byte table
          org 200h         ;place table on an address that ends in 0h
```

```
tableh:  db 00                    ;00 squared is 0000h
         db 00                    ;01 squared is 0001h
         db 00                    ;02 squared is 0004h
         db 00                    ;03 squared is 0009h
         db 00                    ;04 squared is 0010h
         db 00                    ;05 squared is 0019h
;and so on until, at offset location 00FFh in the high byte table
         org 02ffh                ;last entry for number = FFh
         db 0feh                  ;FFh squared is FE01h
;begin low byte table
         org 400h                 ;place table on an address that ends in 0h
tablel:  db 00h                   ;00 squared is 0000h
         db 01h                   ;01 squared is 0001h
         db 04h                   ;02 squared is 0004h
         db 09h                   ;03 squared is 0009h
         db 10h                   ;04 squared is 0010h
         db 19h                   ;05 squared is 0019h
;and so on until, at offset location 00FFh in the low byte table
         org 04ffh                ;last entry for number = FFh
         db 01h                   ;FFh squared is FE01h
code ends
```

NOTE ON EXAMPLE 2

1. Note that the pointer does not have to be doubled, as was done in Example 1, because the pointer of Example 2 is to bytes not words.

Using two tables for our example is not particularly efficient, and one table is clearly easier to program. There may be many instances, however, particularly when accessing multi-byte data, in which multiple tables are easier to write and use.

Example 3: Using CS as a Base Pointer

Using DS to point to the base address of a table involves some inconvenience, particularly the inconvenience of insuring that the tables all start on addresses that end in 0h. Another approach is to use CS as a pointer to the code segment where the table is built (*not* to the base of the table) and use table labels as offsets to point to the base of the table. The program that follows, sqritIII.asm, uses the CS as a segment pointer and labels in the code segment to point to the base of the table.

Two *different* ways of coding Example 3 are shown. The first method puts the *base* of the table in BX and *adds an offset* to it to form a final pointer to the table. The second method places the *offset* in register BX and has *the CPU add the table base* (a constant number) to the offset in the MOV instruction. The MOV operand, in the second case, may be written as:

```
       mov ax,[base+bx]      ;stresses the CPU addition
   or  mov ax,base[bx]       ;another way to write it
```

We shall find that the second method will usually save lines of code, because the offset remains in BX while the assembler keeps track of each base label number.

```
;a program named sqritlll.asm that squares any number from 00 to FFh
;placed in AL, and places the answer (0000 to FE01h) in AX.
;CS is used to point to the base of the table segment.
;when loading AL using D86 be sure AH=0.
code segment                                                        See Note
sqritlll:
        lea bx,table            ;get address of table base to BX
        add ax,ax               ;double the number in AL to point to words
        add bx,ax               ;base plus offset into table points to data
    cs  mov ax,[bx]             ;get answer in AX. CS overrides DS              1
        ;reload AL with same number to demonstrate next code
        mov bx,ax               ;get number in AL
        add bx,bx               ;BX holds offset
    cs  mov ax,[table+bx]       ;CPU adds offset to [bx] during the MOV
    cs  mov ax,table[bx]        ;same result
        jmp sqritlll            ;loop for more demonstration
;
;begin table of squared words at address "table". NUMBERS IN DECIMAL
table:  dw 0000xd               ;00 squared is 0000
        dw 0001xd               ;01 squared is 0001
        dw 0004xd               ;02 squared is 0004
        dw 0009xd               ;03 squared is 0009
        dw 0016xd               ;04 squared is 0016
        dw 0025xd               ;05 squared is 0025
;and so on until, at location 01FEh in the table segment:
        org table + 01feh       ;last entry for number = FFh
        dw 65025xd              ;255 squared is 65025
code ends
```

NOTE ON EXAMPLE 3

1. The data is found at CS:BX, where BX = table + 2*AL.

The same approach used for the second method of Example 3 can be used in our two-table example. By simply loading BX with the offset and using the CPU to add the table base during the MOV to AX, we have sqritIIII.asm.

```
;using CS:BX as pointers to a table, program sqritllll.asm
code segment
sqritllll:
        mov bx,ax               ;get the number from AL into BX
    cs  mov ah,[tableh+bx]      ;get high byte to AH; CS overrides DS
    cs  mov ah,tableh[bx]       ;same operation
    cs  mov al,tablel[bx]       ;the low byte of the answer is in AL
        jmp sqritllll           ;demonstrate again
;
;begin high byte table
tableh: db 00                   ;00 squared is 0000h
        db 00                   ;01 squared is 0001h
        db 00                   ;02 squared is 0004h
        db 00                   ;03 squared is 0009h
        db 00                   ;04 squared is 0010h
        db 00                   ;05 squared is 0019h
;and so on until, at offset location 00FFh in the high byte
```

```
            table org tableh+0ffh  ;last byte in table
            db 0feh                ;FFh squared is FE01h
;begin low byte table
tablel:  db 00h                    ;00 squared is 0000h
         db 01h                    ;01 squared is 0001h
         db 04h                    ;02 squared is 0004h
         db 09h                    ;03 squared is 0009h
         db 10h                    ;04 squared is 0010h
         db 19h                    ;05 squared is 0019h
;and so on until, at offset location 00FFh in the low byte table
            org tablel+0ffh        ;last entry for number = FFh
            db 01h                 ;FFh squared is FE01h
code ends
```

Example 4: A Sine Table

The next lookup table program, sine.asm, will return the sine of any angle between 00 and 90 degrees. The sine is accurate to two decimal places. BX will be used as an offset into the table, as was done in Example 3.

```
;sine.asm, a program that looks up the sine of any angle entered
;by the user. The angle is entered in degrees from 00 to 90. The
;sine of the angle entered by the user is displayed to an accuracy
;of two decimal places. Screen 7 is used by the program.
;                                                                See Note
data segment
            org 0a000h             ;store user keys in data memory
            key1 db ?              ;store ASCII code for user tens key
            key2 db ?              ;store ASCII code for user units key
            answer1 dw ?           ;ASCII first digit and . of sine          1
            answer2 dw ?           ;ASCII tenths and hundredths of sine
            dollar db ?            ;terminate string in memory for INT
                                   ;21H, function 09h DOS string display
data ends
code segment
            textmode equ 0002h     ;set display to 25x80 text display mode
            plump equ 0007h        ;want a fat cursor
            line1 equ 0c08h        ;place cursor at left for first message
            line2 equ 0d08h        ;place cursor at left for second message
            return equ 1ch         ;carriage return scan code
            pagenum equ 07h        ;use page 7 for display
            color equ 9fh          ;white on blue individual characters
            mov sp,0fffeh          ;set the SP
            mov dollar,'$'         ;DOS display function 09h limiting $
            mov ax,textmode        ;set text mode and clear all screens
            int 10h                ;10H is the video interrupt
            mov ah,01h             ;use a plump cursor
            mov cx,plump
            int 10h
angle:
            mov ah,05h             ;switch screen display to page 7
            mov al,pagenum
            int 10h
```

```
                  mov dx,line1              ;position the cursor
                  call _curses
                  mov dx,usermsg            ;display message to user
                  call _display             ;usermsg also erases any earlier entries
                  call _cursorloc
                  sub dx,0007h              ;back cursor up for user entry
                  call _curses
         tenkey:                            ;user may key in ONLY decimal numbers
                  call _getkey              ;get first user key
                  call _number              ;check for ASCII 30h to 39h
                  jc tenkey                 ;if CF=1 then not a decimal number
                  mov key1,al               ;store for later use
                  call _showchar            ;display tens key for user view
                  call _cursorloc           ;move cursor over to the right
                  inc dx                    ;for second user key
                  call _curses
         unitkey:
                  call _getkey              ;get second user key
                  call _number
                  jc unitkey                ;wait for decimal number only
                  mov key2,al               ;store second key
                  call _showchar            ;display units key for user view
                  call _cursorloc           ;move cursor over twice for answer
                  add dx,2
                  call _curses
         entered:
                  call _getkey              ;wait for return key to be hit
                  cmp ah,return             ;look for return key
                  jne entered
                  call _key_to_hex          ;convert entered keys to hex number
                  add bx,bx                 ;multiply BX x 4 to point to 4-byte sine
                  add bx,bx
                  mov ax,w[bx+sinetable]    ;look up first digit and . in table
                  mov answer1,ax            ;store first digit and . in memory
                  add bx,2                  ;point to last two characters
                  mov ax,w[bx+sinetable]    ;get tenths and hundredths
                  mov answer2,ax            ;store last two characters in memory
                  mov dx,offset answer1     ;point to message in data memory
                  call _display
                  mov dx,line2              ;prompt user for next entry or quit
                  call _curses
                  mov dx,promptmsg
                  call _display             ;"q" for quit or "a" for angle
         whatnext:
                  call _getkey
                  cmp al,'a'
                  je begin
                  cmp al,'q'
                  je goback
                  jmp whatnext
         begin:                             ;conditional jump beyond 128 bytes
                  jmp angle
         goback:  mov ax,0500h              ;change active screen to 0 for DOS
                  int 10h
```

2

```
                mov ax,4c00h              ;return to DOS
                int 21h
;END OF PROGRAM

;PROCEDURES
_curses:                                  ;_curses positions the cursor as passed
                mov ah,02h                ;in register DX. The current page number
                mov bh,pagenum            ;is contained in variable "pagenum"
                int 10h

                ret                       ;returns nothing to calling program


_cursorloc:
                mov ah,03h                ;gets current cursor position from the
                mov bh,pagenum            ;page number in variable "pagenum"
                int 10h                   ;returns cursor size in CX
                ret                       ;returns cursor position in DX


_display:
                mov ah,09h                ;DOS int 21h, func. 9 display to $ function
                int 21h                   ;passed pointer to string in DX
                ret                       ;returns nothing to calling program


_getkey:                                  ;wait for key using BIOS INT 16h, func. 0
                mov ah,00h                ;wait for the user to hit key
                int 16h

                ret                       ;returns ASCII in AL and Scan Code in AH


_key_to_hex:                              ;converts two ASCII characters to hex
                mov bl,key1               ;this is the tens key, convert to hex
                sub bl,30h                ;convert ASCII key number to decimal
                mov bh,00h                ;look it up using BX as the table offset
        cs      mov bl,b[bx+hextable]     ;hextable converts decimal to hex
                mov al,key2               ;convert first key to decimal, add to BX
                sub al,30h                ;units number in AL
                mov ah,00h                ;add only AL (also, see AAD)
                add bx,ax                 ;returns hex equivalent of two ASCII
                ret                       ;numbers to calling program in BL


_number:                                  ;checks to make sure key is 0–9
                cmp al,'0'                ;ASCII code must be >= 30h
                jb nonumber
                cmp al,3ah                ;ASCII code must be < 3Ah
                jae nonumber
                clc                       ;clear the carry flag—decimal
                ret                       ;returns with CF = 0 if decimal
nonumber:
                stc                       ;set the carry flag—not decimal
                ret                       ;returns with CF = 1 if not decimal


_showchar:                                ;BIOS INT 10h, func. 9 character display
                mov ah,09h                ;display char on screen for user
                mov bh,pagenum            ;use page found in "pagenum"
                mov bl,color              ;use current "color" attribute
                mov cx,0001h              ;write one character passed in AL
```

```
              int 10h                    ;display character
              ret                        ;returns nothing to calling program
;END OF PROCEDURES

;MESSAGES
usermsg:                     ;first message to user, blanks at end erase old results
       db 'Please enter the angle from 00 to 90 degrees                $'
promptmsg:             ;second message to user
       db 'Enter q to quit or a for next angle $'
;END OF MESSAGES

;LOOK UP TABLES
hextable:     ;changes decimal tens key to hex
;decimal   0  10  20  30  40  50  60  70  80  90
       db 00h,0ah,14h,1eh,28h,32h,3ch,46h,50h,5ah
sinetable:    ;table is organized as X.XX for the angle entered
;degrees  0   1   2   3   4   5   6   7   8   9   10  11  12
       db '0.000.020.030.050.070.090.100.120.140.160.170.190.21'
;degrees  13  14  15
       db '0.220.240.26'
;degrees  16  17  18  19  20  21  22  23  24  25  26  27  28
       db '0.270.290.310.330.340.360.370.390.410.420.440.450.47'
;degrees  29  30
       db '0.480.50'
;degrees  31  32  33  34  35  36  37  38  39  40  41  42  43
       db '0.520.530.540.560.570.590.600.620.630.640.660.670.68'
;degrees  44  45
       db '0.690.71'
;degrees  46  47  48  49  50  51  52  53  54  55  56  57  58
       db '0.720.730.740.750.770.780.790.800.810.820.830.840.85'
;degrees  59  60
       db '0.860.87'
;degrees  61  62  63  64  65  66  67  68  69  70  71  72  73
       db '0.870.880.890.900.910.910.920.930.930.940.950.950.96'
;degrees  74  75
       db '0.960.97'
;degrees  76  77  78  79  80  81  82  83  84  85  86  87  88
       db '0.970.970.980.980.980.990.990.990.991.001.001.001.00'
;degrees  89  90  91  92  93  94  95  96  97  98  99
       db '1.001.00>90D>90D>90D>90D>90D>90D>90D>90D>90D'
;everything above 85 degrees rounds to 1.00
code ends
```

NOTES ON EXAMPLE 4

1. The sine of the selected angle is stored as a series of ASCII bytes from the table into answer1 (X.) and answer2 (XX). A dollar sign ($) is placed in variable dollar.

2. Whatever has been looked up is placed in memory to be later displayed using DOS INT 21h, function 09h.

Tables may be used to do much more than hold numeric data. For instance, graphics can be done using tables of character strings to draw pictures on the monitor screen.

8.5 Structured Programming Techniques

Modular Programs

Most of the example programs in this text are quite small, consisting of a main program that may call a few procedures. The called procedures may themselves call other procedures, and so on. Most beginning programmers are very happy just to have the program work and could care less about such seemingly academic fluff as how the program is organized.

Beginning programmers frequently write what is known as *spaghetti code*. Spaghetti code is one single program that jumps all over, calls procedures that jump all over, and, in general, is impossible to follow or understand, except for very small programs. An early workplace scenario went this way: As beginning programmers became new hires, they continued writing their pasta programs. Deadlines began to be missed. The original programmer would leave for greener pastures. Months would be wasted as a new programmer became acquainted with the program left behind. Moreover, programs began to be so large that no one single individual could finish them in a reasonable time.

Today, the programming industry has evolved strategies to deal with the problem of large, complicated programs. One popular approach is to break a program into many parts, called *modules*. A single programmer can intensely comment on each module, making the program easier to understand. If time is short, many programmers may work on different modules to speed up development.

DEFINITIONS OF MODULAR PROGRAMMING

The *Computer Dictionary*[1] describes modular programming as follows:

> An approach to programming in which the program is broken into several independently compiled modules. Each module exports specified elements (constants, data types, variables, functions, procedures); all other elements remain private to the module. Other modules can use only the exported elements. Modules clarify and regularize the interfaces among the major parts of the program. Thus, they facilitate group programming efforts and promote reliable programming practices.

Another authority, Eric Isaacson,[2] describes modular programming in more concrete terms:

> Isolation of Function. Code that performs one kind of function should be gathered in one place. The number of entry points must be limited . . . so that the interface for those entry points can be cleanly defined.
>
> Isolation of Data Structures. The only data structures communicated between a module and its caller are the parameters passed into the module, and the answers returned. All intermediate data needed by the module should be declared within

[1]*Computer Dictionary* (Redmond, WA: Microsoft Press, 1991), p. 231.
[2]Letter to the author, dated January 1993.

the module and be invisible to . . . other modules, or the main program.

Use of CALL/RET, Not Flags, to Control Program Flow. A hallmark of nonmodular programming is the use of control flags whose meaning is, "From where was this piece of code jumped?" Such flags can always be eliminated by making the piece of code into a subroutine, and CALLing the subroutine from the different contexts.

For our purposes, modular programming is defined as follows:

1. A module is a procedure that performs a *single* program function.

2. Modules are *called* from other modules, or from the main program.

3. All *external* data needed by the module is passed to the module by the caller. Any result is passed from the module to the caller.

ADVANTAGES OF MODULAR PROGRAMMING

Modular programming is popular because it speeds up program development time. Modular programming lets us take a single large programming task and break it into many smaller pieces. Each program piece, or module, can then be attacked simultaneously by a *team* of programmers. Modularity gives the program team these advantages:

a. It is easier to write *and test* small program modules.

b. A library of standard modules can be developed and re-used.

c. Program changes can be confined to a single module.

d. Management can plan, organize, and direct program development as each module is developed, tested, and finished.

Modular programming is usually advantageous, even when the team consists of a *single* programmer. Modular programming that involves more programmers than one is not without some drawbacks, however. Team modular programming requires good management of the team.

TEAM PROGRAMMING MANAGEMENT

Activities that involve many people require good management of the project and the personnel. The project manager must ensure that the following items are addressed:

a. Each module's input and output data are clearly and completely defined. Registers that each module is *allowed* to permanently change are defined.

b. Each module's function is completely defined.

c. The modules work together with no data interference.

d. The modules are carefully and fully documented.

e. Module testing and debugging does not wait on the completion of other modules.

DISADVANTAGES OF TEAM PROGRAMMING

You must "give to get" in all things, and programming is no exception. Among the disadvantages of modular programming are these:

    a. Interchanging module data may result in redundant code, particularly in the main program.

    b. Execution time slows down as a result of calling the modules.

    c. Documentation standards may be difficult to standardize.

    d. Difficulty in testing the overall program increases with the number of people involved.

Despite the disadvantages listed, team programming is proving to be valuable for the rapid development of large programs. Individuals who use modular programming techniques will reap the advantages with very few of the team programming disadvantages.

AN EXAMPLE OF A MODULAR PROGRAM

Suppose that we wish to write a modular program that will display the ASCII code for keys pressed on the keyboard. The program is to produce the following results:

■ If the key is an ASCII printable key, then the display will show the hex ASCII code for the key and the symbol for the key.

■ If the key is an ASCII control key, then the ASCII definition for the key will be displayed (ESC for instance) after the ASCII code for the key.

■ Keys, such as function keys, that have no ASCII definition will cause the message Non-ASCII key to be displayed.

The number of modules needed can be estimated by a short list of program functions. It may be that additional modules will be defined as work progresses, but let us start with the following list.

Module	Function	To Module	From Module
1	Initialize display and cursor shape	Mode in AL Page number in BH Cursor size in CX	No data
2	Position cursor	Position in DX Page number in BH	No data
3	Get keyboard key	No data	Scan/ASCII in AX
4	Find message	ASCII code in AL	Pointer in DX
5	Control key names	ASCII code in AL	Pointer in DX
6	Display message	Pointer in DS:DX	No data
7	Display character and update cursor	Character in AL Page number in BH Color in BL Cursor position in DX	Cursor position in DX
8	Hex to ASCII numbers	ASCII code in AL	ASCII nibbles in BX

Parts of most of these modules have been written in this and previous chapters. The main program and modules follow.

```
;Modular program showascii main part. Variables and constants are
;declared and the various modules called as needed.

data segment
        org 8000h                   ;locate data area above code
        crsrpos dw ?                ;row and column position of cursor
        char db ?                   ;key pressed ASCII hex code variable
data ends
code segment
        color equ 74h               ;red ASCII hex characters, white background
        crsrsz equ 0007h            ;fat cursor
        mode equ 02h                ;Universal color text display mode
        pagenum equ 00h             ;use page 0 for D86 testing
        poscrsr equ 1800h           ;use row 24,column 0 for D86 testing

showascii:
        mov sp,2000h                ;set stack
        call _getkey                ;wait for key
        mov char,al                 ;store ASCII byte code for key
        mov al,mode                 ;text mode
        mov bh,pagenum
        mov cx,crsrsz
        call _initscrn              ;initialize screen and cursor size
        mov bh,pagenum
        mov dx,poscrsr              ;position cursor at posnum
        mov crsrpos,dx              ;save cursor position
        call _poscrsr               ;position cursor
        mov al,char                 ;get ASCII key code
        call _lookupmsg             ;display proper message for key hit
        call _display               ;display appropriate message
        cmp char,00h                ;loop back to start if non-ASCII key
        je showascii                ;screen will be cleared on next key
        mov dx,crsrpos              ;update cursor position
        add dx,0011h                ;17 characters in ASCII key message
        mov crsrpos,dx              ;DOS INT 21h, function 09h, moves cursor
        mov al,char                 ;get ASCII key code
        call _hextoascii            ;ASCII key code into two ASCII characters
        push bx                     ;save characters returned by conversion
        mov al,bh                   ;display first ASCII character
        mov bh,pagenum              ;desired page
        mov bl,color                ;desired color
        mov dx,crsrpos              ;current cursor position
        call _chardisp              ;display ASCII key code first hex numeral
        mov crsrpos,dx              ;save cursor position
        pop bx                      ;get characters
        mov al,bl                   ;display second ascii character
        mov bh,pagenum              ;desired page
        mov bl,color                ;desired color
        mov dx,crsrpos              ;current cursor position
        call _chardisp              ;display ASCII key code first hex numeral
        mov crsrpos,dx              ;save cursor position
        mov al,' '                  ;space after ASCII number for key
        mov bh,pagenum              ;desired page
        mov bl,color                ;desired color
```

```
            mov dx,crsrpos          ;current cursor position
            call _chardisp          ;display ASCII key code first hex numeral
            mov crsrpos,dx          ;save cursor position
            cmp char,20h            ;display control name or printable character
            jb showcontrol          ;if less than 20h then control
            mov bh,pagenum          ;desired page
            mov bl,color            ;desired color
            mov al,char             ;get printable character
            mov dx,crsrpos          ;current cursor position
            call _chardisp          ;display ASCII key code first hex numeral
            jmp showascii           ;wait for next key
showcontrol:
            mov al,char             ;get ASCII code for key
            call _ctrlname          ;get pointer to control name
            call _display           ;display control key type
            jmp showascii           ;get next key
```

;Module 1. Initializes the display for color and text. Sets cursor
;size and active page. Called from main program with mode in AL,
;page in BH, and cursor size in CX. No data returned.

```
_initscrn:
            mov ah,00h              ;select mode function
            int 10h                 ;BIOS sets mode via int 10h
            mov ah,05h              ;set page function
            mov al,bh               ;page in BH
            int 10h
            mov ah,01h              ;set cursor size based on CX
            int 10h                 ;BIOS sets cursor size
            ret
```

;Module 2. Sets cursor position on active page. Position passed in
;register DX, page in register BH. No data returned.

```
_poscrsr:
            mov ah,02h              ;position cursor
            int 10h                 ;cursor now positioned
            ret
```

;Module 3. Waits for key to be pressed on keyboard and returns scan
;(AH) and ASCII (AL) codes in AX. Called from main program.

```
_getkey:
            mov ah,00               ;get the key Scan/ASCII codes in AX
            int 16h                 ;wait for key to be hit
            ret
```

;Module 4. Finds correct message type (ASCII or non-ASCII) based
;on the ASCII key code passed in AH. Called from the main program,
;returns pointer to the message (ending in a $ for DOS INT 21h,
;function 09h) in register DX.

```
_lookupmsg:
            mov dx,offset asciimsg     ;assume the key is an ASCII key
```

```
            cmp al,00h              ;AL = 00 if non-ASCII
            jne ascii               ;assumption correct
            mov dx,offset notmsg    ;pointer to non-ASCII message
ascii:
            ret
asciimsg: db 'The key is ASCII $'   ;beginning of ASCII message
notmsg:   db 'Non-ASCII Key        $'   ;blanks erase any previous
                                        ;displayed message
```

;Module 5. Finishes ASCII message using ASCII control key name after
;displaying ASCII hex code for key on screen. Passed ASCII code in
;AL from main program, returns pointer to 'control name string in DX.

```
_ctrlname:
            mov dx,offset base      ;point to table base
            mov bl,04h              ;each table entry occupies 4 bytes
            mul bl                  ;AL times 4 in AX
            add dx,ax               ;add AX to base to find entry
            ret
base:
            DB 'NUL$'               ;ASCII control key names
            DB 'SOH$'
            DB 'STX$'
            DB 'ETX$'
            DB 'EOT$'
            DB 'ENQ$'
            DB 'ACK$'
            DB 'BEL$'
            DB 'BS $'
            DB 'HT $'
            DB 'LF $'
            DB 'VT $'
            DB 'FF $'
            DB 'CR $'
            DB 'SO $'
            DB 'SI $'
            DB 'DLE$'
            DB 'DC1$'
            DB 'DC2$'
            DB 'DC3$'
            DB 'DC4$'
            DB 'NAK$'
            DB 'SYN$'
            DB 'ETB$'
            DB 'CAN$'
            DB 'EM $'
            DB 'SUB$'
            DB 'ESC$'
            DB 'FS $'
            DB 'GS $'
            DB 'RS $'
            DB 'US $'
```

```
;Module 6. Passed DX by main program. Displays message pointed to by
;DX using INT 21h, function 09h to display messages that end in a $.
;Returns nothing. Assumes that messages are in current data segment.

_display:
        mov ah,09h              ;display message pointed to by DX:DS
        int 21h                 ;display to $
        ret
```

```
;Module 7. Displays a single character at the current cursor
;position and moves the cursor right one space. Passed character
;from main program in AL, color in BL, page in BH, cursor position in
;DX. Returns cursor position in DX.

_chardisp:
        mov ah,09h              ;display character using BIOS
        mov cx,01h              ;one character
        int 10h                 ;write character
        inc dx                  ;move cursor to next position
        call _poscrsr           ;position cursor
        ret
```

```
;Module 8. Converts single-byte ASCII code into ASCII characters for
;each nibble. Passed ASCII hex byte in AL from main program, returns
;equivalent ASCII characters for each nibble in BX.

_hextoascii:
        mov ah,al               ;duplicate byte
        mov cl,05h              ;rotate AH&carry left 5 times
        rcl ah,cl               ;right-justify high nibble
        and ax,0f0fh            ;isolate nibbles
        call cnvrt              ;convert AH nibble to ASCII
        mov bh,ah               ;store high nibble ASCII byte
        mov ah,al               ;convert AL nibble to ASCII
        call cnvrt
        mov bl,ah               ;store low nibble ASCII byte
        ret
cnvrt:
        add ah,30h              ;assume a number from 0 to 9
        cmp ah,3ah              ;if result > 39h then a letter A–F
        jb numb                 ;if result < 40h then a number 0–9
        add ah,07h              ;letter—convert to 41h–46h (A–F)
numb:
        ret
```

As may be clearly seen in the main program, many lines of code are used to pass the same data to several modules, notably _chardisp. Module _chardisp is generic, however. Including the page, color, and cursor location inside _chardisp would make it useful only for those unique parameters. If the caller and the module must interchange a great deal of data, then it may be advantageous to pass pointers to data rather than the data itself.

Some might argue that the procedure named cnvrt in Module 8 should itself be a module because it converts a hex numeral into an equivalent ASCII character. There is, unfortunately, no absolute measure of how *fine* modules must be, or how broadly the word *function* may be defined.

Structured Programming

Structured programming involves applying high-level language programming concepts to assembly code programming. The term *structured* applies to certain standard high-level program statements that can be translated, neatly, to assembly code.

High-level language programmers try to use a few basic program structures to solve all problems. Anyone who has programmed in BASIC, Pascal, or C is familiar with the following standard program structures (or ones that are similar):

```
IF . . . THEN . . . ELSE
DO . . . WHILE
DO . . . UNTIL
CASE
```

Structured programming allows the programmer to think in *pseudocode*. Pseudocode is an English-based descriptive sequence of operations that the programmer wishes to carry out in the program. Programming in pseudocode lets the programmer concentrate on the "big picture" before getting bogged down in the minutia of assembly code mnemonics.

Pseudocode and flow charts are two ways of thinking about the program. Pseudocode is a verbal thinking process; flow charts are a graphical thinking process. Both pseudocode and flow charts are useful to the programmer who is new to a particular assembly code, because it lets the programmer concentrate on essential features of the program.

An example of a small pseudocode program involves searching a finite area of memory until a certain byte of data is found. The program must carry out the following operations, programmed in pseudocode.

```
Point to data area
DO search UNTIL beyond data area
IF data byte THEN exit ELSE continue
```

Once the program is defined in pseudocode the programmer can code the various high-level structures into assembly code.

IF . . . THEN . . . ELSE PROGRAM STRUCTURE

Compare and jump instructions form the basic structure for IF . . . THEN . . . ELSE program structures. From our previous pseudocode example of looking for a certain byte, we can construct an IF . . . THEN . . . ELSE in assembly code as follows.

```
next_byte:  cmp [bx], bytenum    ;compare data with desired byte
            je gotit             ;IF data byte THEN exit
            jmp next_byte        ;ELSE continue
gotit:      nop                  ;the exit point
```

The IF part of the structure is part of the compare and jump opcode sequence. A compare sets the appropriate flags, and the jump is THEN taken IF the conditions are met. ELSE is accommodated by the fact that the program continues at the instruction following the conditional jump.

DO . . . WHILE AND DO . . . UNTIL PROGRAM STRUCTURES

DO . . . WHILE and DO . . . UNTIL both involve loops in the program. A loop is entered under one set of conditions and exited under another set of conditions.

The DO . . . WHILE structure tests a condition when the loop is *entered* and continues looping as long as the condition is satisfied.

The DO . . . UNTIL loop enters the loop and tests a condition at the *end* of the loop. Looping continues until the tested condition is satisfied. Both DO . . . WHILE and DO . . . UNTIL structures generally accomplish the same function.

Using our previous byte search example, a DO . . . WHILE loop that searches through a limited area of memory appears as follows.

```
            mov bx, start-1      ;point to starting point in memory
next_byte:  inc bx               ;increment to next byte
            cmp bx, finish+1     ;DO loop while in desired memory area
            je dunit             ;UNTIL beyond desired data memory area
            cmp [bx], bytenum    ;compare data with desired byte
            je dunit             ;IF data byte THEN exit
            jmp next_byte        ;ELSE continue
dunit:      nop                  ;next program structure
```

The DO . . . WHILE loop begins at memory location start and does the byte search while within the bounds of memory defined by finish. The test for BX pointing to the end of memory, cmp bx, finish+1, is made at the beginning of the DO . . . WHILE structure. The constants start, finish, and byte can all be defined in EQU statements.

The same program using a DO . . . UNTIL program structure can be written as shown next.

```
            mov bx, start        ;DO loop, point to beginning address
next_byte:  cmp [bx], bytenum    ;compare data with desired byte
            je dunit             ;IF data byte THEN exit
            inc bx               ;ELSE continue, point to next byte
            cmp bx, finish+1     ;test at end of loop
            jne next_byte        ;UNTIL out of data area
dunit:      nop                  ;next program structure
```

The test for the end of memory, cmp bx,finish+1, is done at the end of the DO . . . UNTIL loop.

Choosing between a DO . . . WHILE and a DO . . . UNTIL is largely a matter of personal preference.

CASE PROGRAM STRUCTURE

CASE statements are used to select one course of action from among many competing choices. A simple CASE structure can be made up of repeated compare and jump instructions. As an example, suppose that we wished to display a different message for any key pressed on the keyboard. A CASE statement might appear as follows.

```
CASE inkey OF
  'a' : msga
  'b' : msgb
      .
      .
  'z' : msgz
```

The assembly language equivalent of a CASE structure could be written as follows.

```
mov ah, 00h          ;get key
int 16h              ;BIOS to get key, program waits
cmp al, 'a'          ;CASE OF a?
je msga              ;if so, display message number 1
cmp al, 'b'          ;CASE OF b?
je msgb              ;if so, display message number 2
cmp al, 'c'          ;CASE OF c?
je msgc              ;if so, display message number 3
cmp al, 'd'          ;CASE OF d?
je msgd              ;if so, display message number 4
;and so on until all the compares have been done
msga:   nop          ;dummy programs
msgb:   nop
msgc:   nop
          .
msgz:   nop
```

Using repeated compares, as done in the last example, is very time-consuming from both a programmer and CPU standpoint. A better CASE structure can be made by using the register indirect call feature that is discussed in Chapter 6. Register indirect calls are made to procedures whose addresses are pointed to by the contents of a register. The register contents can be changed, as the program runs, to call procedures appropriate to the current register contents.

In our next CASE structure example, a table of procedure addresses is constructed when the program is written, and the key pressed can select the appropriate procedure address by loading a pointing register with a key-derived pointer. Register indirect calls then access the procedure for that key. A simple CASE structure, case.asm, for keys *a through d* is shown next.

```
;a CASE structure program case.asm                          See Note
code segment
nextkey:
        mov ah, 00h     ;get key
        int 16h         ;BIOS waits for key
        cmp al,'q'      ;IF q THEN return to DOS              1
        je xit
        mov ah,00h      ;mask out key scan code
        mov bx,ax       ;point to offset in call table        2
        add bx,bx       ;convert to word address in table
        call table[bx]  ;CASE OF ASCII code in AL             3
        jmp nextkey     ;get next key
xit:
        mov ax,4c00h    ;back to DOS
        int 21h
```

```
table:          ;make CASE OF table
                org table+2*'a'    ;offset for key "a"                              4
;the table of procedure addresses begins here
                dw msga            ;addresses of key procedures                    5
                dw msgb
                dw msgc
                dw msgd
;dummy routines for each key are placed here
msga:           ret                ;dummy procedures, for testing
msgb:           ret
msgc:           ret
msgd:           ret
code ends
```

NOTES ON THE CASE PROGRAM

1. If it is *physically* possible for users to press some other key besides q or a–d, they will. Place a short piece of code after int 16h to make sure that only "legal" keys are pressed.

2. The procedure address table base starts at code address table:. Each procedure address begins on an even word boundary in the table. For our example, characters a through d are ASCII codes 61h to 64h. Procedure addresses are located at offset C2h to C8h in the table.

3. The key offset is added to the table base to find the address, in memory, of the key procedure.

4. The first part of the procedure address table is empty, because we have no entries before a, or 61h×2 words into the table. Originate the address for an a key procedure at 61h×2 = C2h offset from the base of the table. Or, the procedure addresses can be placed starting at the base of table, and every ASCII key code have the ASCII value of a subtracted from it.

5. The assembler places the address of each key's procedure name starting at offset C2h in the table. Although the key procedures are listed next in the example for debugging purposes only, they could be anywhere in the code segment.

Very little programmer effort has been saved by making a CASE table, but the program will find the proper message in one call cycle instead of using repeated compares, as in the first CASE example.

8.6 Port Input and Output

Memory Space and I/O Space

As mentioned in Chapter 3, computer memory can be divided into two distinct "spaces": memory space and input/output space. Memory space is composed of *hundreds of thousands* of byte addresses located in identical DRAM and ROM integrated circuits. Memory space is the only type of memory used up to this point in the book.

Input/output (I/O) space is composed of *thousands* of byte addresses called *ports* located in specialized and nonsimilar hardware circuits. A port is any *hardware* circuit address that is located in I/O space, such as the keyboard, the monitor adapter, or the printer interface ICs. The BIOS interrupt service routines of Chapter 7 deal extensively with I/O ports, although that fact is hidden from the programmer by the interrupt mechanism. Chapters 11 and 12 make use of direct access to I/O ports without using BIOS interrupt services. For the sake of convenience, the I/O space instructions are included here.

IN   Input Data from Port Address

Mnemonic	Operands	Size	Operation	Flags
IN	AL/AX,Port	B/W	AL/AX ← (Port)	None
IN	AL/AX,DX	B/W	AL/AX ← (DX)	None

Addressing Modes
(Destination, Source)

register, I/O direct
register, I/O indirect

Examples:

in al,32h	;move the byte at port number 32h to AL
in ax,32h	;move the word at port number 32h to AX
in al,dx	;move the byte at port number in DX to AL
in ax,dx	;move the word at port number in DX to AX

Operation:

IN moves data from an I/O address (a port number) to the AL or AX register. Byte moves use AL as a destination; word moves use AX. The I/O port address may be specified as 00 to FFh using a direct port number. Ports 0000 to FFFFh may be *indirectly* addressed using DX to hold the I/O address. Note that *word*-length data (signified to the assembler by the use of *AX*) moves the byte at address (port+1) to AH, and the data at address (port) to AL.

OUT   Output Data to Port Address

Mnemonic	Operands	Size	Operation	Flags
OUT	Port,AL/AX	B/W	Port ← AL/AX	None
OUT	DX,AL/AX	B/W	(DX) ← AL/AX	None

Addressing Modes
(Destination, Source)

I/O direct, register
I/O indirect, registerZL

Examples:

out 32h,al	;move the byte in AL to port number 32h
out 32h,ax	;move the word in AX to port number 32h
out dx,al	;move the byte in AL to the port number in DX
out dx,ax	;move the word in AX to the port number in DX

Operation:

OUT moves data from the AL or AX register to an I/O address (a port number). Byte moves use AL as a source; word moves use AX as a source. The I/O port address may be specified as 00 to FFh using a direct port number. Ports 0000 to FFFFh may be addressed using DX to hold the I/O address.

IN and OUT are the only I/O instructions available to us for moving data between the CPU and peripheral devices. I/O is very hardware-oriented. The BIOS and DOS interrupt service routines have permitted us to interface to hardware with no general knowledge of which ports are involved. We shall make use of IN and OUT in Chapters 11 and 12, where the focus of the book shifts from software to hardware interfacing.

8.7 Summary

Various specialized instructions, which augment the basic instructions discussed in Chapter 5, are presented. The specialized instructions include the following major categories and types:

Mathematical

AAA	ASCII Adjust for Addition
AAD	ASCII Adjust for Division
AAM	ASCII Adjust for Multiplication
AAS	ASCII Adjust for Subtraction
CBW	Convert Byte to Word
CWD	Convert Word to Double Word
DAA	Decimal Adjust Addition
DAS	Decimal Adjust Subtraction
DIV	Divide
MUL	Multiply

String Operations

CLD	Clear Direction Flag
CMPSB/W	Compare String Byte/Word
LDS	Load DS and Pointing Register
LES	Load ES and Pointing Register
LODSB/W	Load String Byte/Word
MOVSB/W	Move String Byte/Word

REP   Repeat String Operation
REPNE   Repeat String Operation While Not Equal
REPE   Repeat String Operation While Equal
SCASB/W   Scan String Byte/Word
STD   Set Direction Flag
STOSB/W   Store String Byte/Word

Effective Address Generation
LEA   Load Effective Address of Label

Port Input and Output
IN   Input Data from Port
OUT   Output Data to Port

Several contemporary programming practices and techniques are discussed, as reviewed next.

Lookup tables are used in programs to speed up program execution speed and to enable the program to generate solutions to problems that cannot be mathematically computed.

Modular programming permits long programs to be divided into smaller modules that can be written simultaneously by teams of programmers.

Structured programming involves converting standard high-level language constructs to assembly code. The standard high-level structures include:

IF . . . THEN . . . ELSE
DO . . . WHILE
DO . . . UNTIL
CASE

Input/output space is defined as memory locations that are the addresses of hardware registers called ports. Port access enables program control of external interface circuits. Access to the ports by the program is made possible using IN and OUT instructions.

8.8 Problems

All of the problems that refer to *memory* refer to data segment 6000h, offset addresses 0000h to 0FFFh, for a total of 1000h bytes. Words are defined to start on even addresses. Memory is to be processed from low to high. MAKE SURE YOUR PROBLEM DOES NOT EXCEED THE MEMORY BOUNDS.

Exercise Problems

1. Fill memory with the last two digits of your social security number.

2. Load register AX with the first 2 bytes of program code that you have written to do this problem.

3. Swap the last 500h bytes of memory with the first 500h bytes of memory. The byte at offset address 0000 swaps with the one at offset address 0800h, and so on.

4. Add the ASCII numbers 9 and 9 together, and convert the result to an equivalent ASCII number in AX.

5. Subtract the ASCII number for 5 from the ASCII number for 8, and place the ASCII result in AX.

6. Convert any hexadecimal number from 00 to 63h placed in AX into its equivalent unpacked BCD number in AX.

7. Convert any unpacked BCD number placed in AX into an equivalent hexadecimal number in AL.

8. Add two packed BCD numbers, neither of which is greater than 49d, together in register AL, and convert the answer to packed BCD in register AL.

9. Subtract the packed BCD number 32d from the packed BCD number 87d in register AL, and convert the result to packed BCD in register AL.

10. Multiply the byte in register AL by the byte in register BL. Place the result in register AX.

Intermediate Problems

1. Determine how many times the number 4Ah occurs in memory and place the count in register BX.

2. Assign numbers to the alphabet from 00 for capital A to 25d for capital Z. Offset location 0800h in memory can contain the ASCII code for any letter from A to Z. Convert the ASCII code for the letter at offset location 0800h to an equivalent ASCII code for the decimal number assigned to the letter in register AX.

   Example:

   [0800h] = P ;decimal equivalent is 15d

   AX = 3135 after conversion (ASCII 15)

3. Reverse Problem 2. Take the decimal equivalent of the ASCII numbers found in AX, and convert the numbers into the ASCII letter equivalent for the numbers in offset 0800h.

   Example:

   AX = 3030 ;ASCII for 00

   [0800h] = 41h after conversion (ASCII A)

4. Convert any binary number found in register AX to the equivalent ASCII code for each hexadecimal nibble in AX. Place the ASCII code for the hexadecimal number in AX starting at offset location 0000.

   Example:

   AX = 1010010111000011b = A5C3h

   [0000h] = 41354333h

5. Reverse Problem 4. Convert the ASCII codes for hexadecimal numbers found starting at offset 0000, and convert them to equivalent binary numbers in register AX.

Example:

[0000h] = 35414237h = 5AB7h

AX = 0101101010110111b

6. Registers CX and DX may contain any combination of ASCII-coded characters. Find the first occurrence of the same sequence of ASCII-coded character in memory, and place the address in memory where the sequence begins in register BX.

7. Register AX may contain any number such that AH is greater than AL. Count all of the numbers in memory, using register CX, that are less than the number in AH and greater than the number in AL.

8. Take your last name and convert it into capital letters. Have your program store your name repeatedly in memory from the beginning address pointed to by SI up to the beginning address pointed to by DI. The last entry of your name must be complete.

Example:

SI = 0000, DI = 0100h

Name = AYALA (5 characters)

Store from 0000 to 0103h (260/5 = 52 AYALAs)

9. A transition is defined as a 1 followed by a 0 in a word of data. Have your program count the number of transitions in every word in memory and store the total number of transitions in registers CX (most significant word) and DX (least significant word).

Example:

AAAAh = 1010101010101010b = 8 transitions

0001h = 0000000000000001b = 0 transitions

8888h = 1000100010001000b = 4 transitions

5555h = 0101010101010101b = 7 transitions

10. Find the location of the smallest byte in memory and place its address in register BX. In case of multiple ties, use the address of the first number found.

11. Find the location of the largest byte in memory and place its address in register BX. In case of multiple ties, use the address of the last number found.

12. Convert any byte in memory that contains the hex character A in either nibble of the byte to the ASCII code for A, 41h.

Example:

[0100h] = A3h then change [0100h] = 41h

13. Move all words in memory offset location 0000 to 07FEh that match the word placed in CX to memory offset locations 0800h to 0FFEh.

14. Convert any ASCII bytes in memory that represent a hexadecimal number (30h to 39h and 41h to 46h) to the hexadecimal equivalent of the ASCII code.

Example:

[0100h] = 42h (ASCII "B") then [0100h] = 0Bh

15. Add every word in memory to the random number placed in AX. If the result is more than the number placed in CX, then make the memory word equal to 0000. If the result is greater than or equal to the number in CX, then make the memory word equal to FFFFh.

16. Modify the second example of a CASE structured program and have each key routine display the key pressed on the screen. Screen out all illegal keys.

Advanced Problems

1. Find the address where the largest sum of 10 (decimal) sequential words begins in memory. Put the winning address in register DX, and the winning sum beginning at address 2000h (MSW). If two or more winners are found, put the address of the last winner in register DX.

2. Find the address where the longest consecutively numbered string of bytes begins. Consecutive bytes are defined as those that differ by the number 01. For example, the bytes 39h, 3Ah, 3Bh, 3Ch, 3Dh, 3Eh, 3Fh, and 40h form a string of 7 consecutive bytes. Put the winning address in register DX, and the length of the string (in bytes) in register CX.

3. Find the most numerous byte number in memory and store it in memory location 2000h. Then find the next most numerous byte number and store it in memory location 2001h. Continue this until you have found all pairs and stored them. If two byte numbers occur in equal frequency, then store the numerically larger byte first in memory.

4. Spell as many complete versions of the message "Go Catamounts" as you can using the necessary ASCII characters you find stored for you in memory. Once you have used a character from memory, you may not use it from the same location again. Store the completed messages beginning at offset address 2000h in memory. Do not store incomplete messages. Be sure to include the space between *Go* and *Catamounts*, and use capital *G* and capital *C* where appropriate.

5. A random number is placed in register AH. Sort all the bytes in memory so that those that are closest to the number in AH are at the low end of memory and those that are farthest from the number in AH are in high memory. Closest means that the absolute value of the quantity $|(\text{memory}) - (\text{AH})|$ is smallest. For example, if a 3Dh is placed in AH, then 3Ch and 3Eh are equally close to 3Ch. Numbers do not roll over, that is, FFh is not close to 00. Numbers that are equally distant from AH are stored numerically larger first.

6. Sort all of the bytes from 0000 to 000Fh in ascending order, with the smallest byte in address 0000 and the largest in address 000Fh. Then sort the next 16d bytes, from 0010h to 001Fh in descending order, with the largest byte in address 0010h and the smallest byte in address 001Fh. Continue alternating the sort until you have sorted the bytes from 0FFF0h to 0FFFh.

7. A random number is placed in register AX. Find all pairs of words in memory that, when added together, equal the number in AX with no carry. Store the pairs, in the order found, starting at offset address 2000h. Do not use any word more than once.

8. Imagine that memory is arranged in 256, 16 (decimal) byte rows. Number each row from 00 to FFh. Store the row numbers, beginning at offset address 2000h, that have 1 or more identical bytes in identical places in the row. For example, if rows 12h, 4Ah, and 78h have a 33h in Location 6 in each row, then store 12h, 4Ah, and 78h starting at address 2000h. Do not use any row more than one time.

9. Sort all of the words in memory that can be considered to be decimal ASCII-coded numbers. An ASCII-coded word can vary from 3030h (00d) to 3939h (99d). Sort the ASCII words so that the smallest word is at offset 0000.

10. Sort all of the addresses in memory based on the contents of the byte contained at the address. Sort beginning with the addresses that contain the largest byte. Place the addresses starting at memory offset location 2000h, which contains any addresses that contain an FFH. Addresses smaller than FFh then follow until any addresses that contain 00 are placed last (highest) in memory.

9

DOS, DISKS,
AND FILES

Chapter Objectives

On successful completion of this chapter you will be able to:

- Describe how binary data is stored on a magnetic disk.

- Describe how data is stored on a DOS-formatted disk.

- Describe how DOS locates file data on a disk using directories and file allocation tables.

- Use DOS Interrupts to directly read and write data to disk sectors

- Use DOS file handle interrupts to create, open, read, write, and close files.

- Use DOS interrupts to read and write random file data.

9.0 Introduction

DOS is, of course, an abbreviation for *disk operating system.* Operating systems exist in order to operate the I/O peripheral devices that are connected to the CPU, including the system printer, the system console, and any system mass memory storage devices. Operating systems, as noted in Chapter 7, also serve as an interface between user and computer so that user application programs can be run on the computer.

Operating systems running on PC-compatible computers use the concept of *software* interrupts (INT) to provide application programs with a standard *software* interface to computer peripherals. Chapter 7 discusses a few DOS interrupts that control the system keyboard and monitor peripherals. However, most DOS INT service routines have to do with controlling the system mass memory data storage disks.

Mass memory storage, compared to internal high-speed solid-state memory, is usually cheaper, slower, and magnetic in nature. Magnetic media have been used as mass storage in computer systems for many years, beginning with early adaptations of analog audio magnetic tape drives for binary data recording purposes. Magnetic-media based floppy and hard disks were first developed in the early 1960s and are the preferred mass storage mediums for the majority of contemporary computer systems in use today. Optical disks, however, are now appearing commercially at prices that may challenge magnetic disks in the near future.

Magnetic tape retains a position in computer systems as a backup or *archive* mass storage for data stored on disks. Data stored on hard disks is frequently backed up on tape in case of the *inevitable* hard disk *crash* when the disk bearings wear out, or the disk head positioning mechanism fails.

Magnetic tape is seldom used for "online" storage, because of the time required to access data. Tape stores data sequentially, and the tape must be wound and unwound in order to access data stored at different points on the tape. Magnetic disks, in contrast, may be accessed rapidly because the entire surface of the disk is available at any time. Disks spin on a central axis, much like a circus merry-go-round, and all data areas rotate past a fixed point much faster than does a data area on tape.

DOS is rich with interrupt service routines that can handle all aspects of storing and retrieving data on floppy and hard disks. In order for the DOS disk service routines to work, data on the disk must be *organized* in a rigidly defined manner so that DOS can operate predictably and dependably. We call the way data is organized on the disk a *file* system.

Magnetic data files have very little in common with old-fashioned paper folders stored in metal file cabinets, but the name *file* does conjure up mental images of storing information in some organized manner. All files, magnetic or paper, have *names* that help identify the type of data the file contains. Paper or magnetic files also can be retrieved and stored using some sort of file *indexing* system. Paper files might be indexed alphabetically or by some sequential numbering system. Magnetic files are indexed by *volume, directory, subdirectory,* and *name.*

This chapter is all about disks, how files are stored on disks, and how DOS can be used to manipulate disk files. We shall begin by discussing how bytes of data are stored on magnetic disks.

9.1 Disk Construction

A magnetic disk is a platter covered with a thin coating of magnetic material. Binary data is stored at a spot on the disk by magnetizing the spot in a certain magnetic pole orientation. Spot magnetization is made possible by an electromagnet wire coil that carries current in one of two directions in order to obtain the desired magnetic pole orientation. The coil is in direct contact with the magnetic surface when used with floppy disks, and "flies," on a cushion of air, just above hard disk surfaces.

Data is recovered from the disk by generating voltages in a wire coil as a magnetized spot moves past the coil. The same coil used to magnetize the spot can also be used to generate the voltage, and the coil is thus dubbed a *read/write pickup*. Data recovery is "nondestructive", that is, generating voltages does not alter the magnetic spot and data is stored on the disk until it is erased by magnetic action. The assembly that carries the read/write coil is called a magnetic *head*. The head can move in and out over the surface of the disk as the disk rotates under the head.

The amount of data that can be stored on a disk is a function of disk area (usually indicated by disk diameter) and the smallest spot on the disk surface that can be distinctly magnetized. Disk technology is constantly evolving greater storage densities by using better magnetic disk coatings and smaller heads. The positioning of the head is important also, because data access must be repeatable. The head must be able to reliably return to exactly the same place on the disk, time after time, in order to read and write each spot.

The time required to find a particular spot on the disk, called the disk *access time*, is a function of the rotational velocity of the disk and the time needed to position the head to the desired spot. Speed, as will be noted often, costs money, and how fast do you wish to go? Inexpensive 100-megabyte (100M) hard drives, for instance, currently feature access times of from 12 to 19 milliseconds and cost $2 per megabyte. Disk drives are continuously becoming smaller, denser, faster, and cheaper.

Disks are generally available in two configurations: floppy disks and hard disks. Floppy disks hold less data than a hard disk of the same physical size because of the poorer quality of their magnetic material and the relative sloppiness of floppy drive head positioning systems. Floppy disks do possess one highly regarded quality: They can be removed from the floppy disk drive for safe storage and transportation of data. Hard disks normally are fixed inside the sealed hard disk drive and cannot be removed by the user. DOS views both floppy and hard disk types the same way and stores and retrieves data files on both drive types using the same methods.

Floppy disks are often used to store valuable data and as a distribution medium for application programs. Hard disks are used to store application programs and resultant data from these programs. Despite great leaps in reliability, hard drives are still rotating devices and *must* fail eventually. Current reliability figures for hard drives, specified as *mean time between failures* or MTBF, is 50,000 to 150,000 hours. A MTBF of 100,000 hours yields a mean lifetime of 50 years for a drive used daily—a very long time. Unfortunately, a mean is the average of a set of numbers; half the failures occur before and half occur after the MTBF. Furthermore, the MTBF numbers are calculated, not experimental. We should know in about 20 years how accurate they are. To be fair to the hard drive manufacturers, however, it should be noted that more data is probably lost to human error than to disk failure. *Back up irreplaceable data.*

Physical Disk Organization

A magnetic disk is organized as a set of *concentric* circles that surround the center of the disk. Each concentric circle is called a *track,* and each track is broken up into pieces called *sectors.* Figure 9.1 shows the organization of the disk surface into tracks and sectors.

As shown in Figure 9.1, a disk is organized into *physical* (real, touchable, you-can-see-it) areas of the disk surface that are assigned standard numbers for identification. Tracks on the disk are numbered from 0, the outermost track, up to the highest number of tracks found on the disk. Each track is further divided into *physical* sectors. Sectors, in a most un-binary manner, are physically numbered starting at sector number 1. Track numbers are unique, but sector numbers repeat for each track. Thus each track, for instance, has a sector number 6, but there is only one Track 6 on each disk surface. Using standard *physical* identification numbers, everyone in the disk manufacturing business can discuss their products with some degree of compatibility.

Disks have two sides, and each side is made up of tracks and sectors. Disk sides are numbered Side 0 or Side 1. One or both sides of the disk may be organized into tracks and sectors. Hard disk drives may contain *many* disk *platters,* and like numbered tracks on the stack of platters are called *cylinders* of tracks. Cylinders are numbered starting at 00 to correspond to the tracks that make up the cylinder.

Each track on the disk stores binary data *sequentially* as a circular path of

Figure 9.1 ■ Physical disk organization.

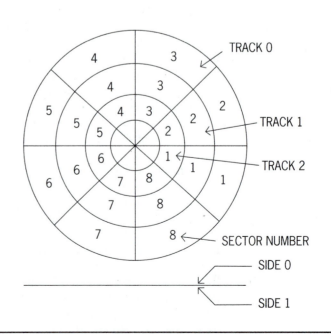

1s and 0s. Depending on the disk type, various numbers of bits are grouped into a single sector. Data on disks are stored sequentially in each sector, and a sector is the smallest disk data surface that can be identified by a unique number. Every track is available to the disk drive head for data access, and, as the disk rotates past the disk drive head, all of the data sectors pass under the head.

Although it would be possible to store data randomly anywhere on the disk surface, it certainly would not be practical. Data access time is important to users. Truly random data access would be very time-consuming as the head waits for 1 byte from each of hundreds of different segments to rotate into its field of view. To make better use of disk head access time, therefore, the head is positioned to the desired track and then an entire sector is read once it rotates under the head. Data is written and read in *sector-sized* batches, even if some sector recording space is wasted in the process.

Logical Disk Organization

Disk manufacturers number data sectors in a certain order, which is the physical arrangement of the sectors on the disk. DOS, however, numbers the sectors in a *logical* order, which does *not* correspond to the physical arrangement of sectors on the disk. DOS numbers sectors on the disk from an efficiency standpoint. Thus, it is not necessary that logically numbered sectors be physically *adjacent* to each other. Logical numbering, particularly when used with double-sided disks, contains physical breaks between sectors when numbering sectors on different sides of the disk.

Figure 9.2 shows how DOS numbers sectors. Track 0 on Side 0 contains the first sector group, and then Track 0 on Side 1 contains the next sector group. Sector numbering then continues at Track 1, Side 0, then Track 1, Side 1; and so on until the last track on Side 1 is used up. DOS *logical* sector identification starts counting at the acceptable binary sector number 0 and continues to the last sector number on the disk (or disks). DOS logical sector numbers start at 0 and count up; they do *not* repeat.

BIOS, it should be noted, uses *physical* numbering (side, track, and physical sector number) to access sectors on the disk. BIOS can access any sector

Figure 9.2 ■ Logical sector numbering.

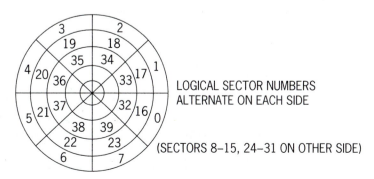

directly, and BIOS is not concerned with how the disk might be arranged for use by DOS. Please refer to Appendix C for a listing of BIOS disk access interrupt service routines.

We shall be concerned with *logical* sector numbering because this is how DOS identifies sectors. Unlike BIOS, DOS can only access disks that have been *formatted* to a DOS standard configuration. Formatting means that information about how data is stored on the disk is prerecorded on the disk by a DOS format program.

Floppy Disk Storage Capacity

Floppy disk storage capacities have been increasing (doubling about every 3 years) since DOS version 1.0 was first introduced. Certain landmark versions of DOS were written to make use of increasing floppy and hard disk capacities. Table 9.1 shows floppy disk capacities and the corresponding versions of DOS that can make use of the higher capacities.

Any DOS version that can access higher density floppy disks can also access disks of lower capacity, so that newer drives can read and write older disks.

Floppy disk capacities have increased by increasing the number of sides, increasing the tracks per side, and increasing the sectors per track. The adjectives used to describe disk density, such as *high* or *extended,* are industry standard names.

Table 9.2 lists floppy disk track and sector formats for disks whose capacity grows from 160K bytes to 2,880K bytes. Note that double-density, single-sided 160K disks are now completely obsolete and may probably be found only in rare instances.

Table 9.2 shows that disk capacities have increased by increasing the number of sides, tracks per side, and sectors per track. Interestingly, the number of bytes per sector has remained *constant* as disk densities have increased. Sector size has largely been frozen at 512 bytes by DOS standardization. Although DOS has the capability of handling sectors of 128 to 1,024 bytes, many parts of DOS *assume* a sector size of 512 bytes. One must assume that 512 bytes per sector has proven to be a convenient size for most DOS applications to date.

To find the capacity of a disk, in bytes, multiply the number of sectors by 512. The total numbers of sectors is found by multiplying together the num-

Disk Size	Bytes	Type	Density	I.D.	DOS Version
5.25	160K	Single-Sided	Double	DD	1.0
5.25	180K	Single-Sided	Double	DD	2.0
5.25	360K	Double-Sided	Double	DD	2.0
5.25	1,200K	Double-Sided	High	HD	3.0
3.50	720K	Double-Sided	Double	DD	3.2
3.50	1,440K	Double-Sided	High	HD	3.2
3.50	2,880K	Double-Sided	Extended	ED	5.0

Table 9.1 ■ DOS Versions and Floppy Disk Capacities

Disk Capacity	Sides	Tracks/Side	Sectors/Track	Bytes/Sector
160K	1	40	8	512
180K	1	40	9	512
360K	2	40	9	512
720K	2	80	9	512
1,200K	2	80	15	512
1,440K	2	80	18	512
2,880K	2	80	36	512

Table 9.2 ■ Floppy Disk Formats

ber of sides, the number of tracks per side, and the number of sectors per track. For example, a 720K floppy disk's byte capacity is found by multiplying the following numbers:

$$2 \times 80 \times 9 \times 512 = 737,280 \ (?) = 720 \times 1,024!$$

In binary usage, K means 1,024. In the metric system, k means 1,000. Thus, 737k bytes become 720K bytes of data. (A survey of sales literature reveals that the K and k are now used interchangeably.) The raw data-holding capacity of a disk exceeds the amounts listed in Table 9.2. For instance, 720k disks are sometimes specified as capable of holding 800k to 1,000k of data. Apparently some disk capacity is "wasted" when the disk is formatted.

Hard Disk Storage Capacity

Floppy disk capacities and physical dimensions are standardized because of the large numbers that are sold and the influence of dominant vendors. Hard disks, because they are not removable, do not conform to exactly standardized data capacities. A certain amount of hard disk capacity standardization has come about as a result of competitive market pressures, but it is quite common to buy a 40M byte hard drive that is actually 42M in capacity. Hard drives have hundreds of tracks per side and may have two or more disk platters per drive. The number of sectors per track are 17 to 63 for contemporary hard disks. Bytes per sector have remained, predictably, at 512.

Hard drive capacities have evolved along with that of floppy drives, and versions of DOS appeared to meet the demands of large hard disk byte capacities. Table 9.3 shows milestone DOS versions that can be used with hard disks up to the capacity listed.

Maximum Hard Drive Capacity	DOS Version
10M	2.0
32M	3.0
256M	4.0
2,000M	5.0

Table 9.3 ■ DOS Versions and Hard Disk Capacities

We shall *not* pursue using DOS for hard drive access for two reasons: First, not everyone is so affluent as to have a hard-disk-drive-equipped computer. Second, and *most* important, a great deal of trouble can come about as we experiment with DOS hard disk access routines. As a matter of fact, it is very possible that we will find ourselves *unexpectedly* using the hard drive on our computer when we actually only wished to access one of the floppy drives.

```
* * * * * * * * * * * * * * * * * * * * * * * * * * * * * * * * * * * * * * * * * * * * * * * * * * * * * * * * *
*                                                                                                               *
*     BE VERY CAREFUL WHEN USING DOS DISK ACCESS ROUTINES                                                       *
*                                                                                                               *
* * * * * * * * * * * * * * * * * * * * * * * * * * * * * * * * * * * * * * * * * * * * * * * * * * * * * * * * *
```

to prevent inadvertent "trashing" of your hard drive.

One way to prevent hard drive problems is to physically disconnect the hard drive controller from the hard drive; another is to use an inexpensive single-floppy PC for disk access experiments. If disconnecting the hard drive is impractical, back up the hard drive. You will very likely trash the hard drive sooner or later as you play around with writing to floppy disks.

9.2 How DOS Organizes Disk Files

DOS Floppy Disk File System

DOS imposes a requirement that any DOS-compatible floppy disk *must* contain *three* special data files that DOS uses to keep track of every file on the disk.

One data file is called the *boot record* file and is the first data area on the disk, located at *logical* sector 0.

The second file is termed the *File Allocation Table* or, somewhat anti-socially, the FAT. The FAT length can vary with the capacity of the disk and starts in *logical* sector 1. The FAT contains information about what *logical* sectors on the disk hold which disk files.

The last data area on a DOS-compatible disk is the *directory*, or the *root directory* for versions of DOS starting with version 2. The directory holds the *name* of every file on the disk and where each file's *first* logical sector address can be found in the FAT. The directory begins at the first logical sector that follows the FAT sectors and uses a predetermined number of sectors (depending on disk size) to hold the names of all the files on the disk.

Before we can study the contents of the boot record, the FAT, and the directory, we must first examine what a file is, and how DOS stores files on a disk.

FILE DEFINITION

A disk *file* is any collection of bytes that can be grouped together under a *single* name. File names are assigned by the programmer under restrictions, such as allowable characters that can be used to write the name, the length of the name, and the number of file bytes that DOS allows.

File names may be from one to eight characters in length, with a one to three character extension name that follows a period (.) after the first name of the file. The extension name is *optional*. Allowable characters for the name include A to Z, the numbers 0 to 9, and the special characters $, _, %, ', @, {, }, ~, `, !, -, and #. File names are *not* case sensitive, which means CAT, cat, and CaT are all the same name. Certain DOS *reserved* names, such as PRN, CON, and the like, are forbidden as file names because they are used for hardware-specific items such as the system printer (PRN) or monitor (CON). The percent sign (%) probably should *not* be used in a file name as it is also used by batch files to indicate replaceable parameters.

The DOS Boot Record

The very first entry, on a DOS-compatible floppy disk, is an area of code and data called the *boot record.* The boot record begins at logical sector 0 and occupies 1 sector, called a *reserved* sector. The boot record contains the following information:

- A jump to a bootstrap routine
- Operating system identification (i.e., MSDOS3.1)
- A BIOS parameter block (BPB)
- The bootstrap routine

The boot record is created when a floppy disk is configured for DOS applications by using the DOS FORMAT.COM program. FORMAT sets up the tracks and sectors on the disk, checks to see which sectors are usable, and creates the boot record, FAT, and directory. If the system option (FORMAT/S) is used with FORMAT, then the DOS system files (typically IO.SYS and DOS.SYS) are placed on the disk as hidden files and entered as the first two files in the directory. The DOS system files contain the stand-alone DOS programs that can boot up the computer and load COMMAND.COM from the disk. (Hidden files are listed using the versions 5 and later DOS command DIR/AH.)

If the formatted disk is in the default drive when the system is reset, then the bootstrap routine is loaded and the jump instruction executed to the bootstrap routine. The bootstrap routine then loads the .SYS files, and DOS is loaded and executed. (Please refer to Appendix D for details on the DOS boot sequence). Should the system have been previously booted up, then the bootstrap routine is ignored, and DOS reads the BIOS parameter block for information necessary to read the disk.

The operating system identification may be used by non-DOS operating systems as a means of identifying the DOS format in use. Conversion programs can be used by other operating systems so that DOS-based disks may be used with non-DOS based computers.

The BPB constitutes the *crucial* part of the boot record. There are now so many versions of DOS that the BPB has been included (starting with version 2) to inform the BIOS disk routines exactly how the disk is configured. Table 9.4 shows the BPB for DOS version 4. Each successive version of DOS is backward-compatible to version 2.

Byte in BPB	Contents	Typical Value
00–01	Number of bytes per sector	512
02	Number of sectors per cluster	02
03–04	Number of reserved sectors	01
05	Number of FAT files	02
06–07	Maximum files in root directory	112
08–09	Total number of sectors on disk	360
0Ah	Type of disk media	FEh
0B–0Ch	Sectors per FAT	02
0D–0Eh	Sectors per track	09
0F–10h	Heads per disk drive	02
11–14h	Number of hidden sectors	00
15–18h	Total number of sectors on disk (use if word at BPB byte 08 is 0)	80000
19h	Drive number of physical drive	01
1Ah	Unused	
1Bh	BPB extends beyond version 2	29h
1C–1Fh	Volume serial number	211B01ABh
20–2Ch	Volume name	DOSVOLUME1

Table 9.4 ■ Bios Parameter Block (BPB) Area of DOS Boot Record

FILE CLUSTERS

Something new can be seen at BPB byte 02: the number of sectors per *cluster*. One or more sequential logical sectors may be grouped together into a *single* cluster. A cluster is important because it is the *smallest* disk storage area that can be assigned to hold any file. Each file stored on the disk must take up at least 1 cluster and can use *all* of the clusters on the disk if necessary. The number of sectors in a cluster is a good indication of how efficiently a disk stores files. If we use 1 sector per cluster, then we can store small files (less than 512 bytes) efficiently. Should we use 8 sectors per cluster, then a 1-byte file would use up 4,096 bytes on the disk. However, large numbers of sectors per cluster may prove efficient if large files are stored.

The number of sectors per cluster varies by DOS version and the type of disk (floppy or hard) in use. Table 9.5 shows the number of sectors per cluster for various floppy disk capacities.

Disk Size	Capacity	Sectors per Cluster
5.25	360K	2
3.50	720K	2
5.25	1,200K	1
3.50	1,440K	1
Hard	10M	8
Hard	32M	4 (Beginning with DOS V3.0)

Table 9.5 ■ Sector per Cluster Assignment from BPB

From Table 9.5 we observe that most modern floppy disks use 1 or 2 sectors per cluster, and that large hard drives use 4 or more sectors per cluster. By increasing the number of sectors per cluster, the chance of wasting sectors on a few bytes of file data increases. Obviously, large drives can afford to waste more storage space than smaller drives, so we should expect sector-per-cluster numbers to rise as hard drive capacity increases. Note that sector numbers per cluster are powers of 2.

The File Directory

DOS keeps track of each file and subdirectory, as they are added to the disk by the user, in the directory data area. DOS uses disk sectors that follow the FAT area to store directory information about each file or subdirectory stored on the disk.

When a disk is first formatted by the DOS format command, the directory is empty, except for any DOS system files that may be present if the disk is bootable. Every time the user adds a file, DOS updates the directory area to show the name of the file, time and date of creation or alteration, and the length of the file. We shall investigate exactly what data is stored in the directory sectors when we discuss the FAT.

THE FILE DIRECTORY AND THE FAT

As briefly mentioned previously, the directory contains important information about files and subdirectories that are stored on the disk. The directory also contains information that tells DOS where to find any disk file by using the FAT.

The FAT is used by DOS to indicate exactly where every cluster, of every file, is logically stored on the disk. The FAT contains addressing information for every data cluster on the disk. Every cluster has an equivalent FAT entry, and every FAT entry corresponds to a unique cluster. The FAT is said to contain a *map* of the cluster locations for each file.

For the sake of accuracy we should point out that there are actually two complete, identical, FAT sectors. DOS, following the example set by earlier operating systems such as CP/M and UNIX, makes two copies of the FAT entries. Two FAT copies might allow file cluster address information to be found in the second FAT, should the first become damaged. A backup FAT may have been the original DOS intent, but it is not obvious that the second FAT has a purpose. Unfortunately, disasters that strike the first FAT usually impact the second also. DOS does not use the second copy of the FAT for any function.

To find a file, DOS first finds the file name in the directory. Directory file name entries occupy the sectors that immediately follow the FAT sectors, (see Table 9.6). The number of FAT and directory sectors increases with disk capacity, because high density disks require more sectors to record FAT and directory information as the number of files grow with disk storage capacity.

The directory information that follows the file name informs DOS of the *first* FAT cluster entry for the file. The first FAT cluster entry then *points* to the second FAT cluster entry for the file. The second FAT cluster entry then points to the third file cluster entry for the file, and so on. The FAT thus holds all of the addresses of all of the clusters that make up a file.

Disk Capacity	Root Directory Files	FAT Sectors	Directory Sectors
360K	112	4	7
720K	112	6	7
1,200K	224	14	14
1,440K	224	18	14
30M	512	122	32

Table 9.6 ■ Directory and FAT Sectors for Various Disk Capacities

The *logical* sector assignment for the boot record, FAT, and directory of a floppy disk can be found from Table 9.6. We know that logical sector 0 holds the boot record, and that the FAT and directory sectors follow in succession. For instance, a 720K disk will have the boot record in Sector 0, the FAT in Sectors 1 to 6, and the directory in Sectors 7 to 13. File storage, for our example 720K disk, begins in logical Sector 14.

DIRECTORY DATA STRUCTURE FOR FILE NAMES

The directory area allocates 32 bytes of information about each file that is stored in the directory. Each 512-byte directory sector can store information on 16 files. File information is stored as follows:

File name	11 bytes
File attributes	1 byte
Unused	10 bytes
File creation time	2 bytes
File creation date	2 bytes
First FAT cluster	2 bytes
File byte length	4 bytes
	32

A file *attribute* is some special characteristic assigned to the file by the file creator, or by DOS. File attributes can be determined by using the DOS attrib command.

Beginning with DOS version 2, *subdirectory* names could be mixed in with file names. The number of file and subdirectory names allowed in the root directory is purely a function of the number of directory sectors multiplied by 16. For instance, a 360K disk has 7 directory sectors, so the root directory may hold as many as 7 times 16, or 112 names.

File and subdirectory information for each file name entry in the directory is structured as shown in Table 9.7.

An inspection of Table 9.7 reveals the origins of many DOS peculiarities:

■ File names in DOS are limited to no more than eight characters and a three-character extension because that is all the space allocated to store a file name in the directory.

■ The "deleted" file character (E5h) makes possible *undelete* file recovery. When a user removes a file with the DOS delete command, an E5h is placed in the first byte of the directory entry for that file.

Byte Numbers	Contents

00–07h	File name in ASCII uppercase
	If the *first* byte in the name is:
	   00h = No file data (empty, never used)
	   E5h = Deleted file (was here)
	   05h = File name starts with E5h (E5h is the
	           DOS extended character *sigma*)

08–0Ah	File extension in ASCII uppercase
0Bh	File attribute:

Bit in Byte		Means
7 6 5 4 3 2 1 0		
1		Read only file
1		Hidden file
1		System file (i.e., DOS)
1		Name is Volume Label
1		Name is Subdirectory
1		File has been updated
1 1		Unused

Byte Numbers	Contents
0C–15h	Unused
16–17h	Time of last update
18–19h	Date of last update
1A–1Bh	Beginning cluster address in the FAT
1C–1Fh	File length, in bytes

Table 9.7 ■ Directory Entry for Each File on the Disk

The remaining 31 bytes in the directory entry are not erased.
Undelete programs can recover the file unless the directory entry has
been used for a new file.

■ A pleasant feature of DOS, the time and date of last file update, is
provided for in the directory entry for each file. Updating a file
involves writing to the file area on the disk. Copies, renames, and
other activities that do not write to the file area are not updates.

■ The update bit in the attribute byte (position number 0Bh) is very
handy for hard disk backup routines—only those files that have been
changed since the last backup need be re-saved on the hard disk
backup media.

The directory sectors that immediately follow the FAT sectors are called
the *root* directory because there are no other directories that exist before the
root directory. The root directory can contain the names of subdirectories,
and a subdirectory is denoted by Bit 4 of directory attribute Byte 0Bh.

Bytes 1Ah and 1Bh of each directory file entry hold perhaps the most
important information of all for DOS after the name of the file—the *beginning* cluster address for the file in the FAT. The beginning cluster address
gives DOS an entry point into the FAT. FAT cluster entries then lead DOS on
a *trail* through the disk that contains all of the cluster addresses for the file in
question.

The File Allocation Table (FAT)

The FAT is a sequence of *nibbles* that identifies the logical *address* of every cluster occupied by any given file. Originally, the FAT table used 3 nibbles (1.5 bytes, or 12 bits) to hold cluster address data. A 12-bit cluster number yields 4,096 discrete cluster numbers, from 000 to FFFh. Each cluster can be made up of 2 to as many as 8 sectors of data. Each sector contains 512 bytes.

Floppy disks up to 4M (4,096 sectors × 1,024 bytes) in size can use 12-bit FAT entries if each cluster holds 2 sectors of 512 bytes per sector. Hard drives up to 16M bytes (4,096 sectors × 4,096 bytes) can use 12-bit FATs if each cluster holds 8 sectors.

As hard disk storage capacity increased, FAT address entries were expanded to 4 nibbles (2 bytes) beginning with DOS version 3.0. Various hard drive capacities can be handled with a 16-bit FAT by varying the numbers of sectors per cluster. FAT cluster numbers of 16 bits allow 65,536 clusters to be identified. Hard drives up to 128M (4 sectors per cluster) or 256M (8 sectors per cluster) are now usable by DOS beginning with version 4.0.

FAT CHAINED ADDRESSES

The FAT exists to inform DOS as to exactly what logical sectors hold a file, or part of a file. Data files are stored on the disk after the boot sector, FAT sectors, and directory sectors.

Sectors used to hold the boot directory, FAT, and directory are part of the *overhead* associated with the disk, and the FAT does not contain any addresses for the overhead sectors. DOS knows from the BPB the disk format in use and skips over the overhead sectors when searching for a data file on the disk.

A file must occupy a *minimum* of 1 cluster and can use as many more clusters as necessary to hold the entire file. Files can be stored in a *sequence* of logical cluster addresses or scattered over the face of the disk in a *random* manner.

The FAT cluster numbers do *not* begin with 000, but with number 002h. The first two FAT entries do not contain cluster 000 and 001h information, rather they contain information as to how the disk is formatted and, indirectly, how many bits (12 or 16) are used for each cluster address. FAT cluster entries 000 and 001h contain the information shown in Table 9.8. As shown in the table for a 12-bit floppy FAT, two cluster entries occupy 24 bits, or 3 bytes.

Every logical cluster, beginning with cluster number 002h, has an equivalent address area in the FAT. FAT cluster address 002h corresponds to the first *active* data file area on the disk.

The initial FAT address for a file cluster is contained in bytes 1A–1Bh of the 32-byte directory entry for that file (see Table 9.7). The FAT entry at the *initial* address contained in the directory holds the address of the *second* cluster occupied by the file. The second cluster address holds the address of the *third* cluster on the disk, and so on until enough clusters have been used to completely store the file. When the *last* cluster address for a file is reached, it will contain the number FFFh. As there are no more clusters allocated to the file, the number FFFh signals DOS that the last file cluster has been found on the disk.

12-bit FAT Cluster Entries 000 and 001h	Disk Capacity	FAT Address
FFFFFFh	160K	12
FDFFFFh	360K	12
FCFFFFh	180K	12
F9FFFFh	720K	12
F8FFFFh	Hard Disk	12/16
F0FFFFh	1.44M	12

Table 9.8 ■ FAT Codes for Disk Size and FAT Address Bits

Each file cluster address holds the address of the next file cluster, starting with the first cluster address in the file directory and ending at the last file cluster address, which contains the "end-of-file" characters FFFh. A scheme in which one address holds the next address is called *chaining*. Like links in a chain, each link (address) is connected to the next link (address).

FAT cluster entries that have *never* been used to hold address chains, or have become *free*, are filled with 000. When the disk is formatted, the DOS format routine assigns all FAT cluster addresses the value of 000. Should a file be deleted using a DOS delete command, then the FAT clusters for that file will also be set to 000. If a file is edited, and becomes smaller so that fewer clusters are needed to hold it than originally, then DOS will also set the *unused* cluster addresses to 000. If the file is edited, and grows, then DOS *skips over* the used clusters and finds expansion room for the newly larger file in clusters far from the beginning file clusters.

The DOS practice of setting re-edited clusters to 000 as files shrink or are deleted explains the disk phenomenon called *fragmentation*. Newly formatted disks will write all original files in *sequential* clusters. Every file follows the last, and all files are stored in sequential clusters.

Sequential clusters can be read by programs very quickly because all file segments are in order and the disk head does not have to waste time seeking random file clusters. But, as files are deleted or edited and change size, clusters are made available on the disk in a *random* manner. New files are assigned these random clusters and disk reads of the new files becomes slower and slower.

One solution to fixing a fragmented disk is to copy the original disk to a backup disk, format the original disk, and re-copy all of the files from the backup disk. Several enterprising writers of PC utilities have also written programs that take a fragmented disk and re-order all the files sequentially.

All FAT entries for unusable (bad) disk clusters, found during formatting, contain the nibbles FF7h. When the disk is formatted, then, all FAT data cluster entries will hold either 000 or FF7h.

A 12-BIT FAT EXAMPLE

Figure 9.3 shows a hypothetical FAT, used on a 1.2M floppy, for two files whose names are contained in the directory. File 1, named TRY.001, has a beginning cluster address of 002h in the directory. TRY.001 is 2,000 bytes long, so it will require 4, 512-byte sectors, or 4 clusters on a 1.2M floppy disk.

Figure 9.3 ■ A 12-bit FAT Example.

Byte number	00	01	02	03	04	05	06	07	08	09	0A	0B	0C	0D	0F	10
Data (actual)	FO	FF	FF	03	40	00	05	FO	FF	FF	OF	00	00	00	00	00..
Cluster number	000		001		002		003		004	005		006		007		008	009
Data(arranged)	FOF		FFF		003		004		005	FFF		FFF		000		000	000

TRY.001 begins at cluster address 002h and ends at cluster address 005h. File 2, TRY.002, which contains 500 bytes, will use 1 cluster and begins and ends at cluster address 006h.

Using 12 bits to hold the address of the next cluster presents some interesting problems for a computer such as the 8086, which can only fetch data in byte or word sizes. When the FAT information is read from the disk to internal RAM, the cluster addresses begin on even-byte boundaries for even cluster numbers and in the middle of a byte for cluster numbers that are odd. Furthermore, the only information that is valid is the first 12 bits for *even* cluster numbers or the last 12 bits for *odd* cluster numbers, of any word of data fetched from memory.

Figure 9.3 shows the FAT address data as it really appears on the disk (to the right of the *actual* heading), and how the data is manipulated to obtain the next cluster address (to the right of the *arranged* heading). File TRY.001 is found in the directory, and the beginning cluster address in the directory is address 002h. Each cluster address uses 1.5 bytes in a 12-bit FAT, so the byte address for the FAT entry that corresponds to cluster 0002h is 0002 × 1.5, or byte number 0003h in the FAT table.

The data word at cluster address 0002h (Byte 03h in the FAT) is read as 4003h using the normal low-byte, high-byte format for all data with the low byte stored first in memory.

An *even* cluster number in a 12-bit FAT occupies the *lower* 3 nibbles of a word, so the 4 in 4003h is thrown away, leaving the 12-bit cluster address 003h. The next cluster address is thus cluster number 003h for file TRY.001. The byte address for cluster number 003 is 003 × 1.5 or byte address 4.5. Clearly, one cannot read in the middle of a byte, so the data word at byte address 04 is read, which is 0040h. Only the *most* significant 12 bits are valid for *odd* numbered clusters, so the last nibble is discarded, leaving the next 12-bit cluster address to be 004h.

Multiplying cluster 004 by 1.5 yields a byte address of 006 in the FAT table, and reading the word that begins at Byte 06h results in data of F005h. Discarding the *upper* 4 bits for the *even* cluster address then yields a next cluster address of 005h.

Finally, cluster address 005 × 1.5 (7.5) lets us read the word that begins at byte address 07 in the FAT, and the resulting data is FFF0h. Keeping only the 12 *most* significant bits for the *odd* cluster address leaves a next cluster address of FFFh, which signals the end of the file.

When file TRY.002 is accessed by DOS, the cluster address 0006h is found in the directory entry for TRY.002. A FAT byte address of 009h is calculated by multiplying cluster 0006 by 1.5, and the data word stored at FAT

Byte 09h is read. The word that begins at Byte 09h contains 0FFFh. Discarding the upper 4 bits for the even cluster address leaves FFFh, the end of the file.

Based on the calculations just illustrated in our example, the rules of finding the next FAT cluster address can be formulated:

1. Multiply the cluster number by 1.5.

2. If the answer is a whole integer, with no fractional part, then read the data word that begins at that byte and keep the last 12 bits as the next cluster number for the file.

3. If the answer has a fractional part, (.5), discard the fraction and read the data word that begins at the integer byte address that is left after the fraction is discarded. Keep the first 12 bits as the next cluster number of the file.

4. Continue until a cluster number of FFFh is read.

Another way to think of the process is to note that even cluster numbers will always yield integer byte numbers, whereas odd cluster numbers will always have a fractional part to discard. Even cluster numbers keep the last 12 bits of the data word, and odd cluster numbers always keep the first 12 bits of the data word.

Using a 16-bit FAT is much simpler than using a 12-bit FAT. The cluster number is multiplied by 2, and the word at the resulting byte is the next cluster number. The first 2 words of a 16-bit FAT are used for formatting information as was done for the 12-bit FAT, and an end of file is signaled by a next FAT number of FFFFh. Free FAT entries are denoted by 0000 and bad clusters by FFF7h.

CONVERTING CLUSTER NUMBERS TO SEGMENT NUMBERS

Once the cluster number is identified, the logical segments that make up the cluster must be computed. Logical sector numbers begin at Sector 0 and count up by 1 for each new sector. Logical Sector 0 holds the boot record, and sectors that follow the boot sector hold the FATs and the directory sectors. As disks grow in capacity, the number of sectors devoted to the FAT and directory entries grows, as is shown in Table 9.6.

Converting from cluster numbers to logical segment numbers involves some knowledge of disk capacity. The formula used to convert from cluster to segment numbers must be altered as dictated by the number of sectors used to store the FAT and the directory. To convert from cluster number to sector number involves adjusting the cluster number back to 0, multiplying by the number of sectors per cluster, and then adding the number of sectors used by the particular disk format for overhead (boot record, FATs, and directory).

The formula for converting from cluster to sector is:

Sector Number = (Cluster Number − 2) × Sectors/Cluster + Overhead

Cluster number 2 is the first cluster number possible in a FAT, because FAT entries 0 and 1 are used for format information.

As an example, the logical sector number that corresponds to cluster number 2 on a 360K floppy disk is:

Sector = (2 – 2) × 2 + 12 = 12

We have used 2 sectors per cluster and added 12 overhead sectors (1 boot record, 4 FATs, and 7 directory from Table 9.6). The last data cluster on a 360K floppy disk is number 355. The sector corresponding to cluster number 355 is:

Sector = (355 – 2) × 2 + 12 = 718

The last cluster on a 360K disk occupies logical sectors 718 and 719, which are the last 2 of the 720 sectors (logically numbered 0 to 719) on a 360K disk.

9.3 Subdirectories

There is only one *root* directory on a disk. The root directory is limited to a certain number of file names because of the restriction of 16 file entries per directory sector. Table 9.6 shows that the number of sectors occupied by the root directory are fixed, thus the maximum number of files in the root directory is equal to 16 multiplied by the number of root directory sectors. The limit on the maximum number of different file names on a disk was greatly increased in DOS version 2.0. DOS V2 introduced the concept of *tree structured* directories by employing the concept of *subdirectories.*

A subdirectory is structured exactly like a directory and differs from the root directory only by the fact that the root directory has no name, whereas a subdirectory *name* can be found in another (sub)directory. Subdirectory names found in the root directory are stored in the same 32-byte format as is done for the file names shown in Table 9.7. The attribute byte (Byte 0Bh) uses Bit 4 to show that the name belongs to a subdirectory and *not* a file.

The starting cluster number in the directory points to a FAT entry for the subdirectory, and the subdirectory clusters are found in the same manner as is used for files. However, instead of file data, subdirectory files contain the same 32-bit file identification format that directories use to locate files on the disk.

Subdirectories can contain file names in exactly the same format as used in the root directory, or the subdirectory can contain the names of additional sub-subdirectories, and so on until the disk data sectors are full of files and subdirectories.

A (sub)directory that contains the name of another subdirectory is called the *parent* (sub)directory, and the named subdirectory is sometimes called the *child* subdirectory. Subdirectory use generates the concept of a *path* name to lead DOS to the final file name contained in a maze of subdirectories. For instance, a word processing file might be found by using the following path:

C:\PFS\B86\BOOK.DSK

The path consists of the following path items:

- The drive letter, C.
- The root directory shown by the first backslash (\).

- The first subdirectory, PFS.
- The second subdirectory, B86
- The file, BOOK.DSK is the text file itself.
- Backslashes (\) to separate directory and file names

In the example, the root directory is the parent of PFS, which is the parent of B86, which holds the cluster address of the file BOOK.DSK.

Subdirectories cannot expand disk capacity beyond the physical number of clusters present. In fact, using subdirectories diminishes the actual usable file storage space on the disk because each subdirectory uses up a cluster, even if the subdirectory is empty.

Subdirectories have proven very popular, particularly on hard disks, as a convenient way to group similar files.

Subdirectory Dot (.) and Dot Dot (..) Entries

One other feature that distinguishes subdirectories from the root directory is the first *two* directory entries in a subdirectory. The first name entered in a subdirectory is the entity named *dot* (.), and the starting cluster number of dot is the cluster address of the subdirectory itself. Use of the DOS command CD . results in the subdirectory remaining the current directory.

The second name entered in a subdirectory is *dot dot* (..). The first starting cluster number of dot dot is that for the *parent* directory of the subdirectory. File name dot dot allows DOS to "back up" from a child to a parent by finding the cluster number of the parent in the second file entry of the child. Using the DOS command CD .. will result in changing the current child subdirectory to the parent (sub)directroy in which the child subdirectory is listed.

9.4 DOS Disk INT Services

At last count, DOS version 6.0 contains over 60 (40 recommended, 20 obsolete) disk INT service routines. In keeping with our traditional (by now) 20/80 rule, we shall examine eight disk INT service routines in detail. Please refer to Appendix D for a complete listing of other DOS INT service routines.

Read Disk Sector: DOS INT 25h

Files on a disk are contained in clusters and each cluster is made up of one or more 512-byte sectors. Sectors are logically numbered beginning at sector number 0. Overhead sectors, such as the boot record, the FATs, and the root directory, occupy the number of sectors shown in Table 9.6. Overhead sectors are established when the disk is formatted.

Any disk sector can be read using INT number 25h, the *absolute* disk read INT service routine. An absolute read means that the sector is copied, exactly as it appears on the disk, into internal RAM. There is no manipulating of the sector data, and the entire sector is read, even if DOS would consider it empty.

Disk data is read from the disk into an area of RAM called a *sector buffer*. Sector buffers are data areas set aside by the program to hold the sector data. Once file data has been loaded into a buffer, D86 can be used to inspect the data.

INT 25h is invoked after loading the following data into the registers listed.

Register	Contents
AL	Drive number. Each drive is assigned a sequential number as follows: A = 00 B = 01 C = 02 D = 03 and so on, up to the drive controller limit.
BX	Offset address into data segment RAM where the sector data begins loading.
CX	Number of sectors to load from the disk into RAM.
DX	Logical sector number, beginning at the boot record (sector 0000) and extending to the last sector.
Returned	Flags: Carry flag is cleared if there was no read error. AX: Error code for disk problem, if CF = 1.

Note that DOS INT 25h is the first INT encountered in this book that returns an *error code*. Error codes are used by DOS to inform other DOS INT routines of I/O problems. If the Carry flag is set on return from interrupt, register AX holds error handling codes. Error codes, their meaning, and the use of error codes is beyond the scope of this book. For more information on error codes, refer to one of the DOS reference books listed in Appendix H, "Sources."

DOS INT 25h does *not* pop the Flag register on return; the user routine must pop the flags to balance the stack. The failure of interrupt type 25h to restore the stack is an unexpected error, and probably the result of an oversight on the part of the programmer who wrote the service routine. But, once INT 25h was released to the world, the world responded as we have by restoring the stack in all application programs that use INT 25h. It would be easy to rewrite INT 25h and fix this bug, but existing programs could not use the fixed INT 25h! Thus, a faulty INT 25h of a very popular DOS operating system will persist for decades. (The most common solution to balancing the stack if CF is used to report an error condition is to end the interrupt routine with a RETF 2 instruction. RETF pops CS:IP from the stack, and the 2 is added to the SP by the CPU to compensate for not popping the flags.)

A DOS SECTOR READ EXAMPLE

The example program, named sector.asm, gets a single-digit sector number from the user (sector number 0 to 9) and loads the sector into a data buffer at offset location 1000h in the data segment. D86 can be used to display the DS:1000h data buffer contents.

The boot record (Sector 0) and the first FAT sectors starting at Sector 1 for a floppy disk can be inspected using sector.com running under control of

D86. Initial directory sectors for a 360K (Sector 5) and 720K (Sector 7) floppy disk may be seen using D86.

If you wish to inspect any sector beyond number 9, simply change the contents of register DX immediately before stepping into INT 25h with the F1 key. You may inspect any sectors that are defined in the FAT table. It has been observed, while using sector.com, that sectors for which there are *no* FAT entries will not be read, and an error is indicated by the Carry flag when the routine returns.

```
;sector.asm, a program that reads a sector and stores the sector ;data in RAM.
data segment                    ;declare data at offset address 1000h
        org 1000h
diskbuff db 512 dup(?)          ;declare a data area named "diskbuff"
data ends
code segment
        mov sp,8000h            ;set the stack
sector:
        mov ah,00h              ;get keyboard key from user
        int 16h                 ;get sector number from 0 to 9
        cmp al,'q'              ;q means quit
        jne dump                ;if not quit then get sector number
        mov ax,4c00h            ;if quit then back to DOS
        int 21h
dump:
        sub al,30h              ;convert ASCII key number to hex
        mov dh,00h              ;get sector number to DX, (0000 to 0009)
        mov dl,al               ;form two-digit sector number
        mov al,00h              ;AL = drive a:
        mov cx,0001h            ;CX = number of sectors to get
        lea bx,diskbuff         ;BX = address of disk data buffer
        int 25h                 ;INT to DOS sector routine
        popf                    ;Int 25h does not pop the flags
        jmp sector              ;loop until done
code ends
```

You may wish to make the diskette in drive A write-protected, just in *case* some error creeps into the program. Figure 9.4 shows the first hex entries from sector dumps from the first FAT sector and the first directory sector of a 1.44M disk containing only two files, try.001 of 2,000 bytes, and try.002 of 500 bytes. See Figure 9.3 for an explanation of the FAT entries.

With the ability to inspect the contents of any sector on a disk, using DOS INT 25h, it is natural to wish to write data to any sector on a disk. DOS INT 26h, which is discussed next, can be used to write data to a sector. You should use a newly formatted disk in drive A: or B: when experimenting with DOS INT routine 26h. Used DOS disks with partially filled FATs and directories will fail to write in sectors that are already defined (used up) in the FAT table.

Absolute Disk Write: DOS INT 26h

INT 26h is the mirror image of INT 25h. Data is copied from a buffer area in RAM (also called the *data transfer area* or DTA) to the specified logical sec-

Figure 9.4 ■ Sample sector display using sector.com program.

```
F0 FF FF 03 40 00 05 F0 FF FF 0F 00 00 00 00 00
```

Beginning of FAT for Files TRY.001 and TRY.002

```
54 52 59 20 20 20 20 20 30 30 31 20 00 00 00 00
|T  R  Y                 0  0  1|
00 00 00 00 00 00 82 5C D2 18 02 00 D0 07 00 00
            | time & date | clstr |   length   |
```

Directory Entry for File Named TRY.001

```
54 52 59 20 20 20 20 20 30 30 32 20 00 00 00 00
|T  R  Y                 0  0  2|
00 00 00 00 00 00 18 58 D2 18 06 00 F4 01 00 00
            | time & date | clstr |   length   |
```

Directory Entry for File Named TRY.002

tor on the disk. The disk directory is *completely* bypassed by INT 26h, and DOS cannot later find any information that may have been stored using this INT, unless the directory is changed by the user. INT 26h uses the following registers to hold information for the service routine.

Register	Contents
AL	Drive number. Each drive is assigned a sequential number as follows:
	A = 00
	B = 01
	C = 02
	D = 03
	and so on, up to the drive controller limit.
BX	Offset address into data segment RAM where the sector data is found for recording to the disk sector.
CX	Number of sectors to load from RAM to disk.
DX	Logical sector number, beginning at the boot record (sector 0000) and extending to the last logical sector on the disk.
Returned	Flags: Carry flag is cleared if there was no read error. AX: Error code for disk problem, if CF = 1.

INT 26h is useful for *efficiently* storing data in specialized applications. Removing the need for all of the DOS overhead sectors—boot record, FATs, and directories—permits the disk to be fully utilized to hold data. Special programs, such as sector.asm, will have to be written to recover data from any disk written to by INT 26h. INT 26h does *not* pop the flags on return; the user routine must pop the flags to balance the stack.

A DOS Sector Write Example

We shall examine a small program that is the reverse of sector.asm to write data to any sector from Sector 0 to Sector 9. You should use a blank, formatted disk to experiment on when using this routine. The program writsect.asm will write a sector full of the ASCII character K when it is run. Program sector.asm can then be run to recover the sector and verify that the program works as intended.

```
;writsect.asm, a program that copies data from DTA area in RAM and
;stores it in user-specified sectors 0 to 9 in RAM.

code segment
        mov sp,9000h        ;set the stack
writsect:
        mov ah,00h          ;get sector number from user
        int 16h             ;get sector number from 0 to 9
        cmp al,'q'          ;q means quit
        jne fill            ;if not quit then sector
        mov ax,4c00h        ;if quit then back to DOS
        int 21h
fill:
        sub al,30h          ;convert ASCII to hex
        mov dh,00h          ;get sector number to DX (0000 to 0009)
        mov dl,al           ;form two-digit sector number
        mov al,00h          ;AL = drive A:
        mov cx,0001h        ;CX = number of sectors to record
        mov bx,diskbuff     ;BX = address of disk data
        int 26h             ;INT to DOS sector routine
        popf                ;Int 26h does not pop the flags
        jmp writsect        ;loop until done
        org 1000h           ;code 512 Ks at offset 1000h
diskbuff:  db 512 dup(4bh)  ;fill DTA area with the K character
code ends
```

Use writsect.asm to fill any sector you wish by changing the sector value in register DX before stepping into INT 26h with the F1 key. Program sector.asm can then be used to display the sectors that have been filled with Ks. You may also experiment with designing your own FAT and directory entries by changing the data in the DTA to strings of your own choosing.

9.5 DOS Files and File Management

DOS is a system for computer disk file storage and retrieval. Data recorded on a disk is organized into entities called files and accessed by means of file directories and file allocation tables (FATs). A disk file is, in general, any consistent method used to group like data on the surface of the disk.

The term *file*, most obviously, had its origin in the concept of a file cabinet. File cabinets are paper mass storage devices that hold individual file folders. Each file folder contains information on a certain subject. Access to the file folders is by means of some standardized storage scheme, such as storing files alphabetically or by a numbering system.

Computer operating systems view files in a broader sense than we view paper file systems. To DOS, a file can be *anything* that can provide or store binary data. The DOS view of a file means that, to DOS, the keyboard, the monitor, the printer, and the serial port are *all* files. Thus a file becomes, in the way DOS uses the term, not only a collection of data in memory but also any device that provides or receives data. DOS assigns *identifiers* to all devices that can store or transmit data. A file identifier is simply a number assigned to a device.

Viewing devices as files results in simplifying DOS based programs. Routines that store data can write the data to the disk, to RAM, to the serial port, or a printer by simply changing the file identifier. A separate routine does not have to be written for every different device that receives data. An identical statement may be made for routines that receive data from a disk, the keyboard, or the serial port.

In this section we shall explore the ways in which DOS identifies files and use some common DOS INT service routines that move data about in the computer system. But, before launching into file manipulation, we must examine how DOS keeps track of file names.

DOS File Management

The term *file management* refers to the ways by which DOS can access files. A hard disk, for instance, may contain thousands of file names organized in many directories and subdirectories. The thousands of files on the hard disk cannot all be accessible at once to DOS, and a system must be established for orderly processing of the data in a file. DOS establishes order in file processing using the concepts of *opening* and *closing* files.

Opening a file refers to the process of *identifying* a specific file name located in a certain device and *finding* the file. The term *opening* originated with getting paper files from a file cabinet. Certainly one must open the file before using it.

Closing a file implies that access to the file has been accomplished, and the file may now be retired from *active* use and returned to its storage device. Again, the vivid image of closing the file drawer comes to mind.

When the computer system is first started, all of the disk files are closed. DOS can open files, under user control, and maintain some number of files that are all open at the same time. Interestingly, DOS may use either of two mechanisms for opening, accessing, and closing files: the *file control block* (FCB), and the file *handle*.

FCBs are a feature of DOS versions prior to version 2.0. The FCB DOS interrupts are retained in all versions of DOS, however using an FCB to access files is now considered *obsolete*. File handles, another concept DOS has borrowed from UNIX, are much easier to use. There are some uses for the FCB method of file access, though, and Table 9.9 shows the merits of file handles and FCBs.

The only legitimate use of an FCB, today, is to create disk volume labels. We shall not pursue the FCB concept further in this chapter. Please refer to Appendix D for a brief description of the FCB interrupts.

File Handle	FCB
Ease of use Makes use of directory structure Redirection of I/O Networking File sharing or locking Enhanced error reporting Random file access	No limit to open files Create volume label names DOS version 1 compatible

Table 9.9 ■ File Handle and FCB Uses

FILE HANDLES

A file handle is a *number*, from 0000 to FFFFh, by which DOS identifies an *open* file. The details of the handle, that is, the information that DOS associates with each file handle number, are hidden from us. We simply know that DOS can identify any open file with a number, and, by using a file number, we may access any file or device. DOS uses the file handle number simply as a pointer to a data area in DOS where all the file details are stored.

And given file handle number may be used for any file name, but only *one* file name at a time. Once a file is opened and a file handle is assigned to the file, the specific handle must be used *until* the file is closed. Once a file is closed, the file handle number is *freed* up for use with *another* file name.

The maximum number of open files, thus the maximum number of file handles that DOS may have in use at one time, is set by the FILES command in the CONFIG.SYS file in the boot disk root directory. When DOS boots up, the FILES command is read, and the maximum number of open files (file handles) is established. A default value of 8 open files is established if no FILES (or CONFIG.SYS) is found on bootup. The maximum number of files that may be opened at the same time, by any *single* program, is 20. The maximum number of all open files, for all programs, is 255 for the system.

Several file handles are *pre-assigned* by DOS for use as handles that access system devices. Pre-assigned DOS file handles are *always* open. Table 9.10 shows the file handle numbers assigned to system peripheral devices.

Obviously, the handles of Table 9.10 are for I/O devices, and care must be exercised so that we do not attempt to write to an input device, such as the

Device	File Handle Number
Keyboard	0
Screen	1
Error output	2
Auxiliary	3
Printer	4

Table 9.10 ■ Device File Handles

keyboard, or read an output device, such as the screen. The auxiliary device is most often a serial port that can be either input or output. The error output device is also the screen.

USING FILE HANDLES

Except for the pre-assigned device file handles, the programmer must have DOS assign a file handle by following a standard file handling procedure. The programmer may create, open, access (read and write), and close a file.

Certainly, the file must first be created before any other actions may take place. To create a file, and assign a handle number to the file, the program performs the following actions:

1. *Name* the file *path* name using ASCII characters.
2. *Terminate* the file *path* name with a 00 byte.
3. *Use* INT 21h, function 3Ch, to have DOS assign a handle.

The file path name, followed by a 00 byte, is called an *ASCIIZ string*. ASCII refers to the ASCII-coded path characters, and the Z comes from the byte, at the end, which is 00. If no directory is specified in the path name, the *current* directory is used by DOS.

Create and Open a File: DOS INT 21h, Function 3Ch

Before a program may access a file for the first time, the file must be created and a file handle must be assigned to the file by DOS. Function 3Ch will return the file handle to the program in register AX. The file is *automatically opened for writing and reading.* Function 3Ch is called from the program as follows.

Register	Contents
AH	3Ch
CX	File Attribute
DS:DX	Address of ASCIIZ file path name
Returned	Flags: Carry flag cleared if no error
	AX: File handle number, if CF cleared
	AX: Error code, if CF set

Register CX may contain any of the following file attribute bytes:

Contents of Register CX	File Attribute
00	Normal
02h	Normal hidden
04h	System
06h	Hidden system

A normal file is the most common file type, and we shall only be concerned with normal attribute files. Hidden and system files are usually DOS operating system files.

Warning: Function 3Ch will create and open a *new* file and assign a file handle if no existing file name matches the ASCIIZ string pointed to by DS:DX. If the ASCIIZ file path name matches *an existing file,* the existing file will be opened and *erased* (set to zero length), and then a file handle number will be assigned to it!

A FILE HANDLE EXAMPLE

To test DOS function 3Ch, the following program, A:trial.try, is created and assigned a handle by DOS.

```
;trial.try, a program that creates the file named A:TRIAL.TRY
code segment
        mov sp,7000h           ;set the stack
trial:
        mov dx,path            ;point to ASCIIZ file path name
        mov ah,03ch            ;assign a file handle with function 3Ch
        mov cx,0000h           ;normal file attribute
        int 21h                ;get DOS
        jmp trial              ;see what is returned in AX and loop
path:   db 'A:TRIAL.TRY',00h   ;file path name in ASCIIZ format
code ends
```

You should note that the same file name, A:TRIAL.TRY, is assigned a different file handle every time the program loops. Should you loop until you have exceeded the allowable number of file handle numbers (depending on what FILES contents are in your CONFIG.SYS file at bootup), the Carry flag will return *set.* AX will contain 04h, which means no (additional) handles are available.

You may have to re-boot your computer after debugging trial.com for a few loops. Trial.com opens file A:trial.try, but does *not* close it. On quitting D86, DOS still has trial.try open.

The DOS Disk File Buffer

DOS maintains an internal RAM storage area, or buffer, to hold data that is in transit to or from the disk. The internal DOS disk buffer can hold any part of a single file up to a file size as large as 4,295M. Four billion bytes is far beyond any sensible internal RAM size or, for that matter, beyond the total storage space available on most hard disk drives.

File data processed between the program and a disk file, using the file handle approach, is completely *transparent* to the program. DOS maintains buffers between the program and the file to handle data transfers. You should not confuse the internal DOS disk buffer with your program's data buffer.

You may create a buffer in your program as large as is needed to store a file. DOS does not need to store your entire file at once, because the DOS buffer is used solely as a temporary stopping place for data in transit between your program and the disk.

Data read from the disk is passed through the buffer to the program. As data from the disk exceeds the capacity of the DOS internal buffer, file contents already read by the program are discarded, and new data from the disk is stored in the DOS buffer. Data written from the program to the disk is stored in the DOS buffer until the buffer is full; the data is then written to the disk.

The only exception to the buffer-full rule is when a file is closed. Closing a file forces DOS to *flush* the disk buffer, a rather inelegant way of saying that any remaining data in the buffer is written to the file. We shall be writing small amounts of data to the disk in the following examples. When the files are closed, DOS will flush the data buffer.

Files should always be closed after the program has completed all operations on the file, or data will be lost.

Close a File: DOS INT 21h, Function 3Eh

Function 3Eh closes an open file and returns the file's handle back to DOS for reuse. Any remaining file data in the DOS internal disk buffer is written to the disk. The FAT entry for the file is marked with FFFh to indicate an *end-of-file* (EOF) condition.

Register	Contents
AH	3Eh
BX	File handle
Returned	Flags: Carry flag set if error
	AX: Error code, if CF set

The Carry flag will be cleared if the file was successfully closed. The only possible error is to use an invalid handle number.

Write to a File: INT 21h, Function 40h

Once a file is *created,* opened, and a handle assigned by function 3Ch, our program can write data to the file. We can read the file also, but remember, if the file already exists then function 3Ch has erased it! Function 40h permits us to write data to the file using the register contents shown next.

Register	Contents
AH	40h
BX	File handle number
CX	Number of bytes to write
DS:DX	Address of data bytes
Returned	Flags: Carry flag set if error
	AX: Error code, if CF set

Should the Carry flag be set, then AX will return an error code indicating an improper file handle. Note that the largest file piece that can be written to the disk, using *one* write function 40h, can be no larger than CX, or FFFFH bytes in length.

A Create, Write, and Close File Example

An example program, write.asm creates a file named trial.try and then uses function 40h to write a string of ASCII characters to the file. Function 3Eh is used to close the file. The file contains that old chestnut, "hello."

You can debug or run the program. Use the DOS command TYPE trial.try to see if the example program works as intended.

```
;write.asm, a program that writes the message "hello" to the file
;named TRIAL.TRY.
data segment
        org 1000h              ;define data area
handle  dw ?                   ;reserve RAM for handle variable
data ends
code segment
        mov sp, 6000h          ;set the stack
        mov dx,path            ;point to ASCIIZ file path name
        mov ah,03ch            ;assign a file handle with function 3Ch
        mov cx,0000h           ;normal file attribute
        int 21h                ;get DOS
        mov handle,ax
        mov bx,ax              ;get file handle to BX
        mov ah,40h             ;use function 40h to write to file
        mov cx,0005h           ;five characters in "hello"
        mov dx,msg             ;point to data
        int 21h                ;write
        mov ax,3e00h           ;close the file and flush buffer
        mov bx,handle          ;get file handle
        int 21h                ;close the file
        mov ax,4c00h           ;return to DOS
        int 21h                ;back to >
path:   db 'A:TRIAL.TRY',00h   ;file path name
msg:    db 'hello'             ;message to be written
code ends
```

It is good programming practice to check that register AX, the number of bytes written by INT 40h, matches CX, the number of bytes to be written. Should AX be less than CX, then the disk file will fill up before the last data can be written to it.

Open a File: INT 21h, Function 3Dh

We can create, open, and write data to a file, using the previous functions. To read data from a file, we must first open the file using function 3Dh and then we may use function 3Fh to read any number of bytes from the disk to a buffer area in RAM.

Remember that the create function, 3Ch, will *erase* any data found in an *existing* file. Files should be created *once,* and opened and closed thereafter as needed. Function 3Dh, which opens a file for read or write, is examined next.

Once a file has been created by function 3Ch, and closed by function 3Eh, it may be opened by function 3Dh for read, write, or read and write operations. To use function 3Dh, load the registers as listed next.

Register	Contents
AH	3Dh
AL	Access mode 00h = Read 01h = Write 02h = Read/write
DS:DX	Pointer to ASCIIZ path name
Returned	Flags: Carry flag is set if errors AX: File handle number if CF cleared AX: Error code if CF set

If the function is in error, the Carry flag will be set and AX will contain error codes that indicate the problem. Note that the file handle assigned by DOS when *opening* a file is not necessarily the same handle assigned by DOS when the file was *created*. File handles are assigned by DOS as needed and are only valid for any given file until the file is *closed*. Once opened by function 3Dh, the file may be read by function 3Fh, which is discussed next.

Read a File: INT 21h, Function 3Fh

Function 3Fh reads bytes from the disk file into a user-specified (program-specific) RAM data area. The data that is read is passed through the DOS disk buffer. To use the read function, preload the registers as indicated below.

Register	Contents
AH	3Fh
BX	File handle number
CX	Number of bytes to read
DS:DX	Address of data buffer in RAM
Returned	Flags: Carry flag clear if no errors AX: Number of bytes read if CF clear AX: Error code if CF set

Read errors will occur if the file is not open, or access is denied.

A CREATE, WRITE, AND READ FILE EXAMPLE

We may now modify our previous program, write.asm, to include a read from the file back to a RAM area we create. The new program is named read.asm.

```
;read.asm, a program that creates, writes, and reads data to the
;file named A:TRIAL.TRY.
data segment
        org 1000h
buff    db 5 dup(?)              ;five bytes in buffer
handle  dw ?                     ;space for handle variable
data ends
code segment
        mov sp,7000h             ;set the stack
        mov ah,3ch               ;open the file
```

```
                mov cx,0000h          ;normal file
                mov dx,path           ;point to file name
                int 21h               ;create file, get handle
                mov handle,ax         ;save handle number
                mov ah,40h            ;write to the file
                mov bx,handle         ;get file handle number
                mov cx,0005h          ;write five characters
                mov dx,msg            ;point to five characters
                int 21h               ;write
                mov ah,3eh            ;close the file
                mov bx,handle         ;get file handle
                int 21h               ;close the file
                mov ax,3d00h          ;open the file for read only
                mov dx,path           ;point to file name
                int 21h               ;open the existing file
                mov handle,ax         ;save file handle
                mov ah,3fh            ;read file
                mov bx,handle         ;get file handle
                mov cx,0005h          ;read five characters
                mov dx,offset buff    ;put the characters in buff
                int 21h               ;read the file
                mov ah,3eh            ;close the file
                mov bx,handle         ;get file handle
                int 21h               ;close the file
                mov ax,4c00h          ;done, back to DOS
                int 21h               ;>
path:    db 'A:TRIAL.TRY',00h         ;define file path name
msg:     db 'hello'                   ;data to be written = hello
code ends
```

Use D86 with memory line 1 set to the data buffer address (buff). You should see line 1 say "hello" when the read interrupt service is executed.

9.6 Sequential and Random File Access

From our discussions concerning the way in which DOS writes data to file clusters, it is apparent that some care must be exercised in how we write data to a file.

The simplest way to write data to a file is to write the file *sequentially*. Sequential file writing means the entire file must be re-written every time any changes are made, even minor ones. Sequential file access is very time-consuming if only small changes to a file are made. Sequential file access is, however, easy to program.

The opposite of sequential file access is *random* file access. Random file access involves changing or reading *only* the desired data, leaving the rest of the file data unprocessed. Random file access is *faster* than sequential file access because changes to the file affect only the clusters where the changes need to be made. Note, however, that random access does *not* allow us to insert more bytes into, or delete bytes from, an existing file. The new data must occupy the same number of bytes as the changed data.

Random file access demands that we have the capability to identify only the data we wish to write or read, while skipping data that is not of immediate importance. INT 21h, function 42h will give us the capability to address, or point to, any area of the file we wish to access.

Random File Pointer: INT 21h, Function 42h

Each time a file is created, or opened, DOS sets a DOS disk buffer byte *pointer* to 0000. Successive reads from or writes to the file, with no intervening closures, advance the pointer by the number of bytes read or written. By using a file pointer, DOS can "keep track" of the data placed in the file buffer and does not overwrite any file data in the disk buffer. The pointer can be manipulated using DOS function 42h and set to any point in the DOS data buffer, as explained next.

Register	Contents
AH	42h
AL	Pointer movement instructions
	00h = Offset from beginning of file
	01h = Offset from current file location
	02h = Offset from end of file
BX	File handle number
CX	Most significant word of offset number
DX	Least significant word of offset number
Returned	Flags: Carry flag clear if no errors
	DX:AX: Current file pointer value if CF clear
	AX: Error code if CF set

Register pair DX:AX contains the current setting of the file pointer. The file pointer is a double-word (32-bit) number that can range from 00000000 to FFFFFFFFh, or from 0 to 4,294,967,295.

RANDOM FILE POINTER EXAMPLE

The next program, pointer.asm, which is a modification of our last program, read.asm, lets us watch the file pointer change as data is written to the disk file buffer. Program pointer.asm should only be run in D86. DO NOT EXECUTE THIS PROGRAM—it never ends.

```
;pointer.asm, a program that gets the DOS file buffer pointer each
;time a 100h byte write is done to the DOS file buffer. Skip over
;the JMP @MORE instruction to leave the write loop and close the file
;when using D86. DO NOT EXECUTE THIS PROGRAM, use D86.

data segment
        org 1000h               ;store "handle" variable here
handle  dw ?
buff    db ?                    ;store read data from disk here
data ends
code segment
        mov sp,8000h            ;set the stack
        mov ah,3ch             ;create file
        mov cx,0000h           ;normal file
```

```
                    mov dx,path              ;ASCIIZ file name at address "path"
                    int 21h                  ;create
                    mov handle,ax            ;save the file handle number
          @more:
                    mov ah,40h               ;write data to file
                    mov bx,handle            ;get file handle
                    mov cx,100h              ;write 256 bytes (1/2 of a sector)
                    mov dx,msg               ;msg is the address of the bytes
                    int 21h                  ;write the 256 bytes
                    mov ah,42h               ;see where the file pointer is
                    mov al,01h               ;move from current position
                    mov bx,handle            ;get file handle
                    mov cx,0000h             ;move 0 bytes from current position
                    mov dx,0000h
                    int 21h                  ;move file pointer
                    jmp @more                ;DX:AX show current pointer number
                    mov ah,3eh               ;skip JMP @MORE when done and close
                    mov bx,handle            ;get file handle
                    int 21h                  ;close the file, flush the buffer
                    mov ax,3d00h             ;open the file
                    lea dx,path              ;name the opened file
                    int 21h                  ;open it
                    mov handle,ax            ;save newly opened file handle
                    mov ah,3fh               ;read the file
                    mov bx,handle            ;get the file handle
                    mov cx,1000h             ;read 1000h bytes (if available)
                    lea dx, offset buff      ;store data in RAM address "buff"
                    int 21h                  ;read file
                    mov ah,3eh               ;close the file
                    mov bx,handle            ;get file handle
                    int 21h                  ;close the file
                    nop                      ;quit out of D86 here
          path:     db 'A:TRIAL.TRY',00H
          msg:      db 100h dup 'a'          ;256 fill characters
          code ends
```

When exercising this program using D86, you may close the file by arrowing over the jmp @more opcode after writing any number of half-sectors. After quitting D86, view the file TRIAL.TRY using the DOS TYPE A:TRIAL.TRY command.

FIND FILE LENGTH EXAMPLE

In our previous program, pointer.asm, we moved the pointer 0 bytes from its current position when function 42h was used. The return value of the pointer then became the current position. We may use a similar "trick" to discover other things about the file.

To determine the length of an unknown file, set AL to 02h (offset from end of file) and request the pointer be moved 0 bytes from the end of the file. The pointer returned will be the length of the file, in bytes. You may use this same technique to position the pointer at the end of the file for appending new data to an existing file.

We shall determine the length of an unknown file using AL set to 02h in the next example program. Use the previous program, pointer.asm, to make a file of any length you wish. Use the next program, howbig.asm, to find the length of your program.

```
;howbig.asm, a program that finds the length, in bytes, of a disk
;file named A:TRIAL.TRY.

data segment
        org 1000h               ;store "handle" variable here
handle  dw ?
buff    db ?                    ;store read data from disk here
data ends
code segment
        mov sp,5000h            ;set the stack
        mov ax,3d00h            ;open the file
        mov dx,path            ;name of file to open
        int 21h                ;open it
        mov handle,ax          ;save newly opened file handle
        mov ah,42h             ;see where the file pointer is
        mov al,02h             ;move from end of file
        mov bx,handle          ;get file handle
        mov cx,0000h           ;move 0 bytes from current position
        mov dx,0000h
        int 21h                ;move file pointer 0 bytes from end
        mov ah,3eh             ;close the file
        mov bx,handle          ;get file handle
        int 21h                ;close the file
        mov ax,4c00h           ;return to DOS
        int 21h
path:   db 'A:TRIAL.TRY',00H   ;ASCIIZ path name for file
code ends
```

Note that an actual read of the file is not necessary. Very large files (greater than a 64K segment) can be gotten from the disk a piece at a time for processing by the program.

Record Length for Random Access Files

Random files must be more *organized* than sequential files. Sequential files are written, from start to finish, by filling as many disk file clusters as are needed to store the file. A sequential file is an amorphous mass, with no sharply defined boundaries within the file. Text files, such as the file that contains this chapter for the word processor, are good examples of sequential files. There is no reason to make a word processor text file random. The word processor program, under human direction, can scan the file until a desired text location is found.

Files are made random when they are composed of many separate *parts*, each of which may be changed *repeatedly*. Random file parts are concerned with the same subject, but each part has a separate identity from all of the other file parts. Examples of random files include a payroll file, a customer order file, or student grade file.

Random files, in order to be random, must be organized into *records*. A record is the smallest identifiable collection of data that a program is designed to operate on. Each record contains all pertinent data about the subject. A record is made up of separate byte groups called *fields*. Each field contains data that is distinctly different from any other fields in the record. Fields may be used to identify the record, or may contain information unique to the record subject. A student grade roster, for instance, might consist of a field for the student *name*, a field for a *social security* number, and a *grade* field for the course mark.

One dilemma posed by records is to fix the record *size*. A student grade file, for instance, contains records for *each* student. Each record is made up of fields that list the student name, social security number, and grade. You can predict the number of bytes for the last two record items, but the name *length* field is extremely variable.

For example, the name field has to be designed to fit the names of Aristotle Demitrius Agorapopolis as well as Amy Li. One solution to the variable field length problem is to fix the name field in the record at some maximum number of bytes and *hope* everyone's name will fit inside. Names that are too long will have to be truncated, so Aristotle may discover his name is now Arist in the file. Names that are shorter than the assigned field length can be filled with zeroes, so that Amy is stored 00Amy.

Once we have designed the record size we must then find a way to identify each individual record so it may be easily found. We need a *key* to the file. A key is any scheme that uniquely identifies a record in the file. A simple key is to assign a number to each file record. The first record is number 0, the second is number 1, and so on. To find any record in the file we simply multiply the file number by the *number* of bytes in a record and position the DOS file pointer to that place in the random file.

A student grade roster is an excellent example of a random file organized into records that contain student information. We shall design a sample student file as follows:

File Bytes	Contents (all in ASCII)
00h–1Fh	Student name as Last, First, MI
20h–28h	Student social security number
29h–2Bh	Student grade (from 000 to 100)
2Ch–2Dh	Line feed, carriage return sequence

Our random file is organized into record lengths of 46 bytes. The name field is 32 bytes long. The student's last name is left justified, followed by a space and the student's first name, followed by a space and the student's middle initial, if there is one. Any bytes that are not used for the name are filled with ASCII zero bytes. The student name field is followed by a SS number and a grade. The last 2 bytes are reserved for a line feed (LF) and carriage return (CR) sequence. Most word processors, and EDLIN in particular, insert the (nonprintable) ASCII characters 0Ah and 0Dh when the Return key is pressed at the end of a line of text.

Aristotle, for instance, appears in the file as:

```
File   0000000000000000111111111111111122222222222 22
Byte   0123456789ABCDEF0123456789ABCDEF0123456789AB CD
       Agorapopolis Aristotle D00000000265882556087LFCR
       |   Last      First   MI|       | Number | Gr |
```

We shall assign every student a student ID number, beginning with student number 0. Each student ID number matches a record number in the file that contains the student grade. As students matriculate they are assigned the next record number. Graduating students' numbers are freed for use by entering students.

RANDOM FILE RECORD SEARCH EXAMPLE

Our last example program will find any student record when the program is supplied a student ID number. The program, random.asm, multiplies the student number by 46 to position the DOS file pointer to the correct student record. Once a record has been located, it is read into a data buffer.

```
;random.asm, a program that will locate any 46-byte record in a
;random file. The record is placed in a data buffer when found.
data segment
        org 1000h
handle  dw ?                    ;store "handle" variable here
buff    db ?                    ;store student record here
data ends
code segment
        mov sp,0a000h           ;set the stack
        mov ax,3d00h            ;open the file
        mov dx,path             ;name of file to open
        int 21h                 ;open it
        mov handle,ax           ;save newly opened file handle
rec:
        mov al,2eh              ;46 bytes per record
        mul bl                 ;***** D86 USER: put student ID number in BL
        mov dx,ax              ;put result in DX
        mov ah,42h            ;position the file pointer
        mov al,00h            ;move from start of file
        mov bx,handle         ;get file handle
        mov cx,0000h          ;move DX bytes from current position
        int 21h              ;move file pointer 0 bytes from end
        mov ah,3fh           ;read the file
        mov bx,handle        ;get the file handle
        mov cx,002ch         ;read 44 data bytes
        mov dx,offset buff   ;store data in RAM address "buff"
        int 21h              ;read file
        jmp rec              ;get next record
path db 'A:STUDENT.REC',00h
code ends
```

To exercise this program, make a student file by using EDLIN, or any ASCII word processor, to make a sample student file with records constructed in 44-byte lengths (plus 2 bytes for LF and CR). Name the file STUDENT.REC and place it on the A: drive so that the correct file will be opened by function 3Dh. Run RANDOM.COM using D86 and supply the student number to BL where indicated in the program.

A sample STUDENT.REC file, using a word processor and then processing by A86, is shown next. A86 will assemble the file as student.bin, because of the org 0000h statement. *Rename* the file student.rec.

```
;sample student.rec file for random processing by random.asm
;0Dh and 0Ah are CR, LF characters
code segment
        org 0000h           ;make A86 assemble as a .BIN file
        db 'Ayala Kenneth J0000000000000000266542346075',0dh,0ah    ;ID 0
        db 'Baldwin Michael J00000000000000234567891100',0dh,0ah    ;ID 1
        db 'Shomaker Heather A0000000000000345689234099',0dh,0ah    ;ID 2
        db 'Martin Fred L00000000000000000349867398090',0dh,0ah     ;ID 3
        db 'Batuwangala Shad-Bantou000000000900234943085',0dh,0ah   ;ID 4
code ends
```

If you are using a word processor, be sure that you save the STUDENT.REC file as a pure *ASCII* file, or problems will arise when you attempt to access the records, because of the "happy faces" inserted by the word processor. EDLIN will automatically save files in pure ASCII.

You may have to adjust the record length 1 or 2 bytes, depending on your word processor's behavior toward inserting LF,CR at the end of each text line. Using A86 to assemble the student.rec file, shown above, is awkward, but gives complete control of the file, including ASCII control characters.

9.7 Summary

Data is stored on floppy disks using the file concept. Disks are organized into sides, tracks, and sectors. A file is composed of one or more disk sectors. Sectors are grouped into clusters.

DOS records the location of each file on the disk by the use of a file allocation table (FAT). The FAT occupies disk Sectors 1 and up and contains file cluster numbers. Each cluster of a file points to the next cluster of the file until an end-of-file (EOF) entry is found in the last cluster.

The root directory, which follows the FAT, lists the name of each file, time and date of creation, and FAT cluster entry point. Files may be organized into groups of directories. Each directory may consist of finer groupings of files into subdirectories. A listing of the drive, directory, and subdirectory by which a file is specified to DOS for access in called the file path, and the text string representing the path is the path name.

File sectors may be directly accessed using the following DOS INT service routines.

INT 25h: Absolute Disk Read

INT 26h: Absolute Disk Write

DOS files are normally accessed using the file handle concept. DOS handle INT service routines that are commonly used to read and write files to a disk are as follows.

INT 21h, Function 3Ch: Create a File

INT 21h, Function 3Dh: Open a File

INT 21h, Function 3Eh: Close a File

INT 21h, Function 3Fh: Read a File

INT 21h, Function 40h: Write to a File

INT 21h, Function 42h: Move the Internal Disk Buffer Pointer

Files may be accessed sequentially, or in a random manner.

9.8 Problems

Use the DOS file access interrupts discussed in this chapter to perform the required file task.

1. Using a text editor, create a file that is between 5,000 and 6,000 bytes long on a newly formatted floppy disk. Calculate the file directory and FAT entries for the type of disk used, and check your calculations using absolute sector program sector.asm.

2. Using a text editor, create three text files of 50, 1,000, and 3,000 bytes in length on a newly formatted floppy disk. Calculate the file directory and FAT entries for the type of disk used, and check your calculations using absolute sector program sector.asm.

3. Take the test disk supplied to you by your instructor and determine the names, dates, and lengths of all the files on the test disk by reading the directory sectors using program sector.asm. Use the DOS DIR command to check your answers.

4. Create a directory and a subdirectory on a newly formatted disk. Predict the contents of the directory entries for the directory and subdirectory and check your prediction with program sector.asm.

5. Write a program that will perform the DOS DIR command for a disk that contains four test files.

6. Create a file named problem5.txt, open the file, write the message "hello, this is a test for problem five" and then close the file. Verify that the message is on the file by using the DOS type problem5.txt command.

7. Write a program that will perform the DOS COPY command by copying one test file on your disk to a new file name.

8. Write a program that will perform the DOS DEL command on a test file on your disk. Verify that the file is deleted by using the DOS DIR command .

9. Construct a random file of your classmates consisting of their names and ages. Assign each person an ID number and write a program that will display the names and ages for each ID number entered from the keyboard.

10. Construct a random file of the sines of all angles from 0° to 90°. Make each entry accurate to three places. Write a program that will display the sine of any angle typed on the keyboard.

10

INTERFACING

TO STANDARD

COMPUTER

PORTS

Chapter Objectives

On successful completion of this chapter you will be able to:

- Describe the difference between memory space and I/O space.
- Use parallel printer port BIOS interrupt services.
- Interface external circuits to the parallel printer port.
- Use serial port BIOS interrupt services.
- Interchange data between two computers using the serial port.

10.0 Introduction

Those who use personal computers to solve problems generally do so on three levels of complexity: pure computing, hardware control using standard peripheral connectors, and hardware control using bus interface cards.

User titles change with the level of usage. Persons most concerned with pure programming are generally named *programmers*, *application programmers*, or *software engineers*. Those who interface the computer to hardware peripherals may be called *system programmers*, *hardware/software engineers*, *real-time programmers*, *computer interface engineers*, or the like.

We shall use the title *system engineer*, in this book, to refer to persons who are primarily machine-language programmers and who also *interface* computers to external hardware circuits. System engineers are concerned with using computers as one *part* of a complete system, such as a robot, a flexible manufacturing cell, a communication network, or an automated factory.

Pure Computing

The first level of computer usage is to employ a computer and its common peripherals, (keyboard, disk drives, monitor, and printer) just as they come from the manufacturer. The programmer can use the personal computer system to run purchased or "canned" programs such as word processors, spreadsheets, databases, CAD programs, and so on. The programmer may also write original programs in high level languages such as BASIC, Pascal, and C, or in assembly code using DOS and BIOS service routines.

The first level of computer use meets the popular conception of a computer as a machine that people use to *run programs* that operate *only* on the computer and its peripherals. All first-level usages use the personal computer as a tool to accomplish a purely programming task.

More detailed knowledge, training, and programming skill are needed as one progresses from using canned programs to writing customized assembly code. But, whether the programs are canned or original, the computer remains a computer and is used to run programs.

Standard Port Interface

The second level of computer usage is to interface the computer to peripherals, other than those *commonly* used in a pure computing environment, using the standard peripheral *ports*. Just as a seaport is an entry and exit point for the goods of a country to other countries, a peripheral port serves as a main point of interchange for the computer with other devices. Ports are generally of two types: *serial* and *parallel*.

Personal computers, originally intended to be used as pure computers, have also found their way into commercial and industrial *control systems*. A personal computer is a *very* inexpensive and very *flexible* control device for machinery of all types, when compared to custom-designed machine control hardware. Competition, and volume production, in the PC industry has driven the price of a typical personal computer *much below* the cost of custom-built control computers.

Interfacing a personal computer to external hardware does require, however, some means of adapting the personal computer so that it can be connected, electrically and logically, to the particular hardware in the system. It is natural, then, to use the common PC peripheral ports as an interface means between the PC and external machinery. The carrot of low price has led to the practice of using personal computer standard peripheral ports as interfaces to nonstandard industrial hardware. By gaining program access to the standard ports, the programmer can use the computer to control external machinery.

A personal computer is normally equipped with *four* standard connectors (ports) to which four common peripherals are plugged. The four common peripheral ports are the keyboard, printer, modem, and cathode ray tube (CRT) video monitor.

All of the standard ports may be used for purposes other than those intended by the manufacturer, however it is *rare* to use the screen connector for any purpose other than for video. The other three connectors—printer, modem, and keyboard—are *often* used to interface the computer to peripheral circuits other than those originally intended by the manufacturer.

Interfacing a PC to a nonstandard peripheral, using a standard peripheral port, requires that the user have a *detailed* knowledge of all port connector *signals,* as well as the BIOS or DOS service routines that *control* the port connector signals. The user must also posses a good understanding of digital *logic design* technology in order to design the *interface* circuitry that will pass signals between the standard port and the nonstandard peripheral. Thus, at the second level of usage, the programmer must be proficient in both software *and* digital hardware design.

Bus Expansion Card

The third level of PC usage is to abandon the standard peripheral ports as interfaces to external hardware and use interface circuits that plug into the computer *bus.* The bus interface circuits, named *expansion cards*, are plugged into one of the empty system bus connectors mounted on the computer motherboard and the external hardware is connected to the expansion card.

A computer bus is any *manufacturer-defined* arrangement of *address, data,* and *control* signals used by the computer motherboard designer. (The term *bus* originated in power plants, where all of the generators are connected to a heavy conductor named a *bus bar*).

Plug-in bus expansion cards are commercially available for data communications, FAX, laboratory instrumentation, logic analyzers, machine sensing and control, bar-code readers, and a host of applications that can, in some sense, be considered "standard." Some bus cards are purchased complete with all software needed to operate them. Other cards, particularly those for machine sensing and control, may require that the user supply any specialized software needed for the project.

To be able to use bus interface cards, the user may need a detailed knowledge of the *internal bus hardware features* of the PC, the ability to *design digital circuits,* and an understanding of the PC *interrupt structure.* The third level of usage, it would seem, requires that the user have the skills needed for second-level usage *and* a good understanding of the internal PC bus and bus interrupt hardware and software.

Hardware Interfacing and Control

Chapters 3 through 8 of this book discuss how to program an X86-based personal computer and use the standard computer keyboard and monitor peripherals for pure, first-level, programming applications. Chapter 11 contains information needed to practice third-level interfacing to the computer internal bus and interrupt system. This chapter investigates the printer and serial connector port interface hardware *and* software as examples of second-level usage.

Discussions of interfacing external circuits to the computer standard ports and bus necessitates involving digital logic circuits. Digital design techniques are beyond the scope of this book and it is assumed the reader has an understanding of digital logic *functions, families,* and, most important, *electrical* and *timing* requirements.

10.1 Memory and Input-Output Address Spaces

As discussed in Chapter 3, the fundamental model of a computer is a CPU and memory. A better model might be: a CPU and CPU *address space.* Address space is the *entire* collection of addresses that the CPU can physically access, by *any* means.

Memory, in the way we have used it in earlier chapters, consists of *high-speed integrated circuits,* such as DRAM, that can store and retrieve thousands of bytes of data at high speeds. We have, for example, written programs that quickly access thousands of memory bytes using MOV- or MOVS-type opcodes. Access to memory can be done as fast as the CPU can execute the memory mode instructions.

A memory address could also be used for a *slow-speed peripheral* device, such as a printer. Peripheral memory has all of the characteristics of integrated circuit DRAM memory, only it is much *slower* (on the order of 100,000 to 1) in response time. In addition to slow speed of response, the *number* of peripheral addresses commonly do not exceed a few hundred as compared to millions of DRAM addresses.

Memory Space and I/O Space

Primarily because of the differences in response *speed* between DRAM semiconductor memory and peripherals and the much more *numerous* semiconductor addresses, overall CPU address space is divided into two parts, as shown in Figure 10.1. One memory space, the numerous high-speed part, is called *memory.* The other memory space, the sparse low-speed part, is called *input/output memory.* (The term *input/output memory* is so universal that it

Figure 10.1 ■ Memory and I/O address spaces.

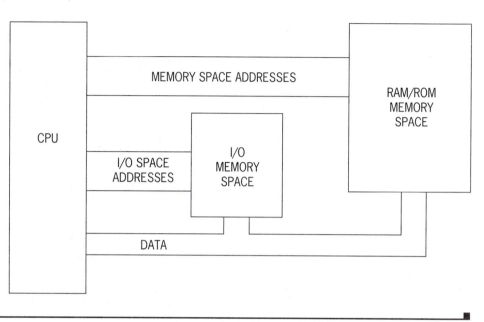

has become the common acronym *I/O*.) Thus, when the term *memory* is used it is *assumed* to be an internal *semiconductor* address, whereas the term *I/O* is *assumed* to mean a *peripheral* address.

The division of overall CPU address space into a memory space and an I/O space is made possible by CPU *hardware* circuits and CPU *software* opcodes. CPU hardware *memory space signals* determine when memory space is addressed and when I/O space is addressed. The CPU memory space enabling signals are, in turn, *generated by opcodes* that cause the memory space or the I/O space to be enabled.

When the CPU is dealing with integrated circuit memory it is said to be *reading* data from or *writing* data to memory. In Chapter 5, we showed how to use the MOV command to read and write data to the data segment in memory. MOV type commands (MOVS, LODS, SCANS, PUSH, POP, etc.) generate CPU *signals* that *access* memory space.

When the CPU is interchanging data with a peripheral it is engaged in *input* or *output* of data with the peripheral. In Chapter 8, we examined the use of the IN and OUT instructions for interchanging data with I/O peripherals. IN- and OUT-type commands are opcodes that generate CPU *signals* that *access* I/O space.

Parallel and Serial Ports

Data interchange by the CPU with internal integrated circuit memory is done via an *internal* bus arrangement of CPU address, data, and control signals. Data interchange by the CPU with external peripherals is done using an arrangement of *external* control and data signals present on the port *connec-*

tor. The two ports most often used for interfacing are the serial and parallel ports. *Serial* ports interchange data on a pair of wires by placing data bits on the wire pair at certain time *intervals.* Serial ports are advantageous when data must be sent over long (greater than 10 meters) distances. Wiring costs are held to a minimum when serial data transmission schemes are employed, because, at a minimum, only two wires are needed to transmit serial data.

Parallel ports have a wire for each data bit; data is present on each of the port wires *simultaneously.* Parallel ports offer greater data speed (bytes per second) than do serial ports, but wiring costs mount when parallel data is transmitted over long distances. The PC printer port is a parallel port and is the subject of the next section.

10.2 The Parallel Printer Port

One advantage of using the printer port for an interfacing application is the *standardization* that exists for the port across PC model lines. The printer port remains the same no matter what PC CPU is employed. All PCs, from those based on the original 8088 to the newest model, are equipped with parallel printer ports that behave in exactly the same manner.

Another advantage of using the printer port is that we may use several *standard* DOS or BIOS INT service routines to interchange data with the port. We do not need to concern ourselves with the exact workings of the circuitry that controls the printer port if we understand how to use the DOS and BIOS printer service routines.

The system programmer is *not* restricted to using only BIOS and DOS printer INT service routines; the printer port can be *directly* controlled by the program. Unfortunately, it *may* be that the integrated circuits (ICs) that control the printer port are not standardized from PC model to PC model, and our program would have to be custom-tailored to each PC make and model. Instead of customizing the program to fit a particular parallel port IC, it may be more feasible to design an expansion card using the techniques discussed in Chapter 11.

A major disadvantage of using BIOS and DOS printer port service routines is a lack of freedom to tailor the parallel port control signals to our application. We must "work around" the available printer port control and data signals and design our peripheral interface circuitry in conformance with signals that were *intended* to control a printer.

The Parallel Printer Port Connector

Figure 10.2 shows the physical configuration of the printer port connector. The connector is an industry standard type, commonly referred to as a *25-pin D type* (DB-XX) connector. D (from the shape) connectors have been standardized for over 30 years by the Electronics Industry Association and can commonly be found in sizes ranging from 9 to 44 conductors. Connectors are configured as male (pins) or female (sockets) and are a constant source of irritation for users who find themselves faced with plugging a male cable into a male connector, or the reverse.

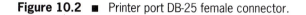

Figure 10.2 ■ Printer port DB-25 female connector.

Good design practice usually dictates that any connector that is the primary source of electrical signals be female, and that any connector that receives signals be male. Female connectors will not short from conductor to conductor as easily as will male connectors. Good design does not seem to happen often, and one observes that parallel printer ports are female and serial ports are male on most personal computers.

Printer Signals

Table 10.1 lists the signals that are present on each pin of the printer port D connector. Signals levels are, nominally, voltages between 0 and +5VDC. Each output signal is capable of driving at least one standard TTL input. Each input signal requires a standard TTL drive level. Note, in Table 10.1, the convention slash (/) has been used to designate *active-low* signals. Signals may be *asserted* active-low or active-high when they are engaged in performing their assigned function, that is, the *condition* the signal represents is *true*. To

Pin	In/Out	Function
1	Output	/Strobe signal, active low
2	Output	Data Bit D0, positive logic
3	Output	Data Bit D1, positive logic
4	Output	Data Bit D2, positive logic
5	Output	Data Bit D3, positive logic
6	Output	Data Bit D4, positive logic
7	Output	Data Bit D5, positive logic
8	Output	Data Bit D6, positive logic
9	Output	Data Bit D7, positive logic
10	Input	/Acknowledge signal, active-low
11	Input	Busy signal, active-high
12	Input	Paper Out signal, active-high
13	Input	Selected signal, active-high
14	Output	/Line Feed signal, active-low
15	Input	/Error signal, active-low
16	Output	/Initialize signal, active-low
17	Output	/Select signal, active-low
18–25		Ground

Table 10.1 ■ Printer Port Connector Signals

avoid semantic confusion, the term *active—when applied to a control signal*—is defined as follows:

Active-High/Low = Signal is +5V/0V when indicated activity is TRUE

Thus if signal A is active-low, it will be 0V when it is doing whatever it is supposed to do, and +5V when it is inactive.

PRINTER HANDSHAKING SIGNALS

The original purpose of the printer port was to send one character (a byte) from the computer to the printer. The computer is very fast; the printer is very slow. To prevent the computer from sending bytes to the printer faster than they can be printed requires that there be some types of control signals *between* the computer and the printer.

The signals that control interaction between the computer and the printer are called *handshaking* signals. Other signals are data bits. Handshaking signals let the computer send data characters to the printer and permit the printer to stop further transmissions from the computer until the last character has been processed.

Printers and computers are built by many different companies, so rigid definitions of the handshaking details (*protocol*) are essential to port *standardization.* One protocol detail is how each handshaking signal is to react, *in time*, as a byte of data is transmitted from the printer port to the printer. Figure 10.3 shows the timing details attendant with sending a character from the computer to the printer. The sequence of printing a character proceeds as follows:

1. The computer *checks* the level of the BUSY signal from the printer. If BUSY is TRUE (active-high), the printer is occupied with printing a previous character. If BUSY is low, then the printer is free to accept the next character. The computer performs no further printing actions until BUSY goes low.

2. The computer *outputs* (latches) a character on the eight character data lines after detecting that BUSY is low.

3. After latching the print character onto the data lines, the computer brings the STROBE line low. The printer uses the STROBE low signal to *accept* the data character present on the data lines as valid, and begins to process the character.

4. The *printer* brings the BUSY line high to signal that it is preoccupied with processing the character and cannot accept another character at the present time.

5. The *computer* brings the STROBE line high and holds it high until the next character is to be printed.

6. When the *printer* has finished processing the character, it will strobe the ACK(nowledge) line low then high, and bring the BUSY line low. The ACK strobe is not strictly needed by the computer but can be of use with circuitry that is edge-triggered, such as flip-flops.

Note that handshaking involves both the computer and the printer, which implies that the printer has limited control circuitry of its own. The

Figure 10.3 ▪ Printer port handshaking signal timing.

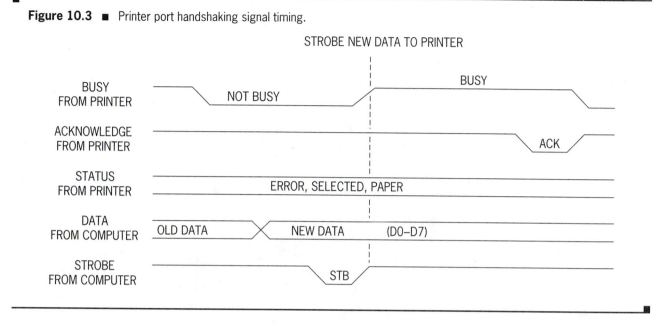

timing of the handshaking signals is *very* important. The computer will respond correctly only to handshaking signals that meet the exact timing specifications just listed.

The printer port character, bits D0 to D7, is normally *latched* at the port connector during the entire print cycle, until a new character is sent. Latching of the data by the parallel port relieves the user from the necessity of having to include a data latch in any parallel port interface circuitry. All of the parallel ports on the PCs used to test programs in this chapter latch the data. It *may* be, if Murphy is still active, that your PC does not latch the parallel port data, and you will have to include a data latch in your designs.

10.3 BIOS and DOS Printer INT Services

In Chapter 7, several DOS and BIOS INT service routines that access the keyboard and the screen peripherals are examined. In this section, we discuss similar BIOS and DOS INT service routines that let us interchange data with the printer port. BIOS interrupt 17h, which permits control of the printer port at the bit level, is examined next.

BIOS Printer Control INT Type 17h

BIOS INT 17h has three functions that output a character to the printer, initialize the printer, or request printer status. As we are in the business of designing a nonstandard peripheral interface, we shall not be concerned with the initialization function, which is *not* used by all printers. We may, however, *test* our program using a printer to verify some of our ideas. Using the printer lets us check out our concepts before building any special peripheral circuitry.

The first INT 17h function, 00h, places a character on the printer port.

Function 00h   Output a Character to the Printer.

Register	Contents
AH	00h
AL	Character to be output
DX	Printer number 00 to 02h
Returned	
AH	Printer status

Operation:

Function 00h of INT 17h writes the character in AL to the printer port. Register DX contains a printer port number, from 00 to 02h. Printer numbers 00 to 02h correspond to standard printer devices LPT1 to LPT3. Most computers have at least one printer port, which is normally set to be LPT1.

After invoking interrupt 17h with an INT 17h opcode, the routine returns a *printer status* byte in register AH. The status byte is the means by which the printer handshaking signals *communicate* with the program. Each bit of the status byte may be interpreted as shown below.

Bit of AH	Printer Port Status
7 6 5 4 3 2 1 0	
– – – – – – – 1	No response from port (Time-Out)
– – – – – – x	Unused
– – – – – x	Unused
– – – – 1	/Error signal low (pin 15)
– – – 1	Selected signal high (pin 13)
– – 1	Paper Out signal high (pin 12)
– 1	/Acknowledge signal low (pin 10)
1	Busy signal low (pin 11)

Note that a printer status bit set to *1* does *not* correspond to the accompanying input signal from the printer in the *high* state. For example, Bit 7 is a 1 when the Busy signal is low, but Bit 5 is a 1 when the Paper Out signal is high.

If the printer Busy signal (pin 11) stays high for 2 seconds or so (resulting in Bit 7 of AH equal to 0), function 00h returns with *no* print action having been done, and Time-Out (Bit 0 of AH) set to 1.

For interfacing purposes, the printer *status* signals permit a limited form of data transmission from a special parallel port peripheral to the program by using Bits 3 through 6 (port pins 10, 12, 13, and 15) as *input* data bits.

Pin 11, the input Busy signal, *must* be kept low so that BIOS assumes the printer is not busy. If Busy pin 11 is not kept low, BIOS will time-out and return AH with Bit 0 set to 1 (Time-Out). When Time-Out Bit 0 is set to 1 by the service routine, it means that the service routine was not able to write a character to the port in a "reasonable" time. Almost all printers can print one character in 2 or 3 seconds, so the time delay involved in setting Bit 0 is reasonable for a printer interface.

If the nonstandard I/O device connected to the printer port requires *more* than 2 seconds to complete an action, then the Busy signal can be used to *slow* down the computer, or a data *buffer* can be included in the I/O device's design. The buffer will have to be "deep" enough (have as much memory as required) to store all of the data from the printer port. Indeed, many contemporary printers have print buffers of 2K and more so that the PC can download printable data to the printer very quickly. Downloading the print data permits the computer to resume other tasks while the printer is occupied. It is generally good system design practice to quickly download data from a computer to its peripherals. If large amounts of data are downloaded, or an equivalent buffer is set up in computer RAM, then the progam does not have to wait for the slower peripheral to act.

An INT 17h function that reads the printer port status, but does *not* write a character to the port, is discussed next.

Function 02h Read Printer Port Status

Register	Contents
AH	02h
DX	Printer number 00 to 02h
Returned	
AH	Printer status

Operation:

Function 02h returns the *same* status byte in AH as does function 00h, but does not write a character to the parallel printer port.

Treatment of a printer port time-out *varies* among different PC BIOS versions. Some PC AT and PC XT 286 models *may* execute an INT 15h, function 90h if the port times-out, in order to generate an error message to the user. When in doubt as to the behavior of the particular BIOS your computer may be employing, do *not* use function 02h. To read the port status bits, write your program using function 00h with the same byte as *last* written. The latched data at the port will not change, but the program will get the latest parallel port status in AH. All of the examples in this section will use function 00h of INT 17h to write and read parallel port data.

FUNCTION 01H: RESET PRINTER

There is an additional INT 17h function, 01h, which is used to initialize *certain* printers. Function 01h will send a reset sequence (*not* recognized by all printers) of characters 08h and 0Ch to the port and return the printer port status in AH.

PRINTING A MESSAGE

We may now try out function 00h and print a message. If the printer you are using *has* an internal buffer, then it will *not* print anything until the internal printer buffer is full or the printer receives a control character from your program to print a line. Typically, a carriage return (0Dh) control byte will *force* a buffered printer to print. A simple print program, printmsg.asm, is written as shown next.

```
;printmsg.asm, a program to print the message "hello"
code segment
        mov sp,1000h        ;set the stack
        lea bx,msg1         ;point to start of message
next:
        mov ah,00h          ;print character
        mov al,[bx]         ;get character to AL
        cmp al,'~'          ;quit if at end of message
        je donit            ;exit to DOS at end
        mov dx,0000h        ;use LPT1
        int 17h             ;just do it
        inc bx              ;point to next character
        jmp next            ;go until done
donit:
        mov ax,4c00h        ;exit back to DOS
        int 21h             ;return to DOS
msg1    db 'hello',0dh,0ah,'~'  ;our favorite message, with CR,LF
code ends
```

Program printmsg.asm is an adaptation of programs used in Chapter 7 to display messages on the screen. The program repeatedly sends characters to the printer until the ~ character is found in the message string.

DATA INPUT PROGRAM

Let us now use a short program to demonstrate using Interrupt 17h to *get* data from an external circuit connected to the printer port. The program shown next, print.asm, will get status bits from the printer port in register AH. Before using the program, a cable can be wired to the printer port for the purpose of placing logic highs (+5VDC) or logic lows (0VDC) on the INPUT signal pins of the printer port connector.

> WARNING: DO NOT TRY TO PLACE VOLTAGES ON *OUTPUT* SIGNALS FROM THE PARALLEL PRINTER PORT CONNECTOR PINS 1–9, 14, 16, AND 17.

We may inspect AH, using D86 to test the program, and verify that the *INPUT* signals on the printer port pins (from the printer to the computer) agree with the settings of the parallel port input switches.

```
;print.asm, a program that gets the status of the printer port pins
code segment
next:
        mov sp,1000h        ;set the stack
        mov ah,00h          ;get status
        mov dx,0000h        ;printer LPT1
        int 17h             ;BIOS printer port status
        jmp next            ;get status repeatedly
code ends
```

To test the program, inspect the contents of AH, using the single-step mode of D86, to see how changing the logic levels on the printer input status pins changes the bits returned in AH.

BIOS Interrupt 17h permits bit-by-bit control of the parallel printer port. DOS has several interrupts that may also be used to write data to the parallel port, as discussed next.

DOS Printer Services

BIOS printer service INT 17h lets the program read all of the input data signals from the parallel printer interface port and output a data byte to the port. DOS is equipped with interrupt functions that write bytes out to the printer port, but the printer port input bits can*not* be read using DOS.

DOS Interrupt 21h, Function 05h   Write Byte to Printer

Register	Contents
AH	05h
DL	Byte to be written to printer port

Operation:

Function 05h outputs a byte to the device STDPRN, which is the same as LPT1, the system printer, unless the user has changed the default device assignments. The function is invoked by loading registers AH and DL and using interrupt 21h.

No printer status data is returned by this INT function. Even worse, the function will *wait* until an acknowledge is received from the printer before returning. Should the printer be faulty, or disconnected, the program will "hang" until the user issues a Ctrl-Break or Ctrl-C keystroke. Function 40h, which is discussed next, may perform in a more satisfactory manner.

In Chapter 9, a discussion concerning the concepts of files and file handles is presented. Briefly, DOS associates a file with a number named a file handle. File handles are usually assigned to files located on disks, but DOS also assigns *permanent* file handles to devices such as the printer. Writing data to a printer is no different than writing data to a disk, as far as DOS is concerned. The first printer port, LPT1, is assigned handle 04h.

DOS Interrupt 21h, Function 40h   Write to File

Register	Contents
AH	40h
BX	04h (permanent printer handle)
CX	Number of bytes to print
DS:DX	Pointer to bytes to be printed

Returned

CF	Cleared if no errors
AX	Number of bytes written, or errors

Operation:

Function 40h of DOS INT 21h can write entire *strings* of data to a file, rather than the single character written by the BIOS printer functions. To use function 40h, the registers AH, BX, CX, and DS:DX are configured prior to using interrupt 21h. Register AX returns the number of bytes written, or an error code if some problems were encountered during the operation.

Let us take an earlier program, printmsg.asm, and use DOS function 40h, INT 21h, to write our "hello" message. The name of the altered program is prntstrg.asm.

```
;prntstrg.asm, a program to print the message "hello" using DOS
;function 40h, interrupt 21h.
See Note

code segment
        mov sp,1000h        ;set the stack
        mov dx,start        ;point to start of message        1
        mov cx,length       ;length of message
        mov bx,0004h        ;file handle for printer
        mov ah,40h          ;print message
        int 21h             ;interrupt to DOS
        mov ax,4c00h        ;return to DOS
        int 21h
start:  db 'hello',0dh,0ah  ;our favorite message
finish: nop
length equ finish-start
code ends
```

NOTE ON THE PROGRAM

1. Note that the data segment register, DS, must be the *same* as the code segment register, CS, for this program to work. The message hello is assembled in the code segment, but the pointer (DX) to the message is combined with DS to locate the message. If you are assembling the programs in .COM format, segment registers CS and DS will be set to the same value by DOS when prntstrg.asm is loaded. Should the file be assembled as an .EXE type, then DS would have to be set equal to CS before printing the message.

Inspection of the DOS-based program prntstrg.asm shows that it is much more compact than the BIOS character-oriented program printmsg.asm. The inclusion of the carriage return (CR) character 0Dh in the message will cause most printers with internal data buffers to print "hello" immediately.

DOS printer interrupts can prove useful when long strings of data are to be sent to a custom peripheral attached to the parallel printer port. The custom peripheral would have to be buffered, using high-speed logic, in order to store long strings of data from the DOS interrupt. BIOS interrupts would be needed to determine the status of any input data from a custom peripheral circuit.

10.4 Custom Printer Port Peripherals

Switch Input/LED Output

We are now prepared to design circuitry that interfaces to the printer port. Figure 10.4 shows a circuit that can sense the state of four switches and illuminate an accompanying LED if a switch is *closed*. If a switch is opened, then the LED associated with that switch is extinguished. Switch status is

Figure 10.4 ■ Parallel port hardware interface.

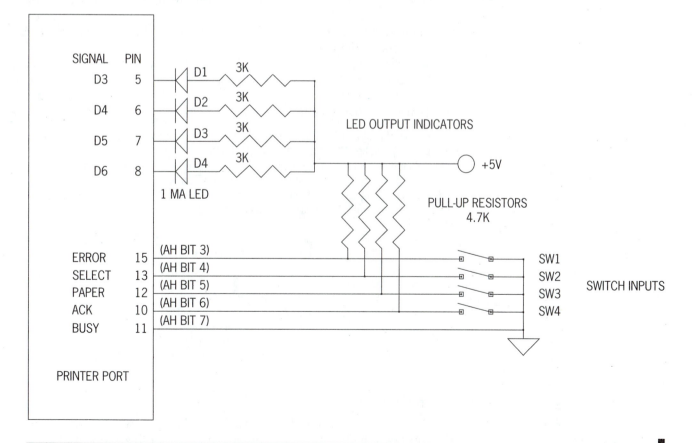

sensed directly from one of the parallel port input data pins. LEDs are lit, via an inverting buffer, from four of the parallel port output data lines. From the circuit diagram of Figure 10.4, note that an LED is turned *off* when its data line is a logical one.

Although the example of lighting output LEDs using the state of input switches may seem trivial, it represents many real-world applications where closing a contact (a switch) causes a motor (an LED) to be energized.

We will use function 00h to *sense* the switches and *control* the LEDs. The very first use of function 00h will output an FFh to the LEDs, turning them *off*. Thereafter, the sensed state of the switches will operate the LEDs. The program associated with the printer port circuit of Figure 10.4 operates as follows.

1. Function 00h is used to sense the state of the switches. An open switch is asserted high, and a closed switch is asserted low. An LED that is assigned to a switch will light when the switch in closed.

2. The switch state is read from AH and corresponding bits are set to enable the appropriate switch LEDs to be lit.

3. Function 00h is then used to write the LED control bits to the port.

4. Steps 1 through 3 are repeated by the program as it loops.

Note that the BUSY signal to the printer port is tied low in the circuit of Figure 10.4, so that the function always returns having written data to the port. AH bits 3 through 6 are used to sense each switch condition, therefore parallel data port bits D3 through D6 are used to drive the LED that matches the switch that sets the corresponding data bit in AH. For example, data Bit D6 (parallel port *output* data pin 8) will drive the LED for the switch that affects bit 6 (parallel *input* data port pin 10) in AH.

The program, switch.asm, that will control the LEDs in response to switch actuations is listed next.

```
;switch.asm will light an LED if a corresponding switch is closed.
;Interrupt 17h, function 00h inputs the switch states by reading
;status bits in the printer port status word. Each switch is
;connected to one of the printer status lines. Switch LEDs are
;controlled by writing a byte to the printer port using function 00h.
;A bit set to zero in the LED control byte will light an LED.
code segment
        mov sp, 1000h       ;set the stack                        See Note
        mov ah,0ffh         ;turn LEDs off at start-up
lup:
        mov dx,0000h        ;select LPT1
        mov al,ah           ;get control byte to AL
        mov ah,00h          ;get status and print character
        int 17h             ;get status and light LEDs
        xor ah,48h          ;convert status bits to active-low      1
        jmp lup             ;drive LEDs and get new status
code ends
```

NOTE ON THE PROGRAM

1. Bit 6 and bit 3 of AH are both a *1* when the corresponding inputs are *low* (switch closed). To illuminate an LED requires a low on the

appropriate LED data line. XORing AH with 48h (01001000b) will turn an active-low input 1 bit into an active-low LED light 0 bit on the output.

The program "switch.asm" can be tested using D86 to observe the individual steps, then run at full speed in order to be able to change the LEDs as each switch is exercised.

Keypad Input/Monitor Output

The final parallel port example involves interfacing the PC to a small 16-key keypad. The program senses which of the keys has been pressed and displays the key symbol on the monitor. The program is to display only *valid* key strokes. Only *one* key may be down on the keypad at any time to be considered valid.

Figure 10.5 shows the keypad connected to the parallel port. The keypad configuration is one of four rows and four columns. Each row is connected to the common input side of four keys. Each key in a row is in a separate column. Each column is connected to the normally open side of four keys. Each column key is in a separate row. As may be seen from the figure, a valid key (for any row) results in only *one* column output connected to a row input.

Figure 10.5 shows that inverters (four NAND gates) have been placed on

Figure 10.5 ■ Four row by four column keypad.

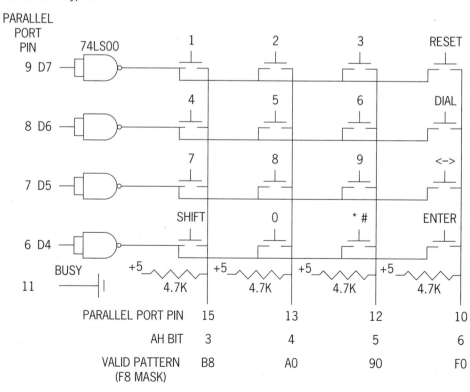

the board between the port data output pins and the keypad. The inverters ensure that shorts between rows (two or more keys down in a column) will not stress internal port circuits. Pullup resistors have been placed on each column so that a reliable logic high will result when no key is down in that column.

The basic operation of the program that drives the keypad, keypad.asm, is as follows.

1. Drive (*strobe*) each row low, one at a time, using four of the port output data bits. Every time a row is strobed, a *matching row lookup table number* is stored in a *temporary* memory location.

2. If a *single* key is pressed in any strobed row, only *one* column will be low. There are only four "valid" column patterns for any strobed row. A valid column pattern consists of one column low (the key down) and the other columns high.

3. After each row is strobed, the program reads the status of the columns using the port input status bits. The program checks for valid column patterns. If a valid pattern is found, the program stores a *matching column lookup table number* in memory. The program also copies the temporary row lookup table number to a permanent memory location. If only *one* valid pattern is seen by the program after all *four* rows have been strobed, then a single key is pressed.

4. The valid key is displayed using a lookup table to find its ASCII value. The lookup table is made by adding the *valid row number* to the *valid column number* stored in memory. If no valid patterns are found (all keys up), or more than one valid pattern is found, then a blank is displayed.

```
;keypad.asm, a program that reads a 4x4 keypad and displays
;the key currently pressed on the system monitor. The keypad is
;strobed by row and read by column, using the parallel port.
data segment                                                    See Note
            org 1000h         ;place data segment above lookup table
temprow     db ?              ;store current row number here
keyrow      db ?              ;valid row number
keycol      db ?              ;valid column number
data ends
code segment
;equates
            row1 equ 80h      ;use data bits D7–D4 to strobe rows
            row2 equ 40h      ;row bits are inverted on keypad circuit
            row3 equ 20h      ;resulting in one row low, all others high
            row4 equ 10h      ;last row pattern
            col1 equ 0b8h     ;valid pattern for only column one low
            col2 equ 0a0h     ;valid pattern for only column two low
            col3 equ 90h      ;valid pattern for only column three low
            col4 equ 0f0h     ;valid pattern for only column four low
;end equates
            mov sp,0ffffh     ;set the stack
            mov ax,0002h      ;set video text mode
            int 10h           ;the video interrupt
            mov ah,02h        ;position cursor in center of screen
```

```
                    mov bh,00h              ;page 0
                    mov dx,0c28h            ;row 12, column 40
                    int 10h                 ;screen blanked, DOS default cursor
;begin to scan each row and test for a valid (one key down) column
;pattern. The valid patterns are B8h, A0h, 90h, F0h.
scan:
                    mov cl,0h               ;CL counts number of valid columns/scan   1
                    mov al,row1             ;start scan with first row activated
                    mov temprow,0h          ;for the lookup table, rows are numbered
                    call _rowscan           ;as 00h, 04h, 08h, and 0Ch                2
                    mov al,row2
                    mov temprow,4h          ;for lookup table, this is row 2
                    call _rowscan           ;scan row 2
                    mov al,row3
                    mov temprow,8h          ;for the lookup table, this is row 3
                    call _rowscan           ;scan row 3
                    mov al,row4
                    mov temprow,0Ch         ;for the lookup table, this is row 4
                    call _rowscan           ;scan row 4
                    call _whatkey           ;determine what to display (blank or key)
                    call _display           ;display whatever was found
                    jmp scan                ;loop again for next key
;_rowscan loads whatever row pattern was passed in AL and sends it to
;the printer port.
_rowscan:
                    mov ah,00h              ;BIOS printer function 00h
                    mov dx,0000h            ;LPT1
                    int 17h                 ;strobe the row and read the columns
                    call _key               ;see if only one column low for this row
                    ret
;_key will compare the AH status bits for a valid column pattern
;(one input low). If a valid pattern is found then CL counts up.
;AL is used to count which column pattern (0-3) is found so that
;the column number may be used in the lookup table.
_key:
                    mov al,00h              ;count valid column compares done in _key
                    and ah,0f8h             ;mask on bits 3–7 in AH                    3
                    cmp ah,col1             ;see if the pattern is valid (column 0)
                    je validcol             ;if so, then leave; else look again
                    inc al                  ;for lookup table, columns are 0,1,2,3
                    cmp ah,col2             ;look for column 1
                    je validcol
                    inc al                  ;look for column 2
                    cmp ah,col3
                    je validcol
                    inc al                  ;look for column 3
                    cmp ah,col4
                    je validcol
                    ret                     ;no valid column found, return
validcol:
                    mov keycol,al           ;save column number (0–3) as valid
                    mov ah,temprow          ;save row number for this valid column
                    mov keyrow,ah           ;latest valid row and column number stored
                    inc cl                  ;count up CL to show valid column found
                    ret
```

```
;_whatkey checks that only one valid column pattern was found for
;four rows. If one valid key is found, then the lookup table
;provides it, else a blank is displayed. The proper character to be
;displayed is returned in AL.
_whatkey:
                cmp cl,01h          ;for a valid key, only one should be found
                je getkey           ;one valid key per four row scan—get it
                mov al,' '          ;not valid—display a blank
                ret                 ;return with AL = ASCII blank
getkey:
                call _lookup        ;lookup the ASCII code for the key
                ret                 ;return with AL = ASCII code for key
;_lookup adds the row number to the column number to form an offset
;into the lookup table of 00 to 0Fh. The key character is returned
;in AL.
_lookup:
                mov bl,keyrow       ;use BX as the offset into the table
                add bl,keycol       ;number in BL is now 00 to 0Fh
                mov bh,00h          ;BX = 0000 to 000F
                mov al,b[table+bx]  ;lookup ASCII symbol for key
                ret                 ;return with ASCII key code in AL
;_display uses BIOS INT 10h, function 0Ah, to display one character
;on screen. Character to be displayed is passed in AL.
_display:
                mov ah,0ah          ;display whatever is passed in AL
                mov bx,00h          ;page 0, white on black—monochrome
                mov cx,01h          ;display one character
                int 10h             ;display it
                ret
;the lookup table is located in code memory
;at this address
table:                              ;key lookup based on row + column numbers
                db '123R456D789>S0#E'                                           4
code ends
```

NOTES ON THE PROGRAM

1. Or, use a data variable to hold the count, *just in case* CL gets changed by INT 17h.

2. Each row must be offset by a count of 4 from the row before it. Thus the first row keys are row 00 + column 0, 1, 2, 3 = 0–3h. The keys in the second row are then row 04 + column 0, 1, 2, 3 = 4h–7h. Third row keys are then 8h–Bh and the fourth row keys are numbered Ch–Fh.

3. The busy bit, AH Bit 7, is included in the masked pattern. It is not necessary to include the busy bit.

4. As a result of using row and column numbers that add up to 00 to 0Fh, the key symbols may be stored from offsets 00 to 0Fh in the lookup table.

We have examined several examples of how the parallel printer port may be used as a data interface. Section 10.9 suggests additional parallel port interface projects. Let us now turn our attention to a much more *universal* computer interface port, the serial port.

10.5 The Serial Modem Port

Serial data transmission techniques are the *most generic* of all computer interface methods for interchanging data between computers and between computers and peripherals. Serial data transmission technology preceded the computer age by many years, and when early computers needed a means to communicate with other devices, serial data transmission methods and equipment were already well developed.

The Development of Serial Data Transmission Technology

Serial data interchange technology had its beginnings in telegraphy, more than 160 years in the past. Early telegraphs worked by sending DC pulses of current from one location to another using a single wire. The telegraph pulses were made long (dashes) or short (dots), with pauses in between, which yields a *trinary* (long, short, nothing) type of signaling. To make the dashes and dots useful, a code had to be invented (Morse code), which assigned a unique pattern of dots and dashes to each letter of the alphabet and each decimal digit.

PULSE DURATION OR TIME INTERVAL SIGNALING

Morse code, which uses pulse *lengths* separated by "dead" time to signal two states, was followed by systems that use pulse *intervals* to signal two binary states. Figure 10.6 shows both pulse length and interval types of binary signaling. As can be seen from Figure 10.6, pulse length signaling is very inefficient, because dead time spaces must be placed between the pulses to differentiate between long and short pulses. Pulse interval signaling uses the presence (or absence) of a signal over each interval of time to signal a binary 1 (or 0). The pulse interval scheme is much more efficient than the pulse length scheme, because there is no wasted time between each bit.

Pulse interval signaling does introduce a complication: The sender and the receiver must measure each time *interval* period the same way. The measurement of equal time intervals means that sender and receiver must both use the same clock *frequency.* The speed of interval serial data transmission is thus directly tied to how *accurately* time can be measured by the sender and receiver.

Figure 10.6 ■ Pulse length and interval codes.

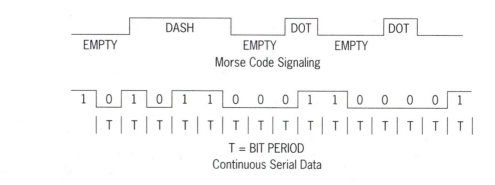

Morse Code Signaling

T = BIT PERIOD
Continuous Serial Data

Early electromechanical systems used electrical clocks to measure time intervals to an accuracy that allowed 50 bits of interval data per second to be transmitted. Later, electronic systems accurately divided time into smaller intervals, so that up to 19,200 bits per second of data could be commercially transmitted. Currently, bit rates exceeding 10 million bits per second are common, using very high frequencies.

Interval data transmission bit rates are no longer dominated by receiver and transmitter clock frequency concerns but are dependent on the transmission *medium*, typically wires, cables, and fiber optics. Speed costs money; the higher the data rate, the costlier the transmission media and all associated terminations.

Serial Data Communication Equipment

The driving force behind the development of serial data interchange techniques was, as usual, economic. Electrical data transmission had its birth in telegraphy using a transmitting key and receiving relay to send dots and dashes over long distances. Each alphanumeric character was assigned a unique Morse code of dots and dashes.

Economics dictated that it was much cheaper to use two wires and serial data, rather than many wires and parallel data, over long distances. Economics also dictated the replacement of Morse code (and human telegraph operators) with electromechanical systems that use fixed pulse time intervals. The overriding goal of all data transmission was, and continues to be, the most data, in the shortest time, with the least errors. In fact, the primary figure of merit for data transmission is how many *bits per second* can be reliably transmitted.

TELEPHONE SYSTEM TYPEWRITERS

The development of the telephone system typewriter or *teletype* was an improvement over telegraph techniques. Instead of using telegraph DC pulses, however, teletype data is transmitted over voice-grade telephone wires as a *modulated* AC sinusoidal wave. Modulation means that an AC waveshape has one of its three unique values (peak, phase, or frequency) changed in response to another modulating signal. One popular binary modulating technique, for instance, is to use one of two frequencies or *tones* for each bit. One tone is a binary 0, and the other tone a binary 1. Conversion from DC bit *pulses* to AC *tones* is done by letting the DC bit signal *level* modulate the AC carrier *frequency*. Each DC-modulated carrier frequency bit persists for a fixed length of time as a tone of fixed frequency. A 1 bit and a 0 bit are equal in time interval, but different in tone, unlike telegraph signals, where the bit intervals (dot and dash) are of different lengths.

The advent of the teletype signaled an era of improved data communications. Teletype operators, sitting at a keyboard, could transmit entire characters with the stroke of a key to a receiving printer that would print the character. The binary code for each teletype character was modified from Morse code so that additional characters, such as punctuation marks, could be added. Control characters, such as line feeds and carriage returns for the receiving printer, also were added. (Control characters, such as a line feed, are also named *nonprintable* characters, because they cause a printer *action* to take place, not a printed character.)

Teletype systems employed two types of data communication lines between communicating units: *full* and *half duplex*. Half-duplex communication means that only *one* communication channel is used for both transmission and reception of data between two terminals. One unit may transmit to the second on the channel, and then the transmission channel is "turned around" so that the second unit may transmit to the first. Full-duplex data communication means that *two* channels are provided for data interchange between terminals; one channel for transmission of data from Unit 1 to Unit 2, and another channel for data transmission from Unit 2 to Unit 1.

Clearly, half-duplex transmission is cheaper than full-duplex transmission, because only one pair of lines is needed for two-way communications. Half-duplex is also slower than full-duplex if data interchange between the two units is balanced in both directions, because there is some delay as each end of the line is turned around. Half-duplex transmission has given rise to many ASCII control characters, such as EOM (end of message) and EOT (end of transmission), so that the receiving unit will know when to turn the line around for transmission.

Teletype characters are coded as differing patterns of 1s or 0s, much like Morse code. One popular early teletype code was the Baudot code. Baudot code is a "five-bit" code because 5 binary bits (32 combinations) are used to encode 26 alphabetic, 10 numeric, and 22 punctuation and control characters. Coding 58 characters with 32 (2^5) bit combinations is made possible only by including two special control characters in the Baudot code. One character (the letter shift) denotes that all characters that follow it are alphabetic characters, or control codes. The second character (the figure shift) indicates that all characters that follow (until the next letter shift character) are numbers or punctuation marks. The Baudot code gave rise to the term *bauds per second* of information, where a *baud* is a piece of information. When the information is coded in binary, then *bauds*, *baud rate*, and *bits per second* (bps) are interchangeable terms.

Teletypes offered faster data interchange than did telegraphs and did not require the skills of a highly trained telegraph operator. By the 1930s, the teletype had become the primary means of nonvoice data transmission. It is interesting to note that one of the first human-computer interface peripherals was the teletype, because of its availability in the 1940s when the modern digital computer was invented. The teletype has, today, given way to CRT terminals and personal computer keyboards and monitors as the main means of human interface to computers. The teletype-telephone data interface *nomenclature*, however, still lives on, and we must become acquainted with some teletype-telephone terms in order to understand serial data interface techniques.

MODEMS

As previously discussed, variable-frequency AC tones may be used to transmit teletype binary data. The choice of frequencies is limited, however, by the bandpass characteristic of the telephone transmission system. At the time data transmission began to be used on the telephone system, so-called "voice grade" lines were the most prevelant. The telephone system was designed to carry human voice frequencies (roughly from 300 to 3,000 Hz), not teletypewriter DC pulses. Some means was thus needed by which teletype pulses could be converted to voice-grade tones, and the reverse.

The standard device used to convert binary data from pulse to tone, and from tone to pulse, is a machine dubbed the *modem*, which is short for *modulator demodulator*. A modulator turns pulses into tones; a demodulator turns tones into pulses. By connecting each teletype to a modem, and then plugging each modem into the phone system, a teletype *network* could be formed.

A complete teletype network, then, consists of teletypes that are directly connected to modems, and modems that are linked together by the telephone system. A teletype operator could simply dial up another teletype and connect the modem when an answering tone was heard. Keyboard data originates at one teletype, is converted to tones in the modem, and the tones are sent through the telephone system. A receiving modem then converts the received tones into pulses that are fed to its accompanying teletype for printing. Today we see the facsimile machine, or *fax*, with its ability to scan and digitize printed material directly into data, as one modern version of the teletype.

When teletypes were adapted from telephone to computer interface use, the teletype was, naturally, equipped with a *standard modem interface*. As a result, early computer interfaces were designed to fit the existing teletype modem standard. The computer port was made to "look like" a modem port. Thus, thanks to the modem, the first really "standard" computer interface was defined.

Modems have steadily become popular as our culture becomes interconnected through the telephone network. Information companies, such as CompuServe, are commonly used to access databases, computer bulletin boards, and reservation and shopping services. Facsimile (fax) transfer has revolutionized the way we exchange printed material. The modem, along with the telephone system, is central to all of these new data transmitting services. To understand the nomenclature of the serial port then, we must understand modem terms and conventions.

10.6 Serial Data Standards

Since the early days of computing, computers have behaved similarly to teletypes in that they both originate and receive data. As a result, most personal computer serial ports are designed to work with a modem as if the computer were a teletype.

When the teletype appeared, telephone equipment was divided into *voice* or *data* categories to differentiate between the two radically different types of transmission. Teletypes were named *data terminals* and modems named *data sets*. The computer is a *data terminal*, the telephone name given to the teletype. Even if the computer serial port is connected to a printer, the printer is still a *data set* in telephonese.

The RS232 Serial Data Physical Standard

As computers began to make serious inroads into our civilization, the need for serial data standardization (of terms, voltages, baud rates, physical features, binary codes, and the like) became paramount if the industry was to grow.

Two standardization *organizations* played a dominant part in serial data transmission standardization efforts, the Electronics Industry Association (EIA) and the American Standards Association (ASA, now ANSI). EIA is the voluntary industry body charged with setting standards for the electronics industry. The serial data transmission standard that evolved from the EIA, in the early 1960s, was named the *RS232* standard. RS232, in turn, was based primarily on existing internal standards used by the Bell System companies. Bell, at that time, was a nationwide telephone monopoly and had many years of experience (and many millions of dollars worth of equipment) involved in data transmission.

The RS232 standard established new equipment names to reflect the realities of the computer age, and renamed the major data communication parts. Data sets became the more generic *data communication equipment* (DCE), and data terminals were renamed *data terminal equipment* (DTE).

It is very important to remember that signals that go *out* from DTE go *in* at DCE, and the reverse. Much, much confusion has been caused by reading an interface specification for a DCE-type serial port and then wiring it as a DTE-type port, or the reverse.

The RS232 standard is a *physical* standard. It defines signal voltage *levels*, standard connector *configurations*, and assigns signal *names* and *functions* to pins in the connector. Any truly compatible RS232 device must be terminated in a standard connector type, have each control or data signal on exactly the correct connector pin, and use signal voltages that meet the RS232 standard.

The RS232 standard does *not* establish the way the data is coded or how fast (in bps) the data is transmitted.

RS232 INTERFACE DEFINITIONS

Table 10.2 lists the standard pin numbers and signal names for an RS232 serial data interface as defined for data communication equipment and data terminal equipment. Note that signal names are *identical* for DTE and DCE signals but that the direction of the signals is generally *opposite* for each equipment type. Table 10.2 is repeated for each type of equipment to emphasize signal direction. Connecting DTE to DCE requires a simple, one-to-one cable that wires corresponding pins together. Typical RS232 serial *cable* connectors are industry-standard 25-pin DB-25P and 9-pin DB-9P types, as shown in Figure 10.7

The voltage levels of RS232 signals are *plus or minus* DC voltage *bands.* RS232 uses −5 to −15VDC as a logic *1*, and +5 to +15VDC as a logic *0.* Personal computers, typically, use 0VDC and +5VDC for logic levels. Special serial port interface circuits are used to convert between computer voltage levels and RS232 signal voltage levels.

Ordinarily, all of the RS232 signals listed in Table 10.2 are not needed by personal computers. Table 10.3 shows the standard RS232 DTE signals that were originally *adopted* from RS232 for use by the PC community for serial data interchange.

The original IBM serial connector included inputs for 4–20ma signals in addition to the more common +/−12 volt signals. Current loop transmission is very popular in industry, as well as in some teletype models, and this may

RS232 DTE (Computer) Connector Signals			
Pin	Signal Name	Signal Function	Direction
1	Shield Ground	Cable shield ground	NA
2	Transmit Data (TxD)	Serial data from computer	Out
3	Receive Data (RxD)	Serial data to computer	In
4	Request to Send (RTS)	DTE request to transmit	Out
5	Clear to Send (CTS)	DCE permission to transmit	In
6	Data Set Ready (DSR)	DCE is on line	In
7	Signal Ground	Common from DTE to DCE	NA
8	Carrier Detect (CD)	DCE senses carrier tone	In
9	No Connection	(May be positive test voltage)	—
10	No Connection	(May be negative test voltage)	—
11	No Connection	—	—
12	Secondary CD	DCE senses second channel	In
13	Secondary CTS	DCE second channel open	In
14	Secondary TxD	Serial data from computer	Out
15	DCE Transmit Clock	DCE transmit data clock	In
16	Secondary RxD	Serial data from DCE	In
17	DCE Receive Clock	DCE receive data clock	In
18	No Connection	—	—
19	Secondary RTS	DTE second channel request	Out
20	Data Terminal Ready (DTR)	DTE is on line	Out
21	Signal Quality	DCE signal quality indicator	In
22	Ring Indicate (RI)	Ring signal detected by DCE	In
23	Baud Rate Select	Select one of two baud rates	Out
24	DTE Transmit Clock	DTE transmit data clock	Out
25	No Connection	—	—

RS232 DCE (Modem) Connector Signals			
Pin	Signal Name	Signal Function	Direction
1	Shield Ground	Cable shield ground	NA
2	Transmit Data (TxD)	Serial data from computer	In
3	Receive Data (RxD)	Serial data to computer	Out
4	Request to Send (RTS)	DTE request to transmit	In
5	Clear to Send (CTS)	DCE permission to transmit	Out
6	Data Set Ready (DSR)	DCE is on line	Out
7	Signal Ground	Common from DTE to DCE	NA
8	Carrier Detect (CD)	DCE senses carrier tone	Out
9	No Connection	(May be positive test voltage)	—
10	No Connection	(May be negative test voltage)	—
11	No Connection	—	—
12	Secondary CD	DCE senses second channel	Out
13	Secondary CTS	DCE second channel open	Out
14	Secondary TxD	Serial data from computer	In
15	DCE Transmit Clock	DCE transmit data clock	Out
16	Secondary RxD	Serial data from DCE	Out
17	DCE Receive Clock	DCE receive data clock	Out
18	No Connection	—	—
19	Secondary RTS	DTE second channel request	In
20	Data Terminal Ready (DTR)	DTE is on line	In
21	Signal Quality	DCE signal quality indicator	Out
22	Ring Indicate (RI)	Ring signal detected by DCE	Out
23	Baud Rate Select	Select one of two baud rates	In
24	DTE Transmit Clock	DTE transmit data clock	In
25	No Connection	—	—

Table 10.2 ■ RS232 DTE and DCE Connector Signals

Figure 10.7 ■ Serial port connectors .

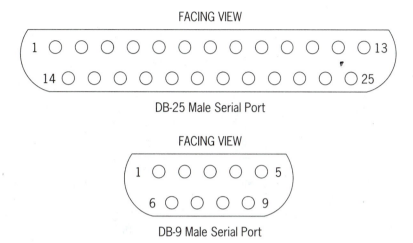

DB-25 Male Serial Port

DB-9 Male Serial Port

Pin	Signal Name	Signal Function	Direction
1	Shield Ground	Cable shield ground	NA
2	Transmit Data (TxD)	Serial data from computer	Out
3	Receive Data (RxD)	Serial data to computer	In
4	Request to Send (RTS)	DTE request to transmit	Out
5	Clear to Send (CTS)	DCE permission to transmit	In
6	Data Set Ready (DSR)	DCE is on line	In
7	Signal Ground	Common from DTE to DCE	NA
8	Carrier Detect (CD)	DCE senses carrier tone	In
9	Transmit Current Out	+20ma signal current source	Out
10	No Connection	—	—
11	Transmit Current In	−20ma signal current sink	In
12	No Connection	—	—
13	No Connection	—	—
14	No Connection	—	—
15	No Connection	—	—
16	No Connection	—	—
17	No Connection	—	—
18	Receive Current In	+20ma signal current	In
19	No Connection	—	—
20	Data Terminal Ready (DTR)	DTE is on line	Out
21	No Connection	—	—
22	Ring Indicate (RI)	Ring signal detected by DCE	In
23	No Connection	—	—
24	No Connection	—	—
25	Receive Current Out	−20ma signal current return	Out

Table 10.3 ■ PC DTE (Computer) Connector Signals

Pin	Signal Name	Signal Function	Direction
1	Carrier Detect (CD)	DCE senses carrier tone	In
2	Receive Data (RxD)	Serial data to computer	In
3	Transmit Data (TxD)	Serial data from computer	Out
4	Data Terminal Ready (DTR)	DTE is on line	Out
5	Signal Ground	—	NA
6	Data Set Ready (DSR)	DCE is on line	In
7	Request to Send (RTS)	DTE request to transmit	Out
8	Clear to Send (CTS)	DCE permission to transmit	In
9	Ring Indicator (RI)	Ring signal detected by DCE	In

Table 10.4 ■ PC DTE (Computer) 9-Pin Connector Signals

account for the IBM current loop connections. Current loops are not now a feature of more recent PC models.

The second generation of PC standard computers reduced the list of signals shown in Table 10.3 to 9, as shown in Table 10.4. Figure 10.2 shows the configuration of a DB-9 serial cable connector.

Clearly, the signals used on DB-9 serial port connectors are the most useful for data interchange between serial devices. The 9 signals and their functions are shown in Table 10.5. DB-25 pin numbers are listed first in Table 10.5, followed by DB-9 pin numbers (in parentheses).

SERIAL PORT HANDSHAKING SIGNALS

The functions listed in Table 10.5, which are the most commonly used RS232 communication signals, are defined as follows.

Transmit Data (TxD)	Binary data *out from* one unit *into* another unit.
Receive Data (RxD)	Binary data *into* one unit *from* another unit.
Request to Send (RTS)	Signal *from* DTE *to* DCE notifying DCE that the DTE wishes to transmit data.

Function	DB-25 Pin	(DB-9 Pin)
Transmit Data (TxD)	Serial data from DTE on pin 2	(3)
	Serial data from DCE on pin 3	(2)
Receive Data (RxD)	Serial data to DTE on pin 3	(2)
	Serial data to DCE on pin 2	(3)
Request to Send (RTS)	Signal from DTE to DCE on pin 4	(7)
Clear to Send (CTS)	Signal from DCE to DTE on pin 5	(8)
Data Set Ready (DSR)	Signal from DCE to DTE on pin 6	(6)
Signal Ground	Ground reference on pin 7	(5)
Carrier Detect (CD)	Signal from DCE to DTE on pin 8	(1)
Data Terminal Ready (DTR)	Signal from DTE to DCE on pin 20	(4)
Ring Indicator (RI)	Signal from DCE to DTE on pin 22	(9)

Table 10.5 ■ Common Serial Data Transmission Port Signals

Clear to Send (CTS)	Signal *from* DCE *to* DTE notifying DTE that data may be transmitted.
Data Set Ready (DSR)	Signal *from* DCE *to* DTE notifying DTE that DCE is online (connected).
Carrier Detect (CD)	Signal *from* DCE *to* DTE notifying DTE that a telephone connection has been made.
Data Terminal Ready (DTR)	Signal *from* DTE *to* DCE notifying DCE that DTE is online (powered up).
Ring Indicator (RI)	Signal *from* DCE *to* DTE notifying DTE that a ring signal from the telephone network is active.

An example of the uses to which the RS232 handshaking signals may be put is outlined below.

- Assume that the DTE and DCE are both powered, and the DCE is connected to the telephone system. DTR and DSR are both active; RTS, CTS, CD, and RI are all inactive.

- The telephone system makes a connection to the DCE, and RI becomes active. The DTE begins to execute a communications program, the DCE connects to the telephone data line, and CD goes active. RI returns to inactive.

- Data is transmitted from the DCE to the DTE. At the end of data transmission from DTE to DCE, the DTE program wishes to respond.

- The DTE program causes RTS to become active, the DCE brings CTS to an active state. The DTE transmits data to the DCE.

- The originator of the telephone connection hangs up. CD and CTS go inactive and the DTE communication program responds by bringing RTS to inactive.

Connecting the serial ports of different equipment is *not* a trivial matter unless they are made by the same manufacturer, as a set. Any time two disparate serial devices are connected to each other you must have available the operation manuals for *both* pieces of equipment. The manufacturers may have both configured their equipment as DTE, or both sets of equipment may have a DCE configuration. The manufacturer's service manuals should specify which configuration (DTE or DCE) is used, and show the signal for each pin of the serial connector.

We have a good example here of how RS232 pin definitions change from one model to another. Note that the 25-pin DTE serial data connector of Table 10.3 uses pin 2 for TxD and pin 3 for RxD. Then note that the DTE 9-pin connector of Table 10.4 uses pin 2 for RxD and pin 3 for TxD! Non-uniformity seems to be the rule for many RS232 devices, rather than the exception.

CONNECTING RS232 DTE EQUIPMENT

Should two identical types of equipment (DTE to DTE, or DCE to DCE) be connected, then the serial data lines must be interchanged between them so that they may communicate. The wires on pins 2 and 3 must be switched

DTE Number One		DTE Number Two	
Pin	**Signal**	**Pin**	**Signal**
2	RxD	3	TxD
3	TxD	2	RxD
4	DTR	6	DSR
5	GND	5	GND
6	DSR	4	DTR
7	RTS	8	CTS
8	CTS	7	RTS

Table 10.6 ■ DB9 DTE-to-DTE Serial Data Cable with Handshaking

between units so that pin 2 of one is connected to pin 3 of the next unit, resulting in TxD of each connected to RxD of the other.

Handshaking signals such as RTS or CTS will usually have to be jumpered to voltage levels that *enable* each function, if needed by the particular piece of equipment. The handshaking signals are commonly active-*high* (+15VDC) at the port pins. Table 10.6 shows a *DB9* cable that may be used to connect DTE equipment to DTE equipment. The cable shown in Table 10.6 connects active output signals from DTE unit Number One to corresponding inputs on unit Number Two, and the reverse. As with all things serial, the cable shown in Table 10.6 is not guaranteed to work in all DTE-to-DTE situations. Use the specific equipment manuals for the units to be connected in order to ensure correct connection.

DTE-to-DTE cabling is the most common interconnect type because of the fact that most computers are configured as data terminal types of equipment. DCE-to-DCE cabling is similar, but would find little use in connecting one peripheral to another.

RS232 defines how to *physically* connect data equipment. The job of defining how to *encode* data is left to the ASCII standard.

The ASCII

The code used to define serial data characters is not part of the RS232 standard. The American National Standards Institute, or ANSI, established a standard data transmission code about the same time that RS232 was established. The standard data transmission code is called the *American Standard Code for Information Interchange* or ASCII. The ASCII code is discussed in Chapter 2 and listed in Appendix E.

Certain large computer companies, notably IBM, did not choose to use RS232 or ASCII for their large mainframe systems. IBM personal computers, however, conform to the EAI and ANSI serial data standards, as have all other personal computers.

Interestingly enough, standard data speeds, in bps, seem to have evolved from earlier technologies. Early teletype baud rates of 110 and 300 bps became 600 and 1,200 bps as modem technology improved. Other "standard" baud rates are 2,400, 4,800, 9,600, and 19,200 bps. All of these standard baud rates differ from each other by a power of 2. Separation by 2 means that a carefully selected oscillator frequency can be divided by a power of 2 to yield many standard frequencies. For instance, an oscillator frequency of 1.2288Mhz. can be divided by the power-of-2 binary number 128 to yield 9,600 baud, by 256 to yield 4,800 baud, and so on. Higher baud rates, up to 100K baud or so, are physically possible for most DOS-based computers. DOS interrupts, however, do not support rates higher than standard values.

Asynchronous Serial Data Transmission

Serial data is generally transmitted using one of two methods: *synchronous* and *asynchronous*. Synchronous data transmission is accomplished in conjunction with a *timing* signal. The receiving unit knows exactly when a bit of data is present because a timing pulse accompanies every data bit. The printer interface is an example of a synchronous parallel data transmission scheme that uses the STROBE line to generate a timing pulse to inform the printer when a character is transmitted from the computer. Synchronous data transmission is not generally applicable to low-volume, low-speed telephone system tone data transmission. Synchronous data transmission is quite common in local area network computer systems where the extra cost of the synchronizing equipment can be justified by high data transfer rates.

Asynchronous data transmission does not feature timing pulses between transmitter and receiver. *Precise* time intervals are used, in asynchronous transmission, to coordinate the actions of the transmitter and receiver. In order for asynchronous serial data transmission to work, both transmitter and receiver must use the *same* time interval (baud rate) for a bit of data.

Asynchronous serial data transmission involves sending a train of electrical voltage pulses, each pulse lasting for a fixed time duration, from the transmitter to the receiver. When no data is being transmitted, the transmitter holds its output at a constant voltage level. When no data is being transmitted, the transmitter is said to be in the *marking* or binary 1 state. When a binary 0 is transmitted, the transmitter is said to be in the *spacing* state. Marking and spacing are terms left over from the early days of telegraphy.

Both receiver and transmitter must have their internal clocks, which are used to measure the bit time interval, at the *same* frequency. Should the clocks not be set to the same frequency, or baud rate, then data transmission cannot be accomplished.

Asynchronous Data Byte

Figure 10.8 shows the waveform used to transmit a byte of data from transmitter to receiver. Note that a total of *10* bits are required in order to transmit 1 *byte* of data. The first and last bit of the byte transmission are needed because of the nature of asynchronous data transmission.

The signal polarities shown in Figure 10.8 *are for internal TTL logic levels, inside the computer.* RS232 port interface circuits convert ground to +15V and +5V to −15V when data is transmitted via the RS232 interface port. The 15V levels are converted back to ground and +5V at the receiving RS232 port. (In practice, most RS232 interface chips will work between 7 and 15 volts.) RS232 line voltage levels are generally of little interest except when troubleshooting an RS232 cable, or observing a cable character on an oscilloscope.

Figure 10.8 ■ Serial data byte with 1 stop bit.

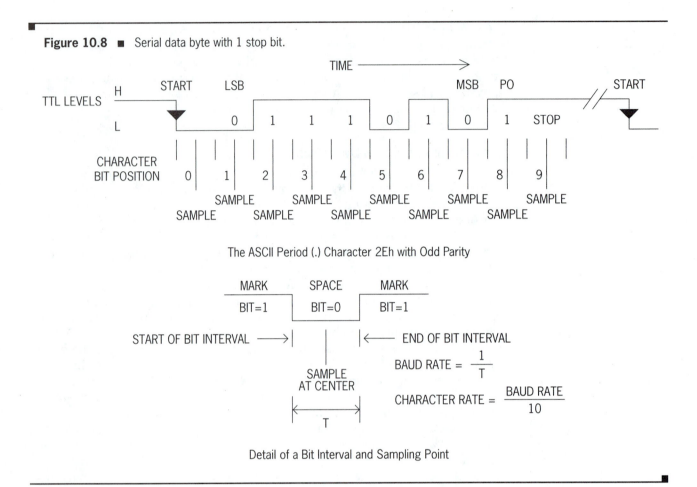

The ASCII Period (.) Character 2Eh with Odd Parity

Detail of a Bit Interval and Sampling Point

    The transmission and reception of an asynchronous data byte proceeds in the following manner. Numbers in *parentheses* ( ) are for the RS232 cable *voltage* levels.

■ When at rest, with no data communication in process, the serial data transmitter maintains its output at a +5V (–15V) marking state.

■ Every time a byte of data is to be transmitted, the marking state changes to a 0V (+15V) spacing state, for *at least one time interval.* The first bit of an asynchronous data byte is *always* a space. The first bit sent is called the *start* bit, because it alerts the receiver that a byte is about to arrive at the receiver.

■ The receiver reacts to the start bit by beginning to keep time (using its internal clock) until *one half* of a bit interval has expired, from the *leading edge* of the start bit. The receiver, *if its internal clock is set to the same frequency as the transmitter*, has now located the approximate *center* of each bit interval.

■ The receiver then samples the state (voltage level) of the start bit to determine if the level is *still* low. If the received data remains low, then the incoming data is probably valid (the falling edge of the start bit was not due to electrical "noise").

■ The receiver then begins to use its internal clock to time complete bit intervals, and samples the received data at the *center* of each bit interval time.

■ After 8 data bits have been sampled, and stored, the receiver *expects* to detect that the last bit (Bit 10) is in the high state. The last bit is named the *stop* bit because at least one high state must *separate* the end of one byte from the high-to-low edge of the start bit for a byte that may immediately follow.

From the preceding discussion, it is clear that the internal receiver clock must operate at a frequency higher than the inverse of a bit interval. If the receiver is to accurately time one half of the initial start bit interval, then the receiver internal measuring frequency must be able to divide a bit interval into at least two pieces. Typically, the internal clock is set to run at a frequency of 8 to 16 times the data baud rate.

Computer serial data circuits commonly employ crystal-controlled oscillators that produce clock periods of hundreds of nanoseconds. Many times the crystal frequency chosen to operate the CPU of a single-board computer is selected based on its ability to be divided evenly by all the standard baud rates, rather than the maximum allowed by the CPU.

The high-frequency clock pulses are capable of timing bit intervals with great accuracy, up to bit rates of tens of thousands of bits per second. Despite the inherently high frequency capabilities of computers, however, most serial data peripherals operate using baud rates of 1,200, 2,400, 4,800, 9,600, or 19,200 bits per second.

THE UART

Serial data transmission is so common that integrated circuits that perform all of the required actions of transmitting and receiving serial data have been commercially available since the early 1970s. Integrated circuits that handle serial data communications are commonly called UARTs, the acronym for *universal asynchronous receiver transmitter*.

An UART will accept parallel bytes, from the CPU, and transmit the byte as serial data according to the rules of asynchronous serial transmission just discussed. The UART will also receive serial data and convert it into a parallel byte for reading by the CPU. Moreover, the UART will generate most of the common handshaking control signals (DTS, RTS, etc.) required by a modem.

Most UARTs may be *programmed* so that baud rates, number of stop bits, modem control signals, hardware interrupt signals, and so on, can be selected by the program. UARTS are sufficiently complex so that you must know the particular model (and have its data sheets) to be able to program one. Several UART types are employed in DOS-based personal computers, so we shall take the more generic approach for our serial data needs and use BIOS INT services to take care of the circuit-peculiar details of whatever computer we happen to be using.

10.7 BIOS Serial Data Transmission INT Services

BIOS is equipped with several INT service routines, most notably type 14h, that facilitate transmitting and receiving serial data via the serial port. It is important to note that all of the BIOS data transmission service routines are written from the *standpoint of DTE*, not DCE, data transmission. The most important consequence of DTE-specific BIOS is that attempting to connect DTE to DTE results in handshaking mismatches. We shall discuss ways to "work around" DTE-to-DTE handshaking as a part of the INT 14h, function 01h discussion.

BIOS ASYNCHRONOUS SERIAL DATA INT TYPE 14H

BIOS Interrupt 14h performs well as an all-purpose serial port interrupt service. Three interrupt 14h functions will let us *initialize* the serial port and then *send* or *receive* a character. We begin with the initializing function.

Function 00h   Initialize the Serial Port

Register	Contents	Function
AH	00h	Select function 00h, initialize port
AL	(see table)	Select serial data modes
DX	0000h–0003h	Select COM1 (0000h) to COM4 (0003h)

Returned

AH	Port status listed in Table 10.7	
AL	Modem status listed in Table 10.7	

AL Register Bit Settings for Serial Transmission

Bit 7 6 5	Baud Rate	Bit 4 3	Parity	Bit 2	Stop Bits	Bit 1 0	Character Length
0 0 0	110	0 0	None	0	1	0 0	7 bits
0 0 1	150	0 1	Odd	1	2	1 1	8 bits
0 1 0	300	1 0	None				
0 1 1	600	1 1	Even				
1 0 0	1,200						
1 0 1	2,400						
1 1 0	4,800						
1 1 1	9,600						

Operation:

Function 00h uses the byte in AL to set the baud rate, parity type, stop bits, and character bit length of serial data transmitted on the selected serial (COM) port. Register DX selects the COM port to be initialized. The status of the serial port and serial port Modem is returned in registers AL and AH.

Return Contents of AH	Serial Port Data Status
Bit 7 6 5 4 3 2 1 0	
x x x x x x x 1	Received Data Present
x x x x x x 1 x	Overrun Error
x x x x x 1 x x	Parity Error
x x x x 1 x x x	Framing Error
x x x 1 x x x x	Break Detected
x x 1 x x x x x	Transmitter Holding Register Empty
x 1 x x x x x x	Transmitter Shift Register Empty
1 x x x x x x x	Time-Out
Return Contents of AL	**Modem (Serial Port Control) Status**
Bit 7 6 5 4 3 2 1 0	
x x x x x x x 1	Change in CTS Status
x x x x x x 1 x	Change in DSR Status
x x x x x 1 x x	Trailing Edge of RI Signal
x x x x 1 x x x	Change in CD Signal
x x x 1 x x x x	CTS Enabled
x x 1 x x x x x	DSR Enabled
x 1 x x x x x x	RI Enabled
1 x x x x x x x	CD Enabled

Table 10.7 ■ Serial Port and Modem Status Bytes

Before any serial data transmission can take place, the serial port *must* be initialized using function 00h. The baud rate selection available with function 00h ranges from 110 to 9,600 bits per second.

The lower rates (less than 600 bits per second) are now *obsolete,* as is the need for 2 stop bits that the lower baud rate devices once required. We shall use a baud rate of *4,800* bps, *no* parity, *1* stop bit, and a character length of *8* bits for our examples. Accordingly, register AL is loaded with the byte C3h before using INT 14h to initialize the port as desired for our examples.

On return from interrupt 14h, function 00h, register AH returns the serial *port* status and register AL returns the *modem* (serial port control line) status, as shown in Table 10.7.

REGISTER AL MODEM STATUS SIGNALS

The meanings of the modem status signals returned in register AL are somewhat obvious and reflect the DTE nature of the computer. A 1 bit in AL bits 7 to 4 means that the corresponding pin on the serial port has been activated by a DCE.

The AL bits that show a status change (Bits 0, 1, and 3) signal that there has been a change (enabled to disabled, or the reverse) in the status of the corresponding modem signal since the last status inquiry. The RI trailing edge bit (Bit 2) means that an RI signal pulse's trailing (low-to-high) edge has occurred. The information returned in register AL is meaningful if we are interfacing the serial port to a modem. We shall not use the serial port for modem control in this book and register AL will not hold our attention.

REGISTER AH SERIAL PORT DATA STATUS

The meanings of the AH data status bits are of great interest to us as we interchange data with other serial devices. The detailed meaning of each status bit in register AH is explained as follows.

Received Data Present (AH Bit 0). A byte of data is present in the serial port receiver memory and has *not* been read by the program. The byte of data should be read by the program before another byte arrives. Bit 0 of AH will be cleared when the received data is read by the program.

Overrun Error (AH Bit 1). The program failed to read a byte of data in the receiver memory *before* the next character arrived. The next character has overwritten the original character in the receiver memory. Bit 1 of AH will be reset after each status read (return of status in AH after an INT 14h).

Parity Error (AH Bit 2). The parity of the received byte (if any type of parity checking is enabled) is in *error*. The parity error bit is set whenever a parity error occurs and is reset after the status of the serial port is returned in AH by INT 14h.

Framing Error (AH Bit 3). A stop bit (marking high state) was *not* detected at the end of the byte. This indicates the byte has been received in error. Electrical noise, or a character that does not match the programmed character bit length, probably has occurred. Bit 3 of AH will be reset after each serial port status byte is returned in AH by INT 14h.

Break Detected (AH Bit 4). The incoming data line has been spacing low for *more* than 10 bit times (one character time). A break can be generated by some serial devices and holds the incoming data in an abnormal (low voltage) state. The normal state of an inactive data line is the marking (high) state. Bit 4 is set by a break condition and reset after each status byte is returned in AH by INT 14h.

Transmitter Holding Register Empty (AH Bit 5). Serial data is transmitted by loading a memory location called a *holding register* with the byte to be transmitted. From the holding register, the byte is loaded into the UART empty shift register and shifted out serially, at the baud rate chosen by the program. After the byte is transferred from the holding register to the shift register, the holding register can then be immediately reloaded by the program. The holding register must then wait until the shift register is empty before loading the second byte. The transmitting registers are said to be *double buffered*, that is, 2 bytes can be sequentially loaded by the program. Once both bytes are loaded, the program must *wait* until the holding register empties before attempting to send any more data. Bit 5 is set to 1 whenever the holding register is empty and cleared to 0 when a byte is written to the holding register by the program.

Transmitter Shift Register Empty (AH Bit 6). The serial transmitting shift register is *empty,* and no data is waiting in the transmitter holding register. (See preceding discussion on transmitter holding register.)

Bit 6 is set to 1 when the shift register is empty and cleared to 0 when a character is transferred to it from the holding register.

Time-Out (AH Bit 7). Bit 7 is a BIOS *error* flag. Depending on the particular BIOS used, BIOS may check to see if the DSR or CTS signals are enabled before attempting to load a character into the holding register or read a character from the receiving register. Should BIOS not be able to write or read a character within a few seconds, the time-out bit is set to 1, and the function returns without transmitting (or reading) a character.

Function 00h should be used to initialize the serial port to the desired baud rate, character length, and parity condition. Note that BIOS does *not* support data rates beyond 9,600 baud, although most PCs are capable of data rates up to 100K baud.

Now that we are equipped to initialize the port, let us see how to transmit and receive characters.

Function 01h   Write a Character to the Serial Port

Register	Contents	Function
AH	01h	Select write to serial port function
AL	Character	Character to be transmitted by serial port
DX	0000h–0003h	Select COM1 (0000h) to COM4 (0003h) port

Returned

AH	Port status shown in Table 10.7	

Operation:

Function 01h transmits the character in register AL via the serial port selected by register DX. Register AH returns the serial port status.

Function 01h writes the character to the serial port data buffer transmitting register and returns the port status in AH as shown in Table 10.7. The port must have been initialized by function 00h *before* attempting to transmit a character from it.

Function 01h will *not* overwrite any previous data in the transmission data buffer; it checks to make sure the data buffer is empty before loading a new character for transmission. Function 01h then returns only *after* any previous character has been transmitted or a time-out (handshaking) failure has occurred.

Function 01h also sets two very important DTE transmitter *handshaking* bits, DTR and RTS, to active (enabled) levels.

DTE-to-DTE Serial Data Transmission

Unfortunately, a DTE will *not* transmit until the DSR and CTS *input* handshaking signals *from* a DCE unit are enabled. A DTE-to-DCE connection presents no handshaking problems for DTE transmit function 01h. However, if a DTE is connected to another DTE (the cable connections shown in Table 10.6 are used), then transmission between units cannot be done until the DTE *transmitter* DSR-CTS input lines have been placed in the active enabled state by function 01h used in the DTE *receiver* to set the matching DTR-RTS signals. Thus, in order to receive a character from the transmitting DTE, the receiving DTE must have transmitted a "dummy" character to enable the DTR-RTS pins. Placing DTR-RTS outputs in the active state on the receiving DTE unit provides a matching active DSR-CTS input state for the transmitting DTE unit. In effect, each receiver *enables* the opposing transmitter to transmit by enabling its own transmit active signals, DTR-RTS. It appears that a unit cannot receive a character until it has transmitted a character, and this is, essentially, the case!

The only sure way to ensure that a receiver DTE can enable a transmitter DTE is to send a dummy character from the receiver to the transmitter after every transmitted character is received. The transmitter program may ignore the dummy character. To relieve the programmer of such tricks, it would appear that some modifications to the cable of Table 10.6 are in order.

An alternative to the serial cable connections of Table 10.6 (and having the receiver send dummy characters) is to connect each DTE unit's handshaking lines *to itself* in order to provide the correct transmission handshaking signals. The only connection between DTE units is then TxD, RxD, and ground. Table 10.8 shows a DTE-to-DTE cable that lets each port supply its *own active level transmit* signals. Every time a unit attempts to transmit, it will supply its own CTS-DSR signals via its RTS-DTR signals, and transmission will be accomplished.

The clear *disadvantage* to using the cable of Table 10.8 is the complete loss of any handshaking *between* terminals. The transmitting DTE unit can merrily send more characters to the receiving DTE unit than can be accepted by the receiver. Data overload in the receiving unit soon results. One solution to data overload is to have the receiving unit transmit a *stop* control character as it approaches overload. The transmitter can check for a stop condition periodically and resume transmission when a *start* control character is sent from the receiver.

If handshaking signals between units is desired, but dummy characters are to be avoided, then the only recourse is to bypass BIOS entirely and directly program the modem control register of the UART used by the partic-

DTE Number One		DTE Number Two	
Pin	**Signal**	**Pin**	**Signal**
2	RxD	3	TxD
3	TxD	2	RxD
4 to 6	DTR to DSR	4 to 6	DTR to DSR
5	GND	5	GND
7 to 8	RTS to CTS	7 to 8	RTS to CTS

Table 10.8 ■ DB9 DTE-to-DTE Serial Data Cable—No Handshaking

ular PC make and model. Programming the UART requires specific knowledge of the particular UART that is used in a given PC model, and program generality may be lost.

The second communication function, that of serial character reception, is handled by function 02h, which is discussed next.

Function 02h Read a Character from the Serial Port

Register	Contents	Function
AH	02h	Select read from serial port function
DX	0000h–0003h	Select COM1 (0000h) to COM4 (0003h) Port
Returned		
AL	Serial data character	
AH	Port status as shown in Table 10.7	

Operation:

Function 02h returns a character in register AL from the serial port receiver selected by register DX. Serial port status is returned in register AH.

The serial port *must* have been initialized using function 00h before attempting to receive a character. Also, a character must be *waiting* in the receiver before it can be read into the program.

The next function, 03h, lets us determine the port *status* only to see if a character is waiting to be read.

Function 03h Determine Serial Port Status

Register	Contents	Function
AH	03h	Select port status function
DX	0000h–0003h	Select COM1 (0000h) to COM4 (0003h) port
Returned		
AX	Port and modem status shown in Table 10.7	

Operation:

Function 03h returns modem and port status, in register AX, from the port selected by register DX. Bit 0 of register AL will be set to 1 by function 03h *only once* for a new character.

Function 03h returns the modem status in register AL and the serial port status in AH, as shown in Table 10.7. Before attempting to read a port data byte, the appropriate serial port status should be determined, and Bit 0 of AH (received data ready) *tested* by the program to be a binary one. If Bit 0 is not a 1, then the program may go on to other tasks or continue to read the serial port status until a byte is received.

Transmit and Receive Serial Data Program Examples

Interrupt 14h lets us use the PC to interface to other *serial* devices. The simplest way to practice using INT 14h is to configure a computer's serial port in a *loop check* mode. A loop check is performed when the transmit pin of a serial port is directly connected to the receive pin of the same serial port. The port may then be configured using function 00h, transmit a character using function 01h, check the port with function 03h, and read the same character via function 02h, with no other outside circuitry. Loop checking is also a handy way to *test* communications programs before connecting outside devices to the serial port.

To help prevent any BIOS-specific handshaking problems, DTR-RTS handshaking signals are connected to the RTS-CTS pins as shown in Table 10.9 for 25-pin and 9-pin connectors.

The loop check connector listed in Table 10.9 is a version of the cable listed in Table 10.8 except only a *single* port is used. Should the computer in use have two serial ports, then the two-port configuration of Table 10.8 can be used to simulate a two-terminal system.

A LOOP CHECK PROGRAM

The next program, comm.asm, checks serial port COM2 by sending and receiving a character using a loop check serial connector that has been configured using Table 10.9.

```
;comm.asm loops a character on COM2

code segment
        mov sp,1000h        ;set the stack
        mov ax,00C3h        ;initialize to 4800, no parity, 8 bits
        mov dx,0001h        ;use COM2
        int 14h             ;COM2 initialized
```

25-Pin Serial Port Connector			
Connect Pin		To Pin	
20	(DTR)	6	(DSR)
4	(RTS)	5	(CTS)
2	(TxD)	3	(RxD)
9-Pin Serial Port Connector			
Connect Pin		To Pin	
4	(DTR)	6	(DSR)
7	(RTS)	8	(CTS)
3	(TxD)	2	(RxD)

Table 10.9 ■ Loop Check Connectors for 25-Pin and 9-Pin Connectors

```
lup:
        mov ah,01h          ;send an 'A'
        mov al,'A'
        int 14h             ;an A transmitted
        mov ah,03h          ;get port status,
        int 14h             ;AH=61h: data ready, xmit reg. empty
        mov ah,02h          ;read the character
        int 14h             ;get the A character
        jmp lup             ;go again?
code ends
```

D86 may be used to exercise the loop check program in order to see exactly how the serial data interrupts operate. The loop check program is *not,* however, suitable for data transmission between two different terminals.

The loop check program enjoys *two* advantages over more general-purpose programs. First, when transmitting, we know that there is *no* previous character in the UART transmitter registers. An empty UART transmitter allows our simple program to transmit a character *without* checking UART transmit status. Second, we know a character is *waiting* when we execute function 02h and can read the character *without* bothering to see if a character has been received.

Data Transmission and Reception Concerns

Transmission of data is relatively simple. Function 01h allows us to transmit *whenever* desired. Function 01h waits until the transmitting register is available before returning to the main program. There is *no* possibility of overwriting the last transmission character when function 01h is used.

Reception is a different story. Function 02h will return *whatever* is in the receiving register. Unthinking use of function 02h results in the return of repetitive (the last character received) or bogus (no character has been received) bytes. A receiving program should *first* check the status of the serial port with function 03h to see if a character is waiting, and then read the character with function 02h.

Function 03h will reset the UART serial port *character received* bit *after* reading and returning the port status in AH, Bit 0. When a character is *first* received, Bit 0 of AH is returned as a 1 by function 03h. The UART serial port *character received* bit is then reset so that *future* uses of function 03h will return Bit 0 of AH as a 0, until the *next* character is received. Normally, after a progam uses function 03h to detect a character, the character should be read by function 02h, to get it before the next character is received.

A small program, recve.asm, shows how function 03h can be used to poll the communication receiver port. When function 03h returns a 1 bit in AH Bit 0, function 02h is used to read the character.

```
;recve.asm, a communications character receive program
code segment
        mov sp,1000h        ;set the stack
        mov ax,00C3h        ;initialize to 4800, no parity, 8 bits
        mov dx,0000h        ;use COM1
        int 14h
receive:
        mov ah,03h          ;get port status, AH bit 0 = 1 if ready
        int 14h             ;data waiting?
```

```
        rcr ah,1              ;rotate AH bit 0 into the Carry flag
        jnc receive           ;poll until character is received (CF=1)
        mov ah,02h
        int 14h               ;get the character
        ;rest of the program
code ends
```

The primary disadvantage of program recve.asm is that the program will *poll* the serial port until a character is received, thus preventing the computer from doing any other useful work. Should there be some fault in the serial data link, the computer will loop forever.

There are two solutions to the polling "hang-up" problem. One solution is to simply loop through the entire computer program and get serial characters when they appear. Characters can be lost using a looping approach, however, if the entire program takes *longer* to loop through than the time it takes to receive a new character.

The loop timing problem is not as dire as may be thought. At a fast baud rate of 19,200 bits per second (bps), a 10-bit character will still require about 500 microseconds to be shifted into the receiver memory. The PC can execute a short (100-byte) program in 500 microseconds. If the program takes longer that 500 microseconds to execute, then the baud rate can be *slowed* down as needed. At a baud rate of 1,200 bps, for instance, the program has over 8 milliseconds to execute between reading characters. Conversely, the PC can be operated at a *faster* clock rate, so that the program runs faster. The obvious drawback to speeding up the computer is the cost of speed.

Should the option of slowing the baud rate or speeding up the PC not be practical, then the program cannot loop and get each character. Instead of looping, the program must become *hardware interrupt* driven *by the serial port*, instead of having the serial port INT driven by the *software*. To be able to make the serial port interrupt the program by hardware action we need to use CPU hardware interrupts, that is, interrupts that are generated by circuit action, and not by a software INT instruction.

Generating hardware interrupts requires detailed knowledge of the serial port's UART *hardware interrupt control* registers. Also, we must have detailed knowledge of the integrated circuit that controls *hardware interrupts* to the computer CPU. The level of hardware detail needed to write interrupt-driven serial data programs is beyond the scope of this chapter, but the concept of interrupt-driven software will be explored in Chapter 11.

Rather than focus on the serial port in Chapter 11, we shall design our own generic interface board for hardware-generated interrupts. By designing our own interrupt ports, using common TTL logic gates and flip-flop functions, we shall have the detailed circuit knowledge necessary to accomplish interrupt-driven applications.

A Computer-to-Computer Communication Program

By using the concepts incorporated in programs recve.asm and comm.asm, a program that enables transmission between two computers is shown next. The example, twoway.asm, programs two computers to transmit and receive messages simultaneously. The non-handshaking cable of Table 10.8 is used to connect the computer serial ports.

The example has each (local) computer send a predefined string of characters to the other (remote) computer. The program allows each local computer user to suspend transmission from the remote computer by pressing the *n* key on the local computer keyboard. Remote transmission can resume when the local user presses the *s* key. Transmission is initiated whenever the user presses the *t* key on the local keyboard, or the *s* key on the remote keyboard. Pressing the *q* key returns control to DOS.

```
;twoway.asm, a program that sends a string from one computer to
;another. Pressing the "t" key initiates local transmission.
;Pressing the "n" key suspends remote transmission; pressing the "s"
;key resumes (or initiates) remote transmission.

data segment
            org 1000h           ;place data above code         See Note
            gostop db ?         ;transmit status flag
            count dw ?          ;number of characters transmitted
            colnum db ?         ;current cursor column location        1
            rownum db ?         ;current cursor row location
data ends
code segment
;use equates to make the program readable and easy to change
;
;character equates
            nak equ 15h         ;NAK = stop transmission               2
            stx equ 02h         ;STX = resume transmission
            cr equ 0dh          ;carriage return
            lf equ 0ah          ;line feed
;
;serial port initialization equates
            slow equ 01000000xb     ;300 baud
            fast equ 01100000xb     ;600 baud
            faster equ 11000000xb   ;4800 baud
            turo equ 11100000xb     ;9600 baud
            eparity equ 00011000xb  ;even parity
            oparity equ 00001000xb  ;odd parity
            nparity equ 00000000xb  ;no parity
            stp equ 00000000xb      ;one stop bit
            length equ 00000011xb   ;8 bit (7 + parity)
;
;program constant equates
            done equ 1800xd     ;25 lines of 72 characters (70 + CR, LF)
            patt equ patterna   ;choose one
            ;patt equ patternb  ;depending on which computer in use
;end equates
;
            mov sp,0fffeh       ;set the stack
            mov ax,0002h        ;select text mode
            int 10h             ;BIOS sets mode via int 10h
            mov ah,00h          ;initialize the serial port
            mov al,00h          ;OR the initialize bits in AL
            or al,fast
            or al,nparity
            or al,stp
            or al,length        ;600 baud, no parity, 1 stop bit, 8 bit
```

```
            mov dx,0000h          ;COM1
            int 14h               ;serial port initialized
            mov gostop,nak        ;disable transmission
            mov count,0h          ;zero character counter
            mov si,0              ;SI points to character lookup table
            call _recv            ;dump any "garbage" in the UART          3
            mov w[colnum],0h      ;initialize cursor location at 0,0
            call _ curses
begin:
            call _recv            ;check for a received character          4
            jc action             ;if CF=1 then character received
            call _getkey          ;get user key and react
            call _recv            ;check again for a received character
            jc action
            cmp gostop,stx        ;see if transmission is enabled          5
            jne begin
            call _status          ;if transmitter is busy then loop        6
            and ah,20h            ;mask off holding register bit
            cmp ah,20h            ;do not attempt to transmit if busy
            jne begin
            call _xmit            ;transmit
            jmp begin             ;loop
action:
            call _action          ;display or react to control character
            jmp begin
;
;received module returns a character received by the serial port
;in register AL. If no character, then returns with the Carry flag
;cleared to 0.
_recv:
            call _status          ;get serial port status
            rcr ah,01h            ;rotate received character bit to C flag
            jnc nothing           ;if AH bit 0 = 0, then no character
            mov ah,02h            ;else, get it
            int 14h               ;get character in AL
nothing:
            ret                   ;CF=0 or character in AL (CF=1)
;
;_status gets serial port status and returns it in AH
_status:
            mov ah,03h            ;get com port status
            mov dx,0000h          ;COM1
            int 14h               ;get port status in AH
            ret
;
;action module determines what should be done with the received
;character. Character passed in register AL.
_action:
            cmp al,' '            ;check for control or display character
            jae showit            ;if 20h (space) or above, display it
            cmp al,cr             ;do not display CR
            je noshow
            cmp al,stx            ;control character
            je go                 ;STX?
```

```
                        cmp al,nak              ;NAK?
                        je stop
                        cmp al,lf               ;if LF, then end of line—reset cursor
                        inc rownum              ;next row
                        mov colnum,00h          ;first column of next row
        noshow:
                        ret
        showit:
                        call _display           ;display character
                        ret
        go:
                        mov gostop,stx          ;go condition
                        ret
        stop:
                        mov gostop,nak          ;stop condition
                        ret
        ;
        ;display module places the character passed in AL to the screen
        ;and increments the cursor to the next column.
        _display:
                        call _curses            ;position cursor
                        mov ah,09h              ;display character function
                        mov bx,0001h            ;page 0, white character, blue background
                        mov cx,0001h            ;one character
                        int 10h                 ;display it
                        mov dl,colnum           ;get current cursor column location
                        inc dl                  ;increment cursor column counter
                        mov colnum,dl           ;save new cursor location
                        ret
        ;
        ;curses module places the cursor at the position in cursnum.
        _curses:
                        mov dx,w[colnum]        ;get current cursor position
                        mov ah,02h              ;position cursor
                        mov bh,00h              ;on page 0
                        int 10h                 ;place cursor
                        ret
        ;
        ;getkey module gets one of the t, s, n, or q keys and reacts
        ;accordingly. Any other key produces no action.
        _getkey:
                        mov ah,01h              ;check for a key
                        int 16h                 ;get key or ZF=1 for no key
                        jz nokey                ;return if no key is waiting
                        mov ah,00h              ;key present, get it
                        int 16h
                        cmp al,'t'              ;act on key action
                        je goxmit               ;t begins transmission
                        cmp al,'s'
                        je restart              ;s resumes transmission
                        cmp al,'q'              ;q quits to DOS
                        je exit
                        cmp al,'n'              ;n stops transmission
                        jne nokey
```

7

```
                        nov al,nak
                        call _sendit
                        ret
        restart:
                        mov al,stx              ;restart remote transmission
                        call _sendit
                        ret
        goxmit:
                        mov gostop,stx          ;start local transmission
                        ret
        nokey:
                        ret
        exit:
                        mov ax,4c00h            ;back to DOS
                        int 21h
        ;
        ;xmit module checks to see if all characters have been sent
        ;before getting and transmitting a character.
        _xmit:
                        cmp count,done          ;see if done < count
                        jb send                 ;if not done, transmit
                        ret                     ;else do nothing
        send:
                        mov bx,patt             ;get character table base to BX
                        mov al,b[bx+si]         ;table offset in SI
                        inc si                  ;point to next character in table
                        cmp si,72xd             ;reset SI pointer after 72 characters
                        jb next                 ;(70 displayable plus CR, LF)
                        mov si,00h              ;reset SI pointer
        next:
                        inc count               ;count transmitted characters
                        call _sendit
                        ret
        ;
        ;sendit module transmits any character passed in AL
        _sendit:
                        mov ah,01h              ;transmit the character
                        mov dx,0000h            ;COM1
                        int 14h                 ;transmit character in AL
                        ret
        ;
        ;lookup tables as determined in the "patt" equates
        patterna:   db  'A123456789A123456789A123456789A123456789'
                    db  'A123456789A123456789A123456789',cr,lf
        patternb:   db  'B123456789B123456789B123456789B123456789'
                    db  'B123456789B123456789B123456789',cr,lf
        code ends
```

NOTES ON THE PROGRAM

1. The cursor moves from left to right until a CR-LF set of control characters is received.

2. Other control characters could have been chosen to stop and start remote transmission.

3. Many computers, on booting up, have *something* (known as *lint*) in the UART.

4. Reception has priority over transmission to minimize the chance of losing received data.

5. Transmission is disabled if a *stop* control character is present.

6. Remember, the transmission interrupt service routine will wait for any *previous* byte to be transmitted before returning. Waiting for a previous byte *wastes time* that could be used to check the receiver.

7. Check key status only; other keyboard INT functions *wait* for a key, thus received data is lost.

Section 10.9 lists several suggested data communication program projects that involve communications between two computers.

10.8 Summary

Interfacing a PC to external circuits may be done using the parallel printer and serial ports in conjunction with BIOS or DOS INT service routines.

Printer ports may be used to interchange parallel data with an external circuit using 8 data pins for output from computer to external circuit and 4 status pins for input from circuit to computer.

The following BIOS and DOS INT service routines may be used to access the parallel printer port pins.

BIOS INT 17h Printer Control

Function 00h: Output a Character to the Printer

Function 01h: Reset Printer

Function 02h: Read Printer Port Status

DOS Interrupt 21h

Function 05h: Write Byte to Printer

Function 40h: Write Data to File Device

Serial 8-bit data transmission and reception from an external circuit may be done using a computer serial port.

The following BIOS INT service routines may be used to initialize, transmit, or receive serial data.

BIOS Asynchronous Serial Data INT Type 14h

Function 00h: Initialize the Serial Port

Function 01h: Write a Character to the Serial Port

Function 02h: Read a Character from the Serial Port

Function 03h: Determine Serial Port Status

External circuits that are connected to the parallel or serial ports must be designed to operate using the port timing and handshaking specifications.

10.9 Questions and Projects

Questions

1. Name each parallel port signal, its function, and whether it is active-high or active-low. Indicate if the signal is an input to the port or an output from the port.

2. Identify the parallel port status bits returned by INT 17h, and indicate if the bit is a logic 1 for an active-high or active-low input signal.

3. Construct a byte that can be XORed with the parallel port status bits so that if an input pin is low, its corresponding status bit will be a 1, and if high, will be a 0.

4. Convert your name, including capital letters and spaces, to ASCII code.

5. Explain the function of the following serial port handshaking signals for *DTE*, and indicate if the signal is from the port or to the port: RTS, CTS, DSR, DTR.

6. Explain the function of the following serial port handshaking signals for *DCE*, and indicate if the signal is from the port or to the port: RTS, CTS, DSR, DTR.

7. Construct a table, similar to Table 10.6, that connects DCE equipment to DCE equipment.

8. Construct initialization bytes for the serial port that will configure the port as follows.

Baud	Parity	Stop Bits	Length
2,400	Even	1	8 bits
9,600	Odd	2	7 bits
300	Odd	1	8 bits
4,800	Even	1	8 bits

9. Draw a picture and identify each bit of the ASCII byte *3* if it is transmitted at 2,400 baud, even parity. Show the time interval for each bit, in microseconds.

10. Define the function of each of the following serial port data status bits returned by INT 14h.

    Bit 0: Received Data Present

    Bit 1: Overrun Error

    Bit 2: Parity Error

    Bit 7: Time-Out

Projects

The projects listed here use the I/O features of the PC parallel and serial ports and associated INT service routines.

Parallel Port

1. Build a 4-switch, 4-LED interface circuit for the parallel printer port and write a program that will illuminate a corresponding LED when any switch is *opened* (input to the printer port is asserted high).

2. Build an interface to the parallel printer port that consists of a 7-segment numeric LED display output and a push-button input. Write a program that initializes the display to 0 and counts up the display from 0 to 9 as the push button is pushed. To get the best grade possible do *not* use a binary to 7-segment type decoder ('7447) chip to drive the display. Instead of using a decoder chip, drive the display from the port using the program to make each segment illuminate.

3. Build an interface to the parallel printer port that consists of 8 input switches and 2 LEDs. Have one LED illuminate if the switch byte has odd parity, and have the other LED illuminate if the switch byte has even parity.

4. Build an interface to the parallel printer port that allows you to do a *loop test* of the printer port. A loop test is a test that uses the program to generate a test pattern on the printer port data bits that are then read in on the port status bits.

   Design the circuit so that the most significant nibble of the output data byte is fed to 4 input status lines when D7 is a logic 0. The least significant nibble is fed to 4 input status lines when D7 is a logic 1. Have the program report on errors if output bits do not match input bits.

5. Design a "game-show" 2-button, 1-LED circuit, and interface it to the parallel port. Have the program illuminate the LED when the *g* key is pressed on the keyboard, and then detect which of the two buttons is pressed first. The winner is shown on the CRT screen as Player 1 or Player 2. Replay the game when the s key is pressed on the keyboard.

6. Use one parallel port data input bit to measure the frequency of an input square wave. Display the input frequency, to the nearest tenth, on the screen.

7. Use one of the parallel port data bits to generate a variable-frequency square wave that ranges from the highest frequency that your PC can generate down to 100Hz. Have the program ask the user to enter the desired frequency, then generate the requested frequency for 1 minute.

8. Use an 8-bit D/A converter and write a program that enables the parallel port to generate a sine wave that may vary from 100Hz up to the highest frequency possible for your computer. Use at least 20 amplitude values per output wave. Have the program query the user as to the frequency desired, then produce the requested frequency for 1 minute.

9. Use an 8-bit A/D converter and write a program that enables the parallel printer port to record the instantaneous amplitude value of a 100Hz input sine wave. Sample the input wave 1,000 times per second, and store 10 seconds worth of data in memory. Write the program so that the user may start recording when the *s* key is pressed and so that the message *sampling done* displays when finished. Enable the user to view the results on the screen, one page at a time, every time the *d* key is pressed.

10. Use the parallel port to read a 16-key X-Y matrix keyboard. Use the data bits to scan each X row of keys, and use the port inputs to read the Y key columns to detect when a key is pressed and released. De-bounce the keys, and display on the screen the sequence in which the keys are pressed and released. Do not display a key if more than one key is pressed

on the keyboard at any one time. Do not miss any keys, no matter how quickly they are pressed and released.

11. Program the parallel port so that it will act as a serial port. Use D0 to transmit serial data, and D1 to receive serial data. Use D3–D6 as RTS, CTS, DSR, and DTR handshaking pins. Connect the parallel port to the serial port, and transmit a character both ways between the ports, displaying the character on the screen as each port receives it. Use software delays or BIOS timing functions to provide timing for the parallel port.

Serial Port

1. Connect the serial port to itself in a loop check configuration, and write a program that takes any *printable* key pressed on the keyboard, transmits it, receives it, and displays the key on the screen.

2. Connect two computers together, via their serial ports, using the full handshaking cable of Table 10.6. Program each computer to operate at your choice of baud rate and parity. One computer is the transmitting computer, the second the receiving computer.

   Write a program for the sending computer that will send any key *scan* code typed on the keyboard to the receiving computer. Write a program for the receiving computer that will receive the character and display it in the center of the screen. Have the receiving computer display ?? whenever a non-ASCII character is received. Have the receiving computer display a Control message whenever an ASCII control character is received.

3. Repeat Project 2 using the non-handshaking cable of Table 10.8.

4. To study the effect of mismatched baud rates, repeat Project 3, but use different baud rates for each computer. Observe the actions of the receiving computer's screen.

5. Connect two computers together, via their serial ports, using the non-handshaking cable of Table 10.8. Write programs for both computers such that any printable ASCII character typed at Computer 1 is displayed on the center of the screen at Computer 2. Computer 2 then re-transmits the character back to Computer 1 for display on the center of the screen of Computer 1. Nonprintable ASCII codes are not to be transmitted by Computer 1.

6. A series of unknown (one of the INT 14h standard baud rate) carriage return (CR) control characters is received by a computer from a second computer. Write a program for the receiving computer that will automatically set the correct receiving baud rate. Have the program display the correct baud rate when it is found. Use the non-handshaking serial cable of Table 10.8.

7. Repeat Project 6, but write the program so that the correct baud rate may be determined in less than 8 tries.

8. Connect two computers together using the non-handshaking cable of Table 10.8. Write a serial terminal program for each PC that performs the following operation.

   ■ Have each program begin by asking the user for the desired baud rate (from 1,200 to 9,600) and parity. Set the desired baud rate and parity, using 8 data bits and 1 stop bit. Position the screen cursor at the upper left corner of the screen.

■ Each user may type in a line at the (local) keyboard. As the line is typed, it is displayed on the local screen. When the Return key is pressed, the string of characters displayed on the local screen is transmitted to the other (remote) computer. The remote computer then displays the received line in reverse video, on the line below its current cursor position.

■ It must be possible to transmit and receive a string at the same time without losing any characters. To test this, users at both computers are to type in a string and press the Return keys on both computers at the same instant.

9. Repeat Project 8 using a split-screen format. Use the left side of the screen (Columns 0 to 38) for the local user. Use the right side of the screen (Columns 40 to 79) for the remote messages.

10. Modify Project 8 so that entire screens (25 rows of 80 characters) may be transmitted. Each user may type in an entire screen, then press the F1 key to transmit the screen. If one user has not finished typing a screen, and the other computer transmits a screen, the received screen is to be stored until the slower user is finished and presses the F1 key. When a remote screen is displayed, blank the local screen and write the received screen in reverse video.

11

INTERFACING

TO THE

EXPANSION

BUS

Chapter Objectives

Upon successful completion of this chapter you will be able to:

- Discuss why computers use standard busses.
- Describe the 8-bit PC and ISA physical busses.
- Describe the 8-bit PC and ISA bus signal groups.
- Calculate bus memory access timing.
- Describe solid-state memory SRAM and EPROM operation.
- Calculate solid-state memory access timing requirements.
- Describe the design of a memory card for the bus.
- Describe the design of a memory-mapped I/O bus card.
- Calculate I/O bus card access time.
- Describe the design of an I/O mapped I/O bus card.
- Describe how the bus hardware interrupt circuitry operates.
- Describe the design of an interrupt-driven I/O bus card.

11.0 Introduction

The material presented in the previous chapter emphasizes using standard ports as the *sole* means by which external hardware interfacing is accomplished. This chapter introduces the idea of *designing* unique PC interface hardware called *bus expansion cards*. We shall still rely on the PC to function as a *host* machine that enables our interface cards to function. The focus of the text, however, shifts from programming to *digital hardware design.*

The selection of the digital logic devices used in the expansion card design examples in this chapter are chosen to be *general-purpose integrated* circuits. Using discrete logic gates is done for illustrative purposes *only*. Modern fixed-logic designs are usually implemented with *programmable* logic devices, notably PALs and GALs. Unfortunately, programmable logic devices tend to hide important circuit logic and timing details that should be brought out in a textbook. You are encouraged to convert the example designs to programmable logic implementations as a vivid demonstration of the space efficiency of such devices.

11.1 Computer Busses

A computer *bus* is a *collection of signals* that is used by the computer system designer to connect system memory and I/O to the central computing core. Almost all computer systems, from the smallest single-board type to the largest mainframe, have made provisions to enable the computer system to be *expanded.* The ability to expand a computer system is desirable from a customer and marketing standpoint. Many customers wish to start with some *basic* computer system configuration and *add* computer resources as their needs grow. A computer bus that allows additional memory and I/O peripherals to be easily "plugged in," is a commercial necessity.

Types of Busses

Personal computer users have need of future expansion capability, and an expansion bus was a prominent feature of the first PC-compatible models. Figure 11.1 shows the basic *hardware* configuration of a personal computer. The PC hardware is shown divided into two main parts: the computing *core*, and the external *expansion* card bus.

The Computing Core and the Local Bus

A basic computing core consists of the CPU, clock generator, memory controller, and hardware interrupt controller. The core may be connected together as desired by the computer hardware designer. Connections between the circuits found in the computing core are sometimes referred to as the *local bus*. The local bus is a *high-speed* bus that can be designed by the manufacturer under few *standardization* constraints. The design of the core can be quite different from manufacturer to manufacturer, and from year to year for the same model by a given manufacturer.

Figure 11.1 ■ Computer core and expansion bus.

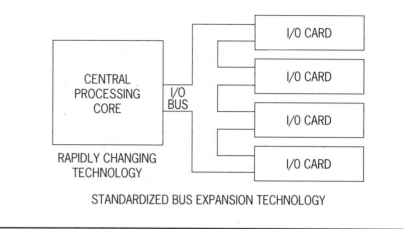

Improvements in integrated-circuit technology drive the design of the computing core, and its associated local bus, toward *fewer* chips and *faster* CPU clock speeds. Local busses may be considered to be nonstandard and proprietary designs.

THE EXPANSION BUS

The *external PC bus,* however, consists of a set of signals and hardware connectors that will *not* change from one PC-compatible manufacturer to another, or from year to year. The reasons for not changing the external bus are market-driven. First, a *standard* external bus encourages second-source vendors to develop products that plug into the bus. Potential customers are very likely to buy a computer that features a popular bus with many optional expansion card sources available to them. Second, once a customer has made a significant investment in bus expansion cards, that customer is much more likely to buy a next-generation computer if the new generation can use the *same* cards as the earlier generation computer.

A Short History of PC-Compatible Busses

The evolution of the PC bus and its successors is a fascinating story—half planning and half accident. IBM, when it first announced its first PC, probably had no idea a dominant future *industry standard* would be established based on the PC bus signal and connector arrangement.

In 1981, when IBM brought forth the first PC, the term *PC bus* began to circulate in the computer industry. Computer busses were certainly not new, having been a feature of most commercial computers beginning with the early mainframes of the 1950s. Every computer manufacturer establishes an expansion bus arrangement for its products, unless the computer is extremely *inexpensive.* The primary advantage of a bus is that it allows expansion of the computer by means of bus plug-in circuit cards that allow the system to "grow."

Busses may be of two general commercial types, *open* or *closed.* Closed busses are considered to be the *proprietary* property of the manufacturer. Second-sourcing of expansion bus cards for closed busses is discouraged by the manufacturer by refusing to share bus information and by actual, or threatened, legal action. Open busses, however, are *available* to all, and the manufacturer makes available whatever information is needed to design expansion cards for the bus. IBM, perhaps unsure of the market for personal computers in 1981, decided to make the *specifications* for the PC bus available to the computer industry .

Bus Standards

When a bus standard is established, every pertinent bus feature that an expansion card designer needs to know is defined. Bus standards are generally composed of the items shown in Table 11.1.

We should note that most industrial standards are *voluntary,* and bus standards are no exception. No knock on the door in the middle of the night will greet the designer who ignores some of the bus standards. Ignoring the standards will, however, mean that the resulting nonstandard card will *not* work with the majority of standard-bus computers in the world.

Personal computer manufacturers are *not* required to publish bus specifications. Apple decided to make the Macintosh bus a *closed* bus and published no information when the Macintosh was introduced.

The PC Bus

By publishing the IBM PC bus specifications, IBM made the PC bus an open bus, available to all other vendors. This decision by IBM to open the PC bus resulted in a host of plug-in circuit cards that allowed PC buyers to add various features to their PC computers. The strategy worked well, and the IBM PC soon became the most popular personal computer ever sold. In the process, the PC bus became a de facto established *bus standard* for second-source vendors.

The AT Bus

IBM declined to publish bus details when the AT class of computers was introduced. The AT was based on the 80286 CPU, which is an enhanced version of the 8086 CPU. To maintain compatibility with its earlier PC model, however, IBM decided to keep *most* of the original PC bus features on the AT bus. New AT bus features were added on a *second, separate* bus connector.

Standardized Feature	Standards Established
Physical dimensions	Length, Width, Mounting, Connector
Electrical quantities	Power, Logic levels, Current loading
Signals	Pin assignments, Signal names, Signal timing, Clock speed

Table 11.1 ■ Standardized Bus Features

Older PC bus cards could be plugged into the new AT bus by using the lower connector.

By the time the IBM AT became a "mature" (commercially popular) model, the PC clone industry was well-established. Vendors of AT-class machines attempted to duplicate the IBM AT bus as closely as possible, but there were the inevitable differences that rendered the term *IBM compatible* an insider's joke. To remedy the incompatibility problems that existed among the various AT clone makers, the clone companies, led by Intel and COMPAQ Corporation, set about to establish a *voluntary industry standard* for the AT bus. The Institute of Electrical and Electronic Engineers (IEEE) acted as the voluntary standards-setting body. IEEE standard P996, better known as the *Industry Standard Architecture* (ISA), was published by the IEEE in 1990. The ISA bus standard covers both 8-bit (8088-based PC-type) and 16-bit (80286-based AT) expansion card busses.

Interestingly, IBM *abandoned* the AT standard about the same time, in favor of their new PS/2 model personal computers and their *micro-channel* bus.

The advent of 32-bit 80386 and 80486 CPUs prompted COMPAQ and others to publish an EISA, or *Extended ISA* bus specification, that covers 32-bit system expansion cards. To further confuse matters, one may hear of the *E-ISA standard*, which is a melding of the 8/16-bit ISA standard and some features of the 32-bit EISA standard.

Fortunately, *all three* bus standards use the IBM AT expansion bus as a starting point, so expansion cards that are designed to ISA, E-ISA, or EISA standards are generally *upwardly* compatible.

The Generic 8-Bit PC Bus

To make the subject of this chapter as useful as possible to a majority of readers (all of whom likely have different computers), I shall use the *8-bit ISA signal names* to design plug-in expansion cards for the *PC bus*. Because the ISA bus grew from the PC bus, 8-bit ISA bus and 8-bit PC bus standards are virtually identical, including timing.

Using the 8-bit ISA bus standard should not be a design burden. As bit widths increase, newer standards add additional connectors to accommodate the new signals. The original 8-bit uses are retained so that an expansion card developed for an 8-bit PC can (*most* of the time) be used in a 32-bit EISA machine. Note the reverse is not true. You can not, physically or electrically, plug a 32-bit EISA card in an 8-bit ISA slot.

All of the designs in this chapter have been *tested* on 8-bit PC and ISA busses connected to 8088, V20, 80286, and 80386 computing cores. It would seem best, mainly for cost reasons, that laboratory exercises based on these designs should be done using PC bus computers.

11.2 The 8-Bit ISA and PC Bus Standard

A computer's overall main bus is composed of *three* smaller busses that are identified in Table 11.2. Each sub-bus contains signals that are grouped together by *function*.

Bus Signals	Function
Address	Address bits for memory and I/O locations
Data	Interchange bits between CPU and memory or I/O
Control	Coordinate operations of address and data busses

Table 11.2 ■ Three-Bus System

The three functional busses all have their origins at the CPU chip. The CPU signals are, in turn, augmented by specialized CPU support chips in the computing core so that the final three-bus signal groups are generated. As Figure 11.1 shows, everything to the left of the external bus line can be designed at the pleasure of the CPU system designer. Everything to the right of the external bus line should meet published bus standards.

The main advantage of bus standards is the fact that expansion cards designed to the standards' specifications will operate in any computer that (honestly) meets the standards. We can see that bus standards are to hardware what BIOS and DOS interrupts are to software: machine independent.

The 8-Bit PC/ISA Bus Physical Standard

Figure 11.2 shows the *physical* standard for an 8-bit expansion card that can plug into any 8-bit PC bus slot connector. Note that no mounting schemes, other than friction with the card edge connector, are established. Those who have used expansion cards will remember that the card is, typi-

Figure 11.2 ■ ISA bus 8-bit expansion card.

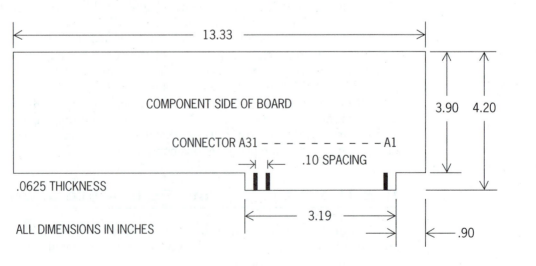
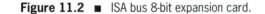

cally, anchored at the connector end by a metal fastener. The mounting position of the fastener is established, but the configuration of the fastener itself is left up to the expansion card maker.

It is not necessary to make expansion cards. Prototype 8-bit expansion cards can readily be purchased from a number of neighborhood and mail-order vendors. Appendix H, "Sources," lists several vendors of ready-made expansion cards. Blank 8-bit expansion cards range from bare cards, to bare cards with address decoding logic mounted on them, to cards with plug-boards mounted on them for rapid prototype circuit construction.

Card *separation*, the spacing from card to card, is not an established standard, although center-to-center connector spacing is typically 0.8 inches. Spacing is very important to keep in mind as it limits the *height* of anything that is mounted on the expansion board to a height less than the card spacing.

The physical standard also determines the *arrangement* of signals on the two-sided (A and B) board edge connector. A very common mistake, especially for the beginner, is to forget which is the A side of the board connector and which is the B side. Fortunately, most blank expansion cards will include etched numbers on the card connector to remind us how the card pins are numbered.

The 8-Bit PC/ISA Bus Signals

The ISA 8-bit standard establishes a connector size of 62 conductors arranged as two groups (one group on each side of the connector) of 31 conductors, as shown in Figure 11.2. The groups are named groups A and B, and pins are identified as Ax or Bx, where x varies from 1 to 31. Table 11.3 lists the signal present on each pin by signal *name*, signal *direction*, and a description of each signal's *function*. Signal direction may be from the card to the system (*Output* to system), from the system to the card (*Input* to card), or bidirectional (I/O) between card and system. Active-low signals are indicated by a slash (/) preceding the signal name.

ISA signals that involve *advanced* features that are beyond the scope of this book (such as DMA transfers) are noted as such using the words "Advanced Feature." Advanced feature signals will be ignored if they are inputs to the card or disabled, if necessary, if they are card outputs.

Differences *between* 8-bit ISA and PC bus standards are noted by the use of an asterisk (*). The PC bus equivalent is preceded by the asterisk and is noted *below* the ISA standard definition.

An inspection of the 8-bit ISA bus standard and the PC bus standard, shown in Table 11.3, reveals that the only differences between the two are the pin *B04 and B08* signals. We shall not use the ISA standard signal /SRDY (pin B08) nor IRQ9 (pin B04). For our purposes then, the 8-bit PC bus and the 8-bit ISA bus pin and signal names are *identical*. PC bus and ISA bus timing *waveforms* are not identical, however, although access times for both busses are virtually the same.

Pin	Name	Direction	Function
A01	/IOCHCHK	Output	General I/O Error—Advanced Feature
A02	SD7	I/O	Data Bit 7
A03	SD6	I/O	Data Bit 6
A04	SD5	I/O	Data Bit 5
A05	SD4	I/O	Data Bit 4
A06	SD3	I/O	Data Bit 3
A07	SD2	I/O	Data Bit 2
A08	SD1	I/O	Data Bit 1
A09	SD0	I/O	Data Bit 0
A10	IOCHRDY	Output	I/O Channel Ready for Data Transfer
A11	AEN	Input	DMA Address Enable
A12	SA19	Input	Address Bit 19
A13	SA18	Input	Address Bit 18
A14	SA17	Input	Address Bit 17
A15	SA16	Input	Address Bit 16
A16	SA15	Input	Address Bit 15
A17	SA14	Input	Address Bit 14
A18	SA13	Input	Address Bit 13
A19	SA12	Input	Address Bit 12
A20	SA11	Input	Address Bit 11
A21	SA10	Input	Address Bit 10
A22	SA9	Input	Address Bit 9
A23	SA8	Input	Address Bit 8
A24	SA7	Input	Address Bit 7
A25	SA6	Input	Address Bit 6
A26	SA5	Input	Address Bit 5
A27	SA4	Input	Address Bit 4
A28	SA3	Input	Address Bit 3
A29	SA2	Input	Address Bit 2
A30	SA1	Input	Address Bit 1
A31	SA0	Input	Address Bit 0
B01	0V	—	Power Supply Ground
B02	RESET	Input	System Reset
B03	+5V	—	Power Supply +5 Volts
B04	IRQ9	Output	Interrupt Request Number Nine
*B04	IRQ2	Output	Interrupt Request Number Two (PC bus)
B05	−5V	—	Power Supply −5 Volts
B06	DRQ2	Output	DMA Channel 2 Request—Advanced Feature
B07	−12V	—	Power Supply −12 Volts
B08	/SRDY	Output	Ready for I/O Data Transfer
*B08	Reserved	—	Unused on PC bus
B09	+12V	—	Power Supply +12 Volts
B10	0V	—	Power Supply Ground
B11	/SMEMW	Input	Memory Space Write
B12	/SMEMR	Input	Memory Space Read
B13	/IOW	Input	I/O Space Write
B14	/IOR	Input	I/O Space Read
B15	/DACK3	Input	DMA Channel 3 Ack.—Advanced Feature
B16	DRQ3	Output	DMA Channel 3 Request—Advanced Feature

Table 11.3 ■ ISA 8-Bit Bus Expansion Card Connector Signals

Pin	Name	Direction	Function
B17	/DACK1	Input	DMA Channel 1 Ack.—Advanced Feature
B18	DRQ1	Output	DMA Channel 1 Request—Advanced Feature
B19	/REFRESH	Input	DRAM Refresh—Advanced Feature
*B19	/DACK0	Input	DRAM Refresh—Advanced Feature
B20	SYSCLK	Input	System Clock Pulse 4.77/8.0 MHz (PC/ISA)
B21	IRQ7	Output	Interrupt Request Number Nine
B22	IRQ6	Output	Interrupt Request Number Six
B23	IRQ5	Output	Interrupt Request Number Five
B24	IRQ4	Output	Interrupt Request Number Four
B25	IRQ3	Output	Interrupt Request Number Three
B26	/DACK2	Input	DMA Channel 2 Ack.—Advanced Feature
B27	TC	Input	DMA Transfer Complete—Advanced Feature
B28	BALE	Input	Bus Address Latch Enable
B29	+5V	—	Power Supply +5 Volts
B30	OSC	Input	System Oscillator Pulse
B31	0V	—	Power Supply Ground

Table 11.3 ■ *Continued*

11.3 Design of a Memory Expansion Card

Additional semiconductor memory is often the first item added to a computer as the user discovers that many application programs and operating systems require ever more memory. The first design, then, will be to add a memory expansion card to the PC bus.

Bus Signals for a Memory Expansion Card

The 62 signals listed in Table 11.3 present an impressive array of address, data, and control quantities. Our first task, in dealing with the ISA bus, is to whittle the 62 signals of Table 11.3 *down* to a number that are truly useful for the design of a memory expansion card.

THE ADDRESS AND DATA BUSSES

All of the address lines can be of use to us when we are addressing one unique memory location on our expansion card. We shall need all of the address pins numbered A12 through A31. We shall also use all eight of the data lines, pins A02 to A09.

THE CONTROL BUS

The memory expansion card we shall design will use several of the *free* address spaces available in the PC 1M memory address space. We shall require memory read and write pulses so that memory accessing is synchronized with the central processing core. Thus, we shall need /SMEMW, pin B11, and /SMEMR, pin B12.

Table 11.4 lists a pared-down group of ISA bus signals that we shall use to design a memory space expansion card. Plus 5V power and ground pins are

Pin	Name	Direction	Function
A02	SD7	I/O	Data Bit 7
A03	SD6	I/O	Data Bit 6
A04	SD5	I/O	Data Bit 5
A05	SD4	I/O	Data Bit 4
A06	SD3	I/O	Data Bit 3
A07	SD2	I/O	Data Bit 2
A08	SD1	I/O	Data Bit 1
A09	SD0	I/O	Data Bit 0
A12	SA19	Input	Address Bit 19
A13	SA18	Input	Address Bit 18
A14	SA17	Input	Address Bit 17
A15	SA16	Input	Address Bit 16
A16	SA15	Input	Address Bit 15
A17	SA14	Input	Address Bit 14
A18	SA13	Input	Address Bit 13
A19	SA12	Input	Address Bit 12
A20	SA11	Input	Address Bit 11
A21	SA10	Input	Address Bit 10
A22	SA9	Input	Address Bit 9
A23	SA8	Input	Address Bit 8
A24	SA7	Input	Address Bit 7
A25	SA6	Input	Address Bit 6
A26	SA5	Input	Address Bit 5
A27	SA4	Input	Address Bit 4
A28	SA3	Input	Address Bit 3
A29	SA2	Input	Address Bit 2
A30	SA1	Input	Address Bit 1
A31	SA0	Input	Address Bit 0
B01	0V	—	Power Supply Ground
B02	RESET	Input	System Reset
B03	+5V	—	Power Supply +5 Volts
B10	0V	—	Power Supply Ground
B11	/SMEMW	Input	Memory Space Write
B12	/SMEMR	Input	Memory Space Read
B29	+5V	—	Power Supply +5 Volts
B30	OSC	Input	System Oscillator Pulse
B31	0V	—	Power Supply Ground

Table 11.4 ▪ ISA Bus Memory Space Expansion Card Connector Signals

included, as is the system RESET pulse and the system oscillator. The last two signals are *not* essential, but they may be needed in future designs.

Bus Memory Cycle Timing

Timing, as many wags have observed, is everything: often in war, occasionally in love, usually in investing, and *always* in logic design. Logic design involves *two* main challenges. Logic design challenge number one is to get the circuit to work *logically,* on paper. Logic design challenge two, the important one, is to get the circuit to work *electrically,* in the real world. Students (under duress from their professors) tend to concentrate on logic design and forget about electrical design.

Electrical logic circuit design hinges on two fundamental tenets: logic *levels*, and response *time*. Proper logic levels ensure that device outputs correctly drive other devices' inputs. A logical 1 voltage output of one device, for instance, must be a logical 1 voltage input to another device, if the whole system is to function reliably. Furthermore, the output source or sink current of the driving device must be adequate enough to supply the input requirements of driven devices.

Proper timing ensures that every circuit can respond, *in the time allowed*, to system signals. Every logic gate has a response time, usually measured in nanoseconds, from input to output. Every flip-flop has a maximum clocking frequency, usually measured in tens of megahertz, at which the flip-flop can toggle. Semiconductor memories, RAM and ROM, are extremely complex devices that, typically, can read or write data in hundreds of nanoseconds under the control of read, write, addressing, and chip-select signals. The *timing requirements* of all of the previously mentioned circuits must be met if our designs are to be successful.

We shall concern ourselves with simple logic designs for the memory space expansion card, while paying particular attention to timing fundamentals.

PC/ISA Bus Memory Access Timing

A memory expansion bus card is *controlled* by the actions of the PC central processing core. The central core establishes memory system timing requirements based on the clock speed of the CPU and the memory control support circuits. The history of personal computers has featured ever-faster CPU clock speeds, accompanied by the need for ever-faster memory.

There is a trade-off inherent in raising core CPU clock speed to gain processing power. "Speed costs money: How fast do you want to go?" is equally applicable to auto racing and computer design. Higher CPU clock speeds translate into more expense for support chips, *particularly* memory. Logic devices are sold by the Hertz, and faster parts are *more* expensive than slower parts. Not only does increased part speed add cost to integrated circuits, it also adds cost to the circuit board that holds the faster ICs. High-speed circuit boards require more expensive electrical noise suppression techniques in their design and manufacture than do slower speed boards. One trend in containing PC cost is to have the CPU run at one speed, while other, more expensive components run at a slower speed. Dual-speed systems may then use slower, less expensive components on the expansion bus. Many designs now use a *cache* of high-speed RAMs on the local bus to hold code and data loaded into them from slower memory.

The original PC, however, linked expansion bus timing *directly* to CPU clock speed. The 8088 CPU that powered the original PC ran at a clock frequency of 4.77 MHz[1], which is considered slow by today's thinking. Later PC designs would *uncouple* ever-increasing CPU clock frequencies from the expansion bus clock.

---

[1]An inexpensive 14.31 MHz crystal was chosen to supply video frequencies for the CGA color monitor. The 4.77 MHz CPU clock frequency was derived from the crystal frequency.

Although we find very little difference between the signal *names* on the PC bus and the ISA bus, we shall find significant differences between PC bus and ISA bus *waveforms*. We begin with PC bus timing.

PC BUS MEMORY ACCESS TIMING

Figures 11.3a and 11.3b show a standard expansion bus memory read and write cycle for a PC based on an 8088 CPU. The signals that control access to the PC bus are loosely synchronized with the CPU clock. Some liberties have been taken with the waveforms of Figure 11.3 in order to not overly complicate the timing diagram. The essential timing facts are presented in Figure 11.3, and a good design always leaves a comfortable margin of error (say 20%) to take care of time, temperature, and vendor drift.

Figures 11.3a and 11.3b show a PC bus read (from the memory) and write (to the memory) timing. Clearly, a read and a write cannot *both* take place simultaneously. Memory access is controlled by the address and read/write control signals. Data is read from the memory as shown for the read cycle. Data is written to the memory as shown for the write cycle.

A PC bus memory access cycle consists of *four* CPU clock pulses, or *T states*, labeled T1 to T4. A T state is the *period* of time used for one CPU clock pulse and is the *inverse* of the CPU clock frequency. Note that 8088 CPU clocks are *not* square waves but have a 33% duty cycle. A bus memory access cycle proceeds as follows.

- Memory address bits SA0 to SA19 become valid from the bus during T1, at the *rising* edge of the positive pulse.

- The controlling signal, /SMEMR (read) or /SMEMW (write), goes active-low from the bus at the *beginning* of T2 to signal that the bus is ready to transfer data bits SD0 to SD7. /SMEMW and /SMEMR persist until the *end* of T3.

- Memory devices have until the *end* of /SMEMR to drive data bits SD7 to SD0 onto the bus during a read cycle, or until the *end* of /SMEMW to get data from the bus during a write cycle.

- The *rising* edge of /SMEMR signals that the data placed on the bus by the memory has been *latched* into the core during a read cycle. The *rising* edge of /SMEMW is used by the memory to *internally latch* data placed on the bus by the core during a write cycle. Period T4 *ends* the memory access cycle.

The total time, from the time the address lines go valid until /SMEMR or /SMEMW goes high, is known as the memory *access time*. The circuitry on the expansion card must have transferred data to the data bus during a read cycle, or gotten data from the data bus during a write cycle, well *within the access time* established by the PC bus timing standard.

The total time the PC bus *allows* for memory to respond can be considered to be two distinct time specifications, *decoding* time and *read/write* time.

For an 8088-based core, the *total* time allowed for accessing memory after addresses go valid is one third of T1 plus T2 and T3, or 2.33T nanoseconds (ns). The time allowed for memory to respond to a read or write cycle is T2 and T3, or 2T ns.

Decoding times 2.33T and read/write times 2T are the times *established* by the bus design standard. Memory chips, however, are designed to respond to control and address signals in certain *minimum* times. The design challenge is to match the right memory speed to the bus timing requirements.

The most critical response time, the access time for memory chips, is the total time needed to internally decode external memory address lines to a single internal address. Thus, for a PC, the access time for the memory chip must be *no greater* than 2.33T. Keep in mind that the timing diagram shown in Figure 11.3a is valid for signals present *on the bus.* Any circuit *delays inside* (between the edge connector and the memory IC) the expansion memory card will *decrease* the available response time for the memory IC on the card.

As an example of memory timing calculations, consider the following: An 8088 running at 4.77 MHz has a T state of 209 ns. Total memory response time then, including all delays through any other circuits, must be comfortably less than 2.33×209, or 488 ns.

ISA 8-Bit Memory Access Timing

The PC bus, discussed in the previous section, relies on the basic 8088 CPU machine cycle for computing core *and* expansion bus timing. The 8088 CPU requires four CPU clock pulses to complete a machine cycle, thus four CPU clock pulses also define an expansion bus cycle. Computers based on the 8088 progressed from 4.77MHz clock to a 10MHz (NEC V20 CPU) clock and did not exceed that frequency. ISA-bus-based computers, however, began with 6-MHz-clock CPUs and now commonly run at clock frequencies of 66 MHz.

Uncoupling CPU and Expansion Bus Timing

The designers of the ISA bus could not extend PC expansion bus clock practice to the ISA expansion bus because the CPUs used for ISA bus computers have machine cycles that were *too fast* for many existing and planned bus expansion cards. Currently, for instance, full-speed 486 CPUs run at two 50MHz T states, or 40 ns for a bus cycle. Most expansion cards designed for a PC bus cycle of 488 ns cannot function at all running at a bus cycle time of 40 ns.

Not only are current ISA bus CPU machine cycles extremely fast, there are *too many different* ISA bus CPU clock frequencies on which to base a standard. The history of the development of 80286, 80386, 80486, and Pentium CPUs has featured ever increasing clock frequencies that have grown from 6MHz to 66MHz. Ever faster clock speeds, if applied directly to ISA bus timing, would mean that expansion cards designed for slower speed busses would *not* work on the latest, higher speed bus.

Decreased machine cycle time, for CPUs beyond the 8088, resulted from a *halving* of the number of clock cycles needed in a machine cycle. The 80286 requires only two CPU clock cycles per machine cycle versus four for the 8088. The first clock time of an 80286 machine cycle is used to address memory, and the second clock accesses the memory. The original 80286 runs at a clock speed of 6MHz, or 166 ns per clock pulse, which approached the access time of the fastest DRAMs then available. Later versions of the 80286 feature clock speeds up to 25MHz. Access times for 80286 CPUs range from 166 to 40 ns. 80386 CPUs, which also use two clocks per machine cycle, fea-

Figure 11.3 ■ PC and ISA bus read and write cycles.

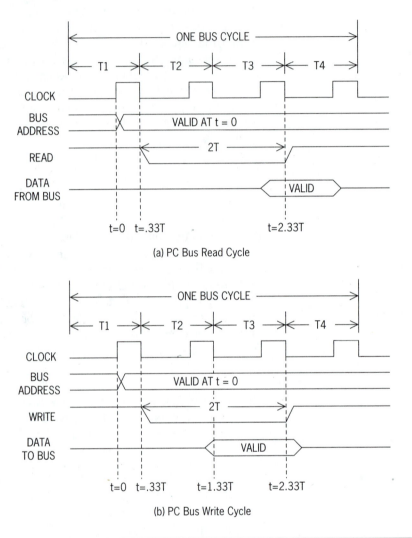

(a) PC Bus Read Cycle

(b) PC Bus Write Cycle

ture clock frequencies of 16 to 40 MHz, which require memory access times of 62 to 25 ns. Processors such as the 80486 and Pentium may run at 100 MHz in the near future, requiring access times of 10 ns.

All of the access times listed for the high-speed processors are *beyond* the capabilities of contemporary inexpensive dynamic RAMs, which need, typically, access times of 80 to 100 ns. Special memory design techniques, such as interleaved memory, can be employed to adapt slow DRAM to fast CPUs, at additional system cost. Because of the special techniques needed to interleave memory, it became clear to designers of high-speed computers that *primary* high-speed system memory would not be expanded to the 8-bit expansion bus as was the case for PC expansion memory. As a result, primary system memory for contemporary ISA bus computers is *part of* the central processing core, where fast local busses and interleaving techniques can employ slow DRAM and very fast SRAM memory.

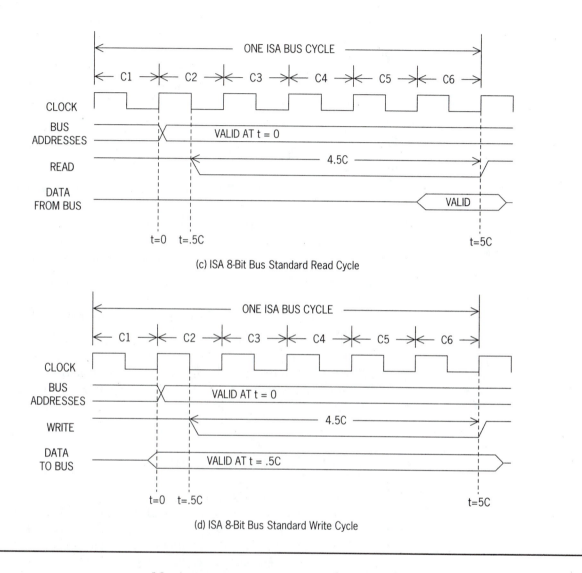

(c) ISA 8-Bit Bus Standard Read Cycle

(d) ISA 8-Bit Bus Standard Write Cycle

Moving system memory to the central processing core, and the great diversity in ISA bus computer CPU clock speeds, has resulted in the adoption of a *standard clock frequency* of 8MHz for the ISA bus. Thus, no matter what the core CPU clock speed happens to be, whether 8MHz or 66Mhz, the ISA bus speed is 8MHz. Standardizing the ISA bus clock greatly simplifies timing problems when designing expansion cards for the ISA bus, because the card designer knows the clock specifications for any CPU-type ISA bus.

ISA BUS 8-BIT STANDARD BUS CYCLE

The ISA bus features three types of memory access cycles: *standard* (six SYSCLK), *ready* (four SYSCLK), and *no-wait* state (three SYSCLK). Two of the ISA bus memory cycles, ready and no-wait state, depend on the bus expansion card asserting one or more of the speed-up control signals, IOCHRDY or /SRDY. If the expansion card does not *request* a faster cycle, then the ISA bus will use the standard memory cycle. We shall examine the standard ISA bus memory cycle, which is shown in Figures 11.3c and 11.3d.

A *standard* 8-bit ISA memory access cycle takes six 8MHz SYSCLK pulses, or a total of 725 ns per bus cycle. (It is interesting to note that the ISA bus standard cycle length of 725 ns is only slightly shorter than the total 872 ns bus cycle of the original 4.77MHz PC.) Of course, the ISA bus cycle can be made as short as 375 ns by using a no-wait state bus cycle, rendering the ISA bus more than *twice* as fast as the PC bus. Much faster data transfer rates are possible on the bus using *direct memory access* (DMA) techniques, but DMA is beyond the scope of this book.

The six SYSCLK pulse standard ISA memory access cycle proceeds as follows.

- Address lines SA0 to SA19 go valid no later than the *end* of clock one.
- Access command (/SMEMR or /SMEMW) goes active low in the *center* of clock two.
- Write cycle data is valid on the bus at the *center* of clock two.
- The /SMEMR or /SMEMW access command goes high at the *end* of clock six.
- Expansion card circuits place data on the bus (read cycle) or latch data from the bus (write cycle) before the *end* of clock six.

The ISA 8-bit memory access standard cycle is a very conservative slow-speed bus cycle. Memory circuits have a minimum of five SYSCLK pulses (625 ns) to respond, which is well within the response time of very inexpensive contemporary memory devices.

11.4 Static Memory Families

The memory board that is designed in this text uses a *static* RAM IC, not a DRAM. Static memory devices, such as SRAM and EPROM, are often used in small memory designs because of their *simplicity* and *speed*. Static RAM is *not* an economical choice for large memories, such as those found in most personal computers, because of its higher cost. Dynamic RAM generally becomes the most cost-effective choice for memory capacities above 64K. For our purposes, however, static RAM lets us investigate the operation of the PC or ISA busses without the distractions of DRAM refresh cycles.

SRAM and EPROM Families

Static memories, both SRAM and EPROM, have evolved over the years as a family of chips with *standard* pin numbers for address bits, data bits, and control signals. As the capacity of the family grew, additional address pins had to be added to the chip package. But, as pins were added to the chip, the original pin assignments of earlier family members were retained.

The orderly growth of static memory address capability has resulted in pin-signal commonality from the earliest 2K family members to 512K family members. Table 11.5 shows the pins and signal assignments for several static

Type: Size:	6264 8Kx8	62256 32Kx8	Type: Size:	6264 8Kx8	62256 32Kx8
Socket Pin	Signal		Socket Pin	Signal	
1	NC	A14	28	Vcc	Vcc
2	A12	A12	27	/W	/W
3	A7	A7	26	E2	A13
4	A6	A6	25	A8	A8
5	A5	A5	24	A9	A9
6	A4	A4	23	A11	A11
7	A3	A3	22	/RD	/RD
8	A2	A2	21	A10	A10
9	A1	A1	20	/E1	/E
10	A0	A0	19	D7	D7
11	D0	D0	18	D6	D6
12	D1	D1	17	D5	D5
13	D2	D2	16	D4	D4
14	GND	GND	15	D3	D3

Table 11.5a ■ 8K and 32K 28-Pin Static RAM Pin Assignments

Type: Size:		2716 2Kx8	2764 8Kx8	Type: Size:		2716 2Kx8	2764 8Kx8
Socket Pin	Pin	Signal		Socket Pin	Pin	Signal	
1	—	*	Vcc	28	—	*	Vcc
2	—	*	A12	27	—	*	Vcc
3	1	A7	A7	26	24	Vcc	NC
4	2	A6	A6	25	23	A8	A8
5	3	A5	A5	24	22	A9	A9
6	4	A4	A4	23	21	Vcc	A11
7	5	A3	A3	22	20	/RD	/RD
8	6	A2	A2	21	19	A10	A10
9	7	A1	A1	20	18	/E	/E
10	8	A0	A0	19	17	D7	D7
11	9	D0	D0	18	16	D6	D6
12	10	D1	D1	17	15	D5	D5
13	11	D2	D2	16	14	D4	D4
14	12	GND	GND	15	13	D3	D3

*Note: 2716 is a 24-pin device aligned at the bottom of the socket.

Table 11.5b ■ 2K and 8K 28-Pin EPROM Pin Assignments

RAM and EPROM family members. Table 11.5a shows the pin assignments for two popular static RAMs, the 6264 8K and the 62256 32K parts. Table 11.5b lists the pin assignments for two EPROM parts, the 2716 2K and 2764 8K models. We shall use an 8K SRAM and a 2K EPROM for bus expansion card memory needs.

Commonality among static memory family members allows a system designer to place a 28-pin socket on a board and use jumpers to let any family member, from 2K to 64K, be placed in the socket. Jumpers can be inserted or deleted by the final user to match the type of memory placed in the 28-pin socket to the user's exact memory requirements. SRAM and EPROM can also be placed in the same socket, with appropriate jumpering of the control signals. If more memory is needed, SRAM and EPROM parts with capacities up to 1M may be used by equipping a bus expansion board with 32-pin sockets.

Static RAM Operation

As listed in Table 11.5a, a static RAM has three sets of *input* pins and one set of *bidirectional* data pins. The input pins are grouped into address, enable, and control sets. The bidirectional pin set consists of the data pins, D0 to D7. The enable and control pins may be active-high or active-low.

The address pins define the size, or number of discrete addresses, of the SRAM. Common static RAM sizes are 2K (A0–A10), 8K (A0–A12), and 32K (A0–A14). The CPU can access any memory location in the RAM by setting the bus address lines to the desired address number.

Depending on the RAM type, there can be one or more chip enable (CE) pins, and they may be active-high or active-low. The enable pins serve to select a *specific* SRAM chip from among the many chips that may make up a physical memory. For instance, if a 64K memory system is made up of eight, 8K memory ICs, then 8 chip enable signals will be needed to individually select one of the memory chips. One addressing scheme for 8K SRAMs is to connect addresses A0 to A12 *directly* to each 8K memory chip, and address lines A13 to A15 are *decoded* to provide eight /CE signals.

The control pins are the read or output enable (OE) pin and the write (WR) pin. The SRAM will read internal data *to* the bus *during* /RD signal, or latch data *from* the bus when /WR goes low to high.

Static RAM Read Cycle

Figure 11.4a shows the timing involved in reading data from a static RAM. The times shown in Figure 11.4a are from the *standpoint of the SRAM*. As memory system designers, we are concerned with the time it will take the static RAM to place valid data on its data pins from the beginning of a read cycle. The RAM read cycle proceeds as follows.

- Addresses are *valid* at t = 0. Valid output data from the SRAM can be available *no sooner* than time T1 after the addresses go valid. The lower address lines may be connected directly to a memory IC. The upper address lines must be decoded to provide the individual chip enable signal, /CE.

- After some upper address decoding delay, the chip enable signal, /CE, is produced by *decoding* logic. Valid output data from the SRAM can be available *no sooner* than time T2 *after* /CE is asserted.

- The read signal, /OE, occurs after allowing time for the decoding logic to produce the /CE signal. Data from the SRAM will not be valid until a *minimum* time of T3 has elapsed *after* /OE is asserted low.

Figure 11.4 ■ Static RAM read and write cycles.

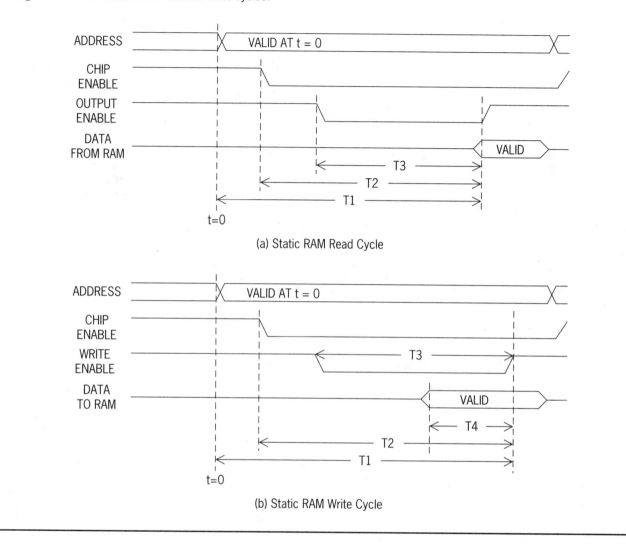

(a) Static RAM Read Cycle

(b) Static RAM Write Cycle

■ A data byte that has been selected by the RAM address lines on the
SRAM chip that is enabled will appear at the data pins of the SRAM
no sooner than T1, T2, or T3, *whichever is latest.*

It is very important to note that all of the times T1, T2, and T3 are *mini-mum* times. The SRAM *cannot* physically respond faster than the times T1,
T2, or T3 after the assertion of the signals that begins the particular timing
action. *All* of the SRAM read timing requirements must be met. For example,
it does *no* good to meet the timing requirements for T1 and T2, but fail to
meet the requirement set by time T3.

EXAMPLE READ TIMING FOR A PC BUS MEMORY EXPANSION CARD

Figure 11.5 shows a block diagram of a PC bus memory expansion card.
The card blocks are arranged as follows:

a. Interface buffers between the bus pins and the card circuits. The assumed delay through the buffers is 20 ns.

b. Address decoding circuits. The assumed delay through the decoding circuit is 50 ns.

c. Memory circuit. The assumed read response times of the memory are: T1, T2 times are 300ns; T3 is 100 ns.

We wish to know how much "safety margin" exists for the PC bus expansion card of Figure 11.5.

Bus access to an expansion card memory location proceeds in two general steps, as listed next. Note that the timing is from the *standpoint* of the bus and is *controlled* by the bus.

1. The desired address is placed on the address lines by the bus. When the address signals have *settled*, they are said to be *valid*. Settling refers to the time it takes for a signal, or group of signals, to become *electrically* stable. The time that the addresses become valid is the

Figure 11.5 ■ Memory card block diagram.

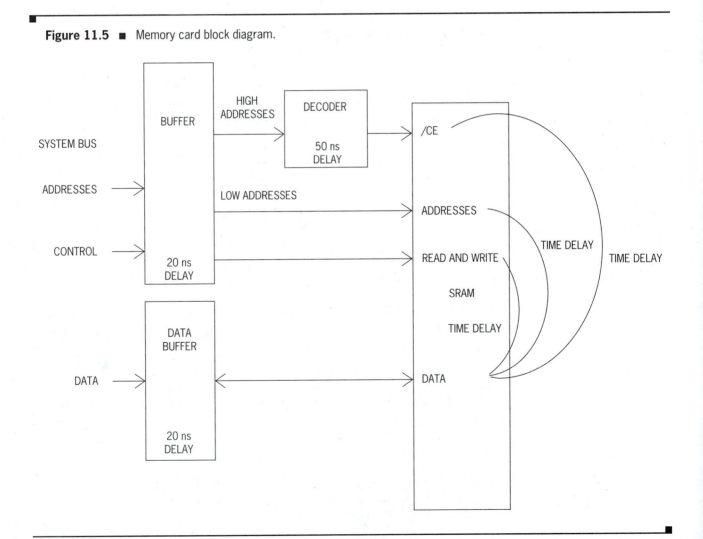

beginning, or time equals 0, of any access cycle. *All times are calculated from t = 0*, the time at which the addresses are *valid* on the *bus* pins.

2. A read or a write signal is issued by the bus. A read signal reads data *from* the RAM *to* the bus. A write signal writes data *from* the bus *to* the RAM.

The expansion card must respond to the bus address, data, and control signals *in the time allowed by the bus*. It is important to remember that the *bus* determines what time it is, not the expansion card. The second point to note is that all the circuits that stand between the bus pins and the memory chip on the card *reduce the time available* for the memory chip to respond. The bus times for memory access are then:

Time from address valid until /SMEMR high

2.33T = 487 ns

Time after addresses valid to /SMEMR low

.33T = 69 ns

Time /SMEMR is low

2T = 418 ns

The card *internal* memory times are less than the bus times because of the buffer and the decoding circuit *delays*. The times available to the SRAM on the expansion card, less card time delays are calculated as follows.

■ The addresses go valid at the SRAM address pins one buffer delay after they are valid on the bus. Data from the SRAM pins reaches the bus pins after one buffer delay. Card T1 time is:

T1 = 2.33T – two buffer delays
T1 = 487 – 40 = 447 ns

■ The /CE signal goes valid one buffer delay and the decoder delay after addresses are valid on the bus. Data from the SRAM reaches the bus after a buffer delay. Card T2 time is:

T2 = 2.33T – two buffer delays – decoding delay
T2 = 487 – 40 – 50 = 397 ns

■ The /RD signal at the SRAM pin goes active-low one buffer delay after it is active-low on the bus. Data from the SRAM reaches the bus after a buffer delay. Card T3 time is:

T3 = 2T – two buffer delays
T3 = 418 – 40 = 378 ns

Note that there are *two* buffer delays that reduce the available bus times. Addresses from the bus lose time as they pass through the address buffer. Data from memory loses time going back to the bus through the data buffer. For our example static RAM we have the following safety margins:

Time	Available on Card	SRAM Specification	Margin
T1	447 ns	300 ns	+117 ns
T2	397 ns	300 ns	+ 97 ns
T3	378 ns	100 ns	+278 ns

The safety margins are adequate. The example points out that for most real-world SRAMs, T1 and T2 are the same. T3 is normally much shorter than T1 or T2. Timing specification T2, because decoding delays must be subtracted, usually determines how much time is available to the SRAM.

Decoding delay is determined by the memory system logic *designer*, and it varies according to the number of gates used and the delay through each gate. Memory logic designs that employ decoding circuits with long delays will slow down the memory access cycles because of the delay from addresses valid to /CE through the address decoder. Decoding is a combinatorial logic function, and thus can be done with a minimum of two gate delays using a sum of products design approach. Once the decoder design is reduced to two gate delays, then choosing a fast logic family becomes the only way to speed up decoding action. As a last resort, the decoder can be *eliminated* by using a single memory IC as large as the *entire* bus addressing space.

EXAMPLE WRITE TIMING FOR A PC BUS MEMORY EXPANSION CARD

Figure 11.4b shows the timing of a cycle that writes data *from* the bus *to* the static RAM. Write cycles are somewhat more complicated than read cycles for static RAM because a write can be controlled by *either* the write-control signals /CE *or* /WR.

A write begins after the *last* write-control signal goes active, and ends with the *first* write-control signal to go inactive. If /WR goes low first, then the write cycle begins when /CE goes low. If /CE goes low first, the write cycle begins when /WR goes low. The write cycle ends whenever /CE or /WR goes high, whichever is *first*.

Reliable operation requires that we choose one signal to control the write cycle, either /CE or /WR. Figure 11.4b shows the latter case, called a *write-controlled* write cycle, which is the type most commonly supported by expansion bus memory cycle timing. A write-controlled cycle is one in which /WR goes low *after* /CE to start the write cycle, and goes high *before* /CE to stop the write cycle.

A static RAM write-controlled cycle proceeds as follows.

■ Addresses are valid at t = 0. Data cannot be written into the SRAM *sooner* than time T1.

■ /CE goes low, after a decoding delay. Data may not be written to the SRAM before T2 nanoseconds.

■ The /WR write signal must be active for at *least* time T3 to enable the SRAM to write data.

■ Valid data must be present on the data pins of the SRAM at *least* setup time T4 *before* /WR goes *high*. When /WR goes high, data is latched from the data pins *into* the RAM address specified by the address lines.

As was the case for a read cycle, all of the times listed are *minimums*, and all must be met. A /CE decoding circuit, for instance, must produce an active /CE signal *before* /WR goes active-low.

Assume our example SRAM has the following write cycle timing specifications: T1 = 300ns; T2 = 300 ns; T3 = 100 ns; T4 = 20 ns.

We shall perform write time calculations using an alternative method to that used for the read time calculations. We shall find the absolute time for all card internal events relative to t = 0 on the bus. Card SRAM write times may then be found from the absolute event times.

■ Addresses are valid on the SRAM pins after one buffer delay.

    Addresses valid = 20 ns

■ /CE is valid after one buffer and the decoding delay.

    /CE active = 20 + 50 = 70ns

■ /WR is active-low at the SRAM .33T plus one buffer delay.

    /WR active = 69 + 20 = 89 ns

■ Data is valid at the SRAM 1.33T plus one buffer delay.

    Data valid at 278 + 20 = 298 ns

■ /WR goes high at 2.33T plus one buffer delay

    /WR high = 487 + 20 = 507 ns

From the absolute times we may find the card's SRAM times.

    T1 = /WR high – addresses valid
    T1 = 507 – 20 = 487 ns
    T2 = /WR high – /CE active
    T2 = 507 – 70 = 437ns

    T3 = /WR high – /WR low
    T3 = 507 – 89 = 418 ns

    T4 = /WR high – data valid
    T4 = 507 – 298 = 209 ns

We see that the cycle is write controlled because /CE is active-low 19 ns before /WR. Note that buffer delays do not affect write times. Because no data has to come out of the card during a write cycle, internal delays simply delay the beginning *and end* of a pulse, while not changing its duration. The safety margins for a write cycle are shown as follows.

Time	Available on Card	SRAM Specification	Margin
T1	487 ns	300 ns	+187 ns
T2	437 ns	300 ns	+137 ns
T3	418 ns	100 ns	+318 ns
T4	209 ns	20 ns	+189 ns

Write cycle safety margins are adequate, and greater than read cycle safety margins. Read cycles are generally the most stringent because card internal delays reduce the time available to the SRAM.

SRAM read and write timing requirements are *minimum worst case* times. It is perfectly all right, and a good thing, if *more* time is used to read and write data to the SRAM than the absolute minimums given by the manufacturer. If the SRAM is not given at least the minimum times specified to do its job, then the circuit will not work reliably.

BUS CYCLES AND RAM SPEED

The time needed by a SRAM to read or write data after the RAM address becomes valid (T1) is called the *access* time of the RAM in manufacturer literature. From our previous example we know that the "real" access time of the RAM is *actually* T2 plus decoding delays for /CE to be asserted. Published RAM access times are equal to real access times only if the /CE delay is 0 ns, a very unlikely prospect. But, manufacturers like to place the best possible times on their memories, because memory is priced by access time. Very fast RAM, with very short access times, is more expensive than slower RAM. Good commercial design practice dictates that RAM access time be fast enough to meet bus read and write cycle requirements with at least a 10% to 20% safety margin.

Note that the bus determines the timing *requirements* for the RAM, not the reverse. If the bus speed is such that the chosen RAM will not operate reliably, then either the bus must be *slowed* down or a *faster* RAM chosen. Speed costs money.

EPROM Operation

An EPROM is a read-only nonvolatile memory that operates in a manner *similar* to that of a static RAM during a read cycle. Table 11.5 shows that EPROM pin assignments are equivalent to static RAM family members of the same memory capacity. The only pin signal differences, between a RAM and an EPROM of the same capacity, is that the /W pin of the SRAM is connected to +5VDC on *the* EPROM when the EPROM is *in* the circuit. At other times, the +5VDC pin is used by EPROM programmers to *program* the EPROM.

EPROMs, of the same technical generation as equivalent static RAMs, are generally *slower* in their response times because of the nature of the EPROM floating gate technology. In all other respects they have the same read timing nomenclature, T1, T2, and T3, as a static RAM.

Bus Buffering

Any logic IC that has *low input* current requirements and *high output* current drive ability is said to be a *buffer*. Buffers are said to be capable of not *loading* the bus with their low input current requirements, and capable of *driving* the bus with high output current abilities. Not loading the bus means that the buffer requires very little input current from the bus. Driving the bus means the buffer can supply significant output current to the bus. How much is a "little" or "significant" depends on the logic family standards employed. ISA bus logic levels are defined to be TTL compatible. A "little" TTL current is anything less than one standard load, or 20 to 40 microamps. Significant TTL current is in the 20 to 40 milliamp range.

We shall exceed TTL standards, however, and buffer signals to and from the bus to the expansion card internal circuits with HCT logic family buffers. HCT is the acronym for *high-speed, CMOS, TTL-compatible* logic. HCT

Voh	Vol	Ioh	Iol	Iih	Iil	tpd	Toggle Rate	Icc	Icc(max)
2.4V	.8V	35/25mA	35/25mA	1μA	1μA	10ns	25Mhz	5μA	70/50mA

Table 11.6 ■ Generic HCT Family Specifications

parts are useful because of their low input drive current and high output drive current abilities.

To aid in standardizing our logic designs, a generic HCT family specification for gates and buffers is shown in Table 11.6. Output currents separated by a slash (/) denote buffer/gate driving currents. Actual data books should be *consulted* for designs that are critically dependent on drive currents and timing.

Inspection of Table 11.6 reveals that very little input current (1μA) is required from the bus by an HCT circuit, and that considerable drive current (35mA) is available from the HCT output to drive the bus. Moreover, very little supply current (5μA) is required to power CMOS logic when it is static. Supply current (Icc) will rise as CMOS logic switches, or supplies heavy driving currents, but rarely exceeds 70 milliamps per chip. Using HCT logic ensures that our expansion card will barely load the bus system signal driving circuits, and that our card can supply adequate signal drive power to the bus system.

For ease of timing calculations, the delay specification for any HCT device is considered to be 10 nanoseconds.

11.5 A Memory Expansion Card Circuit Design

For the sake of illustration, a bus expansion card that adds 8K of static RAM to the system is designed in this section.

PC Bus Expansion Card Memory Addresses

The *first* design issue to be resolved is to determine what memory space has been *allocated* by the ISA standard for expansion cards. Expansion cards must be placed in memory space that has not been reserved for the use of other cards *or* for system memory.

Figure D.1 of Appendix D shows the overall memory map for a 1M PC-based DOS computer. Inspection of Figure D.1 reveals that a large block of memory has been set aside for video memory buffers and special equipment BIOS use. Table 11.7 lists the PC memory space addresses that have been set aside for these functions.

If your system does *not* use one of the addresses shown in Table 11.7, then the memory space *can* be used for a memory expansion card. Based on Table 11.7, expansion RAM from addresses A0000h to A1FFFh is chosen for the bus expansion card memory space.

The *next* design issue is to decide how large the expansion card memory is to be. A memory size of 8K (a 6264 type IC) is chosen for this example.

Memory Address From	To	Kilobytes	Use
A0000h	BFFFFh	128	EGA/VGA video buffer
B0000h	B7FFFh	32	Monochrome buffer
B8000h	BFFFFh	32	CGA buffer
C0000h	C3FFFH	16	EGA BIOS
C4000h	C5FFFh	8	Free
C6000h	C63FFh	1	PGA
C6400h	C7FFFh	7	Free
C8000h	CBFFFh	16	XT hard drive BIOS
CC000h	CFFFFh	16	Free
D0000h	D7FFFh	32	Cluster adapter BIOS
D8000h	DFFFFh	32	Free

Table 11.7 ■ PC Video Memory Map

Memory Expansion Card Design

Based on a memory address starting at A0000h, and 8K in size, the circuit of Figure 11.6 shows an 8K static RAM added to the system using a bus expansion card. The design features these major considerations:

■ The 8K RAM is chosen to be an 8K static CMOS RAM of the 6264 family. Access time for the RAM is determined in the next section.

■ Buffering, into the card and out to the bus, is done using 74HCT245 bus *transceivers*. A transceiver is a device that can pass data in two directions. Although we do not need a transceiver to buffer the address lines (they are unidirectional), it costs no more to make the same part perform several functions. Note that the signal names *change*, as they pass through the buffers, from ISA to circuit names.

■ Address decoding is done using 74HCT00 NAND gates and 74HCT32 OR gates. A more contemporary circuit might use programmable logic, not discrete logic, to perform the address function. Decoding is done after reflecting on the following address decoding considerations:

The 8K SRAM decodes the *lower* 13 address bits to access a particular address in the 8K address space. The upper 7 address lines must be *decoded* in order to select the 8K RAM for address space A0000h to A1FFFh as follows:

$$\begin{array}{ll} \text{A0000h} = \underline{1010000} & 0000000000000\text{b} \\ \text{A1FFFh} = \underline{1010000} & \underline{1111111111111\text{b}} \\ \quad\quad\text{Decoder} & \text{Decoded by 8K RAM} \end{array}$$

■ The design uses OR gates as negative-logic AND gates to detect the 0 bits, and NAND gates to detect the 1 bits. The final /CE output signal, /RAMEN, from the address decoding circuit to the 8K RAM is active-low.

■ The SRAM memory *must* remain tri-stated until the decoding circuit detects a valid address *and* /SMEMR or /SMEMW is active-low.

Figure 11.6 ■ SRAM memory expansion card.

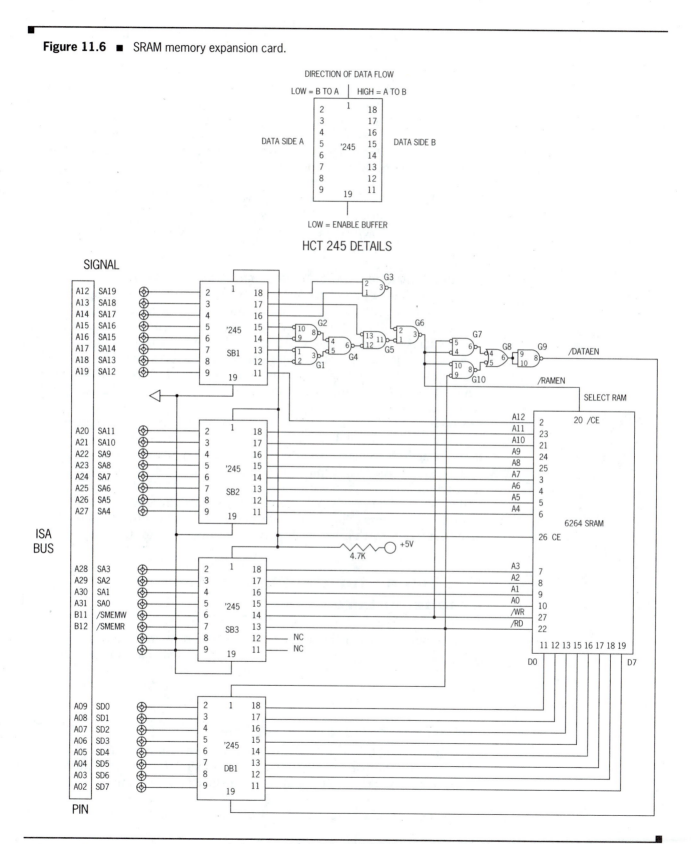

The 8K RAM chip internally ANDs the address valid signal /RAMEN and the read /RD or write /WR signal to ensure that data is not placed on the system data bus at the wrong instant.

■ Bus address and control signals are passed to the remainder of the board circuits by permanently enabling buffers SB1, SB2, and SB3 to pass signals from the bus to the remaining internal circuits. Bus data signals are passed in both directions by controlling the *direction* of data flow through data buffer DB1 using the /RD signal as a direction control. Write operations, when /RD is high, enable data from the bus to pass to the memory. Read operations, when /RD is low, enable data flow from the RAM to the bus. Buffer DB1 remains tri-stated until the /DATAEN signal arrives from the gate circuitry. An active /DATAEN signifies that a valid address has been decoded *and* a read or write cycle is in progress.

MEMORY EXPANSION CARD ADDRESS AND CONTROL DECODING

Positive logic OR and NAND gates are used to decode the upper 7 address bits, A13 to A19. OR gates G1 and G2 are used as negative logic AND gates to detect the 0 bits of the address. NAND gate G3 is used to detect the 1 bits of the address. OR gates G4, G5, and G6 then AND the active-low signals to produce an output, /RAMEN, that enables the RAM. The final decoded address signal (/RAMEN) is active-low to match the *requirements* of the static RAM chip enable input. The RAM internally ANDs the /RD read or /WR write strobes with the address signal /RAMEN to ensure that data is processed only when the address is valid and a read or write is in progress.

To ensure that the data buffer DB1 doesn't react to invalid addresses, the decoded address signal /RAMEN is also ANDed with /RD in gate G7 and /WR in gate G10. The outputs of G7 and G10 are then ORed in gate G8, and inverted in gate G9, producing signal /DATAEN. /DATAEN enables the data buffer to pass data *only* when the memory card address is valid and a read or write is in progress. The data buffer is disabled at all other times so that it cannot *interfere* with other devices that are connected to the bus.

11.6 Timing for the PC Bus Memory Expansion Card

It is possible, even though the bus expansion memory card is *wired* perfectly, that it will not work reliably when tested. The reason it may not work is that the *access time* of the SRAM IC is not fast enough to keep up with the bus.

Typical SRAM and EPROM Access Times

Table 11.8 shows read and write T times for common static RAMs, and read times for equivalent EPROMs. The numbers of Table 11.8 may be used to compare how the PC bus read times of the example compare with equivalent T times for contemporary SRAM. Remember, as long as the SRAM speeds are *less* than those provided by the PC bus, all is well. Should the

SRAM Access Time	Read Cycle			Write Cycle			
(ns)	T1	T2	T3	T1	T2	T3	T4
150	150	150	70	100	100	90	60
120	120	120	60	85	85	70	50
100	100	100	50	80	80	70	40
EPROM Access Time							
(ns)							
250	250	250	100				
200	200	200	75				
180	180	180	65				

Table 11.8 ■ Static Memory Read and Write Times

SRAM speed exceed the times allowed by the bus, a faster SRAM must be found.

Memory Expansion Card Read Cycle Time Calculations

All memory expansion card control and data signals originate from the bus. The bus speed will determine the *maximum* time that the card can take to respond. Any card SRAM chosen must be fast enough to be able to function in *less* time than the maximum allowed by the bus.

From our previous discussion of static RAM read timing (Figure 11.4a), we find that the time needed to read the SRAM is dependent on read times T1, T2, and T3. T2 usually is the most stringent requirement. We must choose a SRAM with timing specifications that are *less* than the times allowed by the bus.

For example, the following bus timing results for times T1, T2, and T3 are calculated for the expansion card inserted into a PC bus. We shall use delay (and absolute time calculations in parentheses) to reach the same results.

■ Time T1, the total access time allowed by the bus from addresses valid until data on the bus pins, for a read cycle, is the PC bus address valid time less two buffer delays.

T1 = 2.33T – (2 × 10) ns
(addresses go valid at the SRAM at 10 ns)

■ Time T2, the time available to the SRAM from /CE to data out, is time T1 *less* any decoding delays in the decoding circuit. Inspection of Figure 11.6 shows that there are four gate delays (40 ns) involved in generating the chip enable signal /RAMEN.

T2 = T1 – 40 ns = 2.33T – 60 ns
(/CE goes active at 50 ns)

■ Time T3 is the SRAM output enable time from /RD active to data ready at the SRAM pins. The /RD signal is generated by the bus

/SMEMR read signal, which is 2T ns in duration, and begins .33T states after the addresses go valid on the bus. Signal /SMEMR is delayed by 10 ns in buffer SB3, and data from the SRAM pins to the bus is delayed by an additional 10 ns in buffer DB1. The time available to the SRAM is thus /SMEMR less the two buffer delays:

T3 = 2T − 20 ns
(/RD goes active at .33T + 10 ns)

There is an additional circuit delay, DB1 enable, that has nothing to do with the SRAM, yet serves to limit bus speed. The data buffer, DB1, will not be enabled until 10 ns after /DATAEN is valid. /DATAEN is valid one buffer and three gate delays (G7, G8, G9) after /SMEMR or /CE, whichever is *latest*. The bus read /SMEMR goes active .33T into a read cycle. If the decoding delay is shorter than .33T, then /DATEN is waiting for the /SMEMR signal. If /SMEMR is faster than the decoding delay, then /DATAEN is generated after all of the buffer and decoding delays. The time at which DB1 becomes enabled is:

DB1 = .33T + 50 ns (/CE valid before /RD)
DB1 = 90 ns (/RD valid before /CE)

Data from the SRAM pins, even if available immediately after /RD goes active (if T3 were = 0), must wait for DB1 to be enabled. If DB1 is enabled in *less* time than T3 for the SRAM, then SRAM data is not held up by the delay to enable DB1. If T3 is less than the time to enable DB1, then the DB1 delay will limit the response time of the memory card.

SRAM READ TIMING SPECIFICATIONS FOR THE PC BUS

If a PC bus clock frequency is set to 4.77MHz, then the T states last for 209 ns. The bus allows the following *maximum* SRAM read times as follows:

T1 = (2.33 × 209) − 20 = 467 ns
(or T1 = Data at SRAM pins − addresses valid = 477 − 10 = 467 ns)

T2 = T1 − 40 = 427 ns
(or T2 = Data at SRAM pins − /CE = 477 − 50 = 427 ns)

T3 = (2 × 209) − 20 ns = 398 ns
(or, T3 = Data at SRAM pins − /RD active = 477 − [69 + 10] = 398 ns)

DB1 enabled = 40 ns after /RD active = [69 + 10] + 40 = 119 ns

T1, T2, and T3 are the *maximum* time requirements *allowed by the bus* for the SRAM to be read reliably. If the SRAM can read faster than the bus-established times, then the circuit is very reliable. The delay associated with enabling DB1 is not practically important at a bus frequency of 4.77MHz, because T3 delay is much longer. DB1 enabling time could become a limiting factor if bus speeds are increased.

As can be seen from an inspection of Table 11.8, time T2 is the most stringent read time, because T3 is usually half of T1 and T2. All of the minimum SRAM read times shown in Table 11.8 are *much* faster than required by a 4.77 MHz PC bus. Any speed SRAM from the group shown in Table 11.8

will function well during a PC bus read cycle. Economics usually dictates that the *slowest*, thus the least expensive, 150-ns SRAM be chosen.

For a 150-ns SRAM from Table 11.8, the design enjoys the following safety margins:

Time	PC Bus Time Allowed	Time Needed by SRAM	Margin
T1	467	150	+317
T2	427	150	+277
T3	398	70	+320
DB1 enabled	398 (T3)	119 (circuit)	+279

Note that a 150-ns SRAM can have data at its data pins 80 ns after /SMEMR while DB1 is enabled 50 ns after /SMEMR. DB1 is enabled a comfortable 30 ns ahead of data from the 150-ns SRAM. Furthermore, data from the SRAM will take 10 ns to propagate through DB1. Should a faster SRAM be used, say the 100-ns part, then DB1 is enabled at the same instant that data is available from the SRAM. Picking a SRAM faster than the 100-ns part will not shorten the read time of the memory card. Data cannot be placed on the data bus faster than 50 ns after /SMEMR goes active.

EFFECTS OF INCREASING BUS CLOCK FREQUENCY

As bus clock speeds rise, the expansion card design begins to experience more stringent SRAM access time requirements. A CPU speed of 12MHz (83.3 ns period), for instance, generates these read times on the PC bus:

$$T1 = (2.33 \times 83.3) - 20 = 174 \text{ ns}$$
$$T2 = 174 - 40 = 134 \text{ns}$$
$$T3 = (2 \times 83.3) - 20 = 146 \text{ns}$$

Speeds of 12 MHz and above on the PC bus will eliminate the 150-ns SRAM from consideration for read cycles. The question then arises: How fast can the PC bus run before the fastest SRAM of Table 11.8 has a time margin of zero? To answer this question, it is only necessary to take the T times of the fastest (100-ns) SRAM of Table 11.8 and equate it to the bus T times. The unknown PC bus period or frequency may then be found. The highest PC bus frequency possible is then the frequency at which one of the T times can exactly be met. For example, using a 100-ns SRAM, we have:

$$T1 = (2.33 \times T) - 20 = 100 \text{ ns} \quad T = 51.5 \text{ ns} \quad f = 19.42 \text{ MHz}$$
$$T2 = (2.33 \times T) - 60 = 100 \text{ ns} \quad T = 68.7 \text{ ns} \quad f = 14.56 \text{ MHz}$$
$$T3 = 2T - 20 = 50 \text{ ns} \quad T = 35 \text{ ns} \quad f = 28.57 \text{ MHz}$$

Thus, a frequency of 14.56 MHz is the fastest PC bus frequency possible at which the 100-ns SRAM timing specification T2 can *just* be met.

Note that if all of the SRAM times are zero, then the DB1 enable time will limit bus frequency. Assuming a perfect SRAM, the bus frequency becomes limited by the total DB1 enable delay time of 90 ns. The limiting frequency is:

$$2.33T = 90 \text{ ns} \quad T = 38.6 \text{ ns} \quad f = 25.88 \text{ MHz}$$

Driving the bus beyond 25.88 Mhz would require that DB1 be enabled in less than 90 ns from the start of a bus read cycle.

Memory Expansion Card Write Cycle Time Calculations

The SRAM speed requirements for a write cycle must also be *checked* for reliable operation. Write cycle times are faster than read cycle times, as an inspection of Table 11.8 shows. Moreover, many card delay times do not penalize write cycles. During a write, addresses and data go into the expansion card, but no data needs to comes *out* of the expansion card. Signals that are delayed on the *leading* edge of the signal are also delayed a similar amount on the *trailing* edge of the signal. Pulses that are *X* ns long on the bus are also *X* ns inside the card, but are delayed by any buffer delay times.

There are four SRAM write cycle times of interest, as shown in Figure 11.4b.

Time T1, the time from address valid to the end of the /WR Pulse at the SRAM, is:

$$T1 = 2.33T$$

One should wonder what became of the buffer delays for write cycle time T1. The reason the buffer delays have vanished is that the delay for valid addresses and the delay for the /SMEMW pulse are the same. The difference in time between the two then becomes:

Bus: T1 = /SMEMW high − Addresses valid = 2.33T
Card: T1 = (/SMEMW high + 10 ns) − (Addresses valid + 10 ns) = 2.33T

T1 is the same, on the bus and *inside* the card. Because all delays through the buffers are equal for all signals, the SRAM just gets all of the signals simply delayed by the buffer delay.

Time T2, the time from /CE to /WR high, is:

T2 = /SMEMW high − /CE = (2.33T + 10) − 50 = 2.33T − 40 ns

Again, there are no buffer delays because the difference, T2, is the difference between two signals that are delayed equally in time. The decoding delays, however, remain. Note that the /CE signal *must* go valid before /WR for a write-controlled SRAM write cycle. For the write cycle to be write-controlled, then, /SMEMW active-low delay time must be slower than the decoding time (.33T must be greater than 40 ns).

Time T3, the minimum allowable /WR pulse width, is:

T3 = /SMEMW pulse width = 2T

Time T4, the time from data valid on the data pins to the end of the /WR pulse, is seen from Figures 11.3b and 11.5 to be:

T4 = /SMEMW + 10 – (1.33T + Data delay through DB1)
T4 = (2.33T + 10) – (1.33T + 10) = T ns (If 1.33T > DB1 enable time)
T4 = (2.33T + 10) – 90 = 2.33T – 80 ns (If 1.33T < DB1 enable time)

SRAM Write Timing Specifications for the PC Bus

For a 4.77 MHz PC bus clock speed, with 209 ns T times, the SRAM write times become:

T1 = 2.33 × 209 = 487 ns
T2 = (2.33 × 209) – 40 = 447 ns
T3 = 2 × 209 = 418 ns
T4 = 209 (1.33T = 278 ns > 90 ns DB1 enable time)

An inspection of Table 11.8 shows that write cycle times for a 4.77 MHz PC bus are well in excess of those required for any of the SRAM speeds listed. There is *one* more check needed, however. The write cycle is /WR controlled, which means that /CE *must* go low before /WR goes low. The /CE delay, from Figure 11.6, is one buffer and four gate delays, or 50 ns. /WR goes low at .33T plus one buffer delay or 79 ns. The /WR pulse *does* occur after the /CE pulse by a *comfortable* 29 ns margin. Should the decoding circuit slow up for any reason, however, or the PC bus use a higher clock rate, the design will become questionable. The safety margins for a 150 ns SRAM and a 4.77 MHz PC bus are as follows.

Time	PC Bus Time Allowed	Time Needed by SRAM	Margin
T1	487	100	+387
T2	447	100	+347
T3	418	90	+328
T4	209	60	+149
/CE enabled before /WR		0	+ 29

At a bus speed of 8 MHz (T = 125 ns), we find the following times for a write cycle:

T1 = 2.33 × 125 = 291 ns
T2 = 2.33 × 125 – 40 = 251 ns
T3 = 2 × 125 = 250 ns
T4 = 125 (1.33T = 166 ns > 90 ns DB1 enable time)
/CE enabled before /WR = (.33 × 125) – 40 = 1 ns

The delay time, for an 8 MHz bus clock, from /CE to /WR for a write-controlled cycle, becomes a *shaky* 1 ns. One solution to the decode speed problem is to speed up /CE by using faster logic, and to reduce the rather sloppy four-gate delay decoder design to a two-gate delay design. Another solution is to *gate* the /WR pulse to the SRAM using the /CE valid level as the gate control. Using a gate, /CE would always *precede* /WR to the SRAM by the gate delay. Alternately, the SRAM could be operated in the /CE control write mode.

The timing examples just discussed for the PC bus may also be applied to the ISA bus. The ISA timing calculations are left as an exercise.

A Test Program for the Memory Expansion Card

The D86 debugger program is invaluable as a debugging tool to see if the memory card works as intended. D86 can display the contents of the memory expansion card addresses A0000h to A1FFFh, in the memory display window, as visual proof that the expansion card functions.

A small test program, memtest.asm, which follows, writes a test word to each expansion card word address in the 8K address space. The test words written to memory start at 0000 and count up by 1 to 0FFFF for the last memory location. D86 can be used to display the contents of the 8K expansion memory after the program runs.

```
;memtest.asm writes a word to each location of the 8K expansion
;memory on the expansion bus.
code segment
        mov ds,0a000h       ;set DS to the 8K memory
        mov bx,0000h        ;set BX to point to the first word
        mov ax,0000h        ;AX counts from 0000h
nxt:    mov w[bx],ax        ;move the contents of AX to word in memory
        inc ax              ;count AX up by 1
        inc bx              ;increment BX to point to next word
        inc bx
        cmp bx,2000h        ;stop when out of memory
        jne nxt             ;go until out of memory
dun:    mov ax,4c00h        ;when out of memory return to DOS
        int 21h             ;back to DOS (if run)
code ends
```

D86, because of its single-step capability, can also help the designer debug the *hardware* on the expansion card. Experience has shown that most construction errors arise from three sources: *miswiring*, using the *wrong* chip, or *inserting a chip inverted* in a socket. The following procedure is recommended if the circuit is built.

- Wire the circuit, insert *no* chips.

- Use a continuity tester and make *sure* +5VDC and ground are *not* connected together.

- Use a continuity tester and check *every* connection.

- Have *another* person continuity-check *every* connection.

- Insert buffers SB1–SB3 and gates G1–G10, and ensure that the *decoder* circuit works using D86 in the single-step mode. A logic probe, placed to detect /RAMEN or /DATAEN, should blink every time D86 writes a byte to a SRAM address.

- Insert the SRAM and verify complete operation by using D86 to display the contents of the SRAM as data is written to it.

D86, in single-step, can be used to generate every control signal to the decoding logic. For instance, every time a write is done to the SRAM, the /RAMEN and /DATAEN lines should pulse low for an instant, and *only* when the card is addressed. A logic probe can be used to trace through the gates to find any decoding or control signal problems.

11.7 A PC Bus Memory-Mapped I/O Expansion Card Design

The 8K SRAM of the memory expansion card can be converted to a general purpose I/O card by *replacing* the SRAM with circuitry that will enable the card to get inputs from external signals and output bits to external devices. The SRAM write function can be replaced with an 8-bit latch, for instance, and the SRAM read function can be replaced by a data buffer. The output data latch enables the program to write a byte of data to peripheral devices. An input data buffer allows the program to read data from an external circuit.

By replacing the SRAM with a latch and buffer, we have defined an I/O address location that happens to be *mapped* in memory space. I/O that is addressed as memory is often referred to as *memory-mapped* I/O. That is, the I/O location is an address in memory space, not I/O space. Memory-mapped I/O does not need many memory addresses, because I/O circuits generally consist of one to a few dozen bytes of input and output data quantities as opposed to the thousands of byte addresses contained in pure memory chips.

Speed Criteria for Memory-Mapped I/O

Placing I/O addresses in memory space is advantageous when the I/O circuitry can respond as fast as, or faster than, conventional memory. Using an IC latch for data output, and an IC buffer for data input, will result in circuit speeds that exceed the response time of conventional memory.

By placing an I/O port in memory space, the programmer gains access to most of the X86 family's *instructions* rather than the standard I/O IN and OUT commands. It is often convenient to be able to use the entire array of instructions that deal with memory addressing and access. Fast I/O is often placed in memory space just to gain access to a rich instruction set.

I/O is not normally placed in the memory space, however, if the I/O device is considerably *slower* than standard semiconductor memory. Slow I/O devices are normally placed in I/O space, with the attendant loss of all but the IN and OUT instructions accepted by the programmer. Slow I/O, placed in the memory space, will so seriously *degrade* system performance that it will not be generally acceptable.

Memory-Mapped I/O Expansion Card Circuit

Figure 11.7 shows our original memory expansion card of Figure 11.6 *adapted* for I/O use. Two changes are readily apparent. Address decoding has been *extended* to a greater number of address bits. The 6264 SRAM has been *replaced* with an 8-bit 74HCT373 output data latch (DL1) and an 8-bit 74HCT245 switch input data buffer (SWB1). For example purposes, data will be input into the I/O card using an 8-bit DIP switch, and output data from the card will be displayed on 8 low-current LEDs.

MEMORY MAPPED I/O ADDRESS

The I/O card has been assigned a not-so-arbitrary memory address range of A1800h to A1FFFh. The address assignment means that the single-byte I/O

Figure 11.7 ■ Memory-mapped I/O card.

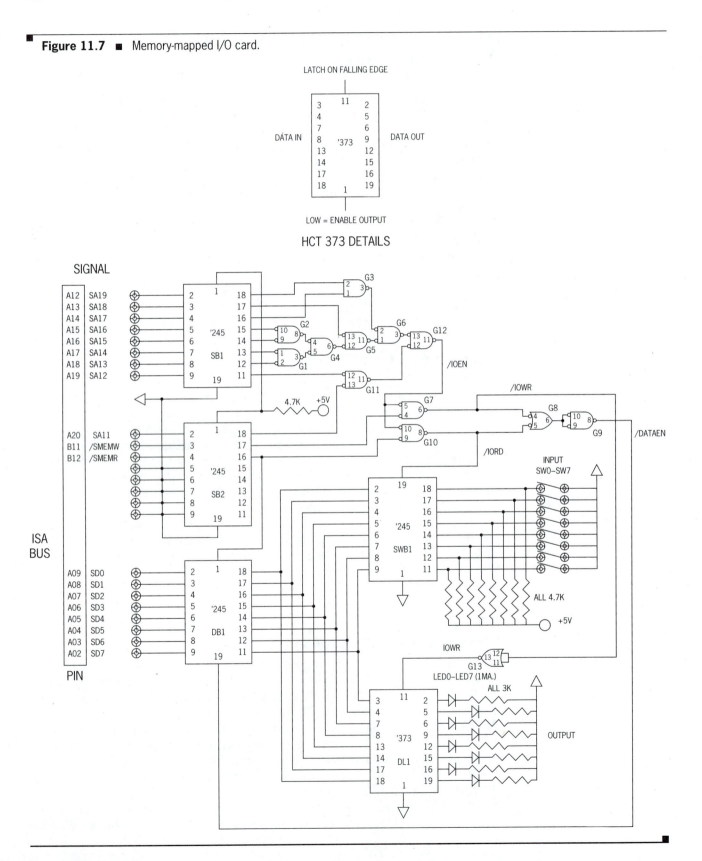

port takes up 800h (2048d) locations in memory! Using an entire block of memory for one memory-mapped I/O port is *not* unusual. Many unused PC memory addresses are available. The convenience of not having to decode the address lines down to a single address outweighs the loss of potential memory addresses. For instance, if we decided to decode the memory mapped I/O port to a single address, we would need about 15 two-input gates. By deciding to "waste" two thousand or so addresses in memory, we can get by with 8 two-input gates of the present design. We shall also find that the unused gates from our 8K memory card can now be fully utilized for the memory-mapped I/O card. The I/O card design features these circuit details:

- Address lines A11 through A19 are decoded by gates G1–G12 to provide a 2K address space.
- Bus address and control signals are buffered by 74HCT245 buffers SB1 and SB2.
- Bus data lines are buffered by 74HCT245 buffer DB1. DB1 is enabled, via gates G7–G10, whenever the bus address equals the I/O decoded address (signal /IOEN valid), and a bus memory read or write cycle is in progress.
- I/O data is buffered and latched by 74HCT245 buffer SWB1 and 74HCT373 data latch DL1. DIP switches SW0–SW7 provide input data to the card, and 1 ma LEDs LED0–LED7 provide visual output for the latched data.

Memory-Mapped I/O Address Decoding

Address bits A11 to A19 are decoded to enable the memory-mapped I/O port. The total address decoded is:

A19	A18	A17	A16	A15	A14	A13	A12	A11	A10–A0
1	0	1	0	0	0	0	1	1	XXXXXXXXXX

The address decoding scheme chosen allows the I/O memory-mapped card to respond to any address from A1800h to A1FFFh.

Reading and Writing I/O Data

As opposed to SRAM action on the memory card, the 74HCT373 data latch, DL1, and the 74HCT245 switch buffer, SWB1, do *not* logically AND the read or write pulses with the address decode signal /IOEN.

The output of gate G10, /IOWR, is used to *latch* data into the 8-bit latch DL1. The output of gate G7, /IORD, is used to *enable* the input data buffer, SWB1, to the internal system bus.

When a byte is written to any address between A1800h and A1FFFh, it will be latched into DL1 to provide a continuous output for external circuits. When a byte is read from any address between A1800h and A1FFFh, any data at the input buffer SWB13 pins will be read into the computer. Bus data buffer DB1 will only be enabled by signal /DATAEN when a valid I/O address is present on the bus *and* a bus memory read or write cycle is in progress.

Memory-Mapped I/O Port Timing

The output data latch and input data buffer can react in 10 ns. Bus memory read and write cycles require hundreds of nanoseconds, as shown for the memory expansion card design calculations. The response times of the I/O card, *using addresses valid on the PC bus at time equal zero,* are as follows.

$$/IOEN = \quad 10\ ns \quad + \quad\quad 50\ ns \quad\quad\quad = 60\ ns$$

Delay: SB1/SB2   G1,G4,G5,G6,G12

$$/IORD = [(.33T\ + 10\ ns)\ or\ /IOEN\ (whichever\ is\ later)] + 10\ ns$$

Delay: /SMEMR   SB2   G10

$$IOWR = [(.33T\ + 10\ ns)\ or\ /IOEN\ (whichever\ is\ later)] + 10\ ns + 10\ ns$$

Delay: /SMEMW   SB2   G7   G13

$$/DATAEN = /IOWR\ or\ /IORD + 20\ ns$$

Delay: G8,G9

Using a 4.77 MHz (T = 209 ns) PC bus, the times needed to read or write a byte of data to the I/O card are as follows.

```
/IOEN = 60 ns
/SMEMW or /SMEMR = .33T = 69 ns (greater than the /IOEN delay)
/IORD = 79 + 10 = 89 ns
IOWR = 79 + 20 = 99 ns
/DATAEN = 89 + 20 = 109 ns
DB1 enabled = /DATAEN + 10 ns = 119 ns
Data from SWB1 to DB1 = /IORD + 10 ns = 99 ns
```

Data cannot pass between the PC bus and the I/O card internal data bus until DB1 is enabled at 119 ns, which is determined by the .33T bus delay and the various circuit delays.

Should the PC bus frequency be increased until the decoding delay is *less* than .33T + 10 ns, then the timing becomes as follows.

```
/IOEN = 60 ns
/IORD = 60 + 10 = 70 ns
IOWR = 60 + 20 = 80 ns
/DATAEN = 70 + 20 = 90 ns
DB1 enabled = /DATAEN + 10 ns = 100 ns
Data from SWB1 to DB1 = /IORD + 10 ns = 90 ns
```

As for the last case, the circuit waits for DB1 to be enabled before data can be passed between the card and the bus.

EFFECT OF INCREASING THE PC BUS FREQUENCY ON THE I/O CARD

At what PC bus frequency could the I/O card prove to be too slow? Let us investigate the I/O card response for any period T of the PC bus. As the last example shows, the time to enable DB1 is the limiting response time if .33T is less than /IOEN.

For an I/O read cycle we have, from Figure 11.3a:

Access Time = 2.33T = DB1 enabled = 100 ns; T = 42 ns; f = 23.3 MHz

The limiting PC bus frequency is 23.3 MHz because of the very slow decoder design. If the decoding delay were reduced to 20 ns, then DB1 could be enabled in 70 ns, and the bus run at 33.3 MHz.

A Memory-Mapped I/O Expansion Board Test Program

The program memio.asm exercises the I/O card in a manner similar to that for the memory expansion card.

```
;memio.asm, a small test program for a memory mapped I/O card that
;uses 8 switches for input and 8 LEDs for output.

code segment
        mov ds,0a000h        ;set DS to I/O memory segment
memio: mov al,b[1800h]       ;read the switches
        mov b[1800h],al      ;light the LEDs
        jmp memio            ;loop forever
code ends
```

The debugging techniques listed for the memory expansion card are equally applicable to the memory-mapped I/O card. A debugging program does not have to be loaded into D86, however.

Using a move command on the D86 command line that reads the switches should produce an /IORD signal to the input data buffer, SWB1, and the data read to AL should match the DIP switch settings.

Similarly, AL may be loaded with a test byte, and a D86 command line move command that writes data to LEDs should produce an IOWR signal to the DL1 data latch, with corresponding LED illumination.

The choice of I/O location A1800h is *arbitrary* in the program; any address between A1800h and A1FFFh will work just as well. All of the rich array of memory access instructions may be used with a memory-mapped I/O card.

11.8 A PC Bus I/O-Mapped I/O Expansion Card

An alternative to memory mapping an I/O bus expansion card is to place the I/O card *address* in I/O space. I/O-space-mapped expansion cards differ in three important aspects from memory-space-mapped expansion cards:

1. I/O space contains *64K*, addressed by the central computing core using address lines SA0 to SA15.

2. I/O space is accessed by bus control signals /IOW (I/O Write) and /IOR (I/O Read). /IOW is *generated* by an OUT program instruction. /IOR is *generated* by an IN program instruction. No other program instructions, except OUT or IN, will generate bus I/O access cycles.

3. Direct memory access (DMA) cycles will use memory space addressing *and* I/O space addressing during a single DMA bus access cycle. DMA transfers, such as the *continuous* DMA access cycles used to *update* video monitors, can interfere with standard I/O bus cycles. The DMA *enable* control signal AEN must be included in the I/O-mapped card design to *disable* the card during DMA cycles.

Figure 11.8 ■ PC Bus I/O access cycle.

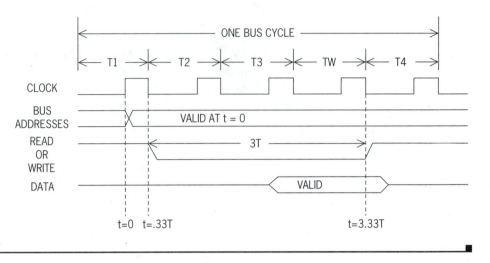

In all other respects, PC bus I/O access behaves exactly as does PC bus memory access. For PC bus I/O space access, substitute /IOW for /SMEMW and /IOR for /SMEMR. Decode address SA0 to SA15 instead of SA0 to SA19. PC bus timing for I/O bus access, shown in Figure 11.8, is seen to be virtually identical to PC memory access cycles, shown in Figure 11.3b, with one exception: I/O cycles are 25% *longer*.

I/O-mapped expansion card PC bus timing for I/O access is *one* clock period longer than for PC bus memory access cycles, as shown in Figure 11.8. Most PC designs add one *wait* state (TW) to the normal four clocks that define a PC bus memory access cycle. Adding the wait state has the effect of relaxing the response time requirements for PC bus I/O expansion cards to 3.33T states, or approximately 1 microsecond for a CPU clock speed of 4.77 MHz.

A Polled I/O Expansion Card Design

I/O space bus expansion cards may request program action by two means, *software* polling or *hardware* interrupt. The I/O expansion card may be polled by the program to determine if any program action is required. Or, the I/O expansion card may *interrupt* the program by *hardware* means when it requires the attention of the program.

We shall begin our study of I/O expansion cards by modifying the memory-mapped expansion card of Figure 11.7 to an I/O-mapped expansion card. The changes to the memory-mapped expansion card circuit that convert it to an I/O-mapped expansion card are shown in Figure 11.9. The most significant change to the memory-mapped I/O card is that address lines SA1 to SA9 replace address lines SA11 to SA19.

Note that the DMA enable signal, AEN, is used to *disable* buffer SB2 during DMA cycles. AEN is active-high during DMA cycles and low at all other times. Pull-up *resistors* R2 and R3 are used to place the card read and write signals in high (nonactive) states whenever SB2 is disabled (tri-stated) by

Figure 11.9 ■ I/O-mapped I/O card.

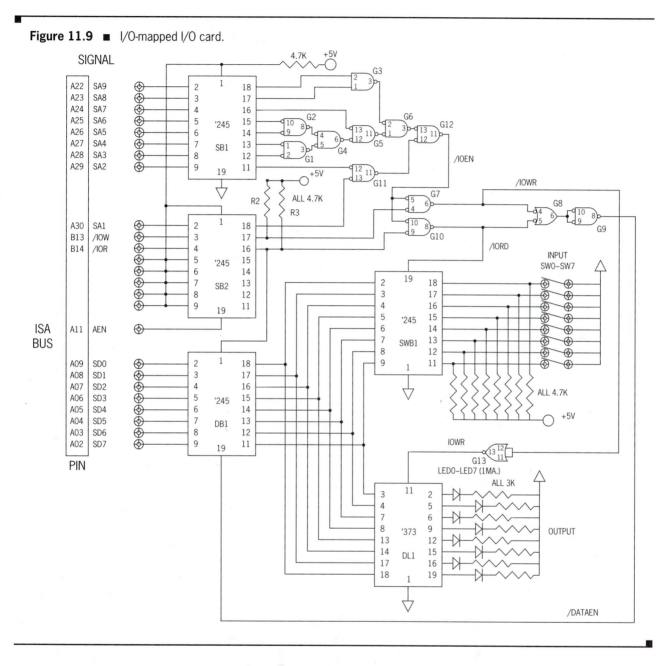

AEN. Failure to incorporate AEN in the decoding design will result in interference as the monitor is refreshed.

PC Bus Expansion Card I/O Addresses

As is the case for selecting unreserved memory addresses for a memory-mapped expansion card, an I/O expansion card must occupy an I/O space address that does *not* conflict with predefined DOS or BIOS I/O addresses. Table 11.9 shows the I/O addresses that have been assigned for DOS or BIOS I/O use on *most* PC computers.

I/O Address (Hex)	Function
0000h–01FFh	Central processing core
0200h–020Fh	Games
0210h–021Fh	XT expansion
0220h–0237h	Not assigned
0238h–023Bh	Bus mouse
023Ch–023Fh	Bus mouse
0240h–0277h	Not assigned
0278h–027Fh	Parallel printer port number 3
0280h–02AFh	Not assigned
02B0h–02DFh	EGA/VGA
02E0h–02E7h	AT general purpose interface bus (GBIB)
02E8h–02EFh	Serial port 4
02F0h–02F7h	Not assigned
02F8h–02FFh	Serial port number 2
0300h–031Fh	Not assigned
0320h–032Fh	XT hard disk
0330h–0377h	Not assigned
0378h–037Fh	Parallel printer port 2
0380h–038Fh	SDLC
0390h–039Fh	Not assigned
03A0h–03AFh	SDLC
03B0h–03BBh	MGA
03BCh–03BFh	Parallel printer port 1
03C0h–03CFh	EGA/VGA
03D0h–03DFh	CGA
03E0h–03E7h	Not assigned
03E8h–03EFh	Serial port 3
03F0h–03F7h	Floppy disk
03F8h–03FFh	Serial port number 1
0400h–FFFFh	Not assigned and not used

Table 11.9 ■ I/O Address Space Map

An inspection of Table 11.9 reveals a startling fact: Only the first 1K of the total 64K of I/O addresses has been used for ISA or PC bus systems! Theoretically, all 64K, except those reserved addresses listed in Table 11.9, are available for I/O expansion card use. In practice, the PC industry *never* used more than the first 1K of available addresses. Thus the PC industry never used more than the first 10 address lines, SA0 to SA9, for I/O address space decoding. Using only 10 address lines reduces the logic needed for I/O address decoding and, conveniently, lets us use our altered memory-mapped card as an (almost) fully decoded I/O-mapped card.

I/O SPACE DECODING

We have available to us, on our memory-mapped I/O card, nine address decoding gate inputs, four active-high and five active-low. Casting about for an I/O address that can use this combination leads us to unused I/O addresses 0300h to 031Fh as one possibility. The bit patterns for this address range are as follows:

A9	A8	A7	A6	A5	A4	A3	A2	A1	A0	
1	1	0	0	0	0	0	0	0	0	= 0300h
1	1	0	0	0	1	1	1	1	1	= 031Fh

If we decide not to decode address bit A0, we find that we may decode the following bit pattern from among those available:

A9	A8	A7	A6	A5	A4	A3	A2	A1	A0	
1	1	0	0	0	0	0	1	1	X	= 0306h (0307h)

Other addresses will work as well, such as 0312h (0313h) or 030Ch (030Dh). The I/O address used on the I/O expansion card of Figure 11.9 is chosen to be 0306h. The card may be addressed at 0306h or 0307h with the same results.

An I/O-Mapped I/O Expansion Card Test Program

Converting from memory-mapped I/O to I/O-mapped I/O results in one important software consideration: only *two* instructions, IN and OUT, can access the I/O-mapped I/O card. The IN and OUT instructions are covered in Chapter 8; this is our first opportunity to use them. A small test program, ioio.asm shown next, demonstrates the use of the IN and OUT instructions.

```
;ioio.asm, a small program to exercise an I/O-mapped I/O card that
;has 8 switches for data input, and 8 LEDs for data output.

code segment
        mov dx,0306h        ;DX points to all I/O addresses
ioio:   in al,dx            ;read the switches
        out dx,al           ;write the LEDs
        jmp ioio            ;loop forever
code ends
```

The debugger, D86, may be used in conjunction with the I/O-mapped I/O expansion card to test the card for proper construction as follows.

- Reading the switches, when single-stepping through the IN command, should produce an /IORD signal at switch buffer SWB1, and the data on the switches should arrive in register AL.

- Writing to the LEDs should produce an IOWR pulse at LED buffer DL1, and the LEDs should illuminate in the bit pattern in AL.

This section has featured using the program to *poll* the expansion card in order to determine I/O operation. Most I/O operations, particularly when many I/O points are controlled by a computer, are done using I/O expansion card *hardware interrupts* to the central processing core. The next section provides an introduction to the fundamental techniques used to interface a PC to hardware interrupt I/O expansion cards.

11.9 Design of Hardware-Interrupt Expansion Cards

Before proceeding to the hardware details of converting the software-polled I/O-mapped I/O card into an interrupt-driven I/O-mapped I/O card, we must re-examine X86 family interrupt concepts discussed in Chapter 7. Information about how the PC organizes and controls hardware interrupts is also necessary in order to be able to write interrupt-driven programs.

Bus Hardware Interrupts

Interrupts are a form of *indirect calls* to subroutines that are written to handle specific system hardware needs. Interrupts may be initiated through *software* means, by including an INT instruction in a program, or through using hardware circuits that *activate an interrupt signal line* on the PC or ISA bus.

A review of interrupt concepts discussed in Chapter 7 follows.

■ The first 1K addresses in system RAM, from address 00000 to address 003FFh, are reserved for interrupt *vectors*. An interrupt vector is a 4-byte group that contains an instruction pointer (IP) word and a code segment (CS) word. The IP is stored first (lower) in memory, followed by CS. The CS:IP combination specifies the address of an interrupt-handling subroutine located somewhere in the total 1M memory space.

■ When an interrupt is activated, the following automatic CPU actions take place:

1. The current CS, IP, and Flags register contents of the original program are pushed on the stack.

2. The interrupt type number is multiplied by 4 to find the address, in low memory, of the interrupt vector for that type number.

3. The CS:IP combination found at the address formed by multiplying type number times 4 is loaded into the CS and IP registers. An immediate jump to the interrupt service routine is done to the subroutine address contained in CS:IP.

4. Program operation begins at the subroutine located at address CS:IP and continues until an IRET instruction is executed in the subroutine.

5. IRET causes the original CS, IP, and Flags register contents to be popped from the stack back to the CS, IP, and Flags registers. The original program resumes.

Computer systems, such as the PC, make use of the 8086 interrupt scheme to predefine many BIOS and DOS system service subroutines. Each subroutine is used to service some common system peripheral such as a keyboard, a printer, or the video display. Each subroutine is assigned a type number, and the collection of subroutines is commonly called the Basic I/O System, or BIOS. BIOS is normally stored in a nonvolatile (ROM) memory so that the computer can organize itself when power is applied. Additional subroutines can be stored on disk and loaded into system RAM after BIOS has initialized the computer. The disk subroutines are grouped into a collection of interrupt-handlers as part of the Disk Operating System, DOS.

We have made use of many BIOS and DOS subroutines throughout this book, particularly in Chapters 7 and 9, by simply including an INT xx instruction in our programs where xx is the particular type number for the BIOS or DOS subroutine we wish to use. By using INT instructions we have employed software interrupts—such as keyboard, video, and port—in many of our programs.

Bus hardware interrupts work exactly as program software interrupts do with one important exception:

> *Circuits connected to the bus supply the interrupt type number* to the CPU, *not* the program.

Hardware interrupts may be generated from circuits within the central processing core, *or* by expansion cards attached to the expansion bus. Bus hardware interrupts are given the signal name *IRQ*, for Interrupt ReQuest.

Many *standard* hardware-generated interrupt type numbers have already been assigned to all of the common system peripherals such as the keyboard, printer, and serial and parallel ports. Each hardware interrupt is activated whenever its associated signal becomes active from a peripheral circuit. Several hardware interrupt signals are also assigned to the PC and ISA bus.

PC Bus 8-Bit Interrupts

The ISA 8-bit bus has been assigned the interrupt request numbers shown in Table 11.10. Each ISA bus IRQ has been assigned a corresponding interrupt type number and interrupt vector address in low memory. The *sole* IRQ assignment difference between the PC and ISA busses is the substitution of IRQ9 on the ISA bus for IRQ2 on the PC bus. The complete (16-bit) ISA bus adds IRQ numbers, up to IRQ15, to the original PC bus IRQ0 to IRQ7. ISA designs added additional interrupt circuitry to the original PC bus interrupt circuitry to allow expanding the number of PC interrupts. To maintain compatibility with existing PC bus expansion cards that use IRQ2, ISA bus IRQ9 can be vectored to the older PC bus IRQ2.

The Interrupt Controller

The interrupt circuitry used in the PC and ISA computer designs that handles hardware interrupts from the bus is called an *interrupt controller*. An interrupt controller is a special-purpose chip whose function it is to *organize*

IRQ	Interrupt Type	Vector Address	Assigned to
IRQ3	0Bh	2Ch–2Fh	Serial port 2
IRQ4	0Ch	30h–33h	Serial port 1
IRQ5	0Dh	34h–37h	XT hard disk, or Parallel port 2
IRQ6	0Eh	38h–3Bh	Floppy disk controller
IRQ7	0Fh	3Ch–3Fh	Parallel port 1
IRQ9	0Ah	28h–2Bh	PC bus IRQ2

Table 11.10 ■ 8-Bit ISA Bus IRQ Assignments

and control the hardware interrupts that are used in the system. An interrupt controller performs these valuable functions for the CPU:

1. *Enables and disables* the various interrupts. Not all of the possible system interrupts may be used by a particular PC configuration. The unused interrupts can be disabled to prevent electrical noise from falsely triggering an interrupt.

2. Organizes the interrupts by *priority*. A higher priority interrupt, such as serial data reception, will be serviced before a lower priority interrupt, such as the keyboard, should they both demand service at the same instant.

3. Expands the *number* of interrupting sources. The 8086 is equipped with a single hardware interrupt input pin. Interrupt controllers can expand the single 8086 interrupt capability up to 64 interrupting sources.

4. *Assigns* interrupt type numbers to each source. Each physical IRQ interrupt signal is assigned one of the 256 possible interrupt type numbers.

5. Handles all interrupt hardware *types* of signals. The interrupt controller can be programed to react to edge-triggered or level-triggered interrupt signals.

All five interrupt controller items listed are *programmable*. The programmer may program the interrupt controller as to the number of interrupts, the type number associated with each interrupt, the priority of the interrupt, whether the interrupt is enabled or disabled, and the type of signal that will trigger an interrupt response.

Programming the interrupt controller is crucial to proper system operation and requires an in-depth knowledge of the exact details of the interrupt controller chip model. The original Intel interrupt controller part number is the 8259. An in-depth discussion of the 8259 interrupt controller is beyond the scope of this book; however, we shall see that using the interrupt controller is simple.

Fortunately for the user, BIOS programs the interrupt controller as part of its *initialization* sequence. BIOS programs each IRQ on the ISA bus to a specific interrupt type number and priority, as well as other interrupt-specific details. There are, however, a few items that must be supplied by the user program. First, we must *enable* the IRQ we wish to use (if BIOS has not already done so), and second, we must *inform* (signal) the interrupt controller when our interrupt subroutine is finished.

PROGRAMMING THE INTERRUPT CONTROLLER

The PC bus interrupts are controlled by a single interrupt controller. The ISA bus interrupts are controlled by two interrupt controllers. On the ISA bus, the second interrupt controller uses IRQ2 of the first controller as an expansion input. To maintain compatibility with the PC bus, IRQ9 of the second controller can be software-vectored to an IRQ2 interrupt handler. The interrupt controller that handles the ISA 8-bit bus IRQ signals (with IRQ9 substituted for IRQ2) is the *same* as the one used for the PC bus.

The interrupt controller IC, originally an Intel 8259, is addressed in I/O space that has been *reserved* in the central processing core. The interrupt controller contains two registers that must be programmed *by the program* at I/O port addresses 20h and 21h.

THE INTERRUPT ENABLE REGISTER

The register at I/O address 21h is the interrupt enable register. The interrupt enable register contains 8 bits and is named the interrupt *mask* register. Masking means to control an activity by setting or resetting a bit in the masking register. Every bit in the mask register is associated with a unique system activity.

Each bit in the interrupt enable register controls the enabling or disabling of one of the ISA 8-bit IRQ signals shown in Table 11.11. Mask register bits 0, 1, and 2 mask IRQ signals that are not assigned to the ISA bus.

For the interrupt mask register, if a bit is set to binary *1,* then the interrupt associated with that bit is *disabled.* If the mask bit is cleared to *0,* then the interrupt associated with that bit in *enabled.*

INTERRUPT ACKNOWLEDGE REGISTER

The interrupt controller register at I/O address 20h is the interrupt controller priority control and interrupt acknowledge register. The interrupt subroutine must write an *end of interrupt* or EOI byte to the priority control register *at the end of the interrupt subroutine* so that the interrupt controller can reset itself to service the next interrupt.

The interrupt controller has no way to determine if a subroutine has completed execution. By writing an EOI acknowledgment byte to the priority control register, the interrupt subroutine *signals* that it has finished and that other interrupts can be serviced by the interrupt controller.

The EOI byte commonly used to signal the end of an interrupt subroutine is constructed as follows:

$$\text{EOI byte} = 01100iiib$$

Binary bits iii are set to the IRQ number (in binary) that has just ended. For instance, iii is 010 for IRQ2, and iii is 101 for IRQ5. The EOI byte can be written to the priority register at any time during the subroutine but is often written to the priority control register immediately before the subroutine issues an IRET command.

Mask Register Bit	IRQ Number	Interrupt Device or Function
0	0	Timer 0 output
1	1	Keyboard interrupt
2	9	Input from second controller
3	3	Serial port 2
4	4	Serial port 1
5	5	Parallel port 2
6	6	Floppy disk controller
7	7	Printer port 1

Table 11.11 ■ ISA Bus Interrupt Enable Register (I/O 21h) IRQ Bits

The Design of a Hardware-Interrupt I/O Expansion Card

We are now ready to modify our I/O card of Figure 11.9 so that it can generate interrupt signals. To be useful for both PC and ISA bus designs, interrupt *IRQ5* is chosen as our interrupt signal. Interrupt signal IRQ5 is unused on the PC bus, used for the hard drive on a PC-XT, and assigned to a second-line printer of the ISA bus. Should your system be an XT type with an original hard drive, or an ISA type with a second printer, please remove the expansion card that controls either conflicting device, to prevent interference with the textbook I/O card.

Changes to the I/O-Mapped I/O Circuits

Very few changes have to be made to the polled I/O-mapped circuit of Figure 11.9 to convert it to an interrupt-driven I/O type card shown in Figure 11.10. The only change is to *add* a hardware-interrupt circuit, as follows:

- The interrupt circuit shown on Figure 11.10 is added. The additional circuitry consists of flip-flop FF1, push button PB1, and gates G14, G15, and G16.

- An IRQ5 interrupt is generated whenever FF1 is set by pushing PB1, which generates a clock to FF1 through inverting gate G14. The output of FF1 brings IRQ5 low to high, signaling the interrupt controller that an interrupt request has been made.

- Flip-flop FF1 is initially reset during power-up by the action of the RESET signal applied to one leg of G15. The RESET signal is issued by the central core any time the CPU is reset. Flip-flop FF1 will also be reset, by /IORD, any time the I/O card is read by the program.

- The IRQ5 active-high positive edge interrupt signal does not have to be buffered; there should be no other circuit connected to the IRQ5 signal on the ISA bus. Interrupts, on the ISA bus, are signaled by a low-to-high transition on the interrupt signal line. Interrupts, on the PC bus, can be triggered by low-to-high transitions or high levels.

An interrupt will take place if the program has enabled IRQ5 in the interrupt enable register. The CPU will perform an interrupt type 0Dh, which is the interrupt type number assigned to IRQ5 by BIOS. The address of the interrupt service routine located at vector address 34h (type 0Dh times 4) in memory will be fetched by the CPU, and the subroutine executed.

An Interrupt-Driven I/O-Mapped I/O Expansion Card Test Program

A short test program, intdriven.asm, can be used to test the interrupt-driven I/O expansion card. The subroutine will read the contents of the switches and then light an appropriate LED based on each switch position.

Interrupt request FF1 will be *reset* by /IORD, whenever the switch is read, by gates G16 and G15. After lighting the LEDs, the subroutine writes an EOI byte, 65h to the priority control register, and then executes an IRET to return to the interrupted program.

It is very important that the interrupt service routine reset the interrupt flip-flop, FF1, *before* returning to the original program. Should FF1 not be reset, it may generate an *immediate* interrupt to the interrupt controller, and

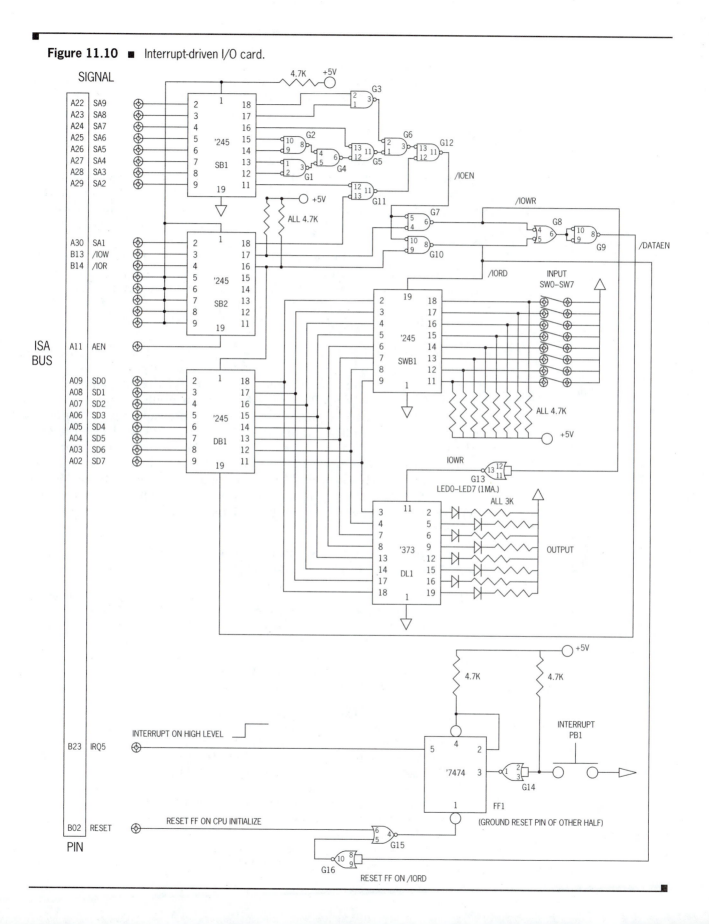

Figure 11.10 ■ Interrupt-driven I/O card.

the interrupt subroutine will be called again. This process of continuous interruption will not end, and the system is trapped in the interrupt routine. About as disastrous, every time the interrupt subroutine is called, the return address and Flag register is pushed on the stack. After several thousand continuous interrupts (a few hundred milliseconds) the stack will underflow, and chaos will result should FF1 finally get reset.

The programmer and system designer should ensure that the device that requests an interrupt can be physically disabled or reset during the interrupt servicing routine. The only alternative to resetting the requesting circuit is to disable the appropriate IRQ bit in the interrupt enable control register prior to issuing an EOI byte.

One more point. The programmer has gross control over all *hardware* interrupts via the CLI and STI opcodes discussed in Chapter 8. CLI clears the interrupt enable (I) flag in the Flag register and prevents *any* maskable interrupt from occurring. The STI opcode sets the I flag and enables maskable interrupts to occur, *if they are enabled in the interrupt controller's interrupt control register.*

STI and CLI give the programmer the ability to prevent *unwanted* hardware interrupts until they are desired. STI and CLI have no power over software interrupts, as these are already under the control of the programmer. The I flag is initially cleared, by CPU reset action, on power-up.

We shall demonstrate the workings of an interrupt driven subroutine using the program intdriven.asm that is listed next. We shall use the same techniques as discussed in Chapter 7 to load the address of our subroutine into the type 0Dh interrupt vector at location 34h–37h in low memory.

Caution: intdriven.asm will *overwrite* whatever vector was in location 34h–37h. Should your system have been equipped with a device that uses IRQ5, *reboot* the computer to load the original device subroutine vector into low memory.

```
;the program intdriven.asm, which initializes an interrupt vector for
;IRQ5, senses switches, and lights LEDs accordingly.

       code segment                                            Note
             mov sp,1000h        ;set the stack
             push ds             ;save current DS
             mov ax,0000h        ;set DS to low SRAM               1
             mov ds,ax
             lea ax,irq5         ;get interrupt subroutine IP      2
             mov w[0034h],ax     ;set interrupt vector IP
             mov ax,cs           ;get CS for the interrupt vector  3
             mov w[0036h],ax     ;set interrupt vector CS
             pop ds              ;restore old DS
             in al,21h           ;read current mask register
             and al,11011111xb   ;enable IRQ5 (bit 5), ONLY        4
             out 21h,al          ;write mask register with IRQ5 enabled
             sti                 ;enable hardware interrupts
wate:        jmp wate            ;loop here until PB1 is pushed
irq5:        mov cx,0ffffh       ;de-bounce PB1                    5
lp:          dec cx
             jnz lp
             mov dx,0306h        ;read in the switches
```

```
        in al,dx
        out dx,al            ;light the LEDs
        mov al,65h           ;EOI byte for IRQ5
        out 20h,al           ;reset interrupt controller        6
        iret                 ;return from interrupt
code ends
```

NOTES ON THE PROGRAM

1. Memory moves use DS as the default register. Interrupt vectors are located in SRAM memory segment 0000h.

2. Program intdriven.asm loads the CS and IP addresses of the small subroutine that begins at label irq5 into interrupt type 0Dh vector address 32h (0Dh × 4). The subroutine has an IP offset (at the address of irq5), in the code segment where DOS loads the program.

3. The interrupt subroutine happens to be in the code segment that is used by DOS when intrdriven.asm is loaded into RAM. The interrupt subroutine could be contained in another code segment, but the programmer would have to have advanced knowledge of the CS:IP for the subroutine.

4. The AND mask will zero Bit 5 (enable IRQ5) and leave all other bits unchanged.

5. As a precaution, a small time delay has been added to the interrupt subroutine to de-bounce PB1. It is quite possible that PB1 can bounce for several milliseconds, causing several successive interrupts. No harm is done when PB1 bounces, but a logic probe may show several IRQ5 pulses when only one pulse is expected.

6. An IRQ5 EOI byte is written to the priority control register. An IRET opcode is then executed to return to the wait loop at label wate.

11.10 Summary

Computer busses are electrical signal and connector physical standards by which a computer system may be expanded by adding peripherals that connect to the bus.

PC-compatible manufacturers have defined the PC, ISA, EISA, and E-ISA bus standards. The PC bus and ISA bus are compatible at the 8-bit expansion card level.

Expansion cards may be designed to plug into the PC bus using standard physical card configurations, and connecting to the bus-defined standard signals.

Several examples of PC bus expansion card designs are demonstrated:

Memory Expansion Card

Memory-Mapped I/O Expansion Card

I/O-Mapped I/O Expansion Card

Interrupt-Driven I/O Expansion Card

11.11 Questions

1. Define the function of a computer bus.
2. List three physical features of a computer bus.
3. Name three reasons why an "open" computer bus may be desirable.
4. List three groups of computer bus signals.
5. Identify the ISA 8-bit (PC) bus signals used for addresses.
6. Identify five control signals used on the ISA 8-bit (PC) bus and their function.
7. Identify the ISA 8-bit (PC) bus data signals.
8. Determine the following times, in nanoseconds, for a PC bus running at 10 MHz.

   a. T state.

   b. Time from addresses valid to control (read or write).

   c. Width of the control pulse.

   d. Time from address valid to data valid during write cycle.

   e. Time from address valid to data valid during read cycle.

9. Repeat Question 8 for a PC bus clock frequency of 5 MHz.
10. Repeat Question 8 for an ISA bus running at 20 MHz.
11. Show a jumpering scheme that will permit an 8K or 32K RAM to be used in the same socket.
12. Show a jumpering scheme that will permit a 2K or 8K EPROM to be used in the same socket.
13. What is the practical result of making T1 and T2 equal for static RAMS?
14. List two desirable features of HCT-type buffers.
15. Design a decoding circuit, made up of two-input gates, that can decode the address A7CE4h.
16. Design a decoding circuit, made up of two-input gates, that can decode the address range E1000h to E13FFh.
17. Calculate all T times, read and write, for the memory expansion card if the decoding delay is 10 ns, and the card is used on an ISA bus running at 20 MHz.
18. Repeat Question 17 for an ISA bus speed of 16 MHz.
19. Repeat Question 17 for an ISA bus speed of 8 MHz.
20. Repeat Question 17 for an ISA bus speed of 50 MHz.
21. What is the function of the AEN signal on the ISA bus?
22. How many I/O addresses are theoretically possible? How many are actually used by PC-type computers?
23. Why is an interrupt controller needed for an X86-type CPU?
24. What is the result of not resetting the interrupting source before returning from the interrupt handler subroutine?
25. Construct the EOI byte for IRQ7.
26. Construct the enable interrupt mask byte for IRQ3.

11.12 Projects

Chapter 12 lists several hardware projects that involve a single board computer (SBC) design. Several projects from Chapter 12 are listed here, modified as appropriate for I/O cards.

1. Control a simple, two-axis, robot arm to pick and place small parts. Control the arm's X- and Y-axis movement with stepper motors, and use digital encoders to feed back arm position. Use a solenoid-actuated gripper.

2. Add a hex keypad and an intelligent LCD display to the I/O card design. Write a program that will display on the LCD each key pressed on the hex keypad.

3. Build an EPROM programmer. Add I/O circuitry, and write a program that enables the card to program a 2K EPROM from an Intel hex file. (Typically, EPROM programmers use D/A converters for programming a range of EPROMs or fixed solid-state switches for one type of EPROM.)

4. Connect a printer to the card. Write a program that will act as a 4K print buffer between the printer and the PC.

5. Add a D/A converter to the card. Write a program that generates sine waves that vary in frequency as the DIP switch is set from 00 to FFh.

6. Add an A/D converter and a three-digit, seven-segment LED display to the card. Write a program that displays input DC voltages from 0 to 9.99V.

7. Add a speech synthesis chip to the card. Write a program that will sing the "Happy Birthday" song.

8. Add a UART, two tone filters, two tone generators and a telephone line interface. Write a program that converts the card into a frequency modulated modem.

9. Add D/A converters to the card so that it functions as a graphic CRT screen controller. Produce a picture of a campus landmark on the CRT screen.

10. Make an LED tic-tac-toe display that interfaces to the card. Use red for an O (player) and green for an X (program). Place a push button in each square. Write a program that plays the game against a human player. The player places an O in a square by pressing the pushbutton in that square. Alternate first entry choice every game.

11. Make a digital oscilloscope. Add A/D and D/A converters to the card. Write a program that will digitize and store an input waveform, and then play the waveform back to an oscilloscope input. Use the card output bits to trigger each oscilloscope trace.

12. Modify Project 11 so that the waveform is displayed on the CRT screen of the PC.

12

SINGLE

BOARD

COMPUTERS

Chapter Objectives

On successful completion of this chapter you will be able to:

- Describe the 8088 microprocessor CPU integrated circuit operation and pinout.
- Describe the 8284 clock generator integrated circuit operation and pinout.
- Describe the 8255 programmable peripheral interface (PPI) integrated circuit operation and pinout.
- Describe the 8250 UART integrated circuit operation and pinout.
- Describe the design of an 8088-based single-board computer.
- Calculate memory and I/O timing requirements.
- Discuss EPROM, RAM, PPI, and UART test program operation.
- Describe an interrupt-driven I/O circuit.
- Discuss an interrupt-driven I/O test program.

12.0 Introduction

A *single-board computer* (SBC) refers to a *physically* small computer built on one circuit board. The term originated as a description for computers with *limited* resources of memory, mass storage, and I/O as compared to a full-size computer system.

Continued development of denser, faster, and cheaper integrated circuits has rendered obsolete the previous definition of a single-board computer. Single-board computers are no longer limited in *capability*. Most modern PC designs use one motherboard for the entire computer. Several vendors have integrated the entire computing *core* of a PC into a *single* chip. Disk drives, once thought indispensable for equipping a computer with an operating system, may now be economically replaced with solid-state equivalents. A modern X86 SBC may be run using DOS. Today, many vendors of single-board computers have models with more power than the original IBM PC.[1] The trend in SBC design is more computing power for less money. We may expect this trend to continue as manufacturers strive to take advantage of the huge *installed* base of DOS-compatible PC software.

There is still a place, however, for the classic, *very low cost*, type of SBC, and Intel, among others, has a family of 8086-based *microcontroller* models available for the SBC designer. Single-board computers are found in applications in which low cost is *paramount*, yet the designer wishes to have the computing power of the X86 family of CPUs. A simple, inexpensive (under $25) SBC typically has *no* DOS or BIOS software programs and programs are assembly-code based. Of course, high-level languages can be used in SBC applications by compiling high-level languages into code that can be stored in ROM or EPROM on the board. Unfortunately, high-level languages tend to drive *up* the cost of a SBC because of their memory requirements.

Very elaborate, and *expensive* ($1,000), SBC designs can include DOS in ROM and nonvolatile RAM that can function as a disk drive as well as system memory. Elaborate SBCs are often found in *rugged* environments that preclude the use of fragile disk drives.

When a project involves using a number of SBCs, the system designer is often faced with an old question: Buy or build? Buying a commercially available SBC is usually the most time- and money-efficient path to take for small volumes, or rapid prototyping. Purchasing off-the-shelf SBCs becomes much less attractive as volume mounts, or the desire for unwanted features diminishes. If SBC *volume* can justify the cost, or the need for *unique* project features eliminates standard designs, then a custom SBC should be designed and programmed to meet the project requirement.

12.1 Issues Involved in Designing an 8088-Based SBC

This chapter studies some of the *concepts* involved in designing an SBC. One SBC concept involves the use of a computer as a *component* part of a

---

[1]A typical 1993 SBC features a 14 MHz totally solid-state 8086-based computer. The unit includes DOS in ROM; serial, parallel, and CGA video ports; 2M of DRAM; and 128K of user code. The entire computer measures less than 4 inches square.

larger system. SBCs are often intended to be placed *inside* a larger system, often electromechanical in nature, as *embedded controllers.*

Embedded controllers are *dedicated* computers that add intelligence to a larger machine. Often the operator of the larger machine is unaware that a computer is part of the overall system.

A second SBC concept involves *real time,* or *interrupt-driven,* computing. Real-time computing involves having the control program(s) actuated by *external* hardware interrupts. After a program has been accessed by an external interrupt, it must control machine operation immediately, within the time limits needed by the machine. Dedicated computers, because they are added to control the machine, are also *I/O rich,* meaning that much of the program is concerned with sensing input data and controlling output signals.

A third concept involves the *lack* of an operating system program to operate the computer system. Dedicated computers typically run a *single* program that has been tailored to meet exactly the needs of *one* application. As there are no keyboards, monitors, or disk drives, the dedicated computer does not need an operating system.

Examples of embedded controller applications range, on the low end, from appliances, toys, entertainment, and automobile controls to robot programmers and aerospace guidance systems, on the high end.

Many *very* inexpensive SBC embedded controllers, such as the one used in all PC keyboards, are based on microcontrollers, *not* microprocessors, for reasons of economy. Often, however, one needs an SBC that offers more programming *power* than the typical microcontroller can offer. Or, the user wishes to maintain software *compatibility* between the SBC and larger computers, such as an X86-based computer. One particular advantage of software compatibility is that programs intended for use in the SBC can first be *exercised* on the PC, then placed into the SBC for final testing.

SBC Hardware Selection

SBC design is *extremely* hardware-intensive. The SBC designer typically goes through these steps in the design process:

■ Decide on CPU type, and whether or not programming is to involve high-level languages. High-level language programming *implies* large ROM memory requirements.

■ Determine the system *specifications* in terms of RAM, ROM, and I/O ports.

■ Determine the final *cost,* physical size, and physical (temperature, shock, pollutant) operating environment.

■ Select appropriate integrated circuits that will *meet* system specification and cost restraints.

Once the board ICs have been selected, the designer must then learn how each IC used in the design operates. Most computer circuits, from the CPU to *peripheral* chips, are complex *programmable* devices, and several days or a week should be set aside to learn how to apply each one.

CPU	8088
Memory	8K RAM (6264), 8K EPROM (2764)
Speed	4.77 MHz clock (8284)
Parallel I/O	Programmable peripheral interface (8255)
Serial I/O	Programmable UART (8250)
Part cost	$25 in small volume

Table 12.1 ■ 8088 SBC Specifications

12.2 Integrated Circuits for an 8088-Based SBC

As an illustration of some of the issues involved in designing an SBC, we shall design an 8088-based SBC that contains limited amounts of RAM, ROM, and I/O. Specifications for the SBC are shown in Table 12.1.

The design involves learning how to use the following integrated circuits:

   8088 Central processing unit

   8284 Clock generator

   2764 8K EPROM

   6264 8K RAM

   8255 Programmable peripheral interface (PPI)

   8250 UART

Chapter 11 discusses the operation of the 6264 SRAM and 2764 EPROM. The remainder of the circuits needed to build the SBC are discussed next.

The 8088 CPU

Figure 12.1 shows the signal *names* for each pin (called a *pinout*) of an 8088 CPU housed in a 40-pin dual inline package (DIP).

The first task involved in learning to use the 8088 is to study the *function* of each pin signal on the 8088 DIP and then, after each pin's function is known, to use those signals that are appropriate for the design and *discard* those signals that are not. We shall investigate the 8088 CPU pin signals in a manner similar to the approach used to design the ISA bus expansion cards in Chapter 11, by grouping the 8088 DIP signals as data, address, and control.

8088 Data Pins

The 8088 is an 8-bit *external* data bus version of the 16-bit 8086 CPU. Both CPUs are *internally* identical in register construction and programming, however the 8088 DIP features *8-bit* data bus and the 8086 DIP has *16-bit* data bus. Using fewer data pins makes the CPU (and surrounding support systems) cheaper, a practice followed by Intel with the 80386SX CPU. We (and IBM) choose to use the 8088 in a first design for its simplicity and *lower* cost.

The data signals are bi-directional, that is, signals may flow in both directions on the pins, between the CPU and memory and I/O chips. The data pin numbers, and functions, are listed in Table 12.2.

Figure 12.1 ■ 8088 CPU pinout.

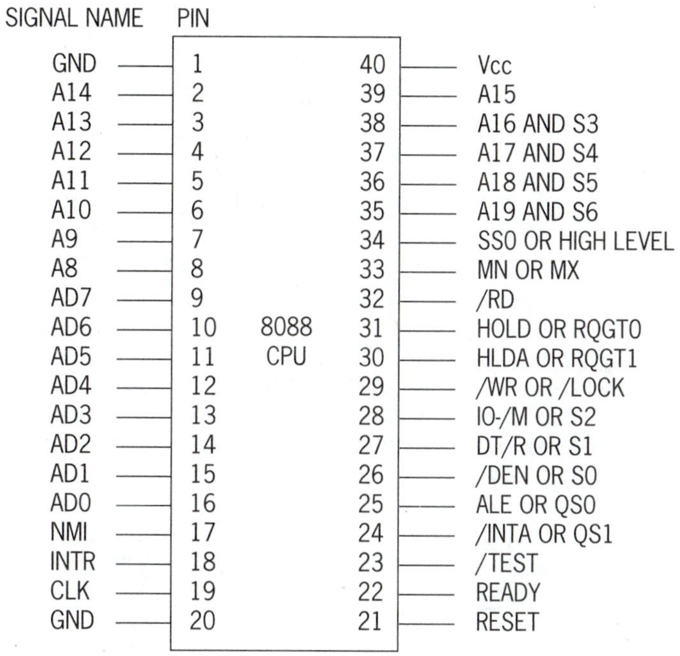

```
SIGNAL NAME    PIN
       GND ───── 1          40 ───── Vcc
       A14 ───── 2          39 ───── A15
       A13 ───── 3          38 ───── A16 AND S3
       A12 ───── 4          37 ───── A17 AND S4
       A11 ───── 5          36 ───── A18 AND S5
       A10 ───── 6          35 ───── A19 AND S6
        A9 ───── 7          34 ───── SS0 OR HIGH LEVEL
        A8 ───── 8          33 ───── MN OR MX
       AD7 ───── 9          32 ───── /RD
       AD6 ───── 10  8088   31 ───── HOLD OR RQGT0
       AD5 ───── 11  CPU    30 ───── HLDA OR RQGT1
       AD4 ───── 12         29 ───── /WR OR /LOCK
       AD3 ───── 13         28 ───── IO-/M OR S2
       AD2 ───── 14         27 ───── DT/R OR S1
       AD1 ───── 15         26 ───── /DEN OR S0
       AD0 ───── 16         25 ───── ALE OR QS0
       NMI ───── 17         24 ───── /INTA OR QS1
      INTR ───── 18         23 ───── /TEST
       CLK ───── 19         22 ───── READY
       GND ───── 20         21 ───── RESET
```

Note: The heading Dir in all of the pin tables shows the direction of pin signals from the standpoint of the chip: I(n) to the chip, O(ut) from the chip, or B(i-directional).

There appears to be some confusion in Table 12.2. The data pin signals are listed as *both* data and address signals. There is no mistake in the table, however: The *same* pin *is* used for both a data bit and an address bit, *but not at the same time.*

TIME MULTIPLEXING DATA AND ADDRESS BITS

Intel has used the concept of *time multiplexing* data and address signals for many of its CPU designs. Time multiplexing means that a pin is used for one function during time *interval* A and *another* function during time *inter-*

Pin	Name	Dir	Function
16	AD0	B	Data bit D0 and address bit A0
15	AD1	B	Data bit D1 and address bit A1
14	AD2	B	Data bit D2 and address bit A2
13	AD3	B	Data bit D3 and address bit A3
12	AD4	B	Data bit D4 and address bit A4
11	AD5	B	Data bit D5 and address bit A5
10	AD6	B	Data bit D6 and address bit A6
9	AD7	B	Data bit D7 and address bit A7

Table 12.2 ■ 8088 DIP Data Pins

val B. Internal CPU circuits *switch* between the two functions during times A and B.

Time multiplexing a pin has the distinct advantage of using fewer pins per circuit package, which makes a lower part cost. Time multiplexing has the disadvantage of adding complexity to the design and can add delay time to a CPU machine cycle.

Figure 12.2 shows how the data pins are time multiplexed during an 8088 CPU *machine* cycle. A typical machine cycle consists of four 33% duty cycle clock pulses, or T states, and proceeds as follows.

Figure 12.2 ■ 8088 time-multiplexed address and data lines.

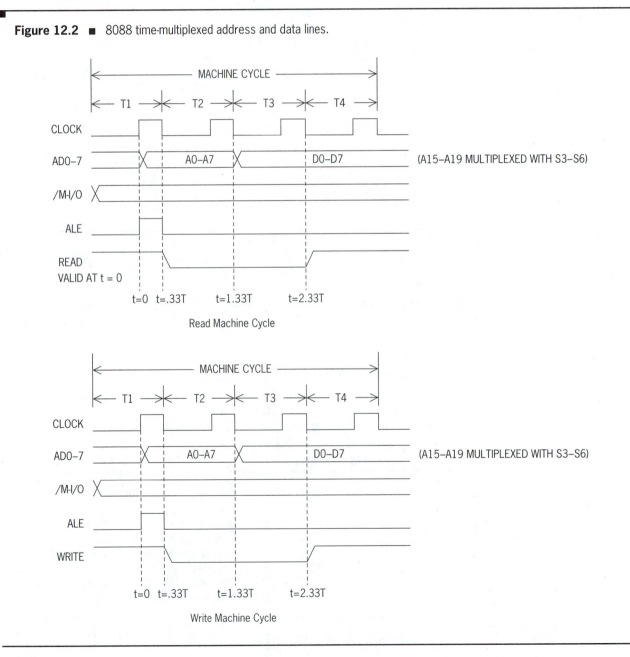

1. During T1, address bits A0 to A7 are placed on DIP pins 9 to 16. The memory-I/O control line goes valid at the beginning of T1 to signal a memory (active-low) *or* I/O (active-high) bus cycle.

2. A CPU control pulse, ALE, is used to latch address bits A0 to A7 into an *external* 8-bit data latch at the end of T1. Address bits A0 to A7 are now *stored* in the external latch, and pins 9 to 16 can be used for *data* transmission.

3. Pins 9 to 16 are internally switched to become data pins D0 to D7. They remain as data pins during the remainder of the cycle.

4. Control signals (/RD *or* /WR) are issued by the CPU at the start of T2 to the start of T4 in order to read or write data to the location specified by the address lines and the /M-I/O lines.

5. Read data is latched into the CPU when the /RD pulse goes high. Write data is latched into the external circuit (memory or I/O) when the /WR line goes high.

As can be deduced from the second step of an 8088 machine cycle, an external latch *must* be wired into the circuit in order to store address bits A0 to A7. Addresses A0 to A7 are often called the *lower* addresses to distinguish them from the partially non–time multiplexed *upper* address signals that are discussed next.

8088 Upper Address Pins

Address signals A8 to A19 are carried on the 8088 pins listed in Table 12.3. Address signals A8 to A19 are placed on the upper address pins at the *same* time the lower addresses are placed on the address/data pins.

Inspection of Table 12.3 shows that addresses A8 to A15 are *not* time multiplexed, whereas addresses A16 to A19 *are* time multiplexed with status signals S3 to S6. Addresses A16 to A19 are time multiplexed with S3 to S6 in a manner similar to addresses A0 to A7 and data bits D0 to D7. Should we wish to use addresses A16 to A19, we would have to latch them into an exter-

Pin	Name	Dir	Function
8	A8	O	Address signal A8
7	A9	O	Address signal A9
6	A10	O	Address signal A10
5	A11	O	Address signal A11
4	A12	O	Address signal A12
3	A13	O	Address signal A13
2	A14	O	Address signal A14
39	A15	O	Address signal A15
38	A16	O	Address signal A16 and status signal S3
37	A17	O	Address signal A17 and status signal S4
36	A18	O	Address signal A18 and status signal S5
35	A19	O	Address signal A19 and status signal S6

Table 12.3 ■ 8088 Upper Address Lines and Status Signals

nal 4-bit latch with the ALE pulse, as must be done with addresses A0 to A7. The remaining addresses, A8 to A15, remain *fixed* during the cycle.

Status signals S3 to S6 are intended to be used in *large* computer systems and are of *infrequent* importance for a simple SBC design. A16 to A19 enable the CPU to address up to 1M of address space, far beyond the 16K memory needs of the example SBC. We shall not make use of signals S3 to S6 or addresses A16 to A19.

8088 CONTROL SIGNALS

The 8088 is equipped with as many control signals as address and data signals. The reason that there are so many control signals included in the 8088 design is the *dual* nature of 8088 applications. The designers of the 8088 CPU envisioned two distinct uses for the circuit: a *minimum* mode and a *maximum* mode of operation.

In minimum operational mode, the 8088 CPU can control all access to memory and I/O locations with the addition of very *few* external circuits. Minimum mode 8088 operation implies, but is not limited to, the use of static memory RAM rather than dynamic RAM.

In maximum operational mode, the 8088 *must* be used with a bus control chip, the 8288, to access memory and I/O. The 8288 memory controller enables multiprocessing (more than one master controller), such as the use of math coprocessors in the system.

Clearly, minimum mode is *appropriate* for a simple SBC design, whereas maximum mode is useful for large computer systems. Table 12.4 lists the control signals found on the 8088. Control signals *used only for maximum mode* are underlined in Table 12.4.

Pin	Name	Dir	Function
17	NMI	I	Nonmaskable interrupt, L-H edge triggered
18	INTR	I	Maskable interrupt, high-level enabled
21	RESET	I	CPU reset signal, active-high level
22	READY	I	Memory or I/O is ready, active-high level
23	/TEST	I	Input bit for WAIT instruction, active-low
24	QS1, /INTA	O	Internal queue status, interrupt acknowledge
25	QS0, ALE	O	Internal queue status, address latch enable
26	S0, /DEN	O	Bus status, data transceiver enable (active-low)
27	S1, DT/R	O	Bus status, data transceiver direction
28	S2, I/O-/M	O	Bus status, I/O (high) or memory (low)
29	/LOCK, /WR	O	Lock out other bus masters, write (active-low)
30	RQGT1, HLDA	B	Bus master request/grant, hold acknowledge
31	RQGT0, HOLD	B	Bus master request/grant, hold request
32	/RD	O	Read (active-low)
33	/MX, MN	I	Maximum mode (low), minimum mode (high),
34	HIGH, SS0	O	High state, bus status bit 0
19	CLK	I	CPU clock pulse, 33% duty cycle
1	GND	I	CPU ground
20	GND	I	CPU ground
40	Vcc	I	CPU power, +5VDC

Table 12.4 ■ 8088 CPU Control Signals

CONTROL SIGNAL FUNCTIONS

The function of each control signal (maximum mode underlined) listed in Table 12.4 is as follows.

CLK Input 33% square wave from *external* clock generator to provide timing for all CPU functions.

HIGH, SS0 *No* maximum function for 8088. May be decoded along with I/O-/M and DT/R to show 8086 bus status in maximum mode.

INTR Maskable external active-high input to request a CPU hardware interrupt cycle. May be masked on or off by STI and CLI opcodes.

/LOCK, /WR Bus control signal in maximum mode. /WR(ite) control signal for memory *or* I/O in minimum mode.

/MX, MN Input signal to *force* maximum mode of CPU operation on reset when low; forces minimum mode of CPU operation on reset when high.

NMI Nonmaskable (by any opcode) external edge-triggered interrupt. Forces an interrupt to type 2 interrupt vector in low RAM.

QS0, ALE Internal BIU queue status for maximum mode. Address latch enable pulse used to latch address data into external latch in minimum mode of operation.

QS1, /INTA Internal BIU queue status for maximum mode. Minimum mode interrupt acknowledge output from the CPU as a response to an INTR signal from an external circuit.

/RD Read memory or I/O.

READY Active-high input signal from memory or I/O signaling that a data transfer *can* take place. Low state of READY causes wait states (Tw) to be inserted by CPU into bus access cycle.

RESET Active-high input signal from external circuit to *force* CPU to reset all internal circuits. Minimum or maximum mode of operation is forced by the state of the min-max pin during reset.

RQGT0, HOLD Bi-directional bus master request and grant signal in maximum mode. Bus master request for CPU to give up control of the bus in minimum mode.

RQGT1, HLDA Bi-directional bus master request and grant signal in maximum mode. Acknowledge that bus will be released by CPU to requesting bus master in minimum mode. CPU tri-states all bus lines immediately after issuing HLDA.

S0, /DEN 8288 bus controller status signal in maximum mode. Enables data transfer buffer (if needed in design) to transfer bi-directional data in minimum mode.

S1, DT/R 8288 bus controller status signal in maximum mode. Data transfer buffer (if needed in design) directional read or write signal in minimum mode.

S2, I/O-/M 8288 bus control status signal in maximum mode. Minimum mode signal for memory (low state) or I/O (high state) bus access cycle. This signal should be ANDed with /WR and /RD to provide memory read and write *and* I/O read and write signals.

/TEST Input pin from an external circuit for use by the WAIT instruction.

As can be seen from Table 12.4, if pin 33, the min-max pin, is connected to a logic high voltage, *then the CPU will power-up in the minimum mode of operation.* Should pin 33 be grounded, the CPU will configure itself in maximum operational mode on power-up. Maximum mode operation requires that a bus controller chip be added to the design. We shall wish to use the CPU in the *minimum* mode of operation for our SBC design, thus pin 33 *must be tied high* for the design to function.

The 8284 Clock Generator

Three timing and control input signals to the 8088 CPU (CLK, RESET, and READY) are most conveniently generated by a special-purpose circuit, the 8284 clock generator and driver. Figure 12.3 shows the 8284 pinout. Table 12.5 lists the 8284 signals and their functions.

We shall use a *non-overtone* crystal for the SBC clock. Overtones, in crystals, refer to using crystal frequencies that are multiples of the fundamental crystal frequency. The tank circuit input, TNK, is used only with overtone crystals, to adjust to the desired overtone frequency.

Bus ready signals, RDY1 and RDY2, can be used by slow peripherals connected to the data bus to signal the CPU that additional time is required for a bus data transfer. RDY1 and RDY2 are gated (enabled) by the accompanying enable active-low 8284 input signals /AEN1 and /AEN2.

If RDY1 and RDY2 are both low, *and* enabled by /AEN1 and /AEN2, then the 8284 READY signal to the 8088 will be low also. The 8088 CPU responds to not READY (READY low) by *inserting* additional clock pulses, called *wait states* (Tw), into the machine cycle. Wait states are added *until* READY goes high. Should *either* RDY signal be asserted high, the 8024 will assert READY to the 8088. The 8088 machine cycle continues after READY is high.

Inserting wait states is the way the 8088 CPU can be adjusted to the response time of *slow* memory and I/O ports. Part of the design task is to determine *if* wait states are needed and, if so, to include circuits that will generate the RDY signals to the 8284. For the example SBC, the memory and I/O chips chosen are sufficiently fast that wait states will *not* be required.

Figure 12.3 ■ 8284 clock generator.

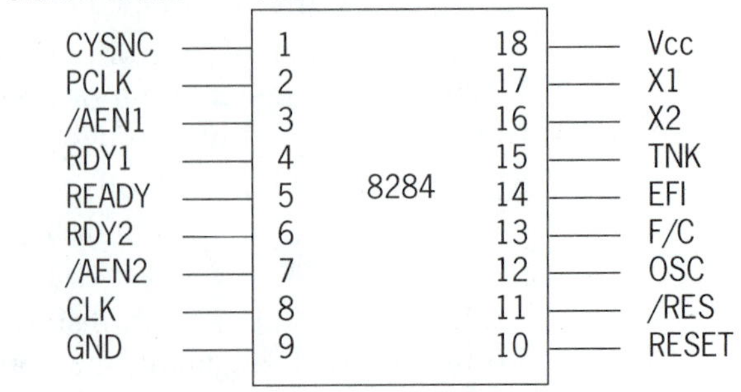

```
SIGNAL NAME      PIN

  CYSNC ——— 1          18 ——— Vcc
  PCLK  ——— 2          17 ——— X1
  /AEN1 ——— 3          16 ——— X2
  RDY1  ——— 4          15 ——— TNK
  READY ——— 5   8284   14 ——— EFI
  RDY2  ——— 6          13 ——— F/C
  /AEN2 ——— 7          12 ——— OSC
  CLK   ——— 8          11 ——— /RES
  GND   ——— 9          10 ——— RESET
```

Pin	Name	Dir	Function
1	CYSNC	I	Clock synchronize for multiple 8284s, active-high
2	PCLK	O	Peripheral clock, 50% duty cycle = CLK/2
3	/AEN1	I	Enable level for bus ready signal RDY1, active-low
4	RDY1	I	Bus ready (for data transfer) signal 1, active-high
5	READY	O	Ready for bus transfer, output to 8088 CPU
6	RDY2	I	Bus ready (for data transfer) signal 2, active-high
7	/AEN2	I	Enable level for bus ready signal RDY2, active-low
8	CLK	O	Clock to 8088 CPU, 33% duty cycle, f = 1/3 crystal
9	GND	I	Power supply ground
10	RESET	O	Active-high reset signal output to 8088 CPU
11	/RES	I	Active-low input reset pulse to 8284 from R/C circuit
12	OSC	O	Output clock pulse, f = crystal
13	F-/C	I	Select external frequency (high), or crystal (low)
14	EFI	I	External frequency, selected when F-/C is tied high
15	TNK	I	External tank circuit input for overtone crystals
16	X2	I	Crystal lead
17	X1	I	Crystal lead
18	Vcc	I	Power supply, +5VDC

Table 12.5 ■ 8284 Clock Generator Signals

The 8255 Programmable Peripheral Interface (PPI)

Many of the 8-bit microprocessor families that were popular in the 1970s featured a "family" of *standard* peripheral ICs. Common standard peripherals include serial data UARTs, interrupt controllers, DMA controllers, DRAM controllers, and parallel port I/O interface circuits. The trend in personal computers has been to integrate all of the interface circuits on one chip. Modern PC *chip sets* consist of a CPU, an optional math coprocessor, one multifunction peripheral chip, a VGA controller, and memory. Many of the older discrete peripheral chips live on, however, in inexpensive SBC designs. Parallel port peripheral chips, in particular, continue to be useful. A very common peripheral chip is the programmable parallel I/O port. Most applications of a microprocessor require some communication with the world outside of the computer. Parallel data flow between computer and other machinery is common as can be seen in the examples in Chapter 10 for parallel data interchange using the printer port.

Most peripheral chips can function in various *modes* of operation, which must be programmed into the chip before it can be used. An operational mode is chosen by the system program to accomplish a particular I/O configuration. Should I/O requirements change, the peripheral can be reprogrammed to meet the new requirements. The flexibility inherent in being able to program the peripherals to meet new and changing I/O situations is one reason the original 8-bit peripherals continue to be popular.

The 8255 PPI is a programmable, three parallel port, I/O peripheral circuit that has been popular for over 10 years. Table 12.6 lists the signal names for each 8255 connection. A pinout diagram of the 8255, along with a simple block diagram of the 8255 internal registers, is shown in Figure 12.4.

The 8255 internal architecture features three data registers, one for each port, and a *Control register* that can be programmed to control the opera-

Pin	Name	Dir	Function
			I/O Port A
4	PA0	B	Programmable Port A Bit 0
3	PA1	B	Programmable Port A Bit 1
2	PA2	B	Programmable Port A Bit 2
1	PA3	B	Programmable Port A Bit 3
40	PA4	B	Programmable Port A Bit 4
39	PA5	B	Programmable Port A Bit 5
38	PA6	B	Programmable Port A Bit 6
37	PA7	B	Programmable Port A Bit 7
			I/O Port B
18	PB0	B	Programmable Port B Bit 0
19	PB1	B	Programmable Port B Bit 1
20	PB2	B	Programmable Port B Bit 2
21	PB3	B	Programmable Port B Bit 3
22	PB4	B	Programmable Port B Bit 4
23	PB5	B	Programmable Port B Bit 5
24	PB6	B	Programmable Port B Bit 6
25	PB7	B	Programmable Port B Bit 7
			I/O Port C
			Lower Nibble
14	PC0	B	Programmable Port C Bit 0
15	PC1	B	Programmable Port C Bit 1
16	PC2	B	Programmable Port C Bit 2
17	PC3	B	Programmable Port C Bit 3
			Upper Nibble
13	PC4	B	Programmable Port C Bit 4
12	PC5	B	Programmable Port C Bit 5
11	PC6	B	Programmable Port C Bit 6
10	PC7	B	Programmable Port C Bit 7
			Control Inputs
9	A0	I	Internal registers Address Bit 0
8	A1	I	Internal registers Address Bit 1
6	/CS	I	Chip select, active-low
5	/RD	I	Read internal register, active-low
36	/WR	I	Write internal register, active-low
35	RESET	I	Reset, active-high
			Data
34	D0	B	Data Bit 0
33	D1	B	Data Bit 1
32	D2	B	Data Bit 2
31	D3	B	Data Bit 3
30	D4	B	Data Bit 4
29	D5	B	Data Bit 5
28	D6	B	Data Bit 6
27	D7	B	Data Bit 7
26	Vcc	I	Power supply voltage, +5V
7	GND	I	Power supply ground

Note that the 8255 does *not* use pins 40 and 20 for Vcc and ground.

Table 12.6 ■ 8255 PPI Pin Signals

Figure 12.4 ■ 8255 PPI.

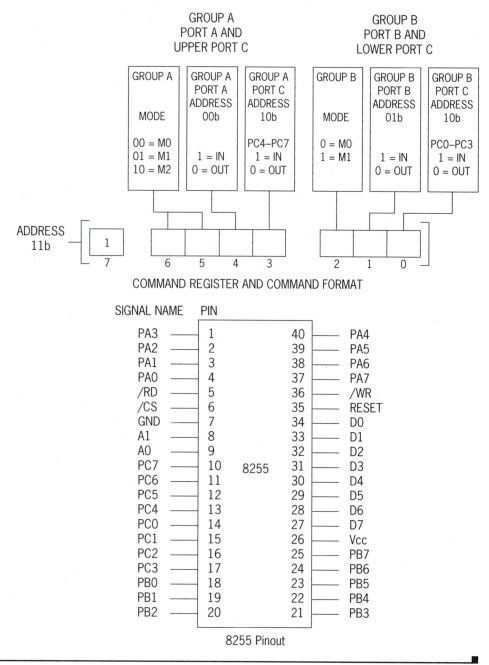

port, and a *Control register* that can be programmed to control the operational *mode* of the 8255.

The 8255 features *three* parallel I/O ports, named Ports A, B, and C. Each port can be programmed to be either an input or an output port, or in the case of Port A can be made bi-directional. Port C is further divided into upper and lower nibbles, which can be *individually* programmed to be input, output, or control.

In addition to programming each port for input or output, the 8255 can be operated on one of *three modes* by programming the control register. The three operational modes are as follows.

Mode 0: Basic I/O for Ports A, B, and C

Mode 1: Strobed I/O for Ports A and B using C for control

Mode 2: Bi-directional bus for Port A using C for control

Mode 0 is the most readily understood I/O mode and is used in the first SBC interface example.

PROGRAMMING THE **8255** PPI

Before the 8255 PPI can be used it *must* be programmed by the user program. The 8255 is programmed by writing *control* bytes from the CPU to the 8255 internal control register. Each control byte, written by the user's program to the 8255 control register, sets the 8255 mode and the input-output configuration of each of the I/O ports. Table 12.7 shows how a control byte to select an 8255 operational mode can be calculated by the programmer.

Note that *only* Port A can be programmed into operation mode 2. It is not necessary to program all of the ports to the same operational mode; the operational modes may be mixed.

The programmer can inspect Table 12.7 to determine the makeup of a control byte that is to be written to the 8255. For instance, the control byte 10010001b will program the 8255 as follows.

Ports A and C (upper) in mode 0

Port A is an input port, Port C (upper) an output port

Ports B and C (lower) in mode 0

Port B is an output port, Port C (lower) an input port

Control Byte Bit	Bit Function
7	1 = active (program the control register) 0 = Port C bit set/reset command
	Port A and C (upper nibble)
6,5	00 = Operation mode 0 01 = Operation mode 1 1X = Operation mode 2
4	1 = Port A is an input port 0 = Port A is an output port
3	1 = Port C (upper nibble) are inputs 0 = Port C (upper nibble) are outputs
	Port B and C (lower nibble)
2	0 = Operation mode 0 1 = Operation mode 1
1	1 = Port B is an input port 0 = Port B is an output port
0	1 = Port C (lower nibble) are inputs 0 = Port C (lower nibble) are outputs

Table 12.7 ■ 8255 Control Byte Format

8255 Register Address A1 A0	Addressed Register
0 0	Port A
0 1	Port B
1 0	Port C
1 1	Control register

Table 12.8 ■ 8255 Internal Register Addresses

ADDRESSING THE 8255

The 8255 contains one Control register and three port registers, as shown in Figure 12.4. The 8255 registers are addressed by Address Bits A0 and A1 on pins 9 and 8. The addresses for each of the four internal registers in the 8255 are shown in Table 12.8.

Note, for mode 0, that when a port register is written, the data byte written is *latched* into the port for *output*. When a port register is read, the port pin logic levels are read from the port pins at that *instant*. Modes 1 and 2 latch both output and input data. Note also that the control register should *only* be written, not read.

The instructions used by the program to read and write the 8255 registers depend on how the 8255 has been *mapped*. If the 8255 is designed *into memory* space, then data *moves* will program and operate the 8255. If the 8255 is mapped into *I/O* space, then *input and output* instructions are used to access and program the 8255. We shall I/O map the 8255 for the SBC design, and use IN and OUT instructions to access the internal registers.

The 8250 Universal Asynchronous Receiver/Transmitter (UART)

The UART chosen for the SBC design is the 8250, a very popular 1980s design used in many PC-compatible computers. A faster design, the 16450, is also available today.

A pinout and programmable register block diagram of the 40-pin 8250 DIP is shown in Figure 12.5. Table 12.9 lists the 8250 pins, signal names, and pin functions. Note that some of the features of the 8250 are the multiple chip select (CS0, CS1, /CS2) and read/write (/RD, RD, /WR, WR) pins. Multiple active-low/high control pins let the hardware designer use whichever active level is most convenient.

As was the case for the 8255 PPI, the 8250 UART must be programmed by the program before it can be used. In Chapter 10 we discussed using PC serial port BIOS interrupt services to initialize the PC UART for serial data transmission. The SBC, of course, is not equipped with a BIOS; therefore, we shall have to write initialization programs for the UART. Refer to Chapter 10 to review serial data transmission concepts and terms that are used in the following sections.

PROGRAMMING THE 8250 UART

Before the UART may be used to receive or transmit serial data, the data word length, parity type, number of stop bits, and data baud rate *must* be programmed. Modem handshaking bytes may also be programmed, if desired.

Figure 12.5 ■ 8250 UART 40-pin DIP and internal registers.

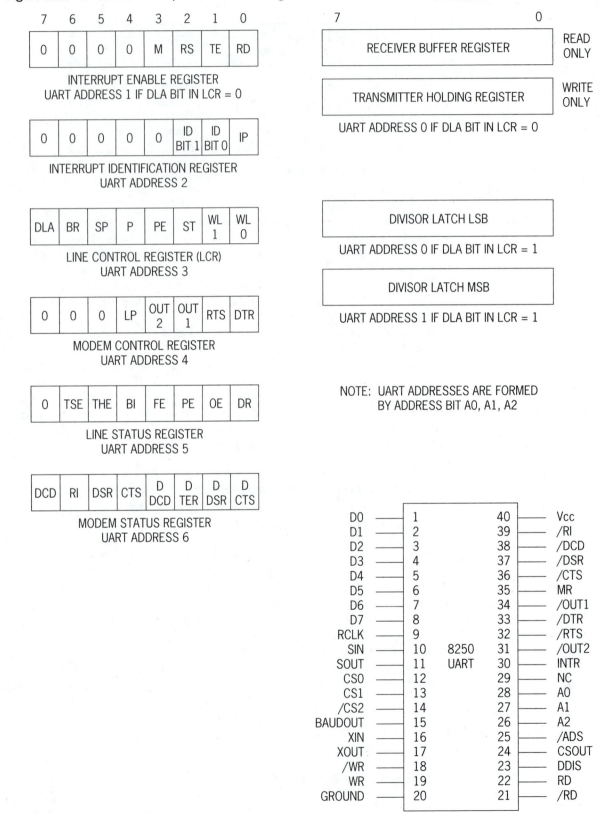

7	6	5	4	3	2	1	0
0	0	0	0	M	RS	TE	RD

INTERRUPT ENABLE REGISTER
UART ADDRESS 1 IF DLA BIT IN LCR = 0

7							0
		RECEIVER BUFFER REGISTER					

READ ONLY

| | | TRANSMITTER HOLDING REGISTER | | | | | |

WRITE ONLY

UART ADDRESS 0 IF DLA BIT IN LCR = 0

0	0	0	0	0	ID BIT 1	ID BIT 0	IP

INTERRUPT IDENTIFICATION REGISTER
UART ADDRESS 2

DLA	BR	SP	P	PE	ST	WL 1	WL 0

LINE CONTROL REGISTER (LCR)
UART ADDRESS 3

| | | DIVISOR LATCH LSB | | | | | |

UART ADDRESS 0 IF DLA BIT IN LCR = 1

0	0	0	LP	OUT 2	OUT 1	RTS	DTR

MODEM CONTROL REGISTER
UART ADDRESS 4

| | | DIVISOR LATCH MSB | | | | | |

UART ADDRESS 1 IF DLA BIT IN LCR = 1

0	TSE	THE	BI	FE	PE	OE	DR

LINE STATUS REGISTER
UART ADDRESS 5

NOTE: UART ADDRESSES ARE FORMED
BY ADDRESS BIT A0, A1, A2

DCD	RI	DSR	CTS	D DCD	D TER	D DSR	D CTS

MODEM STATUS REGISTER
UART ADDRESS 6

D0	1	40 — Vcc
D1	2	39 — /RI
D2	3	38 — /DCD
D3	4	37 — /DSR
D4	5	36 — /CTS
D5	6	35 — MR
D6	7	34 — /OUT1
D7	8	33 — /DTR
RCLK	9	32 — /RTS
SIN	10	31 — /OUT2
SOUT	11	30 — INTR
CS0	12	29 — NC
CS1	13	28 — A0
/CS2	14	27 — A1
BAUDOUT	15	26 — A2
XIN	16	25 — /ADS
XOUT	17	24 — CSOUT
/WR	18	23 — DDIS
WR	19	22 — RD
GROUND	20	21 — /RD

8250 UART

Pin	Name	Dir	Function
Data			
1	D0	B	Data Bit 0
2	D1	B	Data Bit 0
3	D2	B	Data Bit 0
4	D3	B	Data Bit 0
5	D4	B	Data Bit 0
6	D5	B	Data Bit 0
7	D6	B	Data Bit 0
8	D7	B	Data Bit 0
Address			
26	A2	I	UART register Address Bit A2
27	A1	I	UART register Address Bit A1
28	A0	I	UART register Address Bit A0
Control			
12	CS0	I	Chip select 0, active-high
13	CS1	I	Chip select 1, active-high
14	/CS2	I	Chip select 2, active-low
18	/WR	I	Write, active-low
19	WR	I	Write, active-high
21	/RD	I	Read, active-low
22	RD	I	Read, active-high
23	DDIS	O	Bus driver disable
24	CSOUT	O	Chip select out
25	/ADS	I	Address strobe, active-low
30	INTR	O	Interrupt request
35	MR	I	Master reset
Serial Data Transmission and Modem Control			
9	RCLK	I	Receiver clock
10	SIN	I	Serial data input
11	SOUT	O	Serial data output
15	BAUDOUT	O	Transmitter clock
31	/OUT2	O	User-programmable output bit
32	/RTS	O	Request to send, active-low
33	/DTR	O	Data terminal ready, active-low
34	/OUT1	O	User-programmable output bit
36	/CTS	I	Clear to send, active-low
37	/DSR	I	Data set ready, active-low
38	/DCD	I	Data carrier detect
39	/RI	I	Ring indicator
Power and Clock Crystal			
16	XIN	I	Crystal circuit
17	XOUT	O	Crystal circuit
20	Ground	—	Power supply ground
29	NC		No connection
40	Vcc	—	Power supply +5 VDC

Table 12.9 ■ 8250 UART Pin Signals

PROGRAMMING THE WORD LENGTH, PARITY TYPE, AND NUMBER OF STOP BITS

The 8250 UART has three address lines (A0, A1, and A2), that may form UART register addresses from 0 to 7. Actually, 10 UART registers are addressed, because bit 7 in the Line Control register (LCR = UART address number 3) controls access to *five* registers. UART register addresses 0 and 1 may be used by the Transmitter, Receiver, Interrupt Enable, and Divisor registers.

Bit 7 (bit DLA) of the Line Control register may be set to 1 by the programmer to address the divisor latch registers at UART addresses 0 and 1. Should DLA be set to 0 by the programmer, the divisor latch registers are *not* addressed. UART address 0 is used to *read* from the received data buffer, or *write* to the transmit data holding register when *DLA is set to 0.* UART address 1 accesses the Interrupt Enable Register when DLA is set to 0.

The 8250 Line Control register is programmed as shown in Table 12.10 to obtain the desired serial data structure and to control access to the five divisor/data registers.

PROGRAMMING THE BAUD RATE

The transmission baud rate is determined by dividing the input clock on XIN pin 16 by the *word* contained in the divisor latch registers. The resultant *baud rate clock* is internally connected to the transmitter shift register and made externally available on BAUDOUT pin 15. BAUDOUT is usually connected to RCLK pin 9 so that data reception is done at the same baud rate as data transmission.

An external clock oscillator may be connected to XIN, or a crystal clock oscillator circuit connected between XIN and XOUT. The clock frequency is divided by the word (the *divisor*) programmed into the divisor latch registers to generate a *serial* data transmission *baud rate clock.* The baud rate clock is *16* times the actual data baud rate. Making the baud rate programmable

Bit	Set	Function
7 (DLA)	0	Address Receive/Transmit and Interrupt Enable registers
	1	Address divisor latch registers
6 (BR)	0	Clear break output, set SOUT to logic high
	1	Set break output, set SOUT to logic low
5 (SP)	0	Use parity bit set by Bits 3 and 4
	1	Set parity bit to 0 if even parity selected
		Set parity bit to 1 if odd parity selected
4 (P)	0	Odd parity enabled
	1	Even parity enabled
3 (PE)	0	No parity enabled
	1	Parity enabled
2 (ST)	0	1 stop bit (1.5 for 5-bit words)
	1	2 stop bits
1 0 (WL)	0 0	5-bit character
	0 1	6-bit character
	1 0	7-bit character
	1 1	8-bit character

Table 12.10 ■ Line Control Register Program Bits

Baud Rate	Baud Rate Clock	Divisor
300	4,800	384
600	9,600	192
1,200	19,200	96
2,400	38,400	48
4,800	76,800	24
9,600	153,600	12
19,200	307,200	6

Table 12.11 ■ 1.8432 MHz Clock Divisors for Standard Baud Rates

enables the programmer to choose many standard (or nonstandard) serial data baud rates.

A review of Chapter 10 reveals that the highest BIOS data baud rate is 9,600. The minimum clock frequency needed to generate a ($16 \times 9,600$) baud rate clock is 153,600 Hz, if the divisor is 1. Baud rates of 19,200 and above are often used in modern systems, however, so that the clock frequency is often chosen to be in the MHz range. Table 12.11 shows divisors that may be used to generate standard baud rate clocks from a 1.8432 MHz clock frequency.

The divisor latch registers may be accessed at UART addresses 0 (MS byte) and 1 (LS byte) *if bit 7 (DLA)* of the Line Control register is set to 1 by the programmer.

MODEM HANDSHAKING BITS

Please refer to Chapter 10 for a discussion of modem handshaking signals. The 8250 UART may function as a DTE or DCE part, however the programmable handshaking signals are given DTE names. The Modem Control register (accessed at UART address 4) may be programmed as shown in Table 12.12.

DATA TRANSMISSION AND RECEPTION

The program may address the 8250 Transmitter Holding register, Receiver Buffer register, and Line Status register. The transmitter and receiver registers are *both* addressed at UART address 0, *if* DLA is set to 0 in the

Bit	Set	Function
7, 6, 5	0	Not used
4 (LP)	1	Loop check UART by internal connection SIN to SOUT
	0	Normal operation, disable loop check
3 (OUT2)	1	User handshaking bit /OUT2 set logic low
	0	User handshaking bit /OUT2 set logic high
2 (OUT1)	1	User handshaking bit /OUT1 set logic low
	0	User handshaking bit /OUT1 set logic high
1 (RTS)	1	DTE handshaking bit /RTS set logic low
	0	DTE handshaking bit /RTS set logic high
0 (DTR)	1	DTE handshaking bit /DTR set logic low
	0	DTE handshaking bit /DTR set logic high

Table 12.12 ■ Modem Control Register Bits

Bit	Set	Function
7	0	Not used
6 (TSE)	1	Transmitter Shift register empty
	0	Transmitter Shift register busy
5 (THE)	1	Transmitter Holding register empty
	0	Transmitter Holding register busy
4 (BI)	1	Break interrupt
	0	No break interrupt
3 (FE)	1	Framing error
	0	No framing error
2 (PE)	1	Parity error
	0	No parity error
1 (OE)	1	Overrun error
	0	No overrun error
0 (DR)	1	Data ready (New data in receiver)
	0	Data not ready

Table 12.13 ■ Line Status Register

Line Control register. The receiver may be accessed at address 0 when the address is *read*, whereas the transmitter is accessed at address 0 when the address is *written*.

The Line Status register is accessed at UART address 5 and *must* be checked by a *polling* program before reading or writing data to the UART. The Line Status register contains bits that show the status of the receiver and transmitter, as shown in Table 12.13.

In particular, a polling program *must* check Bit 0 (data ready) *before* attempting to read the receiver register. Failure to check for new received data results in reading the *previous* character. *Note that Bit 0 is reset to 0 whenever the receiver register is read* by the program. Resetting Bit 0 ensures that new data will be read once.

A polling program *must* also check Bit 5 (Transmitter Holding Register Empty) *before* writing new data to the Transmitter Holding register. Failure to check for an empty Transmitter Holding register will result in the program *overwriting* a previous character that has not yet been transmitted.

GENERATING INTERRUPTS FROM THE UART

The 8250 is capable of generating an interrupt (INTR, pin 30) if it has been programmed to do so. Interrupts are enabled by programming the Interrupt Enable register at UART address 1 (if DLA is set to 0). The cause of the interrupt may be read by an interrupt service routine program in the Interrupt Identification register, UART address 2. Tables 12.14 and 12.15 show the bit assignments for the interrupt registers. We shall not connect the 8250 as an interrupting device on the SBC, so the interrupt features will not be used.

MODEM STATUS REGISTER

The status of handshaking signals connected to the modem status pins of the 8250 may be read in the Modem Status register, UART address 6. The modem status bits are listed in Table 12.16.

Bit	Set	Function
7,6,5,4	0	Not used
3 (M)	1	Enable interrupt on modem status change
	0	Disable modem status change interrupt
2 (RS)	1	Enable interrupt on receiver line status error
	0	Disable receiver line status error interrupt
1 (TE)	1	Enable interrupt on transmitter holding register empty
	0	Disable interrupt on THRE
0 (RD)	1	Enable interrupt on received data ready
	0	Disable interrupt on received data ready

Table 12.14 ■ Interrupt Enable Register

Bit	Set	Function
7,6,5,4,3	0	Not used
2 1 0	0 0 0	Modem status: CTS, RTS, DSR, RI, DCD active
		Reset interrupt by reading Modem Status register
2 1 0	0 1 0	Transmitter Holding register empty
		Reset by writing data to Holding register
2 1 0	1 0 0	Received data available
		Reset by reading Receiver Buffer register
2 1 0	1 1 0	Receiver line status error
		Reset by reading the Line Status register
2 1 0	X X 1	No interrupt currently active

Table 12.15 ■ Interrupt Identification Register

Bit	Set	Function
7 (DCD)	1	Data carrier detect
	0	No data carrier
6 (RI)	1	Ring signal indicator
	0	No ring signal
5 (DSR)	1	Data set (modem) ready
	0	Data set not ready
4 (CTS)	1	Clear to send (transmit) data to data set
	0	Not clear to send
3,2,1,0		Change (D) in the associated modem status bit
		1 = Change, 0 = No Change
		Change bits are set to 0 when the
		Modem Status register is read
3 (D DCD)		
2 (D TER)		Indicates trailing edge (low-to-high) RI signal
1 (D DSR)		
0 (D CTS)		

Table 12.16 ■ Modem Status Register

RS232 Level-Shifting Gates

As discussed in Chapter 10, RS232 transmission levels are positive and negative voltages that typically are +/–15VDC. There are many RS232 level-shifting gates available on the market. We shall use the common 1488 (logic levels to RS232 levels) and 1489 (RS232 levels to logic levels) gates in the design.

12.3 An 8088 SBC Design

A design for the 8088 SBC, based on the original specifications of Table 12.1, is shown in Figure 12.6. The SBC design is based on the following considerations.

Figure 12.6 ■ 8088-based single-board computer.

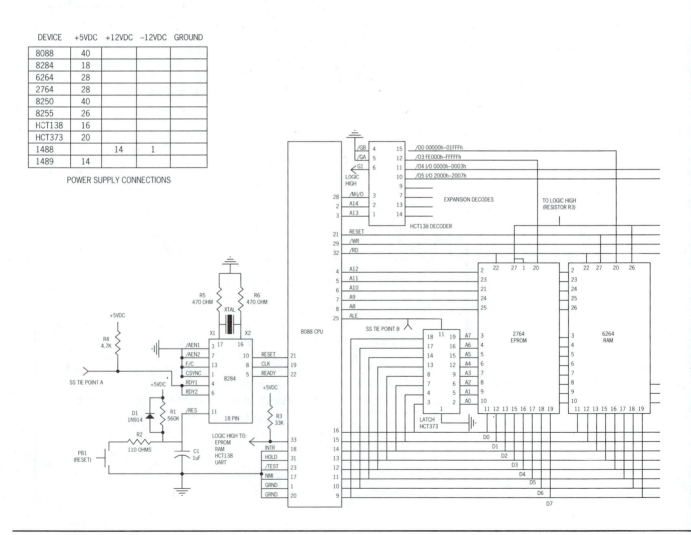

DEVICE	+5VDC	+12VDC	–12VDC	GROUND
8088	40			
8284	18			
6264	28			
2764	28			
8250	40			
8255	26			
HCT138	16			
HCT373	20			
1488		14	1	
1489	14			

POWER SUPPLY CONNECTIONS

8284 Clock Generator

■ Ground /AEN1 and /AEN2 to enable RDY1 and RDY2.

■ Tie RDY1 and RDY2 high through R4 to enable READY CPU signal.

■ Ground F–/C to enable a crystal frequency source.

■ Ground CSYNC to disable any multiple 8284 synchronization.

■ Connect the reset circuit of R1, R2, C2, and D1 to /RST. On power-up, or when PB1 is pushed, an active-low /RST is generated until C2 charges to +5V. The 8284 /RST *input* generates an 8284 RESET *output* to the CPU.

■ Connect a 14.318 MHz fundamental mode crystal to the crystal

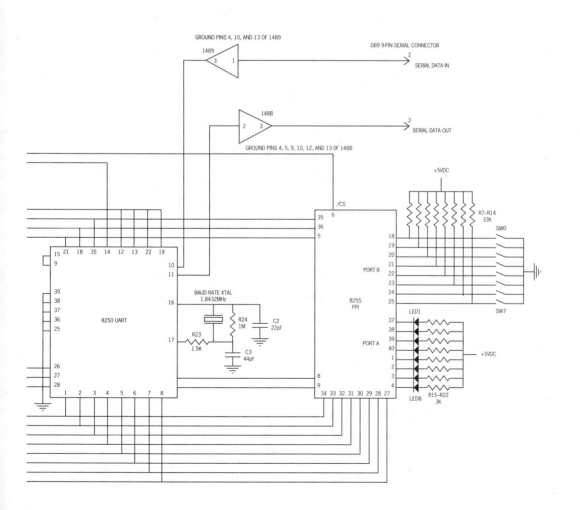

input pins X1 and X2. Resistors R5 and R6 complete the crystal circuit.

8088 CPU, HCT373 Lower Address Latch, RAM, and EPROM

■ Connect AD0–AD7 to the external HCT373 data latch and the RAM, EPROM, UART, and PPI data pins. Use the ALE pulse from the CPU to store A0–A7 into the HCT373 latch during T1.

■ Connect A8–A12 to the 8K RAM and EPROM A8–A12 address lines.

■ Connect A13, A14, and /M-I/O to the HCT138 address decoder.

■ Connect A0–A7 from the HCT373 latch to RAM and EPROM address lines A0–A7.

■ Connect A0 and A1 from the HCT373 to the PPI internal register address lines A0 and A1.

■ Connect A0, A1, and A2 from the HCT373 to the UART address lines.

■ Ground HOLD, /TEST, INTR, and the NMI inputs to prevent unwanted CPU operation.

The inclusion of the CPU input signals NMI, INTR, /TEST, and HOLD in our list of signals is the result of an old digital design adage: *Leave No (Control) Input Floating.* We may *safely* ignore output signals, such as A15–A19, that we do not *wish* to use. But we may *not* ignore input control signals to the CPU, whether we wish to use them or not.

Table 12.17 lists the 8088 CPU signals that are used in the SBC design.

8255 PPI

■ Connect the 8255 /CS to the HCT138 decoder I/O output 4.

■ Connect the 8284 RESET signal to the 8255 RESET input.

■ Connect the 8255 data lines to the 8088 data lines.

■ Connect eight input switches to Port B.

■ Connect eight output LEDs to Port A.

■ Connect HCT373 address lines A0 and A1 to the 8255 address pins.

■ Connect the /RD and /WR control lines to the 8255 control pins.

The 8255 is mapped into I/O space by *physically* connecting it to an HCT138 decoder output that is active-low only when an I/O instruction is used by the program. It is important to note that the space (memory or I/O) that an output device uses depends solely on the hardware design. The programmer is forced to use the correct type of instruction (memory or I/O) by the hardware-mapping design.

Pin	Signal Name	Function
16	AD0	Data bit D0 and address bit A0
15	AD1	Data bit D1 and address bit A1
14	AD2	Data bit D2 and address bit A2
13	AD3	Data bit D3 and address bit A3
12	AD4	Data bit D4 and address bit A4
11	AD5	Data bit D5 and address bit A5
10	AD6	Data bit D6 and address bit A6
9	AD7	Data bit D7 and address bit A7
8	A8	Address signal A8
7	A9	Address signal A9
6	A10	Address signal A10
5	A11	Address signal A11
4	A12	Address signal A12
3	A13	Address signal A13
2	A14	Address signal A14
17	NMI	Nonmaskable interrupt, L-H edge-triggered
18	INTR	Maskable interrupt, high-level enabled
21	RESET	CPU reset signal, active-high level
22	READY	Memory or I/O is ready, active-high level
23	/TEST	Input bit for WAIT opcode, active-low
24	/INTA	Interrupt acknowledge
25	ALE	Address latch enable
28	I/O-/M	I/O (high) or memory (low)
29	/WR	Write, active-low
31	HOLD	Hold request, active-high
32	/RD	Read, active-low
33	MN-/MX	Minimum mode (high)
19	CLK	CPU clock pulse, 33% duty cycle
1	GND	CPU ground
20	GND	CPU ground
40	Vcc	CPU power, +5VDC

Table 12.17 ■ 8088-Based SBC CPU Signals

8250 UART

- Connect a 1.8432 MHz crystal circuit to XIN and XOUT.
- Connect BAUDOUT to RCLOCK.
- Connect /CS2 to the HCT138 decoder I/O output 5 (I/O).
- Connect RESET input to the 8088 reset signal.
- Connect data lines D0–D7 to the 8088 data lines.
- Connect HCT373 address outputs A0, A1, and A2 to the 8250 addresses.
- Connect the 8088 /RD and /WR signals to the 8250 read/write pins.
- Disable the unused 8250 RD, WR, CS0, CS1, and /ADS inputs.

■ Ground modem signals /DSR, /RI, /DCD, and /CTS.

■ Connect SIN and SOUT to the RS232 level-shifting gates.

74HCT138 Octal Address Decoder

Decode A13, A14, and I/O-/M in an HCT138 3-to-8 decoder as follows.

8088 Address			HCT138-Decoded	Hex Address	Space
I/O-/M	A14	A13	Output	in Program	
0	0	0	0	X0000–X1FFF	Memor y
				X8000–X9FFF	(RAM)
0	0	1	1	X2000–X3FFF	Memory
				XA000–XBFFF	
0	1	0	2	X4000–X5FFF	Memory
				XC000–XDFFF	
0	1	1	3	X6000–X7FFF	Memory
				XE000–XFFFF	(EPROM)
1	0	0	4	0000–1FFF	I/O
				8000–9FFF	(PPI)
1	0	1	5	2000–3FFF	I/O
				A000–BFFF	(UART)
1	1	0	6	4000–4FFF	I/O
				C000–DFFF	
1	1	1	7	6000–7FFF	I/O
				E000–FFFF	

Note that the SBC design of Figure 12.6 does *not* decode any address line from A15 to A19. The memory specifications of Table 12.1 call for 8K each of RAM and EPROM. Memory sizes of 8K require 13 address lines, A0 to A12. Address lines A13 and A14 are decoded by the HCT138 decoder to provide *four* memory and four I/O chip *select* signals. Because all address lines *above* A14 are ignored, *any* address in memory or I/O space can be specified as shown.

MEMORY AND I/O ADDRESSES

EPROM is selected whenever a memory address contains a 1 in address bits A14 and A13. Addresses FE000h–FFFFFh will access the EPROM. RAM is selected whenever a memory address contains a 0 in bits A14 and A13. Low RAM address 00000h–01FFFh will access the RAM. The PPI is selected whenever an I/O cycle is in progress and the I/O address has bits A14 and A13 set to 0. I/O addresses of 0000h–1FFFh access the PPI. The UART is accessed at any I/O address from 2000h to 3FFFh.

When the CPU is reset, it will automatically fetch its first instruction from memory location FFFF0h. We *must* decode our EPROM in a memory space such that all instructions are fetched from it. Address lines A14 and A13 are both 1 for a reset address of FFFF0h, so the program in the EPROM is accessed by the HCT138 decoder on reset. As long as the program does not try to jump out of the range of FE000h to FFFFFh (a 1FFFh ROM space), A14 and A13 will both remain a 1, and the 8K EPROM will be accessed by the decoder for all program instructions.

If interrupts are to be used by the SBC design, the interrupt vectors *must* be placed in low RAM, from 00000h to 003FFh. The HCT138 decoding is chosen so that any memory address from 00000h to 01FFFh will make A14 and A13 both 0 so that the RAM will be accessed for any address between 00000h and 01FFFh.

Actually, *any* address that makes A14 and A13 both 1 will access the EPROM, and any memory address that makes A14 and A13 both 0 will access the RAM. Program addresses from FE000h to FFFFFh, however, describe a 2000h-byte ROM space. Program addresses from 00000h to 01FFFh describe a 2000h-byte block of RAM space.

The 8255 I/O chip is chip-enabled by the HCT138 whenever I/O/M goes high, signifying an I/O cycle is in progress. The 8255 *may* be mapped anywhere in I/O space that has address lines A14 and A13 both 0. The 8255 is *chosen* to be addressed at I/O locations 0000h to 1FFFh, which makes address lines A13 and A14 both 0.

The 8250 UART is enabled by the HCT138 for any I/O address that makes A14 a 0, and A13 a 1. I/O addresses from 2000h to 3FFFh will generate the A14 and A13 bits that enable the UART from the HCT138.

Additional memory and I/0 chips may be added to the SBC design of Figure 12.6 simply by connecting unused outputs from the HCT138 decoding chip. Typically, one *chooses* program addresses for memory and I/O at convenient numbers, such as 02000H to 03FFFh for memory and C000h to D1FFFh for I/O.

12.4 SBC Timing

The SBC designer *must* check the timing available for memory read and write bus cycles so that RAM and EPROM parts are chosen that can respond in the available time.

Chapter 11 discusses the timing requirements of static RAM and EPROM memories. The timing available to static memories on the SBC are very similar to memory cycles on the PC bus. PC bus timing is based on 8088 CPU cycle timing.

Using the *simplified* 8088 CPU memory access timing diagram shown in Figure 12.2, the SBC bus timing events are described as follows.

- The I/O-/M signal goes active at the beginning of T1.
- Memory address bits A0 to A7 become valid when T1 goes high $(t = 0)$, and persist to the end of T2 $(t = 1.33T)$.
- ALE rises and falls in phase with the positive part of T1. A falling-edge triggered latch will latch A0 to A7 at the end of T1 $(t = .33T)$.
- The controlling signal, /RD (read) or /WR (write), goes active-low from the bus at the beginning of T2 to the end of T3.
- Memory devices have until the end of /RD to drive data bits D0 to D7 onto the data bus during a read cycle, or until the end of /WR to get data from the bus during a write cycle. The rising edge of /RD signals that the data placed on the bus by the memory has been latched into the 8088 CPU during a read cycle. The rising edge of

/WR is use by memory to latch data placed on the bus by the 8088 CPU during a write cycle.

■ Period T4 ends the memory access cycle.

Note that the term *memory* refers to devices decoded into memory space *and* I/O space. The total time, *from the time the address lines go valid at the memory pins* until /RD or /WR goes high, is known as the memory *access* time. The memory and I/O circuits on the SBC *must* have interchanged data with the CPU by the *end* of /RD or /WR. Any circuit delays that occur between the CPU and memory will reduce memory access time. For instance, using the SBC design shown in Figure 12.6, the address lines do not go valid until A0 and A7 have propagated through the HCT373 latch. The HCT373 latch is a *transparent* latch, which means the data will flow through it, from input pins to output pins, when ALE goes high.

SBC Memory Timing Requirements

For our timing calculations we shall assume all HCT time delays are 10 nanoseconds, as is done in Chapter 11. Remember, t = 0 when the addresses go *valid at the CPU pins.*
Addresses are valid at memory pins:

Addresses valid = Addresses valid + 10 ns = 10 ns

$$\underbrace{\hspace{4cm}}_{\text{HCT373 delay}}$$

The chip enable signals to each memory chip are generated from the HCT138 decoder. The HCT138 decoder receives the decode signals /M-I/O at the beginning of T1. Addresses A14 and A13 are available when addresses go valid at the CPU, as no delay through the latch is involved with the high address lines. The chip select (enable) signals are then available at:

Chip enable (/CE) = Addresses valid + 10 ns = 10 ns

$$\underbrace{\hspace{4cm}}_{\text{HCT138 delay}}$$

/RD or /WR go active-low at the end of T1 (t = .33T) and persist until the end of T3 (t = 2.33T). Memory chips have until the end of the /RD or /WR pulse to interchange data with the 8088 CPU.

RAM and EPROM Read Times

Using the memory times defined in Chapter 11, we have:

T1 = End of /RD – addresses valid at memory = 2.33T – 10 ns
T2 = End of /RD – /CE = 2.33T – 10 ns
T3 = End of /RD – beginning of /RD = 2.33T – .33T = 2T

For an 8088 CPU clock frequency of 4.77 MHz, a T state lasts 209 ns. The read access times for memory are as follows:

T1 = 2.33T – 10 ns = 487 – 10 = 477 ns
T2 = 2.33T – 10 ns = 487 – 10 = 477 ns
T3 = 2T = 418 ns

Comparing the SBC memory read times with those required by the SRAMs and EPROMs of Table 11.8 shows the following for the slowest memories listed. All times are nanoseconds.

Read Time	Available on SBC	SRAM Memory Requirement	Safety Margin
T1	477	150	327
T2	477	150	327
T3	418	70	348

Read Time	Available on SBC	EPROM Memory Requirement	Safety Margin
T1	477	250	227
T2	477	250	227
T3	418	100	318

The SBC read times are more than adequate for the slowest memories now commercially popular. Note that EPROMs are *slower* than SRAMs of the same generation. EPROMs are slower than equivalent-technology SRAMs because of the nature of EPROM floating-gate technology versus SRAM flip-flop technology. Fast SBC designs will typically be constrained by EPROM read speeds.

RAM Write Times

Using the definitions of Chapter 11, we have the following SRAM write times.

T1 = End of /WR – addresses valid at memory = 2.33T – 10 ns
T2 = End of /WR – addresses valid at memory = 2.33T – 10 ns
T3 = /WR pulse width = 2T
T4 = End of /WR – data valid at memory = 2.33T – 1.33T = T

For T states of 209 ns, the SBC actual write times are:

T1 = 487 – 10 = 477 ns
T2 = 487 – 10 = 477 ns
T3 = 418 ns
T4 = 209 ns

Based on the SRAM write times listed in Table 11.8, a 150–ns SRAM will have the following safety margins.

Write Time	Available on SBC	SRAM Memory Requirement	Safety Margin
T1	477	100	377
T2	477	100	377
T3	418	90	328
T4	209	60	149

Again we see that a relatively slow SRAM has generous write timing safety margins.

Timing Parameter	Type	PPI Timing Requirement	
		8255	8255A-5
Tcs = Chip select to data valid	R/W	250	200
Tar = Address valid to data valid	Read	250	200
Trd = /RD low to data valid	Read	250	200
Trw = Read pulse width	Read	300	300
Taw = Address valid to data valid	Write	100	100
Tww = Write pulse width	Write	400	300

Table 12.18 ■ 8255 PPI Timing Requirements

8255 PPI Timing

Timing for all memory and I/O devices on the SBC must be checked. The 8255 PPI is mapped in I/O space. 8088 CPU I/O cycles are identical to 8088 CPU memory access cycles. The 8255 PPI has read and write access times (in nanoseconds) as shown in Table 12.18.

From the timing requirements shown in Table 12.18, the 8255 timing safety margins are as follows. All times are nanoseconds.

Time	Available on SBC	PPI Requirements		8255 Margin	8255A-5 Margin
		8255	8255A-5		
Tcs	477	250	200	227	277
Tar	477	250	200	227	277
Trd	418	250	200	168	218
Trw	418	300	300	118	118
Taw	477	100	100	377	377
Tww	418	400	300	18	118

The faster 8255A part should be chosen to leave a generous write pulse width safety margin.

8250 UART Timing

The timing requirements for the 8250 UART are listed in Table 12.19. All times are in nanoseconds.

Time	Type	UART Requirement
Tcs = Chip select to data valid	R/W	285
Tar = Address valid to data valid	Read	285
Trd= /RD low to data valid	Read	175
Trw = Read pulse width	Read	175
Taw = Address valid to data valid	Write	235
Tww = Write pulse width	Write	175

Table 12.19 ■ 8250 UART Timing Requirements

Based on the timing requirements shown in Table 12.19, the 8250 safety margins are shown next.

Time	Available on SBC	UART Requirements	Margin
Tcs	477	285	192
Tar	477	285	192
Trd	418	175	243
Trw	418	175	243
Taw	477	235	242
Tww	477	175	302

The timing margins for the 8250 are adequate.

12.5 A Single-Step Test Circuit for the SBC

A single-step circuit for the SBC is shown in Figure 12.7. Such circuits are useful for hardware debugging when first testing a newly built design.

The single-step circuit may be connected to the tie points *A* and *B* shown in Figure 12.6. The single-step circuit operates by bringing the 8284 RDY1 and RDY2 signals *low* after each machine cycle ALE pulse clears the 7474 flip-flop, FF1. The 8284 responds to /RDY1 and /RDY2 by generating a READY low signal to the 8088 CPU, which then enters a *wait* state until READY goes high.

Figure 12.7 ■ Single-step circuit.

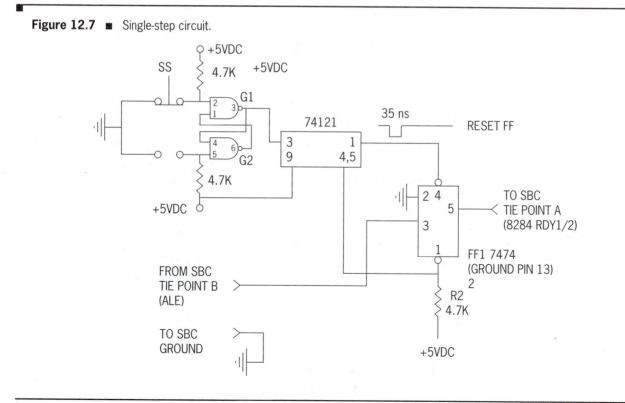

An instruction step is initiated by pushing the SS switch. RS flip-flop G1-G2 will then set flip-flop FF1, via the active-low 74121 one-shot output. FF1 asserts the 8284 RDY1/2 signals high when set. The 8284 brings 8088 CPU input pin READY high, and the 8088 completes one bus cycle. The next ALE pulse then clears FF1. When FF1 is cleared, the 8284 RDY1/2 inputs are low. The 8284 then brings the 8088 READY signal low, and the CPU enters a wait state until SS is pressed again.

The pulse length of the 74121 one-shot is chosen to be much shorter (35 ns) than a machine cycle (209 ns) so the flip-flop is not DC set during ALE. Switch SS must *not* bounce (or be de-bounced by an RS latch as shown), or several flip-flop sets may take place for each SS switch actuation.

Remember that each bus cycle fetches a *byte* of program opcode, *not* an entire opcode byte sequence. A logic probe may be used to check the address and data lines to determine if the proper code addresses and code bytes are fetched by the computer. Remember also that the opcode bytes are stored in the 8088 CPU internal 4-byte instruction cache, and opcode execution will not take place until an entire opcode has been fetched.

12.6 Test Programs for the SBC

Once the SBC is built, it must be tested to ensure that no construction errors exist. Testing any computer should begin with the simplest possible test, one that ensures the computer will fetch and execute code, and then proceed to more advanced tests. I shall present a series of simple programs that test all of the SBC functions.

Before beginning any program testing, *every* SBC connection should be checked with an ohmmeter or other continuity-checking meter. The most important continuity check is to ensure that +5V and ground are *not* connected to each other. Next, you should check to make sure that all chips have +5V and ground connected to the appropriate power pins. Finally, all interconnections should be checked for accuracy. Then, *someone else* should make the same checks.

Keep in mind that the SBC has no operating system, such as DOS, to load and run our programs. On reset, the CPU simply fetches an instruction from the EPROM at memory space address FFFF0h. It is the designer's *responsibility* to make sure there is a valid opcode in the EPROM at address FFFF0h.

EPROMs are most commonly programmed from files that are pure binary in format, thus we *must* assemble our programs with the extension .BIN instead of our normal .COM extension. Binary files begin the program at 0000h, not 0100h as is done for COM files. A86 will produce .BIN files from .ASM files automatically *if* an ORG 0000h directive is found in the .ASM file.

EPROM Function Test

The computer must be able to fetch code from the EPROM and execute the code before any advanced tests can be run.

Begin program testing by inserting the 8088 CPU, the 8284 clock generator, the HCT373 latch, and a programmed EPROM into the SBC. The EPROM should contain the test program codetest.asm, which is listed next.

```
;Test program codetest.asm to verify that the CPU can reset and
;fetch code from the EPROM. A series of jumps is done to test all
;EPROM address lines. The CPU will reset to address FFFF0h in memory
;space, or address 1FF0h on the 8K EPROM.

code segment
        org 0000h               ;bottom of EPROM – address FE000h in memory
        jmp 0002h               ;take jumps at addresses = power of 2
        org 0002h
        jmp 0004h               ;0004
        org 0004h
        jmp 0008h
        org 0008h               ;0008
        jmp 0010h
        org 0010h               ;0010
        jmp 0020h
        org 0020h               ;0020
        jmp 0040h
        org 0040h               ;0040
        jmp 0080h
        org 0080h               ;0080
        jmp 0101h
        org 0101h               ;0101 *** SEE NOTE AT BOTTOM ***
        jmp 0200h
        org 0200h               ;0200
        jmp 0400h
        org 0400h               ;0400
        jmp 0800h
        org 0800h               ;0800
        jmp 1000h
        org 1000h               ;1000
        jmp 1ff0h
        org 1ff0h               ;CPU resets here
        jmp 0fe00h:00h
code ends
```

NOTE: A86 *assumes* a .COM file if an ORG 100h statement is found and a .BIN file if an ORG 0000h statement is found. A86 acts on the *last* of these two ORG directives found in the program, so an ORG 101h was used to test address line A9, not ORG 100h, and thus a .BIN file is assembled.

All of the jumps in program codetest.asm are relative jumps except the last jump instruction at memory space address FFFF0h (8K EPROM address 1FF0h). The jump instruction, JMP 0FE00h, will cause the CPU to jump to memory space address FE000h, or address 0000h in the 8K EPROM.

Program codetest.asm exercises every address line used to address the EPROM. On reset, the CPU fetches the first instruction from memory space address FFFF0h, which is address 1FF0h, in the EPROM. A jump instruction at EPROM address 1FF0h then jumps to address FE000h in memory space, or address 0000h in the EPROM. A jump at EPROM address 0000h then jumps to EPROM address 0002h, the code there jumps to EPROM address 0004h, and so on until all individual address lines have been checked. Using the single-step circuit, the tester can check the EPROM address pins after every jump to ensure that the jumps are correctly taken.

RAM Test

Once we have confidence that the computer will fetch and execute a program by running codetest.asm, we can test the RAM. Insert the RAM in the SBC, and run the following program, ramtest.asm.

```
;RAM test program ramtest.asm writes data to every RAM
;location and then reads it back and checks for any error. Correct
;operation causes the program to loop at label "pass." Failure
;causes the program to loop at label "failed."

code segment
        org 0000h              ;assemble as a .BIN file for EPROM
start:  mov ax,0000h           ;set DS to point to the RAM segment 0000h
        mov ds,ax
        mov cx,2000h           ;check all 2000h RAM locations
        mov bx,0000h           ;BX points to each RAM byte
next:   mov al,0aah            ;use a checkerboard byte test
        mov b[bx],al           ;write the RAM
        mov ah,b[bx]           ;read the RAM
        cmp ah,al              ;compare read and write bytes
        jne fail               ;fail loop if not equal
        inc bx                 ;point to next RAM byte
        loopnz next            ;loop until CX = 0000
        jmp pass               ;all tested good
fail:   jmp failed
        org 200h               ;fail address is 0200h
failed: jmp failed             ;loop here if failed
        org 400h               ;pass address is 0400h
pass:   jmp pass               ;loop here if pass
        org 1ff0h              ;CPU resets here
        jmp 0fe00h:00h         ;jump to beginning of program
code ends
```

RAM test program ramtest.asm can be run at full speed, and a logic probe used to see where the program loops by checking A9 active (fail) or A10 active (pass). (A0 and A1 will be active in any case as the CPU fetches the three jump opcode bytes.)

8255 PPI Test

The next test program, iotest.asm, exercises the 8255 by reading the input switches and lighting the LEDs. First, the 8255 must be programmed for simple I/O, then the switch Port B is read and LED Port A is written. The test verifies that a closed switch will light a LED for that bit position.

```
;program iotest.asm, which exercises the 8255 I/O PPI by programming
;the 8255 to operate in mode 0, then reading port B switches and
;writing port A LEDs.

        prta equ 00h           ;port A address is xxxxxx00b
        prtb equ 01h           ;port B address is xxxxxx01b
        conreg equ 03h         ;contol register address is xxxxxx11b
        mod0oi equ 8bh         ;mode 0, A output, B and C input
```

```
code segment
        org 0000h              ;assemble as a .BIN file for EPROM
        mov al,mod0oi          ;program the 8255 control register mode 0
        out conreg,al          ;out to 8255 in I/O space
        in al,prtb             ;read the switches on port B
        out prta,al            ;write the LEDs on port A
        org 1ff0h              ;reset CPU here
        jmp 0fe00h:00h         ;jump to beginning of program
code ends
```

8250 UART Test

A simple UART test involves connecting the SBC serial port to a PC serial port, using the nonhandshaking serial cable of Chapter 11. The PC program sends a character to the SBC and then displays whatever character is received from the SBC. A PC serial data transmission/reception program similar to those listed in Chapter 11 may be used.

The SBC test program, uart.asm, listed next, involves programming the UART and then waiting for a character to be received from the PC. The received character is then retransmitted by the UART test program.

```
;The UART test program uart.asm for testing the SBC 8250 UART.
;Data transmission is programmed to be at 9,600 baud, even parity,
;1 stop bit, and an 8-bit character.

        code segment                                                   Note
                divlsb equ 0ch     ;divide by 0012d for 9,600 baud from a
                                   ;1.8432 MHz crystal
                divmsb equ 00h
                divl equ 2000h     ;Divisor register LSB address            1
                config equ 1bh     ;9,600 baud, PE, 1 stop bit, 8 bits/char 2
                lcr equ 2003h      ;Line Control register at UART address 3
                lsr equ 2005h      ;Line Status register at UART address 5
                mcr equ 2004h      ;Modem Control register at UART address 4
                rtran equ 2000h    ;Transmitter Holding register, Receiver  3
                                   ;Buffer register at UART address 0
        ;end equates

                mov al,80h         ;set LCR to address divisor registers
                mov dx,lcr         ;set DLA bit to 1 in LCR
                out dx,al          ;Divisor registers now addressable
                mov dx,divl        ;set LSB of divisor for 9,600 baud
                mov al,divlsb
                out dx,al
                inc dx             ;set MSB of divisor
                mov al,divmsb
                out dx,al
                mov dx,lcr         ;now program LCR for character format
                mov al,config      ;configure data character
                out dx,al          ;receive/transmit registers now addressable
                mov dx,mcr         ;set Bit 4 of MCR for normal operation
                mov ax,00h
                out dx,al
```

```
            mov dx,rtran        ;get power-up character from UART
            in al,dx            ;dump any power-up UART lint              4
            mov dx,lsr          ;wait for PC character to be received
    rcve:
            in al,dx            ;get receiver status
            rcr al,1            ;wait for data reception bit to be 1
            jnc rcve
            mov dx,rtran        ;read character from receiver buffer
            in al,dx
            mov bl,al           ;save character
    tran:
            mov dx,lsr          ;make sure transmitter is not busy
            in al,dx
            and al,20h          ;mask for Bit 5 (THE) status
            cmp al,20h          ;wait for transmitter to be empty
            jne tran
            mov dx,rtran        ;tranmitter empty, send byte
            mov al,bl           ;retrieve character
            out dx,al           ;send
            jmp rcve            ;repeat as desired
```

NOTES ON THE PROGRAM

1. If DLA bit in the LCR is a 1.
2. Refer to Chapter 10, Interrupt 14h.
3. If DLA bit in LCR is 0.
4. On reset, the Line Status register is set to 60h, indicating that no data is ready, and the transmit registers are empty. Power-on noise may, however, give a false data-received state in LSR Bit 0. The code will empty the receiver register.

The UART test program is a polling type. Connecting the UART as a real-time interrupting source is left as a project.

12.7 Interrupt-Driven I/O

The 8088 will respond to an INTR active high level signal by executing what is known as an *interrupt acknowledge* bus cycle. An interrupt acknowledge bus cycle is one in which the CPU reads an interrupt *type* byte from the data bus, *not* memory or I/O data. The CPU interrupt acknowledge signal, */INTA*, is used to read the type byte, instead of the /RD pulse normally used to read data.

The external circuit that caused INTR to go active *must* be designed to place its interrupt type byte on the data bus in response to an /INTA pulse. If more than one interrupting source is present on the SBC, then the interrupting sources must be designed in such a way that only *one* source can supply an interrupt type byte to the CPU at any given time. A special interrupt control IC, such as the 8259 interrupt controller discussed in Chapter 11, may also be incorporated in the design to handle multiple interrupt sources.

Figure 12.8 ■ 8088 interrupt acknowledge cycle.

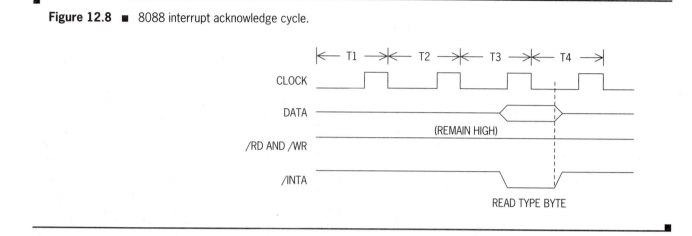

An interrupt bus cycle timing diagram is shown in Figure 12.8. An interrupt bus cycle appears very similar to a standard data read bus cycle, *except* the /INTA pulse is used as a read control signal. The sequence of events, for an interrupt cycle, is as follows.

- The external interrupting circuit brings the INTR line to the CPU to the high state.

- The CPU completes whatever instruction is in progress, stores the current program's return address and flags on the stack, and executes two interrupt acknowledge bus cycles.

- The 8088 CPU converts the address-data lines to data lines and issues the /INTA read pulses.

- The interrupting device *must* place its interrupt type byte on the data bus in response to the last /INTA pulse.

- The CPU reads in the interrupt type byte during the second interrupt acknowledge cycle, multiplies the type by 4, finds the interrupt vector in low RAM, and jumps to the interrupt service routine associated with the interrupt type.

- The original program proceeds after the interrupt service routine returns.

Note that there are *two* interrupt acknowledge bus cycles per interrupt. The first interrupt cycle may be used by the interrupting circuitry to reset (turn off) the INTR signal, so that it does not re-interrupt the CPU. The second interrupt bus cycle can then used by the interrupting circuit to place its interrupt type number on the data bus.

The interrupt type byte associated with the interrupting device can be permanently wired into the device, or programmed by the CPU during system initialization.

To illustrate the operation of using interrupt-driven hardware with the SBC, the circuit of Figure 12.6 will be modified to enable a hardware-interrupt cycle. The modifications to Figure 12.6 are shown in Figure 12.9.

Figure 12.9 ■ Interrupt circuit.

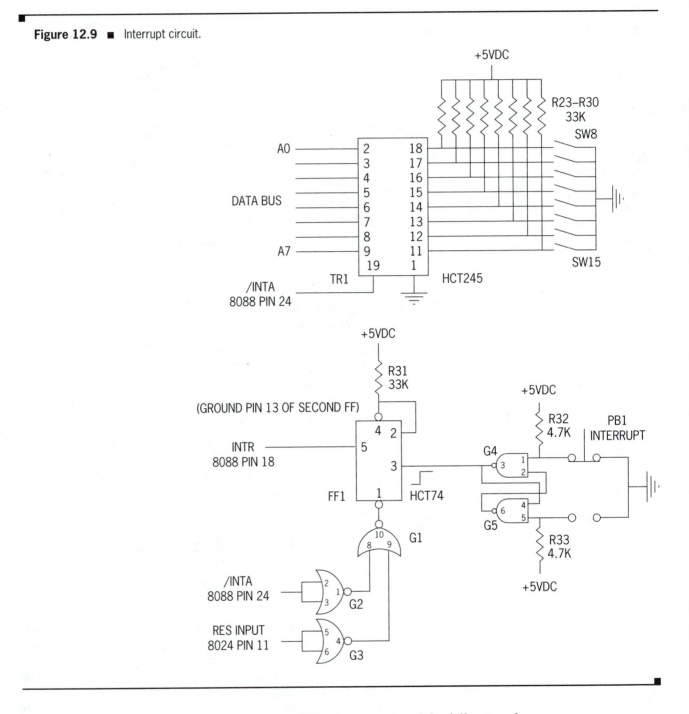

Interrupt modifications consist of the following changes.

■ An interrupt flip-flop, FF1, is added. FF1 is set by the clock edge generated when push button PB1 is pressed. FF1 is reset by the /INTA pulse. PB1 must be bounce-free, to prevent multiple interrupts, or de-bounced using an RS latch, as shown.

- Each /INTA cycle enables the HCT245 transceiver, TR1, which places the interrupt type number, set by switches SW0–SW7, on the data bus. The CPU executes the interrupt service routine associated with the type number.

- The user may set the interrupt type byte into switches SW0–SW7 and then cause a real-time interrupt to the program to occur by pushing PB1. The user program must have loaded the address of an interrupt service routine in low RAM and placed the interrupt service routine in ROM.

A Real-Time Interrupt Test Program

A program that exercises the circuit of Figure 12.9, int.asm, is shown next. The example program is itself a modification of an earlier interrupt service program from Chapter 11, intdriven.asm, along with the previous PPI I/O program iotest.asm. The program *first* initializes low RAM so that the interrupt associated with INTR is executed. When PB1 is pressed, the interrupt service routine is called, which proceeds to light the LEDs of PPI Port A based on the switch settings of PPI Port B. For demonstration purposes, the interrupt type is chosen, randomly, to be type 3Ah, and the interrupt type switches, SW8–SW15, are set to 3Ah. The SBC interrupt-driven program int.asm is written as follows.

```
;program int.asm, which exercises the 8088 interrupt capability by
;responding to a type 3Ah interrupt by reading the PPI Port B switches
;and writing the PPI Port A switches.

            prta equ 00h          ;port A address is xxxxxx00b
            prtb equ 01h          ;port B address is xxxxxx01b
            conreg equ 03h        ;contol register address is xxxxxx11b
            mod0oi equ 8bh        ;mode 0, A output, B and C input
code segment
            org 0000h             ;place in bottom of EPROM
            mov sp,1fffh          ;set the stack to RAM address
            mov ax,0000h          ;set DS to low RAM
            mov ss,ax             ;set SS to low RAM
            mov ds,ax
            lea ax,ppi            ;get interrupt subroutine IP
            mov w[00E8h],ax       ;set interrupt vector IP
            mov ax,cs             ;get CS for the interrupt vector
            mov w[00EAh],ax       ;set interrupt vector CS
            sti                   ;enable hardware interrupts
wate:   jmp wate                  ;loop here until PB1 is pushed
;interrupt service routine follows next.
ppi:    mov al,mod0oi             ;program the 8255 control register mode 0
            out conreg,al         ;out to 8255 in I/O space
            in al,prtb            ;read the switches on port B
            out prta,al           ;write the LEDs on port A
            reti                  ;back to wate
            org 1ff0h             ;reset CPU here
            jmp 0fe00h:00h        ;jump to beginning of program
code ends
```

Note that the stack was set to the *top* of the RAM, which is mapped at 00000h to 01FFFh.

12.8 Summary

Single-board computers (SBCs) are typically employed as intelligent controllers embedded within a larger system. SBCs may range from very simple and inexpensive single-program single-function models to complex multi-programmed and expensive PC equivalents using solid-state memory in lieu of disk drives.

A simple 8088-based SBC is designed, at a parts cost of less than $25, using the following major parts.

  8088 CPU

  8284 Clock Generator

  8255 PPI

  8250 UART

  6264 8K SRAM

  2764 8K EPROM

  74HCT373 Latch

  74HCT138 3 to 8 Decoder

  1488 Logic Level to RS232 Level

  1489 RS232 Level to Logic Level

  Switches and LEDs

Programs that test the SBC are developed for ROM, RAM, and I/O. A single-step circuit may also be added to aid in testing the SBC.

Parallel I/O is accomplished by programming the 8255 PPI. Parallel I/O may be done using software polling or may be interrupt-driven using an external 74HCT245 buffer to supply an interrupt type during a CPU interrupt acknowledge cycle. The interrupt cycle is generated when a 74HCT74 flip-flop is set via a push button.

Serial I/O may be done by programming the 8250 UART and using it in a polled mode.

12.9 Questions

For questions 1 through 20, supply the name and pin number of the 8088 CPU signal that best matches the description given in the question. All questions apply to the 8088 CPU operated in minimum mode.

1. Addresses that are not time-multiplexed.
2. Selects memory space or I/O space.
3. Reads data.
4. Addresses that are time-multiplexed.
5. Pulse that triggers external address latch.

6. Reads interrupt type byte from external interrupt source.
7. Writes data.
8. External interrupt signal.
9. Causes CPU to insert wait states when low.
10. External clock signal.
11. Causes CPU to set CS:IP to FFFF:0000h.
12. Executes a type 2 interrupt.
13. Selects CPU mode of operation.
14. Selects data direction.
15. Bus release request by bus master.
16. Used by WAIT opcode.
17. Ground.
18. Vcc.
19. Bus release acknowledge by CPU.
20. Enables data transceivers.

12.10 Projects

The SBC designed in this chapter may be used, as is, for a variety of real-time control applications or may be expanded by the addition of peripheral chips. The following projects are listed as suggestions for a capstone design project.

1. Control a simple, two-axis, robot arm to pick and place small parts. Control the arm X- and Y-axis movement with stepper motors, and use digital encoders to feed back arm position. Use a solenoid-actuated gripper.

2. Add a hex keypad and an intelligent LCD display to the SBC. Map the keypad and LCD display in I/O space. Write a program that will display on the LCD each key pressed on the hex keypad.

3. Connect the UART so that it can interrupt the 8088 CPU. Write a serial data program similar to uart.asm that is interrupt-driven.

4. Add a hex keypad and an intelligent LCD display to the SBC. Map the new components in I/O or memory space. Write a monitor program that will load a program into RAM from a PC serial port. Have the SBC run the program when the *g* key is pressed on the PC.

5. Write a monitor program for the SBC that permits programs to be loaded and debugged from a PC using the PC serial port. Have the monitor program perform the same functions as the D86 debugger.

6. Write a program that remotely controls the PPI parallel ports from a PC serial port. The PC user must be able to program the PPI to any operational mode, and be able to read and write all the PPI ports.

7. Build an EPROM programmer. Write a program such that EPROM data may be loaded from a PC file into the SBC RAM, via the serial port. Add I/O circuitry to the PPI, and write a program that enables the SBC to pro-

gram a 2K EPROM. (Typically, EPROM programmers use D/A converters for programming a range of EPROMs or fixed solid-state switches for one type of EPROM.)

8. Connect a PC Printer Port to one PPI parallel port, and a printer to another PPI parallel port. Program the SBC to act as a 4K print buffer between the printer and the PC.

9. Add a D/A converter to one of the PPI ports, and a DIP switch to another PPI port. Have the SBC generate sine waves that vary in frequency as the DIP switch is set from 00h to FFh.

10. Add an A/D converter and a 3-digit, 7-segment LED display to the SBC. Interface the A/D and LED display to the PPI. Program the SBC so that input DC voltages from 0 to 9.99V are displayed.

11. Add a speech synthesis chip to the SBC via the PPI. Write a program that will sing the "happy birthday" song.

12. Make the SBC into a frequency modulated modem by adding two tone generators, two tone filters, and a telephone line interface. Use PPI port pins to interface to all components, including the PC serial port.

13. Add D/A converters to the PPI and use the SBC as a graphic CRT screen controller. Produce a picture of a campus landmark on the CRT screen.

14. Make an LED tic-tac-toe display that interfaces to the PPI. Use red for an *O* (player) and green for an *X* (program). Place a pushbutton in each square. Write a program that plays the game against a human player. The player places an *O* in a square by pressing the pushbutton in that square. Alternate first entry choice every game.

15. Make a digital oscilloscope. Add A/D and D/A converters to the SBC. You may map the converters in memory or I/O space, or connect them to the PPI. Write a program that will digitize and store an input waveform, and then play the waveform back to an oscilloscope input. Use one of the PPI ports to trigger each oscilloscope trace.

16. Remove the microcontroller from a PC keyboard, and use the SBC as a keyboard controller. Use the PPI as an interface to the keyboard. Have the SBC perform all functions of the original keyboard controller.

A

ASSEMBLY

CODE AND

THE A86

ASSEMBLER

PROGRAM

A.0 Introduction

Assembly language programming, for any particular CPU design, implies that programs are written using assembly language instruction mnemonics that are unique for that CPU. Normally, the CPU assembly language instruction mnemonics are defined by the CPU manufacturer. The manufacturer also supplies an assembler program that converts programs in mnemonic form (assembly language) to programs in binary form (machine language). Mnemonics are intended for human use; hexadecimal code is intended for use by the CPU.

Intel Corporation, the original 8086 manufacturer, supplied the first 8086 assembler, ASM86. Other manufacturers, notably Microsoft (MASM) and Borland (TASM), have written 8086 assemblers.

The assembler used for all the examples in this book is Eric Isaacson's *A86* assembler. A86 has the distinct advantage of being very compact (27K) and very fast (1000 lines of code per *second*). Because of its compact size, A86 can run on systems equipped, at a minimum, with 320K dual-floppy-drive PCs using DOS version 2.0 or later. More advanced DOS-based computers will also execute A86.

Learning to use A86 involves having to cope with very few *magic words.* Magic words are assembler terms that must be used to get things to work, but you don't know how they do it (or why you need them). The lack of magic words should not imply that learning to use A86 is obvious, however. The 8086 family CPU structure and addressing modes are not simple, and some complication must creep into any 8086 assembler in order to use all of the family's power.

A.1 A Guide to Using A86

This guide is meant to be used by those learning to program in assembly code, perhaps for the first time or maybe the first time using the 8086 CPU family. A86 is a very powerful assembler and has all of the features normally employed by professional programmers. Beginners are not professionals, however, so we shall study the *basic* operations of A86. Those who wish to know more about the full capabilities of A86 should register their copy of A86/D86 with Eric Isaacson in order to get a manual. Registration instructions are in file A01.DOC of the A86 disk manual. See READ.ME to unpack the manual.

Also, in keeping with our 80/20 rule of life (see Chapter 5), we shall study those 20% of assembler operations that we shall be using 80% of the time. Please note also that many of the commands used to get A86 to do something can be entered into the computer program in a variety of ways. The examples given in the text are guaranteed to work, but other commands may work just as well.

Using A86

The A86 assembler program exists to turn programs written in 8086 instruction mnemonics into machine code that can be executed by an 8086 CPU. The instruction mnemonic program is often called the *source* program.

Programs to be assembled are written in 8086 mnemonics using an ASCII text editor and stored on a disk. 8086 ASCII text programs can have any DOS legal name and extension except that the extensions .OBJ, .BIN, and .COM *must* not be used.

The source program (the extension .ASM is highly recommended) is prepared using any of a variety of word processing or text editing programs. The only restrictions on the source file are that it *must* be saved in pure ASCII form on the disk and *each* text line *must* end by pressing the Return key on the keyboard *after* each line is typed.

Pure ASCII format programs have no "strange" characters (happy faces), often introduced by more complicated word processing programs. Most word processing programs have the ability to save files in ASCII format. Ending each text line in the file with a Return key ensures that A86 can separate one line of text from the next line of text.

Once the mnemonic source file is written it can be processed by the assembler program simply by typing the name of the assembler program followed by the *path name* of the source file to be assembled.

USING A86 ON A FLOPPY DRIVE

Typically, the disk containing the assembler and the program to be assembled is in drive A. The assembler will fetch the source program, assemble it, and put the resulting assembled program back on drive A with a different extension so that the assembled program can be identified. For instance, suppose that the source file is named yourfile.asm. The A86 assembler and yourfile.asm are both on a disk in drive A, and DOS has the A prompt on the screen. You type one of the lines below, *depending on what type of output file you wish to have.*

> A> a86 yourfile.asm to yourfile.com

or: A> a86 yourfile.asm to yourfile.obj

or: A> a86 yourfile.asm to yourfile.bin

- A86 is the name of the assembler program. By naming it in the line you type, you cause the computer to load it from disk into RAM and begin running it.

- yourfile.asm is any program you have created and stored in pure ASCII format on the A drive.

- yourfile.com will cause A86 to produce an output file in .COM form. A .COM file is used for single-segment programs and can be loaded and run under DOS with no linking. *Single-segment .COM files are the primary format for all programs used in this book.*

- .OBJ files must be linked together using linker programs such as DOS LINK, to produce an .EXE file that can then be loaded and run under DOS. .EXE files are usually multi-segment programs and are not featured in this book.

- yourfile.bin will cause A86 to produce an output file that can be placed in an EPROM for use by board-level computers that have no DOS operating system.

A86 will *also produce* a file named yourfile.sym. .SYM files are known as *symbol* files and are used by the D86 debugger (see Appendix B) to enable you to run and troubleshoot your programs.

USING A86 ON A HARD DRIVE

Assuming that A86 is on hard drive C, and the user source program is on floppy drive A, the following procedure is recommended for hard drive use.

- Be sure the *path* to the A86 assembler is part of the PATH command in the boot drive autoexec.bat file.

- Go to drive A and type:

    A:> A86 yourfile.asm to yourfile.com

A86 will place all of its output files (yourfile.com, yourfile.sym, etc.) on the *default* DOS drive. If you issue the assembler command while in the C drive, for instance, then all of the assembled files will end up on drive C, not on drive A. Most instructors do not want the users' files "filling up" drive C, so try to make sure the assembled files end up on drive A, the user drive.

A.2 A86 Keywords: Mnemonics, Names, Numbers, and Directives

Once you have chosen an assembler program, such as A86, you become bound by the "rules of the game" for that assembler.

The first rule is that you may use no words, of any kind, in your source program that the assembler cannot "understand." Assemblers are programs, and you may only use *words, numbers, and punctuation,* in any source program, that are *legal.* Legal numbers, words, and punctuation, are those items that are defined by the assembler rules. The rules for the assembler are referred to as the assembler *syntax.* Chapters 4 and 5 contain a discussion of the basic syntax, or rules of grammar, that must be used for writing lines of assembly language instructions for the A86 assembler.

There are four groups of legal words (also called *keywords*) that may be used in your source file for processing by A86. The legal words are CPU *instruction* mnemonics, *names, numbers,* and assembler *directives.*

Instruction Mnemonics

Instructions consist of the mnemonics defined by the person who writes the assembler program. In order not to commit financial suicide, the assembler program writer usually chooses the same mnemonics as defined by the original manufacturer of the CPU. The assembler program can have additional features, however, that extend the manufacturer's original instructions to include mnemonics or syntax that simplify programming chores.

An essential set of mnemonics is presented in Chapter 5, and additional instructions are included in later chapters. Appendix F contains a complete list of 8086 instructions.

Names

Names are symbols that the programmer invents in order to have the assembler take care of program details, such as code addresses. Symbols enable the programmer to write programs much faster, and more clearly, than could be done otherwise.

Numbers

There are two types of numbers: *variables* and *constants.* Variables are addresses to data in memory. Whenever a variable is used in a program, the assembler will use the *contents* of the variable address, not the variable address number itself. Constants are pure numbers. When a constant is used in a program the assembler will use the constant number itself.

Numeric *constants* can be expressed in decimal, hexadecimal, binary, and octal. Most of the numbers used in this book and numbers used in programs are *hexadecimal.* Decimal numbers are used in the text where their meaning is obvious, such as page numbers, example numbers, and the like. If there may be any confusion, a small *d* will be placed at the end of decimal numbers in the text. Thus, Example 10 means the decimal number 10, whereas memory location 400h means the hexadecimal number 400.

All numbers used in a program must begin with a *decimal* digit of 0 to 9. The number *should* end in an *abbreviation* for its base, as shown next.

> 1234 is decimal number 1234
>
> 1234h is hexadecimal number 1234
>
> 0abch is hexadecimal number ABC
>
> 1234q is octal number 1234
>
> 1234xd is decimal 1234
>
> 0100xb is binary number 0100

Making the first digit of a number a *decimal* digit ensures that the assembler will not mistake a number for a name. The number 0ABCDh is clearly a hexadecimal number, whereas ABCDh is a name. Using a leading 0 for hexadecimal numbers that begin with the numbers A–F will not confuse the assembler.

If the base abbreviation is omitted, then A86 makes the following assumptions:

> Any number that starts with a 0 is hexadecimal.
>
> Any number that starts with a 1–9 is decimal.

Using the tail *xd* for the decimal number ensures that the decimal number 123xd will not be mistaken (by the *programmer)* for the hexadecimal number 123D. The same reasoning applies to the *xb* tail for binary numbers. By using the xb tail, binary 01001xb will not be confused with hexadecimal number 01001B.

Hexadecimal numbers are the most natural numbers to use with multibyte binary machines. You may as well use them for everything and memorize their decimal and binary equivalents now.

Directives

Directives are words that are not instruction mnemonics or program names. Directives are *assembler instructions,* not CPU instructions. Directives may be written as a separate program line or incorporated as part of a CPU mnemonic. For instance, some directives help the assembler determine the *size* of data that is to be used in a mnemonic. Because the 8086 has byte, word, and double-word size data, size directives *must* be added to certain CPU mnemonics so that the assembler will produce the proper code.

A.3 Using Symbols in a Program

A86 is what is known as a *symbolic* assembler. Symbols are words we use to *identify* things. *Dog* is the symbol, for instance, for a domesticated four-legged animal. Symbols are handy because they let the programmer use words, which are very readable, rather than depend entirely on boring numbers. Legal symbols are of two general types, *label names* and *data names.*

Legal Names

A legal name is any combination of letters from A(a) to Z(z), numbers from 0 to 9, and ?, _, $, or @. A name *cannot begin with a decimal digit 0–9,* because the assembler might mistake a name for a number. For instance, the symbol ABCDh is a *name,* whereas the symbol 0ABCDh is a hexadecimal *number.*

RESERVED NAMES, ILLEGAL NAMES

Almost any name that can be invented may be used as a name in a program. Names may contain up to 127 characters. Names *must* begin with a *letter, underline* (_), *dollar* sign ($), *question* mark (?), or an *"at"* sign (@). Names must *not* begin with a *decimal* digit, and may not be any *reserved* keyword. Reserved words are those words used by A86 as mnemonics, or assembler-defined directives. A list of reserved keywords is given at the end of this Appendix.

Label Names

Labels are names followed by a colon (:) that are used in a program to give a *name* to an *address* constant. A label is normally used as a name for the *address* number of an opcode in code memory. For instance, the name _open: is a label, as well as the names lp2:, put:, here:, and table1:. You place the colon after the label name when it is used at the beginning of a line. You do not use a colon when the label is used as an instruction operand. For instance,

```
loop3:  mov ax,bx
        jmp loop3
```

Use the colon only when the label marks the line of code. Using the jmp opcode with loop3 does *not* use the colon.

Labels may be used alone on a line, and A86 will assign the label to the next opcode's address number. For instance, the example listed above may be written as follows.

```
loop3:
        mov ax,bx
        jmp loop3
```

Isolated label lines are easy to change, because you do not have to rewrite any instructions on the label line when changing the line that *follows* the label.

Labels make the work of writing assembly language programs easier for the programmer, and make the program more *readable*. The assembler program keeps track of what absolute number in memory belongs to each label name and lets the programmer concentrate on writing instructions. For instance, the mnemonic jmp loop4 is much easier to write than jmp 0acdfh. The assembler knows (keeps a .SYM table of all names) that the label loop4: represents location ACDFh in the program. In order for us to know where loop4: is, we would have to count the bytes of every instruction in the program up to the label loop4: to find out where it is located in the program. Even worse, every time we add a line of code in the program ahead of loop4:, the address of loop4: changes, and we would have to count the code bytes again.

Variable and Constant Data Names

Variable and constant symbols are names that are *not* followed by a colon when used by the programmer as the first symbol in a program line. Thus loop4: is a label and loop4 is a variable name.

VARIABLE NAMES

Variable names stand for the *contents* of an address in memory, not the *address* of the data. For instance, suppose offset address 1000h in the data segment contains the byte 32h, and that the variable name varble has been assigned to the contents of offset address 1000h. Any program reference to the name varble, as an operand, will use the contents of address 1000h, or the byte 32h in this example.

Because the contents referred to by variable names are often in data memory, the contents represented by the variable name may be changed by the CPU as the program runs. For the previous example, should the program write an F3h to offset address 1000h, then the next operand reference to the name varble will use the contents of offset address 1000h, or F3h.

Variable names may also be used in code memory and be associated with fixed data. The important thing to keep in mind when using variable names is that they refer to the *contents* of the variable address, not the address itself.

CONSTANT NAMES

Constant names are programmer-assigned symbols to *absolute* constant numbers in the code segment. Absolute numbers are numbers that the programmer types into the program. Absolute numbers will not change unless the programmer changes them. Labels are also constant numbers, but they are assigned by the assembler, not the programmer. Label constants may change each time the program is assembled (as lines are added or deleted).

Constant names used to represent fixed numbers use the EQU directive (see the next section) to equate the constant name with an absolute number.

A.4 Program Numbers: Constants, Labels, and Variables—More Detail

The following section is also found in Chapter 5.

Assembly language instructions are made up of action mnemonics and operands. The assembler program encodes the mnemonics and the operands into binary code. Thus every mnemonic and every operand become real, fixed, binary numbers in code memory. There are no names in code memory, just numbers. Assemblers, however, let the programmer use *symbols* (names) in place of *code numbers.*

Instructions are encoded by the assembler depending on the mnemonic used and the operands associated with the mnemonic. Operands, no matter what the addressing mode, must be numbers, symbols, or the name of a register.

Register names are assigned by the CPU designers. The programmer, then, has no choice but to use the register names (AX, DI, BP, etc.) chosen by the manufacturer. Numbers in a program, however, are chosen by the programmer as *absolute constant* numbers or *symbolic* numbers.

Absolute constant numbers are numbers that are entered as part of the program and do not change. Absolute constant values can be seen when the program is read. For instance, the following lines all use an absolute constant as one of the operands.

```
mov ax,1000h          ;immediate data constant of 1000h
mov w[1000h],bx       ;constant memory address 1000h
mov w[2000h],1234h    ;move the constant 1234h to constant memory
                      ;location address 2000h
```

As can be seen in the preceding example, absolute constant numbers are just plain numbers that do *not* change each time the program is assembled.

Symbolic constants, however, are numbers that are assigned names *by the programmer* and become numbers when the program is *assembled.* There are *three types* of symbolic names available to the programmer: constants, labels, and variables.

Constant names are assigned by the programmer using the EQU directive. There is very little difference between an absolute constant and a symbolic constant, except for the convenience of using meaningful names in place of obscure numbers. For instance, the following constant name may make the resulting program easier to understand.

```
code segment
        beernum equ 99        ;99 bottles of beer
        fall equ 1            ;1 less bottle of beer
        mov cx,beernum        ;count the bottles of beer
one_less:
        sub cx,fall           ;if one happens to fall
        jnz one_less          ;continue until no more
code ends
```

The names used with the EQU directives are not variable names (although they do not end in a colon), they are the names of constants. A86 assumes that any undefined name in a program (except jump operands) is the name of a constant.

Labels are names assigned to address location numbers in code or data memory. The assembler knows a name is a label because it ends in a colon (:). The absolute number that A86 assigns to a label is determined when the program is assembled into code. For example, the label one_less: in the preceding example is assigned to memory location 0103h by A86 when the program is assembled. The label symbol becomes a fixed number only during assembly. If the program is rewritten, and a new line of code inserted before the label, then the number assigned to one_less: will change during assembly. For instance, if a 1-byte instruction, such as NOP (no operation), is inserted in the program before the label, the assembler will assign the number 0104h to the symbol one_less. Thus, as the program is re-written, and lines are added or deleted, numbers assigned to labels may change each time the program is assembled.

Variables are names assigned to address location numbers in data memory. The assembler knows a name is *not* a label because it *does not* end in a colon. Whenever A86 encounters a variable name in a program, it assumes that the programmer wants to *use the contents* of the variable location, *not the address number* associated with the variable name. Variable names are assigned by the programmer using the various *define* directives such as DB, DW, and so on. For example, the following program assigns a variable name to a data memory location in the program.

```
data segment
        org 1000h              ;begin assigning data locations starting at 1000h
        can_vary dw ?          ;reserve a word space at 1000h in data memory
data ends
code segment
        mov can_vary, 1234h    ;move constant to data memory location 1000h
        mov ax,can_vary        ;move contents of variable to AX
        mov ax,w[1000h]        ;does exactly the same thing
```

The assembler uses the correct opcode for "contents of" whenever it encounters a variable name as one of the instruction operands. The address number assigned to a variable name may change as we alter the program. Should we modify the last program, and ORG the data segment at some other location, say 2000h, then the assembler will assign the address number 2000h to the variable name can_vary when the program is assembled.

The programmer can *override* A86, however, and force it to use a label as a variable, or a variable as a label, in an instruction. The override directive that makes A86 use the variable address number is the *"offset" operator*. The override directive that makes A86 treat a label number as the "contents of" is the bracket [ ] operator. For instance, in the next program we shall define a variable and a label and use them interchangeably.

```
data segment
        org 1000h              ;start data memory at 1000h
        label_name: dw ?       ;A86 assigns the number 1000h to the label
        vary_name dw ?         ;A86 assigns the number 1002h to the variable
data ends
```

```
code segment                    ;COM programs are assembled beginning at 0100h
start_here:                     ;the label "start_here" = 0100h
        mov ax,label_name       ;the constant number 1000h is moved to AX
        mov vary_name,ax        ;the 1000h in AX is moved to contents of 1002h
        mov ax,offset vary_name ;AX = 1002h (number assigned to variable)
        mov w[label_name],ax    ;move 1002h to contents of 1000h
        mov bx,codeptr          ;BX = 0117h
        mov cx,[bx]             ;CX = 4567h
        mov cx,w[codeptr]       ;CX = 4567h–brackets force contents of
        jmp start_here          ;jump to location 0100h in program code
codeptr:   dw 4567h             ;codeptr: is a label = 0117h
code ends
```

A.5 Commonly Used Assembler Directives

Directives are keywords that *help* the assembler convert mnemonics into machine code. Directives enable the assembler to resolve problems with instructions that can have more than one meaning.

A typical example arises when decrementing a memory location. The DEC command applies equally well to bytes and words, and the assembler must be told what data entity you wish to have decremented. The keywords b and w are used by the assembler to decrement the byte at a memory address, or a word at a memory address.

Comment Directive

Comments *must* begin with a semicolon (;). The assembler will ignore anything typed after a semicolon. Comments may be placed anywhere on a line and are also a handy way to "remove" instructions while debugging a program. For instance, if the following opcode:

```
mov ax,bx
```

is suspected of causing some problem, then typing a semicolon in the .asm file in front of the line results in the following correction:

```
;mov ax,bx
```

which will be ignored by the assembler when the program is re-assembled. The semicolon may be edited out later, if desired.

Equate Directives

Often it is handy to be able to assign a name to an absolute constant number in the code segment. For instance, you might be writing a program that counts a number down from some initial value to zero. The number represents, say, the number of beer bottles to be counted on a wall. You could know the number, which might be 99, and use that number in your program. But, if you used a name, such as beernum, then your program is much more readable. Not only that, but if you decided to change to 100 bottles of beer on the wall, you wouldn't have to search all over your program looking for every opcode that used the old number, 99. For these reasons, clarity and convenience, a keyword named EQU can be used. EQU is used to *equate a name* with an absolute *constant* number in the code segment. The syntax for EQU is shown next.

```
beernum equ 99xd              ;number of bottles of beer on the wall
```

Whenever you use an opcode that could use the number 99xd, you can substitute the name beernum. For instance:

```
        mov al,99xd
    or, mov al,beernum
```

performs the same operation, moving the number 99 into AL.

ORG Directive

The *originate* directive org is used to place code or data at exact offset addresses from the *start* of the segment. An org is followed by a number that specifies the offset address of any data or code that *follows* the org directive. For instance, the following org directive:

```
    org 400h
```

will place whatever comes next (data or code) exactly at offset address 400h from the *start* of the segment.

An org directive is *normally* used to place the data area of memory well above any program code memory that is in the code part of the same segment. The assembler assembles the code segment first in *lower* segment memory, and provides space for the data memory at the org address declared in the data segment. Forgetting to org your data will result in an error code (*complaint*) from A86.

Constant data may also be placed in *exact* offset addresses in the code segment using org.

■ *But, beware!* If you org constant data at the end of your program, and then *add* to your program, it may grow into the constant data area. Should the program and the constant data overlap, the assembler will assemble your program first and then write over it with the constant data.

■ *Beware again!!* When ORGing constant data in you *code* segment, be sure that each org address is *larger* than the previous one. The assembler code counter stops at your *last* org address. If the last ORG address is *smaller* than those before it, the assembler stops at the last (lower) org address. (Everything at org addresses greater than the last one is lost.)

Segment Directives

Often it is desirable to include constant or variable data in a program. Exact, known, *constant* numbers and addresses can be placed in the program segment, and *space* for variable data (that will be generated by the program as it runs) can be *reserved* in a data segment. Reserving data space means that the programmer names each variable address, and the assembler associates the contents of the address with the variable name.

Name	Bits	Bytes	Words	A86 Directive
Byte	8	1	—	B
Word	16	2	1	W
Dword	32	4	2	D
Qword	64	8	4	Q
TByte	80	10	5	T

Table A.1   ■   Data Sizes

Only two segment areas need to be defined for a single-segment .COM program: data and code. The data segment is usually declared at the start of a program by using a data segment keyword. The end of the data segment is signaled to the assembler by the keyword data ends. In a similar manner, the code segment is declared by starting with the keyword code segment, and the code segment is ended with code ends. An example is shown next that uses the code and data segment declarations.

```
data segment
        org 1000h
                                ;variable names defined
data ends
code segment

                                ;constants equated to names
                                ;program instructions
                                ;constant data defined
code ends
```

Number Size Directives

Data may be stored in memory using a single byte (8-bits), a word (2 bytes), a double word (2 words), a quad word (4 words), or 10 bytes. Table A.1 shows the relative size of each type of data storage.

When we wish to *make sure* that the assembler knows exactly the *size* of data we have in mind as an operand we include, with the operand, one of the *key letters* B, W, D, Q, or T. If the size of the operand is obvious we need do nothing. For instance, loading register AL with a byte from memory is obvious, as in:

```
mov al,[400h]
```

which moves a byte from location 400h to AL. The assembler knows that AL is byte-sized and moves a byte from memory. But any mnemonic that does not make the data size obvious must have a size designation added. For instance, in the opcode that follows, it is not clear if a byte or a word is to be decremented:

```
dec [400h]
```

This could mean decrement the byte at location 400h, or decrement the word

```
DB defines Byte(s) of data
DW defines Word(s) of data
DD defines a Double Word of data
DQ defines a Quad Word of data
DT defines Ten Bytes of data
```

Table A.2 ■ Data Define Directives

at 400,401h. To make sure the assembler knows what you want, specify the data size as follows:

```
dec b[400h] ;decrements a byte
dec w[400h] ;decrements a word
; or
dec [400h]b
dec [400h]w
```

B and W size indicators are the most *commonly* used size directives because most data that can be accessed by the CPU is byte- or word-sized. Alternates to B and W directives are byte ptr for b and word ptr for w.

Define Data Directive

The D *define* directive, in conjunction with a size specifier, can be used to define a data constant (byte, word, etc.) in the code segment, or to reserve data space of the proper size for a variable in data memory. There are five define-size designators, as shown in Table A.2.

When D is used in a data segment, only *space* is *reserved* (usually with a variable name) by the assembler. No actual, constant numbers are placed in RAM by the assembler (it has no means to do so, only the CPU can do that).

If D is used in a code segment, then the *actual constant numbers that follow the D directive are assembled by the assembler and placed in the code segment.* That is what assemblers do: convert instruction mnemonics to code bytes in the code segment.

Some examples of using D and size specifiers include the following.
Used in data segment:

```
varname    db ?  (= leave space for a byte of data)
           dw ?  (= leave space for a word of data)
           dd ?  (= leave space for two words of data)
           dq ?  (= leave space for a quad word of data)
           dt ?  (= leave space for ten bytes of data)
```

The example variable name varname is used to associate the contents of the variable data in the data segment with the variable name.

A question mark (?) used after the D directive in a data segment *emphasizes* that data defined in a data segment has no real value and only space for future data is reserved in data memory.

Used in code segment:

```
coname    db 5ah
          dw 1234h
```

```
dd 12345678h
dq 123456789abcdef0h
dt 1234567891bcdef01234h
db 32h,34h,35h,36h,
dw 1234h,5678h
db 10h dup 33h
```

The example variable symbol *coname* is used to associate the contents of the fixed data in the code segment with the variable name.

The *constant* number specified by the programmer after the DX directive is assembled at the *current* location of the program counter. The assembler keeps track of where the next code byte is to be placed in the code segment by maintaining its own version of an instruction pointer counter. Every time a code byte is assembled in the code segment, the assembler updates the pointer to point to the next empty address in the code segment.

You may separate items to be defined by commas (,) as shown in the code segment examples, to avoid having to write a DX for every item. For example, to code the first three ASCII numeric characters, you could write:

```
db 30h
db 31h
db 32h
```

or: `db 30h,31h,32h`

Using commas tells A86 that the entire string of bytes is to be placed in code memory.

A.6 Convenient Assembler Features

Math Directives

The assembler is capable of performing four common mathematical operations on program *constants*. Program constants are *integer* numbers such as address label constants and absolute constant data used in your program. Using A86 to do math for you in your program often helps document exactly the meaning of the program.

The assembler can do addition, subtraction, division, and multiplication. The symbols for each math operation are the following:

```
+ addition
– subtraction
/ division
* multiplication
```

Examples of using assembler math in programs include the following:

```
org table + 0ffh     ;ORG the assembler at address table + 0FFh
mov bx, 12+13        ;BX = 25xd
mov bx, 12*13        ;BX = 156xd
mov bx, 21/5         ;BX = 4xd (integer/integer = integer)
```

Remember that the numbers used can be absolute constants (such as 1234), or label address constants (nmbr:), or names that are absolute constants (nmbr equ 1234).

DUP Duplicate Directive

The DUP directive serves to save a place for as many data items as are specified before the DUP keyword. The structure of DUP is as follows:

DX nn dup yy

where: nn = number of X things to duplicate
yy = the value of each X quantity

For example, to duplicate 100h bytes of data, each byte containing a 33h, the following line can be used:

db 100h dup 33h

Single Quote ASCII LookUp

ASCII data is the byte equivalent of all the keyboard printable characters and control characters. Printable characters are those that can be displayed on the screen or printed on the system printer. Control characters cause actions such as line feeds (LF) on the screen or printer to take place. An *A* is a printable character; an LF is a control character. (See Appendix E for a listing of all of the ASCII codes.)

It is convenient to be able to have the assembler do the drudgery of looking up the ASCII equivalents of characters, rather than using Appendix E. To let A86 do the work, use the single quote (') mark on both ends of a string of characters. For instance, to code the message "hello":

db 68h,65h,6ch,6ch,6fh
or: db 'hello'

will convert the string of characters h,e,l,l,o into the equivalent ASCII code bytes 68h, 65h, 6Ch ,6Ch ,6Fh.

Anything you can possibly type between the single quote marks will be converted to ASCII, including space (ASCII 20h). Control characters, such as LF, can be looked up in Appendix E and included after the single quote. For instance, to end the message "hello" with an LF:

db 'hello',0ah

Be sure to include the *comma* after the quoted string.

A.7 Mistakes

When you make a mistake in your source code, A86 will notify you where the error has occurred. Your original program, with no messages, is copied to a file named yourfile.old and error *messages* are placed, by A86, in your original yourfile.asm file.

Error messages are placed in, or right after, the line of code where the error has been found by A86. Error messages begin and end with the tilde (~) mark. You can inspect your source code file and determine the nature of the problem. Error messages can be cryptic at times, so regard error messages as "hints" from A86. The most common errors are as follows:

- *Misspelling* an opcode
- Using *different* spellings of the same name or label (see *Note*)
- Using a *reserved* assembler keyword
- *Forgetting* a comma
- *Forgetting* a semicolon before a comment

Note: A *special* kind of error occurs when you spell the *same* name in *different* ways in the program. For instance, suppose your program has used a label in one place and misspelled the label in a jump statement, as follows:

```
loop1:  mov ax,bx
        ;more opcodes
        jmp loopel        ;ERROR: loopel should be loop1
```

A86 will inform you that a symbol has been found in your program that is *not* defined. A86 will type the names of the symbols that are not defined in your program after it returns from assembling your program. All symbol errors are also placed in a yourfile.ERR file. Have DOS type out the contents of yourfile.ERR to see what symbols are undefined in the program, and correct any errors in symbol use found in the file.

A.8 Conclusion

This appendix documents enough of Eric Isaacson's A86 assembler to enable the student programmer to write and assemble all of the problem assignments in this book. A86 has many, more powerful features, and you are encouraged to contact Eric to register your copy of A86.

A.9 Reserved Symbols in the A86 Language

Note: Not *all* of these reserved symbols are mnemonics or other keywords used in 8086 programs. Many of the reserved symbols pertain to the 808186 or 80286 CPUs, or to 8087 or 80287 math coprocessors. Use of *any* of these reserved words will draw an error message from A86: ~Misplaced Built-In Symbol~.

AAA	ASCII adjust addition	AT	SEGMENT specifier
AAD	ASCII adjust division	AX	Word register
AAM	ASCII adjust multiply	B	Byte memory specifier
AAS	ASCII adjust subtract	BH	Byte register
ABS	EXTRN specifier	BIT	Bit-mask operator
ADC	Add with carry	BL	Byte register
ADD	Instruction	BOUND	Instruction
ADD4S	NEC instruction	BP	Word register
AH	Byte register	BX	Word register
AL	Byte register	BY	Bytes-combine operator
AND	Instruction/operator	BYTE	Byte memory specifier
ARPL	286 Prot instruction	CALL	Instruction
ASSUME	Ignored, compatibility	CALL80	NEC instruction

CBW	Convert byte to word		F	Far specifier
CH	Byte register		F2XM1	87 instruction
CL	Byte register		FABS	87 instruction
CLC	Clear carry		FADD	87 instruction
CLD	Clear direction		FADDP	87 instruction
CLI	Clear interrupt		FAR	Far specifier
CLRBIT	NEC instruction		FBLD	87 instruction
CLTS	286 Prot instruction		FBSTP	87 instruction
CMC	Complement carry		FCHS	87 instruction
CMP	Compare		FCLEX	87 instruction
CMP4S	NEC instruction		FCOM	87 instruction
CMPS	Compare string		FCOMP	87 instruction
CMPSB	Compare string byte		FCOMPP	87 instruction
CMPSW	Compare string word		FDECSTP	87 instruction
CODE	Segment name		FDISI	87 instruction
COMMENT	Directive		FDIV	87 instruction
COMMON	SEGMENT specifier		FDIVP	87 instruction
CS	Segment register		FDIVR	87 instruction
CWD	Convert word Dword		FDIVRP	87 instruction
CX	Word register		FENI	87 instruction
D	Dword specifier		FFREE	87 instruction
DAA	Decimal adjust add		FIADD	87 instruction
DAS	Decimal adjust sub		FICOM	87 instruction
DATA	Segment name		FICOMP	87 instruction
DB	Define bytes		FIDIV	87 instruction
DD	Define Dwords		FIDIVR	87 instruction
DEC	Decrement		FILD	87 instruction
DH	Byte register		FIMUL	87 instruction
DI	Word register		FINCSTP	87 instruction
DIV	Divide		FINIT	87 instruction
DL	Byte register		FIST	87 instruction
DQ	Define Qwords		FISTP	87 instruction
DS	Segment register		FISUB	87 instruction
DT	Define Twords		FISUBR	87 instruction
DUP	Duplicate operator		FLD	87 instruction
DW	Define words		FLD1	87 instruction
DWORD	Memory specifier		FLDCW	87 instruction
DX	Word register		FLDENV	87 instruction
ELSE	Conditional term		FLDL2E	87 instruction
ELSEIF	Conditional term		FLDL2T	87 instruction
END	Start specifier		FLDLG2	87 instruction
ENDIF	Conditional term		FLDLN2	87 instruction
ENDP	End of procedure		FLDPI	87 instruction
ENDS	End of segment		FLDZ	87 instruction
ENTER	Instruction		FMUL	87 instruction
EQ	Equals operator		FMULP	87 instruction
EQU	Equate directive		FNCLEX	87 instruction
ES	Segment register		FNDISI	87 instruction
EVEN	Coerce to even address		FNENI	87 instruction
EXTRN	Ignored, compatibility		FNINIT	87 instruction

FNOP	87 instruction		JBE	Jump below equal
FNSAVE	87 instruction		JC	Jump on carry
FNSTCW	87 instruction		JCXZ	Jump on CX zero
FNSTENV	87 instruction		JE	Jump on equal
FNSTSW	87 instruction		JG	Jump on greater
FPATAN	87 instruction		JGE	Jump greater equal
FPREM	87 instruction		JL	Jump on less
FPTAN	87 instruction		JLE	Jump less equal
FRNDINT	87 instruction		JMP	Jump unconditional
FRSTOR	87 instruction		JNA	Jump not above
FSAVE	87 instruction		JNAE	Jump not above equal
FSCALE	87 instruction		JNB	Jump not below
FSETPM	87 instruction		JNBE	Jump not below equal
FSQRT	87 instruction		JNC	Jump not carry
FST	87 instruction		JNE	Jump not equal
FSTCW	87 instruction		JNG	Jump not greater
FSTENV	87 instruction		JNGE	Jump not greater equal
FSTP	87 instruction		JNL	Jump not less
FSTSW	87 instruction		JNLE	Jump not less equal
FSUB	87 instruction		JNO	Jump not overflow
FSUBP	87 instruction		JNP	Jump not parity
FSUBR	87 instruction		JNS	Jump not sign
FSUBRP	87 instruction		JNZ	Jump not zero
FTST	87 instruction		JO	Jump overflow
FWAIT	87 instruction		JP	Jump parity
FXAM	87 instruction		JPE	Jump parity even
FXCH	87 instruction		JPO	Jump parity odd
FXTRACT	87 instruction		JS	Jump on sign
FYL2X	87 instruction		JZ	Jump on zero
FYL2XP1	87 instruction		L2E	Real constant
GE	Greater/equal operator		L2T	Real constant
GROUP	Group of segments		LABEL	Declaration
GT	Greater than operator		LAHF	Load AH flags
HIGH	High byte of word op		LAR	286 Prot instruction
HLT	Halt		LDS	Load into DS
IDIV	Integer divide		LE	Less equal operator
IF	Skip/conditional term		LEA	Load eff address
IMUL	Integer multiply		LEAVE	Instruction
IN	Input from port		LES	Load into ES
INC	Increment		LG2	Real constant
INCLUDE	Ignored, compatibility		LGDT	286 Prot instruction
INS	Input string		LIDT	286 Prot instruction
INSB	Input string byte		LLDT	286 Prot instruction
INSW	Input string word		LMSW	286 Prot instruction
INT	Interrupt		LN2	Real constant
INTO	Interrupt on overflow		LOCK	Instruction
IRET	Interrupt return		LODBITS	NEC instruction
JA	Jump on above		LODS	Load string
JAE	Jump above equal		LODSB	Load string byte
JB	Jump on below		LODSW	Load string word

LONG	Operator	REP	Repeat prefix
LOOP	Instruction	REPC	NEC instruction
LOOPE	Loop on equal	REPE	Repeat while equal
LOOPNE	Loop not equal	REPNC	NEC instruction
LOOPNZ	Loop not zero	REPNE	Repeat not equal
LOOPZ	Loop on zero	REPNZ	Repeat while zero
LOW	Operator	REPZ	Repeat non zero
LSL	286 Prot instruction	RET	Return
LT	Less than operator	RETF	Far return
LTR	286 Prot instruction	ROL	Rotate left
MACRO	Directive	ROL4	NEC instruction
MEMORY	Segment specifier	ROR	Rotate right
MOD	Operator	ROR4	NEC instruction
MOV	Instruction	SAHF	Store AH to flags
MOVS	Move string	SAL	Shift arith left
MOVSB	Move string byte	SAR	Shift arith right
MOVSW	Move string word	SBB	Subtract with borrow
MUL	Multiply	SCAS	Scan string
NAME	.OBJ module name	SCASB	Scan string byte
NE	Not equals operator	SCASW	Scan string word
NEAR	Operator	SEGMENT	Directive
NEG	Instruction	SETBIT	NEC instruction
NIL	No code instruction	SGDT	286 Prot instruction
NOP	No operation	SHL	Instruction/operator
NOT	Instruction/operator	SHORT	Operator
NOTBIT	NEC instruction	SHR	Instruction/operator
OFFSET	Operator	SI	Word register
OR	Instruction/operator	SIDT	286 Prot instruction
ORG	Directive	SLDT	286 Prot instruction
OUT	Output to port	SMSW	286 Prot instruction
OUTS	Output string	SP	Word register
OUTSB	Output string byte	SS	Segment register
OUTSW	Output string word	ST	EQU Zero for
PAGE	Ignored, compatibility		compatibility
PARA	Segment specifier	STACK	Segment specifier
PI	Real constant	STC	Set carry
POP	Instruction	STD	Set direction
POPA	Pop all	STI	Set interrupts
POPF	Pop flags	STOBITS	NEC instruction
PROC	Procedure directive	STOS	Store string
PTR	Ignored, compatibility	STOSB	Store string byte
PUBLIC	Ignored, compatibility	STOSW	Store string word
PUSH	Instruction	STR	286 Prot instruction
PUSHA	Push all	STRUC	Structure directive
PUSHF	Push flags	SUB	Instruction
Q	Qword specifier	SUB4S	NEC instruction
QWORD	Memory specifier	SUBTTL	Ignored, compatibility
RADIX	Directive	T	Tbyte specifier
RCL	Rotate carry left	TBYTE	Memory specifier
RCR	Rotate carry right	TEST	Instruction

TESTBIT	NEC instruction	WAIT	Instruction
THIS	This-location specifier	WORD	Word specifier
TITLE	Ignored, compatibility	XCHG	Instruction
TYPE	Operator	XLAT	Translate byte
VERR	286 Prot instruction	XLATB	Translate byte
VERW	286 Prot instruction	XOR	Instruction/operator
W	Word specifier		

Appendix

B

THE D86 DEBUGGING PROGRAM

B.0 Introduction

Debugging is a term that can trace its roots back to the prehistoric era of computing, the 1940s. In those days long ago, some computers actually used electromechanical relays as part of the computer circuitry. One day, so the story is told[1], the computer suddenly ground to a halt. After a long and thorough search, it was discovered that an insect had become caught inside the contacts of one of the relays. The technician, one Bob Rose by name, recounted that he had "removed the bug" that had caused the problem. The term *debug* thus was born. The very first computer bug may now be found in the Smithsonian Museum. Unfortunately, the species is alive and well to this day, a constant menace to all computer programmers.

A *debugger* is a program that lets you run your program, under *controlled* conditions, in order to ensure that it actually works. Some programmers, now mostly unemployed, have been known to write programs that were not subsequently adequately checked to see if they were in proper working order. It is absolutely all right, and unavoidable, to make mistakes when you write a program. It is not all right to write a program and not check to see if it works.

There are limits to how completely a program can be verified for proper operation. Large programs, that deal with an almost infinite set of possibilities, cannot be completely checked. Large programs are often checked using other programs, named *test* programs, that attempt to simulate all conceivable situations.

Commercial firms conduct initial testing of a new program, by the programmer or other company personnel, at the *alpha* test site. The alpha test site is usually the place where the programmer works. After testing at the alpha site, the program is given to a *very* sympathetic customer who agrees to try it out. The friendly customer is referred to as the *beta* test site. After beta testing (and, usually, more debugging), the program is released to an unsuspecting world.

Programs are normally released as *versions*. The first release might be version 1.0, the second could be numbered version 1.1, and so on. Radically new versions may be identified using a new first number (before the period), as version 2.0, then 2.1 for the next incremental version, and so on. Versions that have the same first number, but different second numbers (after the period), generally represent a program that is functionally unchanged but has had various bugs "fixed." Versions with new first numbers not only have the bugs of the previous program fixed but offer "new and improved" features. Some cynics believe that program manufacturers issue new versions periodically in order to continue to profit from well-known earlier versions.

Generally, a program becomes "mature" about version 3.0 or so, and any improvements beyond that point may be marginal for the average user. Users who buy version 1.0 are considered to be adventuresome.

---

[1]Admiral Grace Hopper. "Reflections: An Oral History of the Computer Industry" (*Computer System News*, November 1988 p. 26).

B.1 Debugger Features

Debuggers should offer three features: *display* the contents of CPU registers and external memory; *change* the contents of CPU registers and external memory; and a *run* the program in a variety of ways.

When you are debugging a program, you generally want to be able to see *everything* that is happening as the program operates. You should never "guess" or *assume* that the program has done something; use the debugger to show you *proof* of operation. Most debuggers supply the user with three types of information about the program.

Display

First, you need to know *where* you are in the program, that is, what are the IP contents and what instruction is next to be executed. Second, you want to *see* the contents of all of the CPU *registers,* and how they change as the program runs. Third, you need to see any areas of *memory* that the program affects.

Changing Data

The purpose of debugging is to *test* the program. To test most programs you must have the ability to *alter* registers and memory data so that the program deals with a variety of data. A debugger should let the user easily change the contents of any register in the CPU and any location in program memory.

Running The Program

Testing a program involves *executing* program instructions and *inspecting* program results in the CPU registers and memory locations. The debugger should permit the programmer to *single-step* the program, one instruction at a time, so that the results of the program can be inspected at human speed.

B.2 Testing a Program

First, get a *printout* of the program to be tested. The debugger does not display comments (they are not assembled), so a printout of a well-commented program will serve to remind the tester exactly what is *supposed* to happen in the program.

Very inexperienced program testers run a new program, full speed, and then try to "guess" what went wrong. Based on the guess, the program is rewritten and run, full speed, again. The process usually proceeds past the date the program is due to be finished.

Wise, and battle-scarred, program testers know that patience is, indeed, a virtue, and that the best way to test a program is to *make it show you* exactly what it is doing at every instruction. Debuggers allow the tester to *single-step* the program. Single-step means that one instruction is done, and the debugger screen updated to shown the tester the result of the instruction. The tester can then *compare* the results of the step against the results the tester *expects* to see. If the results of the step are not as expected, the tester *should not* proceed until the inconsistency is *resolved.*

Sometimes single-stepping is not practical, as when in a long loop. Not many of us wish to single-step a program through a 10,000-step loop. Debuggers allow the user to run the program full speed until a user specified *label,* or *breakpoint,* is reached in program code. Once the particular instruction is reached, the program stops so that the user can inspect the program results.

Eric Isaacson has provided an efficient, fast, and easy-to-use debugger called *D86.* It will do all of the normal debugging tasks just outlined, and quite a few more. It is called a *symbolic* debugger because you can refer to names of things, instead of the numbers the names represent. For instance, if you have labeled an instruction loop3:, you can set a breakpoint at loop3, instead of the absolute address number in the program that loop3 represents. Take advantage of symbolic debugging by liberally scattering labels throughout your program.

As was done for the A86 assembler user instructions contained in Appendix A, the D86 user instructions presented in this appendix are adequate for student use. If you wish to obtain all of the power of D86, you should order a manual and a registered copy of the A86 and D86 programs from Eric Isaacson. See text file A01.DOC of the A86 manual on the disk for registration instructions. See the READ.ME file for manual unpacking instructions.

B.3 Using the D86 Debugger

How to Start D86

Assume you have written a program named yourfile.asm and assembled it with A86 to yourfile.com on the A floppy drive. Go to the drive and directory that contain D86 and type:

D86 A:yourfile.com

and D86 will load your .COM file and show the screen that is reproduced in Figure B.1. An alternative to going to the drive and directory that contain D86 is to include the path to D86 in the DOS PATH command contained in the AUTOEXEC.BAT file. Type the D86 start line, while using drive A, as follows:

D86 yourfile.com

If the screen of Figure B.1 comes up filled with nothing but NOP instructions, D86 could *not* find yourfile.com. Get out of D86, and be sure that the exact *path* name of the file to be tested is typed correctly.

Getting Out of D86

To stop debugging, press the letter *q* (for quit), followed by the Enter key. The screen will return to the drive from which D86 was called.

Figure B.1 ■ D86 screen

```
START:
#  0100      MOV AX,01234          D86 debuffer, V3.2
   0103      MOV BX,05678            Copyright 1990 Eric Isaacson
   0106      MOV CX,AX
   0108      MOV DX,BX             All rights reserved, see documentation
   010A      MOV DI,-1               for permission/restrictions
   010D      PUSH DI              To register for D86 only send $50 to:
   010E      POPF
   010F      MOV SI,0             Eric Issacson          Visa/MC call
   0112      PUSH SI              416 E. University Ave.  (812)339-1811
   0113      POPF                 Bloomington IN 47401-4739
   0114      PUSH AX                    or call 0297 24088 in England
   0115      PUSH BX                           A86+D86 is $80
   0116      POP AX               F10 key toggles windows
                                  Alt-F10 key toggles HELP mode

AX  0000    i          1:
BX  0000    IP  0100    2:
CX  00FF    CS  2714    3:
DX  2714    SS  2714    4:
SI  0100    DS  2714    5:
DI  FFFE    ES  2714    6:
BP  091C    SP  FFFE    0:
```

The D86 Display Screen

The screen shown in Figure B.1 has loaded a program named sample.com. The program sample.asm is reproduced in Figure B.2. You should type sample.asm into drive A, assemble it with A86, and load sample.com into D86 if this is your first try at using D86.

The D86 screen, shown in Figure B.1, is divided into *six* parts as follows.

INSTRUCTION DISPLAY

The upper left column is the program instruction display area. Because sample is a .COM program, the program begins at offset location 100h in program code memory. .COM programs are assembled with the first 100h bytes used by the assembler to pass program information to DOS.

The right column shows a copyright message by Eric Isaacson. As the program is exercised, the copyright area will disappear and be replaced by an additional program display area. Long programs will "swap" between the left and right column program display areas as the program is single-stepped.

REGISTER CONTENTS

The lower left corner of the screen displays all of the CPU registers. All of the CPU registers are identified, by name, except the Flags register.

The Flags register is displayed to the right of the AX register contents. Flags are *blank when cleared* (the flag is a 0) and *displayed* as lowercase let-

Figure B.2 ■ The sample.asm program

```
                  ;sample.asm, a first program for trying out D86
                  code segment
                  start:  mov ax,1234h        ;load some registers
                          mov bx,5678h
                          mov cx,ax
                          mov dx,bx
                          mov di,0ffffh       ;set DI to FFFFh
                          push di             ;push DI and the pop to flags
                          popf                ;all flags set to 1
                          mov si,0000h        ;set SI to 0000h
                          push si             ;push SI and pop to flags
                          popf                ;all flags cleared to 0
                          push ax             ;interchange ax and bx via the stack
                          push bx
                          pop ax
                          pop bx
                          mov sp,2000h        ;set the stack to a new place
                  loop1:  inc dx              ;a small loop
                          cmp dx,5680h
                          jne loop1
                  dun:    jmp start           ;want to go again?
                  code ends
```

ters *when* they are *set* (the flag is a 1). Flags are displayed, from *left to right,* as follows:

o d i s z a e c

where: o = Overflow
         d = Direction
         i = Interrupt
         s = Sign
         z = Zero
         a = Auxiliary Carry
         e = Even Parity
         c = Carry

THE STACK

The *current* stack area is shown at the bottom of the display, to the right. The contents of the stack area of memory are shown on a line next to the contents of the SP, right after the stack item counter, which is set to 0: on start-up. The stack item counter shows how many words are on the stack. As the program has not yet been run, there are no items on the stack (0:), and nothing is shown on the line to the right of the 0:.

DOS sets the SP to FFFEh when the program is loaded, and shows no items on the stack. Any subsequent change of SP will display whatever is on the stack at the new stack location, and show some number of items on the stack. Press *Ctrl-t* if you wish to reset the stack item count to 0 *after* changing the stack pointer.

As *words* are pushed on the stack the stack item counter will change to show the number of words pushed, and the words pushed on the stack will be displayed. Words on the stack are displayed in *true* form (as we humans read them), not in the low-byte, high-byte format of *actual* memory. Only one line of stack data is displayed. The last item pushed on the stack (at the location of the stack pointer) is at the far left of the displayed stack line. The stack item counter can count from 0 to 32K words.

MEMORY DISPLAY

A memory display area is placed to the right of the numbers 1: to 6: located to the right of the CPU register display area. Any line, from 1: to 6:, may display *any* area of memory that exists in the computer.

COMMAND LINE FOR D86 OPERATION

The command line area is located at the bottom line of the D86 active display screen. The command line is used by the program tester. Commands typed by the user are shown on the command line. If you err in typing a command, a correction message will be displayed immediately after the command. Retype the command if an error is made.

PROGRAM SCREEN OUTPUT DISPLAY LINE

The very bottom line of the screen (not a part of the lighter, D86 display area) shows a cursor in column 0. Any programs that output data to the screen can use the very bottom line to display short, one-line messages.

Note that there are instructions shown beyond the end of program sample.com. These instructions are nonsense and are the disassembled contents of memory beyond the program, in an area of data or "no-man's-land."

B.4 D86 Debugging Commands

Immediate Instructions

One of the novel *features* of D86 is the ability to *immediately execute* any legal 8086 *instruction* typed on the command line. Using mnemonics as commands lets you, at any time, control your program as if the command you type is the next program instruction. For instance, if you wish to jump to any near location in your program, simply type jmp loop3 or jmp 0200h to have the program immediately go to label loop3 or the instruction at location 200h.

You may jump to any area of memory, not just your program code. You may, for instance, go to a high address in memory by typing jmp A000h:0000h. If you wish to inspect the BIOS ROM you can jump to individual BIOS routines. When you jump to areas outside of your program, D86 will try to disassemble whatever you have jumped into.

Single-Stepping Your Program

When D86 begins, a reverse video cursor (a # sign) is shown next to the first instruction in your program. Press *F1* to single-step your program. Each press of F1 will execute the instruction at the # cursor, and advance the cursor to the next instruction.

Any time you single-step at an INT mnemonic (interrupt request to DOS or BIOS), your program will *run* through the interrupt service routine and resume single-stepping at the instruction that follows the INT mnemonic. The # will go to normal video while the interrupt is executing (waiting for a key to be pressed, for instance).

String move operations that feature one of the repeat-until-done prefixes (REP) may be stepped through, at once, by pressing *F2*. Key F2 will also run any CALLed procedure to completion.

Moving Around the Program

If you wish to move the # to an instruction, *without* single-stepping through every instruction on the way, use the *up* and *down* arrow keys. Pg Up and Pg Dn keys will move # rapidly through the program.

The Up arrow should be used with *care*. If you move up in your program, especially near the beginning of your program, D86 may back up too far (before your program starts), and find itself in the middle of an instruction. It will disassemble as best it can but the following code may be jumbled. Use *Ctrl-D* to step D86 down a byte, or *Ctrl-U* to step D86 up a byte. When the disassembler gets your code right, then resume.

Use of the Up arrow near the beginning of the program is not recommended.

Running the Program

SETTING BREAKPOINTS

You may set up to two breakpoints, or stopping places, in your program. Breakpoints are set by typing b,xxxx. The xxxx quantity can be a *constant number*, such as 1234h, or a label name, such as loop3. If you know the address of the instruction where you wish to stop, you can enter xxxx as a number. If the stopping point happens to have a label, such as loop3, then you can use loop3 instead of the IP contents that loop3 represents.

Typing *b* followed by a *Return* will *reset* all breakpoints that are currently set. If you wish to see what breakpoints are set, press *Ctrl-s* to display any breakpoints that are set. Breakpoints remain set until cleared

RUN TO A BREAKPOINT

Type g,xxxx where xxxx is the IP contents (a *number)* or the *label* (a name) of the instruction where you wish to stop program execution.

If you issue only g, *and no breakpoint has been set,* then your program will run to completion. If the program loops continuously then it will never end. Or, if it is not meant to be operated without using D86 it may run right past its end into no-man's-land and crash. Set a breakpoint if you wish to just go with no stopping point specified. The program will stop at the next breakpoint. When the program stops, it stops just before it executes the instruction at the stopping point.

The program will also stop, after a go command, if the trap flag in the

Flags register is set by a POPF command. Re-issue the go command if your program uses the trap flag.

RE-RUN THE PROGRAM

Press the *Home* key to restart your program from the beginning. The Home key will alternate between the last program instruction executed and the beginning of the program. If you used the Up arrow to excess, and find yourself with strange code, use the Home key to return to the start of the program.

B.5 Changing Register and Memory Contents

Changing Register Contents

Use a MOV instruction typed on the command line to change register contents. D86 can instantly execute any legal 8086 instruction you type on the command line. To load register AX with 1234h, for instance, type mov ax,1234h on the command line. Don't forget the h for hex, or you will get the decimal equivalent of 1234 (04D2) in AX.

If you want hex, end each number with an *h*. If you don't you will get decimal, and this may cause you much grief.

Displaying Memory

To see a *line* of memory, type a memory line number (1–6) on the command line. The D86 cursor will shift from the instruction display to the line number you have typed and await further instructions.

You may choose to display memory in one of the following formats:

> b = bytes
> w = words
> t = ascii text

The format specifier is followed by a comma and the address in memory you wish to see. Memory addresses are specified as:

> segment,offset

Segment is optional, but, if you do not specify a segment, then *DS* will be used as the *default* segment register display. If you want another segment then type its name (CS, ES, or SS) as an override for the automatic selection of DS, or type a constant segment number.

The following are all legal memory line display commands:

> b,400h = DS:0400h as bytes
> t,400h = DS:0400h as ASCII text
> w,400h = DS:0400h as words
> b,1000h,0100h = Segment 1000h, offset 0100h as bytes
> w,cs,200h = CS:0200h as words

NOTE THAT ADDRESSES ENTERED ON MEMORY DISPLAY LINES USE COMMAS (,) TO SEPARATE THE SEGMENT FROM THE OFFSET. DO NOT USE COLONS (:) IN A MEMORY DISPLAY SPECIFIER.

You may use *register names* or symbols to specify an offset memory location. For instance, to display the area of memory pointed to by SI, you may type ds,si as your memory identifier. Memory locations that have names may be conveniently viewed also. If a variable is named foobar, then the memory specification ds,foobar will display the variable foobar in the data segment.

MEMORY UPDATE

Once you have chosen an area of memory for display, it is *updated* anytime the contents of that memory area change. Also, if a memory pointing register (BX, BP, SI, DI) is used as a memory address, the memory window changes as the pointing register contents change.

CLEAR MEMORY LINE

To *clear* the memory display, type the memory line number (1–6) you wish to blank out followed by the *space* bar.

VIEWING LARGE AREAS OF MEMORY

If you wish to see memory beyond that displayed on a single line, press *Ctrl-n* to observe memory that starts at the next address after the last memory display *line* you have chosen. For instance, say Line 1 displays text bytes from offset 0000h to 0020h. Pressing Ctrl-n will cause a memory window to appear in the upper right hand of your screen. The new window begins just beyond the last item shown in the original memory line. For our example, the window begins at offset address 0021h.

Pressing *Ctrl-p* will back up the memory window so that you can see memory data at addresses below the last address of the last memory display line. The window disappears when no memory display line is in use.

Changing Code Memory

Whenever the contents of any code memory area are altered, the alteration is called a *patch*. The term goes back to the days of using punched paper tape to hold program code. The tape could literally be "patched" with another piece of tape to insert new program code into existing program code.

Function key F7 is used to patch memory. Any time the cursor is opposite a memory address number, pressing F7 puts D86 in the patch mode. When in the patch mode, you can type in new instructions, just as if you had written them in the original program and assembled them.

> *One warning:* The code you patch in *must occupy* exactly the same number of bytes as the old code, or you will overwrite code you do not wish to change. Patching code should be reserved for small operand changes, such as when you meant to write 1234h and wrote 1234 (decimal) instead.

If you type the mnemonic incorrectly (move for mov, for instance), you will get a beep and can try again. You may *exit* the patch mode by pressing *F7* again.

Changing Data Memory

CHANGING SMALL AREAS OF MEMORY

A command line move instruction, such as mov w[400h],1234h, will place a 1234h in memory starting at location DS:400h. Command line moves are useful for changing single items in memory.

CHANGING LARGE AREAS OF MEMORY

Individual moves are fine for 1 or 2 words of data, but they become cumbersome if large areas of data are to be altered. To fill large areas of data, use a JMP instruction to get to the area of memory you wish to change, and then use DB, DW, DD, and the like to define data type instructions to fill the data area.

For instance, to fill data from location DS:0400h to DS:0500h, type jmp 0400h on the command line. The # cursor will jump to that address. Then press the F7 key to enter the patch mode. Use any legal D command, such as db 100h dup 00h, to set all the bytes from 400h to 4FFh to an initial value of 00.

NOTE THAT THE COMMAND TO PATCH MEMORY ENDS IN A COMMA (,).

Use the Return key to finish patching data in memory. Every time the Return key is pressed, memory patching continues beyond the last area filled. To stop patching data into memory, press *F7* again.

If you wish to change the data in a data segment *different* from the one in current use, jump to the desired data segment, and then change the data. For instance, the command line jump command jmp 6000h:0400h will cause a jump to physical address 60400h for patching. Press the Home key to return to your program after changing data memory.

B.6 Displaying Program Output to the CRT Screen

If your program includes interrupts to DOS or BIOS that display characters on the screen, you need to take the following points into consideration.

1. D86 uses screen (page) *number 0*. If your program uses another screen, then the D86 screen will disappear when your program changes screens, and your program screen will (hopefully) appear. D86 is *still* running, however, and your program should include an interrupt to change back to screen 0 when you wish to regain access to a visible D86.

2. If your program outputs only a single line, then you can use display Row 24 (the blank space just below the D86 screen) to show program output. Have your program position the cursor at Row 24, and then you may exercise program output. If you should attempt to use a D86 screen 0 line, other than Line 24, you will find that D86 *redraws* the screen after every step, and you will not see your output.

3. If you are running D86 on a color monitor, you will need to make sure that your *attribute* byte is correct. Using an attribute of 00h, which is the normal attribute for a monochrome monitor, will result in a black foreground on a black background on a color monitor—an *invisible* character.

B.7 Disclaimers and Words of Warning

Pay Attention

D86 is sometimes very unforgiving of operator mistakes. For instance, if you JMP to some weird place in the computer from which there is no return, or execute a g(o) instruction with no stopping place, then D86 may lock up. The only remedy for lockup is to reboot the computer.

If your program outputs data to the screen by using DOS or BIOS calls, then you may not like the results on a color monitor. Use Alt-F9 to clean it up.

Saving Changes You Have Made

This works *only* for .COM files. Press a *w* (for write) to save any programs you have patched heavily. Saving patched programs is *not* recommended. You should find your mistakes and re-edit and re-assemble your program. Patch minor items, such as operands and numeric constants, test your program, and then change the .asm file.

B.8 Control Key Summary

Place this handy summary near the computer when first learning to use D86. <R> = Return key on keyboard.

- Clean Up Color Display = Alt-F9
- Display Memory = Line Number (1 to 6), format, segment, offset <R>
- Display More Memory = Ctrl-n(ext) or Ctrl-p(revious)
- Get Out = q <R>
- Move Through Program = Up arrow, Down arrow, Pg Up, Pg Dn
- Move Through Program by Bytes = Ctrl-u(p) or Ctrl-d(own)
- Patch = F7
- Remove Breakpoints = b, <R>
- Reset Stack = Ctrl-t
- Run = g,yyyy <R> [yyyy = label or number]
- Run to Next Breakpoint = g, <R>
- Run through REP or CALL = F2

- Save Changes = w
- See Breakpoint Symbols = Ctrl-s
- Set Breakpoint = b,xxxx <R> [xxxx = label or number]
- Single-Step = F1
- Start Over = Home

Appendix

C

C.0 Interrupt Type Low Memory Table

Table C.1 shows the interrupt type number addresses in low memory. The address listed in the table is the location of the interrupt vector IP address word for the interrupt type listed. The CS word for an interrupt address follows the IP word for a total of 4 bytes of address vector per interrupt type.

BIOS interrupts are listed in this appendix by interrupt type and, where applicable, by function. Interrupt function listings show values to be loaded into CPU registers before using a multifunction interrupt type, and any return register values. Many functions use the Carry flag (CF) set to 1 to indicate an error status.

A brief description of each type and function is included. Interrupts that are externally generated (automatically by hardware action) or are reserved should generally not be used in application programs.

The contents of this appendix are intended to be used as a quick lookup table for BIOS interrupt type service addresses. For complete descriptions of each BIOS function, please refer to one of the references listed in Appendix H, "Sources."

C.1 Hardware-Generated Interrupts

The following hardware-generated interrupts are not generally invoked via the software INT instruction. Hardware interrupts are normally automatically generated by CPU action or CPU action in conjunction with external signals. Hardware-initiated interrupts are termed *pure* if they are the direct result of CPU action or an external hardware signal. Several pure hardware interrupts, notably the keyboard interrupt, invoke other interrupt types listed in this section.

00h   Divide Error

Automatically generated by the CPU on a divide error. Normally vectored to a DOS handler that displays a divide error message on the screen.

01h   Single Step

Automatically generated by CPU (after executing an instruction) if the trap flag is set. The trap flag is normally used by single step (monitor) programs to interrupt a user program. The monitor program gains control of the CPU in order to display CPU and memory contents.

02h   Nonmaskable

Automatically generated when an external circuit brings the NMI input pin high. Most PCs use NMI for parity error signals; single-board designs may use NMI for any external interrupt purpose.

04h   Arithmetic Overflow

Automatically generated by the CPU if the OF is set, and the INTO instruction is executed. This type may be used to handle overflow errors in BIOS.

Address	Type	Address	Type	Address	Type	Address	Type	Address	Type	Address	Type
00000*	00	000AC	2B	00158	56	00204	81	002B0	AC	0035C	D7
00004*	01	000B0	2C	0015C	57	00208	82	002B4	AD	00360	D8
00008*	02	000B4	2D	00160	58	0020C	83	002B8	AE	00364	D9
0000C	03	000B8	2E	00164	59	00210	84	002BC	AF	00368	DA
00010*	04	000BC	2F	00168	5A	00214	85	002C0	B0	0036C	DB
00014	05	000C0	30	0016C	5B	00218	86	002C4	B1	00370	DC
00018	06	000C4	31	00170	5C	0021C	87	002C8	B2	00374	DD
0001C	07	000C8	32	00174	5D	00220	88	002CC	B3	00378	DE
00020*	08	000CC	33	00178	5E	00224	89	002D0	B4	0037C	DF
00024*	09	000D0	34	0017C	5F	00228	8A	002D4	B5	00380	E0
00028	0A	000D4	35	00180	60	0022C	8B	002D8	B6	00384	E1
0002C*	0B	000D8	36	00184	61	00230	8C	002DC	B7	00388	E2
00030*	0C	000DC	37	00188	62	00234	8D	002E0	B8	0038C	E3
00034*	0D	000E0	38	0018C	63	00238	8E	002E4	B9	00390	E4
00038*	0E	000E4	39	00190	64	0023C	8F	002E8	BA	00394	E5
0003C*	0F	000E8	3A	00194	65	00240	90	002EC	BB	00398	E6
00040	10	000EC	3B	00198	66	00244	91	002F0	BC	0039C	E7
00044	11	000F0	3C	0019C	67	00248	92	002F4	BD	003A0	E8
00048	12	000F4	3D	001A0	68	0024C	93	002F8	BE	003A4	E9
0004C	13	000F8	3E	001A4	69	00250	94	002FC	BF	003A8	EA
00050	14	000FC	3F	001A8	6A	00254	95	00300	C0	003AC	EB
00054	15	00100	40	001AC	6B	00258	96	00304	C1	003B0	EC
00058	16	00104	41	001B0	6C	0025C	97	00308	C2	003B4	ED
0005C	17	00108	42	001B4	6D	00260	98	0030C	C3	003B8	EE
00060	18	0010C	43	001B8	6E	00264	99	00310	C4	003BC	EF
00064	19	00110	44	001BC	6F	00268	9A	00314	C5	003C0	F0
00068	1A	00114	45	001C0*	70	0026C	9B	00318	C6	003C4	F1
0006C*	1B	00118	46	001C4	71	00270	9C	0031C	C7	003C8	F2
00070*	1C	0011C	47	001C8	72	00274	9D	00320	C8	003CC	F3
00074*	1D	00120	48	001CC	73	00278	9E	00324	C9	003D0	F4
00078*	1E	00124	49	001D0	74	0027C	9F	00328	CA	003D4	F5
0007C*	1F	00128	4A	001D4	75	00280	A0	0032C	CB	003D8	F6
00080	20	0012C	4B	001D8	76	00284	A1	00330	CC	003DC	F7
00084	21	00130	4C	001DC	77	00288	A2	00334	CD	003E0	F8
00088	22	00134	4D	001E0	78	0028C	A3	00338	CE	003E4	F9
0008C	23	00138	4E	001E4	79	00290	A4	0033C	CF	003E8	FA
00090	24	0013C	4F	001E8	7A	00294	A5	00340	D0	003EC	FB
00094	25	00140	50	001EC	7B	00298	A6	00344	D1	003F0	FC
00098	26	00144	51	001F0	7C	0029C	A7	00348	D2	003F4	FD
0009C	27	00148	52	001F4	7D	002A0	A8	0034C	D3	003F8	FE
000A0	28	0014C	53	001F8	7E	002A4	A9	00350	D4	003FC	FF
000A4	29	00150	54	001FC	7F	002A8	AA	00354	D5		
000A8	2A	00154	55	00200	80	002AC	AB	00358	D6		

* = Pure hardware interrupt

Table C.1 ■ Interrupt Type Low Memory IP:CS Vector Addresses

05h   Print Screen

Invoked by INT 09h, the keyboard handler, when a Print Screen key sequence is detected.

08h   System Timer

Automatically generated by the system clock every 54.925493 milliseconds, or 18.20648193 times per second, via the INTR input pin. Used to drive the time of day DOS service program. There are 1,573,040 interrupts in a 24-hour day.

09h   Keyboard

Automatically generated by the keyboard, via the INTR input pin. PC keyboards contain a microcontroller that transmits key data information to BIOS. BIOS converts the microcontroller key data into scan and ASCII codes and stores the codes in a 16-key input buffer for user programs.
    This routine also will invoke other interrupt types in response to the Ctrl and Print Screen keys. The routine also detects the Ctrl-Alt-Del key combination, and invokes the POST software to reset (warm boot) the computer

0Bh   Communications
0Ch

Automatically generated by communication port(s) via the INTR pin.

0Dh   Peripheral

Automatically generated by peripherals via the INTR pin.

0Eh   Floppy Disk

Automatically generated by floppy disk controller(s) via the INTR pin.

0Fh   Line Printer

Automatically generated by LPT1 via the INTR pin.

1Bh   Ctrl-Break

Invoked by INT 09h, keyboard interrupt, if the Ctrl-Break key combination is detected from the keyboard.

1Ch   Time

Invoked by INT 08h, timer interrupt. User programs that use the system timer should be accessed by this type so that INT 08h may be reset for the next clock INTR input.

1Dh   Video Controller Initialization (RESERVED)

Not a true interrupt. Used as a pointer to a video controller initialization data lookup table.

1Eh   Disk Controller Initialization (RESERVED)

Not a true interrupt. Used as a pointer to a disk controller initialization data lookup table.

1Fh   Video ASCII Character (RESERVED)

Not a true interrupt. Used as a pointer to a video ASCII character bit-mapped lookup table. Each ASCII screen character may be made up of an 8×8 array of bits (pixels) found in the table.

70h Real Time Clock

Automatically generated 1,024 times per second by a real-time clock (if the PC is so equipped) via INTR input pin. This interrupt permits finer time control than does timer interrupt 08h.

C.2 BIOS Interrupt Services

The interrupt services listed here are meant for general use by user programs. Many of these BIOS handlers may have many possible functions. Functions are shown under the "Load" heading, usually referring to the value to be loaded into register AH in order to invoke the desired function. If the service routine returns some data values to the user program, the registers used to return data are listed under the "Return" heading.

03h Single byte INT opcode (CCh).

Often used by debugging programs to insert a single-byte opcode (CCh) in the user code to return control to the debugger when a certain code address (a breakpoint) is reached.

06h RESERVED

07h RESERVED

0Ah RESERVED

10h CRT Video Functions
Set Video Mode

Load	Returned
AH = 00h	None
AL = Mode number:	

Mode Number	Display Format	Adapter Type	Colors
00h	Text, 25 rows of 40 characters	CGA, EGA, VGA	16
01h	Text, 25 rows of 40 characters	CGA, EGA, VGA	16
02h	Text, 25 rows of 80 characters	CGA, EGA, VGA	16
03h	Text, 25 rows of 80 characters	CGA, EGA, VGA	16
04h	Graphics, 320 by 200 pixels	CGA, EGA, VGA	4
07h	Text, 25 rows of 80 characters	MDA, EGA, VGA	Mono
08h	Graphics, 160 by 200 pixels	PCjr	16
09h	Graphics, 320 by 200 pixels	PCjr	16
0Ah	Graphics, 640 by 200 pixels	PCjr	4
0Bh	Reserved for future		
0Ch	Reserved for future		
0Dh	Graphics, 320 by 200 pixels	EGA, VGA	16
0Eh	Graphics, 640 by 200 pixels	EGA, VGA	16
0Fh	Graphics, 640 by 350 pixels	EGA, VGA	Mono
10h	Graphics, 640 by 350 pixels	EGA, VGA	16
11h	Graphics, 640 by 480 pixels	MCGA, VGA	2
12h	Graphics, 640 by 480 pixels	VGA	16
13h	Graphics, 320 by 200 pixels	MCGA, VGA	256

Set Video Cursor Height

Load	**Returned**
AH = 01h	None
CH = Top line (0–1Fh)	
CL = Bottom line (0–1Fh)	

Set Video Cursor Screen Position

Load	**Returned**
AH = 02h	None
BH = Page number (0–7)	
DX = Position (row,column)	

Read Video Cursor Position

Load	**Returned**
AH = 03h	BH = Page number (0–7)
BH = Page	CX = Cursor size (top,bottom)
	DX = Position (row,column)

Read Position of Light Pen

Load	**Returned**
AH = 04h	AH = Pen on
	BX = Pixel column
	CX = Pixel row
	DX = Character position

Select Displayed Video Page

Load	**Returned**
AH = 05h	None
AL = Page number (0–7)	

Scroll Active Video Window Up

Load	**Returned**
AH = 06h	None
AL = Number of scrolled lines	
BH = Attribute of scrolled lines	
CX = Left corner row,column	
DX = Right corner row,column	

Scroll Active Video Window Down

Load	**Returned**
AH = 07h	None
AL = Number of lines	
BH = Attribute of scrolled lines	
CX = Left corner row,column	
DX = Right corner row,column	

Read Character Code at Cursor Position

Load	**Returned**
AH = 08h	AH = Attribute
BH = Page number	AL = ASCII

Write Character and Attribute

Load	**Returned**
AH = 09h	None
AL = ASCII	
BH = Page number	

BL = Attribute:

Monochrome Display Attribute	**Attribute Byte Bit Settings** **Bits 7 6 5 4 3 2 1 0**
Normal character	0 x x x x x x x
Blinking character	1 x x x x x x x
Dark background	x 0 0 0 x x x x
Light background (inverse)	x 1 1 1 x x x x
Normal intensity	x x x x 0 x x x
High intensity	x x x x 1 x x x
Light foreground	x x x x x 0 0 0
Underlined light foreground	x x x x x 0 0 1
Dark foreground	x x x x x 1 1 1

Color Display Attribute	**Attribute Byte Bit Settings** **Bits 7 6 5 4 3 2 1 0**
Normal character	0 x x x x x x x
Blinking character	1 x x x x x x x
Background color	x b b b x x x x
Foreground color	x x x x f f f f

Background color	**bbb**	**Color**
Foreground color	**ffff**	
	0000	Black
	0001	Blue
	0010	Green
	0011	Cyan
	0100	Red
	0101	Magenta
	0110	Brown
	0111	White
	1000	Gray
	1001	Light blue
	1010	Light green
	1011	Light cyan
	1100	Light red
	1101	Light magenta
	1110	Yellow
	1111	High-intensity white

CX = Number of copies

Write Character and Graphics Mode Color

Load	**Returned**
AH = 0Ah	None
AL = ASCII	
BH = Page number	
BL = Color (graphics mode only)	
CX = Copies	

Set Color for Graphics Display

Load	**Returned**
AH = 0Bh	None
BH = Palette	
BL = Color	

Write Graphics Mode Pixel

Load	**Returned**
AH = OCh	None
AL = Color	
BH = Page number	
CX = Pixel column	
DX = Pixel row	

Read Pixel Color

Load	**Returned**
AH = ODh	AL = Color
BH = Page number	
CX = Pixel column	
DX = Pixel row	

Teletype Write Mode

Load	**Returned**
AH = OEh	None
AL = ASCII	
BH = Text page number	
BL = Graphics color	

Read Display Mode in Use

Load	**Returned**
AH = OFh	AH = Number of screen columns
	AL = Mode
	BH = Current page

Set Video Controller Registers

Load	**Returned**
AH = 10h	BX = Data (varies by AL contents)
AL = Control	CX = Data (varies by AL contents)
BX = Data	DH = Data (varies by AL contents)
CX = Data	ES:DX = Pointer (varies by AL contents)
ES:DX = Pointer	

Customize Graphics Character Generator

Load	**Returned**
AH = 11h	Varies

Additional EGA Control

Load	**Returned**
AH = 12h	Varies
Varies	

Write ASCII Character String

Load	**Returned**
AH = 13h	None
AL = Write mode	
BH = Page number	
BL = Attribute	
CX = Length	
DX = Row, column	
ES:BP = Pointer to string	

PS/2 Read or Write Display Codes
Load	**Returned**
AH = 1Ah	Varies
AL = Control	
BX = Code	

PS/2 Read State of the Video Display
Load	**Returned**
AH = 1Bh	AL = Status
ES:DI = Pointer to buffer	

PS/2 Save or Restore Display State
Load	**Returned**
AH = ICh	AL = Status
AL = Action	BL = Blocks
CX = Map	
ES:BX = Pointer to buffer	

11h Report Peripheral Status
Load	**Returned**
None	AX = Status of standard peripherals

12h Report Memory Size
Load	**Returned**
None	AX = Size

13h Disk Controllers

Reset Disk Controller
Load	**Returned**
AH = 00h	CF = Status
DL = Drive number	AH = Errors

Read Disk Controller Status
Load	**Returned**
AH = 01h	AH = Status
DL = Drive number	

Read Disk Sectors
Load	**Returned**
AH = 02h	CF = Status
AL = Number	AH = Errors
ES:BX = Pointer	AL = Number
CX = Track, sector	
DX = Head, drive	

Write Disk Sectors
Load	**Returned**
AH = 03h	CF = Status
AL = Number	AH = Errors
ES:BX = Pointer	AL = Number
CX = Track, sector	
DX = Head, drive	

Verify Sector CRC Code

Load	**Returned**
AH = 04h	CF = Status
AL = Number	AH = Errors
CX = Track, sector	
DX = Head, drive	

Format Track on Disk

Load	**Returned**
AH = 05h	AH = Status
ES:BX = Pointer	
CH = Track	
DX = Head, drive	

Format Cylinder on Disk

Load	**Returned**
AH = 06h	CF = Status
AL = Factor	AH = Errors
CX = Cylinder, sector	
DX = Head, drive	

Format Cylinders on Disk

Load	**Returned**
AH = 07h	CF = Status
AL = Factor	AH = Errors
CX = Starting cylinder, sector	
DX = Head, drive	

Drive Information

Load	**Returned**
AH = 08h	CF = Status
DL = Drive number	CX = Track, sector
	DH = Sides
	DL = Number of drives
	ES:DI = Pointer
	BL = Drive type
	AH = Errors

Initialize Fixed Disk Table

Load	**Returned**
AH = 09h	CF = Status
DL = Drive number	AH = Errors

Read Sector from Fixed Disk

Load	**Returned**
AH = 0Ah	CF = Status
AL = Number	AH = Errors
ES:BX = Pointer	
CX = Track, sector	
DX = Head, drive number	

Write Sector to Fixed Disk

Load	**Returned**
AH = 0Bh	CF = Status
AL = Number	AH = Errors
ES:BX = Pointer	
CX = Track, sector	
DX = Head, drive number	

Seek Fixed Disk Cylinder

Load	**Returned**
AH = 0Ch	CF = Status
CX = Track	AH = Errors
DX = Head, drive number	

Reset Fixed Disk Controller

Load	**Returned**
AH = 0Dh	CF = Status
DL = Drive number	AH = Errors

Read PC/XT Sector Buffer

Load	**Returned**
AH = 0Eh	CF = Status
ES:BX = Pointer	AX = Errors

Write PC/XT Sector Buffer

Load	**Returned**
AH = 0Fh	CF = Status
ES:BX = Pointer	AX = Errors

Test Fixed Disk System

Load	**Returned**
AH = 10h	CF = Status
DL = Drive number	AX = Errors

Initialize Fixed Disk

Load	**Returned**
AH = 11h	CF = Status
DL = Drive number	AX = Errors

Diagnose PC/XT Controller RAM

Load	**Returned**
AH = 12h	CF = Status
	AX = Errors

Diagnose PC/XT Fixed Disk

Load	**Returned**
AH = 13h	CF = Status
	AX = Errors

Diagnose PC/XT Controller

Load	**Returned**
AH = 14h	CF = Status
	AX = Errors

Return Drive Information

Load	**Returned**
AH = 15h	CF = Status
DL = Drive Number	AH = Information
	CX:DX = Sectors

Disk Change Status

Load	**Returned**
AH = 16h	CF = Status
DL = Drive number	AH = Information

Initialize for Disk Format

Load	**Returned**
AH = 17h	None
AL = Format type	
DL = Drive number	

Disk Format Information

Load	**Returned**
AH = 18h	CF = Status
CX = Tracks	AH = Errors
CL = Sectors	ES:DI = Pointer
DL = Drive number	

Park Fixed Disk Heads

Load	**Returned**
AH = 19h	CF = Status
DL = Drive number	AX = Information

Format PS/2 ESDI Drive

Load	**Returned**
AH = 1Ah	None
AL = Table	
ES:BX = Pointer	
CL = Modifier	
DL = Drive number	

14h   Serial Communications Port

The serial ports are based on programmable devices. Before serial data may be received or transmitted, the ports must be programmed. Many serial port functions return the status of the serial port modem handshaking signals and serial data. Port status is returned in AX, as follows.

Return Contents of AH

Bit 7 6 5 4 3 2 1 0	**Serial Port Data Status**
x x x x x x x 1	Received data present
x x x x x x 1 x	Overrun error
x x x x x 1 x x	Parity error
x x x x 1 x x x	Framing error
x x x 1 x x x x	Break detected
x x 1 x x x x x	Transmitter holding register empty
x 1 x x x x x x	Transmitter shift register empty
1 x x x x x x x	Time out

Return Contents of AL

Bit 7 6 5 4 3 2 1 0	**Modem (Serial Port Control) Status**
x x x x x x x 1	Change in CTS status
x x x x x x 1 x	Change in DSR status
x x x x x 1 x x	Trailing edge of RI signal
x x x x 1 x x x	Change in CD signal
x x x 1 x x x x	CTS enabled
x x 1 x x x x x	DSR enabled
x 1 x x x x x x	RI enabled
1 x x x x x x x	CD enabled

Initialize Port

	Load							**Returned**

AH = 00h   AX = Status

AL = Program port:

Bit	**Baud Rate**	**Bit**	**Parity**	**Bit**	**Stop Bits**	**Bit**	**Character Length**
7 6 5		4 3		2		1 0	
0 0 0	110	0 0	None	0	1	0 0	7 bits
0 0 1	150	0 1	Odd	1	2	1 1	8 bits
0 1 0	300	1 0	None				
0 1 1	600	1 1	Even				
1 0 0	1,200						
1 0 1	2,400						
1 1 0	4,800						
1 1 1	9,600						

DX = Port number (0, 1, 2, 3)

Write Character to Port

Load	**Returned**
AH = 01h	AH = Status
AL = ASCII	
DX = Port number	

Read Character from Port

Load	**Returned**
AH = 02h	AH = Data status
DX = Port number	AL = ASCII

Port and Modem Status

Load	**Returned**
AH = 03h	AX = Status
DX = Port number	

PS/2 Port Initialization

Load	**Returned**
AH = 04h	AX = Status
AL = Break	
BH = Parity	
BL = Stop	
CH = Length	
CL = Baud	
DX = Port	

PS/2 Port Control

Load	**Returned**
AH = 05h	AX = Status
AL = R/W	
BL = Data	BL = Data (AL = Read)
DX = Port number	

15h   PC Tape Cassette Control

Turn Cassette Motor On

Load	**Returned**
AH = 00h	CF = Status
	AH = Errors

Turn Cassette Motor Off

Load	**Returned**
AH = 01h	CF = Status
	AH = Errors

Read Cassette Data (obsolete)

Load	**Returned**
AH = 02h	CF = Status
ES:BX = Pointer	AH = Errors
CX = Number	DX = Number read
	ES:BX = Pointer

Write Cassette Data

Load	**Returned**
AH = 03h	CF = Status
ES:BX = Pointer	AH = Errors
CX = Number	ES:BX = Pointer

PS/2 Format Interrupt

Load	**Returned**
AH = 0Fh	CF = Status
AL = Code	

PS/2 POST Log Update

Load	**Returned**
AH = 21h	CF = Status
AL = R/W	AH = Information
BX = Data	BX = Number
	ES:DI = Pointer

Intercept Keyboard Characters

This routine is invoked by hardware INT 09h, the keyboard interrupt. A programmer may point to a custom scan code translation program with this interrupt vector.

Load	**Returned**
AH = 4Fh	CF = Status
AL = Code	CF = 1, then AL = Scan code
	CF = 0, INT 09h rejects key

Multitasking Device Open

Load	**Returned**
AH = 80h	CF = Status
BX = Device	AH = Errors
CX = Process	

Multitasking Device Close

Load	**Returned**
AH = 81h	CF = Status
BX = Device	AH = Errors
CX = Process	

Multitasking Program Terminate

Load	**Returned**
AH = 82h	CF = Status
BX = Process	AH = Errors

Multitasking Event Wait

Load	**Returned**
AH = 83h	CF = Status
AL = Arm	AH = Errors
CS:DX = Delay	
ES:BX = Pointer	

Read Joystick Position/Switches

Load	**Returned**
AH = 84h	CF = Status
DX = P/S	DX = 0:
	AL = Switches
	DX = 1:
	AX = Position w
	BX = Position x
	CX = Position y
	DX = Position z

System Request Key Pressed (May Be Invoked by INT 09h)

Load	**Returned**
AH = 85h	CF = Status
	AL = 00, Key Down
	AL = 01, Key Up

Operating System Delay

Load	**Returned**
AH = 86h	CF = Delay status
CX:DX = Delay time	

808286/80386 Block Move

Load	**Returned**
AH = 87h	CF = Status
CX = Count	AH = Errors
ES:SI = Pointer	

Size of Extended Memory

Load	**Returned**
AH = 88h	CF = Status
	AX = Size

Switch to Protected Mode

Load	**Returned**
AH = 89h	CF = Status
BX = Vectors	
CX = Offset	
ES:DI = Pointer	

Multitasking Device Wait

Load	**Returned**
AH = 90h	CF = Status
AL = Device	
ES:BX = Pointer	

Multitasking Device Interrupt Complete

Load	**Returned**
AH = 91h	CF = Status
	AL = Device Code

Read System Type and BIOS Data

Load	**Returned**
AH = C0h	CF = Status
	ES:BX = Pointer to data table

PS/2 Extended BIOS Data Address

Load	**Returned**
AH = C1h	CF = Status
	ES = Data segment address

PS/2 Auxiliary Pointing Device

Load	**Returned**
AH = C2h	CF = Status
AL = Control	AH = Errors
BH = On/Off	

PS/2 Watchdog Timer Control

Load	**Returned**
AH = C3h	CF = Status
AL = On/Off	
BX = Time-Out count	

PS/2 Expansion Board Register Select

Load	**Returned**
AH = C4h	CF = Status
AL = Board	BL = Slot number
	DL = Register address

16h   Keyboard Control

The keyboard interrupts the system every time a key or key combination is activated by the keyboard user. (See hardware INT 09h.) INT 16h reads the keyboard buffer that is automatically filled by INT 09h.

Read Keyboard Buffer

Load	**Returned**
AH = 00h	AH = Scan code
	AL = ASCII code (AL = 00 if no ASCII code)

Determine if Key Waiting

Load	**Returned**
AH = 01h	ZF = Status
	ZF = 0, then key waiting
	AH = Code
	AL = ASCII
	ZF = 1, then no key available

Read Control Keys Status

Load   AH = 02h

Returned   AL = Status:

7 6 5 4 3 2 1 0	
x x x x x x x 1	Right Shift down
x x x x x x 1	Left Shift down
x x x x x 1	Ctrl down
x x x x 1	Alt down
x x x 1	Scroll Lock on
x x 1	Num Lock on
x 1	Caps Lock on
1	Insert key toggled

Adjust Key Repeat Rate (Not All PC Models)

Load	**Returned**
AH = 03h	None
AL = Control	
BX = Rate	

PCjr Key Click Sound On/Off

Load	**Returned**
AH = 04h	None
AL = On/Off	

Enhanced Keyboard Buffer Write

Load	**Returned**
AH = 05h	AL = Full
CX = Scan code, ASCII	

Read Key from Enhanced Keyboard

Load	**Returned**
AH = 10h	AH = Scan code
	AL = ASCII

Check Enhanced Keyboard Status

Load	**Returned**
AH = 11h	ZF = Status
	ZF = 0, then key waiting
	AH = Code
	AL = ASCII
	ZF = 1, then no key waiting

Enhanced Keyboard Status

Load

AH = 12h

Returned

AL = Control key status (See Function 02h)
AH = Enhanced status:

7 6 5 4 3 2 1 0	
x x x x x x x 1	Left Ctrl down
x x x x x x 1	Left Alt down
x x x x x 1	Right Ctrl down
x x x x 1	Right Alt down
x x x 1	Scroll Lock down
x x 1	Num Lock down
x 1	Caps Lock down
1	SysRq down

17h   Printer Control

Printer status is returned in AH. Each bit of AH, *when set to 1,* indicates a printer port active signal level condition.

AH Bit	**Status**
7	Printer not busy
6	Acknowledged
5	Paper out
4	Printer is selected
3	I/O error
2	Not used
1	Not used
0	Time-out error

Write Character to Printer

Load	**Returned**
AH = 00h	AH = Status
AL = ASCII character	
DX = Printer number	

Initialize Printer

Load	**Returned**
AH = 01h	AH = Status
DX = Printer number	

Get Printer Status

Load	**Returned**
AH = 02h	AH = Status
DX = Printer number	

18h   Start BASIC Program in IBM Machines

19h   Warm Boot

Restart (warm boot) the computer without loss of memory.

1Ah   System and Real-Time Clock Control

PC systems contain a system clock that interrupts the CPU every 54.936 milliseconds. (See hardware INT 08h.) Many PC systems also contain a high-resolution clock called the *real-time* clock, which operates at a frequency of 1,024 counts, or *ticks* per second. (See hardware INT 70h.)

Read System Clock Tick Counter

Load	**Returned**
AH = 00h	AL = 01h, if past midnight
	CX:DX = Tick count since midnight

Set System Clock Counter

Load	**Returned**
AH = 01h	None
CX:DX = Count	

Read Real-Time of Day

Load	**Returned**
AH = 02h	CF = Status
	CH = Hours (BCD)
	CL = Minutes (BCD)
	DH = Seconds (BCD)
	DL = Daylight Savings Time flag

Set Real-Time of Day

Load	**Returned**
AH = 03h	None
CH = Hours (BCD)	
CL = Minutes (BCD)	
DH = Seconds (BCD)	
DL = Varies (BCD)	

Read Real-Time Date

Load	**Returned**
AH = 04h	CF = Status
	CH = Century (BCD)
	CL = Year (BCD)
	DH = Month (BCD)
	DL = Day (BCD)

Set Real-Time Date

Load	**Returned**
AH = 05h	None
CH = Century (BCD 19 or 20)	
CL = Year (BCD)	
DH = Month (BCD)	
DL = Day (BCD)	

Set Real-Time Alarm

Load	**Returned**
AH = 06h	CF = Status
CH = Hours (BCD)	
CL = Minutes (BCD)	
DH = Seconds (BCD)	

Turn Real-Time Alarm Off

Load	**Returned**
AH = 07h	None

Read Real-Time Alarm Status

Load	**Returned**
AH = 09h	CH = Hours (BCD)
	CL = Minutes (BCD)
	DH = Seconds (BCD)
	DL = Alarm status

Get Number of Days Since January 1, 1980

Load	**Returned**
AH = 0Ah	CX = Number of days

Set Number of Days Since January 1, 1980

Load	**Returned**
AH = 0Bh	None
CX = Number of days	

Set PCjr Sound Source

Load	**Returned**
AH = 80h	None
AL = Source	

Appendix

DOS AND

DOS

INTERRUPTS

D.0 A Short History Of DOS

Origins of DOS

DOS first saw the light of day, (commercially) in 1980 as an operating system program named *86-DOS*. 86-DOS was written by Tim Paterson as an operating system for *S100* bus 8086 CPU cards manufactured by Seattle Computer Products. (S100 was the name for one of the first personal computer bus standards.) S100-based computers helped launch the whole personal computer movement. The operating system for most 8-bit microprocessor S100-bus systems was Digital Research's CP/M. 86-DOS was, essentially, a 16-bit version of the Digital Research 8-bit CP/M.

Microsoft Corporation, also of Seattle, bought all the rights to 86-DOS in 1981 (and got Tim Paterson also) and released MS-DOS (MicroSoft-DOS) version 1.0 that year.

IBM adopted MS-DOS, rewrote parts of it, and named it plainly DOS— the operating system for its new personal computer, the IBM PC. To differentiate IBM DOS from Microsoft DOS, IBM DOS became known as PC-DOS. Soon the IBM PC became the dominant personal computer, and PC-DOS tagged along. In the years since 1981, new versions of DOS have been introduced to take advantage of rapidly growing PC CPU and disk drive capabilities.

DOS is inexorably tied to developments in disk storage technology. Floppy disk drives have evolved from an original storage capacity of 160K to today's 2.88M, and higher capacities are planned. Hard disk drives, which originally offered 10M of storage capacity, are now commonly offered at 120M and up. CD drives are now available with 600M of storage.

DOS evolved to handle the ever increasing size of disk storage and to meet competitive challenges from other operating systems. There are many versions of DOS (for example, 3.2, 3.21, 3.3, 3.3 Plus) that were sold by Microsoft to various PC manufacturers. A short history of PC-DOS (up to PC-DOS 6.2) is listed in Table D.1, with the caveat that many variations of the later releases (beyond version 3.0) are possible.

IBM licensed MS-DOS and made various modifications to produce PC-DOS. Microsoft continued to develop MS-DOS for non-IBM customers, and most versions of PC-DOS listed in Table D.1 have one, or more, MS-DOS equivalents. If you need to know the specifics of the DOS version you are using, please refer to the DOS manual that accompanies your particular computer.

DOS Versions, 1981 to 1993

Variations in DOS programs are called "versions", and are numbered (so far) from 1 to 6. Each number represents a major new version, with many changes from previous versions. If the version number is followed by a period and another number (as in DOS 3.2), the second number represents minor changes to the version. Another number with or without a preiod (as in DOS 3.2.1) represents very minute changes to the version.

The significant versions of DOS, their dates of release by the manufacturer, and major features are listed in Table D.1.

The different versions of DOS have a wide variation in popularity. Many people feel that each version of DOS is larger and slower than the one that preceded it. You may wish to use the earliest version that will support your

Version	Date	Features
PC-DOS 1.0	August 1981	DOS for IBM PC. Support for 180K single sided 5.25-inch floppy drives. Added AUTOEXEC.BAT files for system initialization.
PC-DOS 1.1	June 1982	Supports 360K 5.25-inch double sided floppy drives. Time and date for files.
PC-DOS 2.0	March 1983	DOS for the IBM PC-XT. Hard drive support, directories and subdirectories, installable device drivers, pipes, filters, print spooling, disk volume labels, I/O redirection, and file handles.
PC-DOS 2.1	October 1983	Fixes bugs in version 2.0.
PC-DOS 3.0	August 1984	DOS for IBM PC-AT. Supports high-density 5.25-inch floppy drives. File sharing.
PC-DOS 3.1	November 1984	Fixes bugs in version 3.0. Network support.
PC-DOS 3.2	January 1986	Supports 720K 3.5-inch floppy drives.
PC-DOS 3.3	April 1987	DOS for IBM PS/2 series. Supports 1.44M 3.5-inch floppy drives. Hard drive partitions up to 32M. Supports up to four COM serial ports.
PC-DOS 4.0	August 1988	Hard drive partitions greater than 32M. Expanded memory support.
MS-DOS 5.0	June 1991	Supports 2.8M 3.5-inch floppy drives. Supports up to eight hard drives. Improved full-page editor. QBASIC. Frees more RAM for applications. Command shell for graphic interface to user.
MS-DOS 6.0	March 1993	Disk compression, antivirus, file transfer, file backup, and memory management. Power management (for battery-powered applications).
MS-DOS 6.2	October 1993	Fixes bugs in version 6.0.

Table D.1 ■ Major Versions of DOS

computer. Popular versions of DOS are V2.1 (for older floppy drive PCs), V3.3 (hard drive PCs), and V5.0. Most computers, new or old, come with a suitable version of DOS that matches the computer hardware.

MS-DOS version 5.0 features a command *shell*. A shell program is a user interface program, usually featuring a *mouse* as a screen pointing device. For MS-DOS 5.0, the shell is a graphics (CRT) interface that features extensive screen menus, icons, and "point-and-click" ease of use. Other shell programs, such as DesQview, Windows, GEM, TopView, and GeoWorks also perform operating system tasks, using DOS only for disk access.

MS-DOS version 6.0 adds many contemporary features such as antivirus protection and disk compression. IBM plans to have a version called IBM DOS 6.0. The popularity of MS-DOS V6.0 has not yet been fully measured, but it was a best-seller only 1 month after commercial introduction.

Microsoft and IBM are not the only sources for PC-compatible DOS programs. Digital Research (now part of Novell) has produced several excellent DOS versions under the title *DR-DOS*; the latest is version 6.0.

Most early 360K floppy drive systems will work very well with DOS 2.1; high-capacity hard drive systems will generally use version 3.3. If you are purchasing a new system, it will probably be equipped with an original equipment manufacturer (OEM) copy of version 5.0 or 6.2.

D.1 DOS Structure

A DOS operating system is not one single, large program but a collection of programs that operate the computer. Many DOS programs, such as FORMAT and EDLIN, are not continuously in computer RAM. When needed by the user they are loaded from disk to RAM. Such DOS programs are considered to be *transient*. Transient means that they are not retained in RAM but are overwritten when they have finished executing. Other programs, called *terminate and stay resident* or TSRs, remain in RAM and can be executed without having to be reloaded from disk.

PC Memory Organization

We have previously noted that the area of memory first accessed by the CPU on power-up is ROM BIOS. BIOS is a collection of hardware interface programs that are fixed in ROM. PC memory also contains RAM chips, and Figure D.1 shows how PC memory is organized.

Figure D.1 ■ Generic PC memory map.

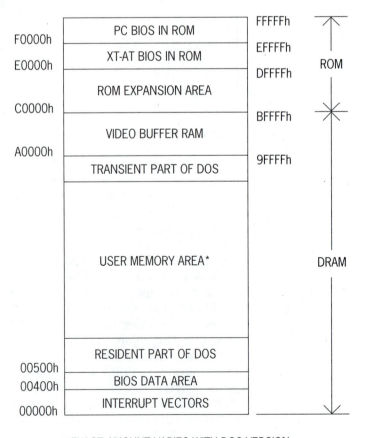

*EXACT AMOUNT VARIES WITH DOS VERSION

Reserved Memory	Address Range Function
00000h–001DFh	BIOS and DOS interrupt vectors
001E0h–002FFh	User interrupt vectors
00300h–003FFh	BIOS stack, user interrupt vectors
00400h–004FFh	BIOS data buffers
00500h–005FFh	DOS and BASIC use
A0000h–BFFFFh	Video (CRT Screen) data buffer
C0000h–EFFFFh	BIOS ROM expansion
F0000h–FFFFFh	BIOS

Table D.2 ■ Reserved Memory Area

PC memory is organized so that BIOS and DOS use certain *reserved* memory addresses. Reserved means that no application program should use memory addresses meant for DOS and BIOS use. DOS has no memory "protection" features, which means that the user can write programs that will interfere with DOS, and thus crash the system.

The major *reserved* areas of memory addresses, and their functions, are listed in Table D.2. The list is generic; some versions of DOS may not use all of the reserved memory.

The "user" area of RAM, from memory addresses 00600h to 9FFFFh, is set aside for the computer user.

Part of the user space is occupied by the operating system. Typical values for the amount of user RAM occupied by DOS vary from 16K for early DOS versions up to 115K for versions beyond 4.0. More DOS user memory may be required if numerous device drivers are loaded by DOS from CONFIG.SYS files.

Although the CPU can address 1M of memory, or more, DOS cannot use more than 640K of memory, from addresses 00000h to 9FFFFh. When DOS was first written, 640K of personal computer memory was thought to be beyond the needs of any future user. Programs tend to grow to fit the memory available however, and RAM has become so cheap that personal computers with 4M of memory are common. We shall not discuss the "tricks" that can be employed to access memory beyond the DOS limit of 640K. 640K of memory is more than adequate for our needs.

Actually, we cannot make use of all 640K for our application programs. BIOS and DOS will need some of it for their use.

D.2 Booting DOS

Booting a computer means the actions taken to get the computer to load a program that will operate the computer. The term *booting* comes from the old saying to lift oneself "by one's own bootstraps." Lifting yourself up by grasping and lifting the tops of your shoes is, of course, impossible. Computers, however, can be booted. Booting a computer involves getting an elementary program into the computer code space that will allow the computer to load another, smart program from disk into RAM. The disk-based program in RAM can then load smarter programs from disk into RAM, and the smarter programs can run the computer.

In the stone age of personal computing, around 1975, boot programs were placed into computer memory by using switches and push buttons on the front panel. The invention of ROMs let us put the boot program inside the ROM and have the CPU boot itself up.

The sequence of events that transpire from the moment you turn power on to the computer until the DOS prompt (>) shows up on the screen are discussed next.

DOS Load Sequence

When power is applied to the CPU (or the CPU RESET pin is brought high), the CPU fetches an instruction from location FFFF0h in memory. Resetting the CPU causes the CS register to be set to FFFFh, and the IP to 0000h. Intel chose reset address FFFF:0000h for the 8086 family because low memory (0000:0000h and up) is used to hold RAM-based data.

Resetting the CPU to fetch the first program instruction from FFFF:0000h leaves the programmer 16 bytes of code space before hitting the last addressable memory byte at FFFF:000Fh. Sixteen bytes of code is adequate for a far jump to a lower memory address, say the beginning of ROM BIOS at F0000h.

Programs in ROM may reside in memory from addresses C0000h to FFFFFh. BIOS occupies addresses F0000h to FFFFFh, thus the CPU resets to the last 16 bytes in the BIOS program. A far jump located at address FFFF0h directs the program to the first BIOS service routine in segment F000h.

BIOS proceeds to test the system hardware (memory, disk drives, printer, etc.) checking for proper operation. BIOS will skip this test, called a *power on self test* or POST, if the reset was caused by pressing Ctrl-Alt-Del on the keyboard instead of a power-on reset.

After POST, BIOS executes a bootstrap routine that configures lower memory to support system hardware. Low memory refers to RAM that begins at segment address 0000h, and, in particular, the RAM from offset addresses 0000h to 0600h. High memory is the area at the top of the 640K (9FFFFh) RAM memory that DOS can directly access, in particular, the last segment that begins at 9000h.

Once low memory is initialized, the BIOS bootstrap routine searches for a disk drive that contains a disk with a disk bootstrap routine. BIOS starts searching at drive A, skips drive B, and proceeds until the first disk with a disk bootstrap routine is found. BIOS loads the disk bootstrap routine into RAM and jumps to the disk bootstrap routine in RAM.

The disk bootstrap program loads an initialization program from the system disk: IBMBIO.COM, or IO.SYS (MS-DOS). IBMBIO.COM completes the initialization of any hardware devices not done by BIOS.

One part of IBMBIO.COM, a procedure named SYSINIT, loads IBMDOS.COM (PC-DOS) or MSDOS.SYS (MS-DOS) into RAM and then reads any CONFIG.SYS file that may exist on the system disk. SYSINIT configures the system according to the directives in the CONFIG.SYS file. Mouse drivers, and similar device drivers, are loaded into RAM for later use by the devices that require them. A CONFIG.SYS file is a flexible way for users to add hardware devices, such as a mouse, to a system that has no BIOS provisions for the hardware built into the computer.

Once the system is fully configured, SYSINIT uses a DOS function to load and run a program that we constantly use: COMMAND.COM. COM-MAND.COM is a program that interprets user keystroke sequences given in response to the DOS command prompt. COMMAND.COM then attempts to translate each sequence into some DOS action.

COMMAND.COM is broken into two parts when it is loaded into RAM. The resident part is stored in lower RAM, and contains all of the code necessary to reload COMMAND.COM at the end of some command action. Temporary parts of COMMAND.COM are held in high memory and serve to interpret command instructions from the user, or batch files. The resident part of COMMAND.COM can reload the temporary part after a user program returns the computer to DOS control.

When it first starts operating, COMMAND.COM executes any directives found in the AUTOEXEC.BAT file (if one exists) found on the system disk, and then displays the system prompt (>) on the screen for us to see. COMMAND.COM then waits for us to react.

Many people assume COMMAND.COM is, in fact, DOS. DOS is a group of programs, also called functions, that are loaded into low RAM beginning at address 00600h. Collectively, the DOS functions are named the DOS *kernel*. Properly, DOS consists of ROM BIOS, the bootstrap loader, IBMBIO.COM, IBMDOS.COM, and COMMAND.COM. Several DOS programs are not permanently resident in RAM (they are overwritten by later DOS modules), so that the permanent DOS that remains in RAM after boot-up is IBMDOS.COM and COMMAND.COM.

DOS is included on a system disk as a *hidden* file. The files that we see displayed when looking at the contents of a DOS system disk are really utilities included by the DOS manufacturer.

COMMAND.COM is one of many programs that are included with DOS. As previously mentioned, COMMAND.COM processes DOS commands that are entered by the user in response to the DOS prompt. Most common user directives (COPY, TYPE, DEL, etc.) are called *internal* and are located in that part of COMMAND.COM loaded into high memory. Internal commands are executed immediately. Other user directives, (FORMAT, CHKDSK, DISKCOPY, etc.) are called *external* and are .COM programs that are included on the DOS disk. External commands are loaded from the disk to RAM by DOS and run as application programs.

D.3 Running Programs Under DOS

When we wish to run a program, we simply type the name of the program opposite the COMMAND.COM prompt (>). Our program is loaded from the disk file where it resides, and our program executes.

When our program is finished it exits back to DOS, and the COM-MAND.COM prompt reappears on the screen. If the application program has overwritten high memory part of COMMAND.COM, the resident portion of COMMAND.COM in low memory reloads the entire COMMAND.COM version.

In order to run our program, DOS has, somehow, found our program, loaded it from the disk to internal RAM, turned control of the computer over to our program, and run our program. When our program exits, control of the computer mysteriously returns to COMMAND.COM. What, exactly, is going on?

Before we answer that question, we must examine the two types of programs that can be run under DOS control, COM and EXE.

COM Files

Chapter 3 discusses the segmented nature of 8086 memory. A memory segment is any linear address array of 64K size. 8086 segment addresses are held in the segment registers CS, DS, SS, and ES. Segment registers are 16 bits long, so there can be as many as 64K different segments. Segments can begin at address intervals of 16 bytes, yielding a total memory size of 1M (64K segments × 16 bytes) for the 8086 family of CPUs.

8086 programs (that run under DOS) fall into two types; .COM and .EXE, identified by the first two assembler-coded bytes of the program. COM programs are programs that occupy only 1 segment of memory. COM programs use all of the segment registers—CS, DS, SS, and ES—set to the same segment address.

COM programs also have another distinguishing feature: All program code begins at an IP offset address of 100h. The first 100h bytes of the code segment are used by DOS for writing a short preamble to the program named the *program segment prefix* or PSP. The PSP is not part of a COM program and is not stored on disk. When COM programs are loaded by DOS into RAM, DOS always loads the COM program starting at an offset of 100h into the program segment. DOS uses the PSP to store program operation details.

COM program code is limited to 64K, the maximum size of the code segment. COM programs can, however, access any area of memory simply by changing the DS, SS, or ES segment registers as the COM program runs.

Careless memory access is a dangerous practice, because it is possible that memory area the COM program accesses holds COMAND.COM, DOS, or other resident programs that you depend on for smooth operation of the computer. A sure sign that your COM program has *trashed* an essential area of memory is the refusal of the computer to respond to the usually reliable Ctrl-Alt-Del reset keys. The only remedy for trashed memory is to turn the computer off, wait 10 seconds, and re-power it.

COM programs are the predominant type of programs used in this book. There are three primary reasons for using COM programs:

1. COM programs are easy to write, and involve a minimum of esoteric assembler directives in order to produce usable programs. COM program format lets the beginner concentrate on writing code for the 8086 without getting bogged down in the details of assembler directives. See Appendix A for details on how to use the A86 assembler.

2. COM programs will run on PCs with limited amounts of memory, and can be easily converted to ROM for use in board-level computers.

3. 64K of code, data, and stack memory is more than adequate for any student program.

EXE Files

Programs that use more than a single memory segment must have the extension .EXE added to the program file name. EXE files may extend over many code and data segments. The segments do not have to be contiguous, that is, they may be scattered all over memory.

Programs written for EXE files may be very large, and are usually written in high-level languages. Several high-level language programs may be combined into a single program by a linker. Linkers can combine, or link, many program modules into a single EXE file.

Multi-segment EXE programs require many more assembler directives to write than do COM programs. A very simple, 3-line EXE program may require 10 lines of assembler directives to assemble. Assembler directives used by our assembler, A86, for creating EXE files are extremely few in number compared to other popular assemblers. However, a study of the A86 assembler directives in the A86 manual will reveal that EXE files require considerably more assembler directives to write than do COM programs.

EXE code segments that are larger than 64K require some special handling by the programmer. The only way to progress from one code segment to the next is to perform a far jump from the original code segment to the next code segment. The programmer must be sure that calls and jumps do not extend from one code segment to the next without the proper short, near, or far assembler directive.

Another way to build large programs, without resorting to multi-segment code, is to call far procedures from a main program code segment. If one segment contains the main program, and the remainder of the program consists of far procedures of less than 64K called from the main program, then enormous programs may be written.

Alternatively, the main program may temporarily suspend operation and use DOS to initiate the execution of a second program. The second program will terminate, and DOS will restart the original program. The original program is named the *parent* program, and the second program is named the *child* program. DOS can be thought of as the parent of every program executed on the computer.

EXE programs are stored on the disk with an assembler-generated program header that enables DOS to construct a PSP when the EXE file is loaded from disk to memory. The EXE header contains information to help DOS determine how much memory is needed by the EXE program, and how it should be located in memory. EXE programs do not have to exist as one single block of memory but can be broken into pieces and placed in any free memory areas DOS finds.

Loading Programs Under DOS

DOS is designed to load programs from disk, run them, and regain control after the program terminates. To do this, every program stored on disk must contain information that will enable DOS to compute where to put the program in memory.

DOS may, theoretically, put a program anywhere within a RAM memory of 640K. Actually, not all the 640K are available to DOS. Figure D.2 shows the memory map for system RAM, from 00000h to 9FFFFh. As can be seen in

Figure D.2 ■ System RAM map.

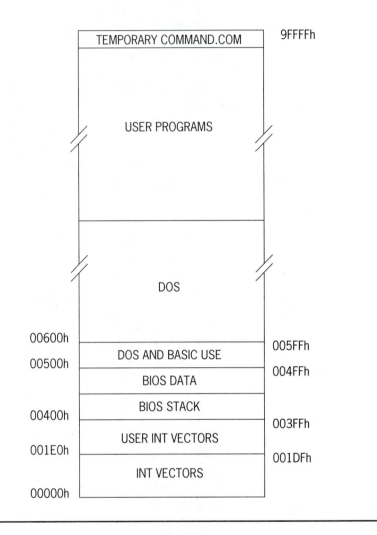

Figure D.2, the interrupt vector table, BIOS tables, COMMAND.COM, any CONFIG.SYS files, and DOS itself require up to 200K of RAM. The memory area above the resident part of COMMAND.COM, dubbed the *transient program area* or TPA, is the only memory space available to DOS for locating application programs.

DOS remembers the address of the first empty block of memory in the TPA, and loads application programs starting at that block address. Programs that require more memory than is available in the TPA will not be loaded, and DOS will display an error message to that effect. The programmer must then divide the large program into separate pieces and run them separately.

To return to DOS after execution, the application program executes an interrupt to DOS, and DOS regains control of the CPU via the interrupt mechanism. The resident part of COMMAND.COM determines if the transient part of COMMAND.COM still exists. If DOS did load an application program over transient COMMAND.COM, then COMMAND.COM is reloaded by DOS. Should COMMAND.COM not exist on the disk (you changed disks to run your application program), DOS will request that you insert a system disk in the current drive.

THE PROGRAM SEGMENT PREFIX: PSP

How does DOS keep track of a program that is to be executed? The answer is found in a 256-byte (100h) data area that contains all of the information needed by DOS to execute, control, and terminate a program. The data area is named the *program segment prefix* or PSP. DOS constructs the PSP when a program is commanded to be run by the user. The PSP is placed in available memory immediately ahead of the DRAM memory used to hold the program code.

Programs assembled as .COM types are loaded by DOS in the first available memory segment with the PSP occupying offset bytes 0000h to 00FFH. The .COM program begins at IP offset 100h, immediately after the PSP. All of the segment registers are set by DOS to the beginning segment of the PSP, the IP is set to 100h, and the SP is set to the top of the segment, FFFEh. All .COM type programs occupy one segment, thus DOS can allocate a segment of memory to hold them.

Programs assembled as .EXE types are loaded in as many segments as required by the program, and a PSP is built from information contained in a linker-generated *header* that is stored on the disk with the .EXE-type program. DOS constructs a PSP for the .EXE program, sets the DS and ES registers to the segment containing the PSP, sets CS:IP to the starting address of the first byte of code, and sets SS:SP as directed in the header.

DOS has not been known to "lose" an application program, but the excuse "DOS ate my disk" has replaced "The dog ate my homework" as a high-tech excuse for not having homework assignments completed on time. All users, all the time, should back up their valuable application programs to prevent losing them.

D.4 DOS Memory Management

DOS manages the free area of RAM named the transient program area (TPA) part of RAM. TPA RAM exists just above resident COMMAND.COM and extends to the top of RAM at 9FFFFh. Every time DOS loads an application program into RAM, it must decide where to place the application program, and how much RAM to let it use.

TPA memory is organized into contiguous groups called *memory blocks*; all of the blocks combined form a group named the *memory arena*. Some of the blocks may contain programs, and others are empty, or *free*. Each memory block is equipped with a header section that is called an *arena header*. Figure D.3 shows a diagram of the TPA and a group of memory blocks.

Arena headers tell DOS how much memory is available in its block, and if the block is *in use* or free. Free blocks of memory are available to DOS as a space available for loading application programs. Each header also contains information as to the address of the next memory block. The headers form

Figure D.3 ■ TPA memory organization.

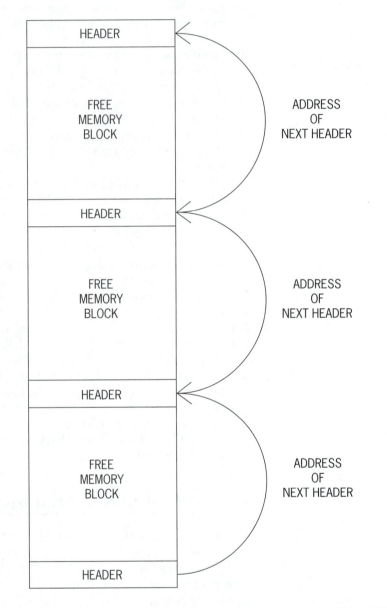

what is known as a *chain* or a *linked list*. Chaining implies that each part points to the next part. DOS need only keep track of the address of the first arena header, and then follow the chain to find all of the rest of the free memory blocks.

A free memory block may be as small as 1 paragraph (16 bytes) or may hold the entire TPA. One paragraph, we should note, is the smallest possible segment size. Should 1 free block come in contact with another free block, DOS will combine them into a single block in the chain. By combining free blocks, DOS creates a flexible memory structure that has free blocks as large as possible.

When DOS is called on to load an application program, it begins at the first block arena header and follows the chain of memory blocks until a free block is found that can hold the application program. Finding the first available block large enough to hold the application program is a strategy known as a *first fit.* First fit does not guarantee that memory is used efficiently. The first free block found that can hold the application program may be much larger than absolutely necessary, and thus waste a lot of RAM by using an entire block to hold a small program.

DOS operates the PC for the user. The remainder of this appendix investigates how assembly-language programmers make use of DOS service routines (interrupts) in an assembly language program.

D.5 DOS Interrupt Services

DOS interrupts are listed by interrupt type and, where applicable, by function. Interrupt function listings show values to be loaded into CPU registers before using a multi-function interrupt type, and any return register values. Many functions use the Carry flag (CF) set to 1 to indicate an error status—CF cleared to 0 indicates proper interrupt operation. If CF is 1 on return from an interrupt, AX usually contains an error code to assist the program in corrective action.

A brief description of each type and function is included. Interrupts that are reserved should generally not be used in application programs.

The contents of this appendix are intended to be used as a quick lookup table for DOS functions and not as a detailed reference. An inspection of this appendix will indicate, however, the scope of the services that an operating system may provide for application programs. For a complete description of each DOS function, please refer to one of the references listed in Appendix H, "Sources."

20h　Terminate Program (Also see Interrupt 21h, Function 4Ch)
Load	**Returned**
None	None

21h　Primary DOS Interrupt Type

V1　Terminate Child Program and Return to Parent Program (Obsolete)
Load	**Returned**
AH = 00h	None
CS = PSP segment address	

V1　Read Key and Write to Screen
Load	**Returned**
AH = 01h	AL = ASCII character

V1　Display Character on Screen
Load	**Returned**
AH = 02h	None
DL = ASCII character	

V1   Read Serial Character from Port 1

Load	**Returned**
AH = 03h	AL = ASCII character

V1   Write Serial Character to Port 1

Load	**Returned**
AH = 04h	None
DL = ASCII character	

V1   Write Character to Printer

Load	**Returned**
AH = 05h	None
DL = ASCII character	

V1   Read Keyboard or Write Screen

Load	**Returned**
AH = 06h	ZF = 1:
DL = Read or write	No character available
control code	ZF = 0:
	AL = ASCII character

V1   Read Keyboard Character and Ignore Ctrl-C and Ctl-Break

Load	**Returned**
AH = 07h	AL = ASCII character

V1   Read Keyboard Character and Handle Ctrl-C and Ctrl-Break

Load	**Returned**
AH = 08h	AL = ASCII character

V1   Display String Ending in $

Load	**Returned**
AH = 09h	None
DS:DX = Pointer to string	

V1   Read Keyboard Input Characters to Buffer

Load	**Returned**
AH = 0Ah	None
DS:DX = Pointer to buffer	

V1   Check Keyboard for Character Ready

Load	**Returned**
AH = 0Bh	AL = Keyboard status

V1   Clear Keyboard Buffer and Process Key

Load	**Returned**
AH = 0Ch	AL = Processing result
AL = Key processing action	

V1   Flush Disk Buffers

Load	**Returned**
AH = 0Dh	None

V1   Select Disk Drive

Load	**Returned**
AH = 0Eh	AL = Last drive number
DL = Drive number	
[0(A) to 25(Z)]	

V1   Open File
Load	**Returned**
AH = 0Fh	AL = Status
DS:DX = Pointer to FCB	

V1   Close File
Load	**Returned**
AH = 10h	AL = Status
DS:DX = Pointer to FCB	

V1   Search for First File Entry
Load	**Returned**
AH = 11h	AL = Status
DS:DX = Pointer to FCB	

V1   Search for Next File in Current Directory
Load	**Returned**
AH = 12h	AL = Status
DS:DX = Pointer to FCB	

V1   Delete File
Load	**Returned**
AH = 13h	AL = Status
DS:DX = Pointer to FCB	

V1   Read Sequential File
Load	**Returned**
AH = 14h	AL = Status
DS:DX = Pointer to FCB	

V1   Write Sequential File
Load	**Returned**
AH = 15h	AL = Status
DS:DX = Pointer to FCB	

V1   Create File
Load	**Returned**
AH = 16h	AL = Status
DS:DX = Pointer to FCB	

V1   Rename File
Load	**Returned**
AH = 17h	AL = Status
DS:DX = Pointer to FCB	

V1   Determine Current Drive Number
| **Load** | **Returned** |
| AH = 19h | AL = Drive number [0(A) to 25(Z)] |

V1   Write Disk Transfer Area (DTA) Address
Load	**Returned**
AH = 1Ah	None
DS:DX = Pointer to new DTA	

V1   Read Current Disk Information
           Load   **Returned**
           AH = 1Bh   AL = Sectors per cluster
                                             CX = Bytes per sector
                                             DX = Clusters per disk
                                             DS:BX = Pointer to media descriptor byte

V2   Read Disk Information
           Load   **Returned**
           AH = 1Ch   AL = Sectors per cluster
           DL = Drive number   CX = Bytes per sector
                                             DX = Clusters per disk
                                             DS:BX = Pointer to media descriptor byte

V2   Get Disk Drive Parameter Block (DPB) Address
           Load   **Returned**
           AH = 1Fh   AL = Status
           DS:BX = Pointer to DPB

V1   Read Random Records to File
           Load   **Returned**
           AH = 21h   AL = Status
           DS:DX = Pointer to FCB

V1   Write Random Records to File
           Load   **Returned**
           AH = 22h   AL = Status
           DS:DX = Pointer to FCB

V1   Get Size of File into File Control Block (FCB)
           Load   **Returned**
           AH = 23h   AL = Status
           DS:DX = Pointer to FCB

V1   Update File Control Block (FCB) for Random File Oprations
           Load   **Returned**
           AH = 24h   None
           DS:DX = Pointer to FCB

V1   Initialize Interrupt Vector
           Load   **Returned**
           AH = 25h   None
           AL = Interrupt type number
           DS:DX = Pointer to interrupt handler address

V1   Create New Program Segment Prefix (PSP)
           Load   **Returned**
           AH = 26h   None
           DX = PSP Segment address

V1   Read Random Records to Disk Transfer Area (DTA)
           Load   **Returned**
           AH = 27h   AL = Status
           CX = Number of records   CX = Number of records read
           DS:DX = Pointer to FCB

V1   Write Random Records to File from Data Transfer Area (DTA)
                      Load   **Returned**
              AH = 28h   AL = Status
              CX = Number of records   CX = Number of records written
              DS:DX = Pointer to open File Control Block

V1   Parse File Name to a File Control Block (FCB)
                      Load   **Returned**
              AH = 29h   AL = Status
              AL = Flags   DS:SI = Pointer to parsed file name
              DS:SI = Pointer to file name   ES:DI = Pointer to updated FCB
              ES:DI = Pointer to FCB

V1   Read Date
                      Load   **Returned**
              AH = 2Ah   CX = Year
                                                DH = Month
                                                DL = Day
                                                AL = Day of week

V1   Set Date
                      Load   **Returned**
              AH = 2Bh   AL = Status
              CX = Year
              DH = Month
              DL = Day

V1   Read Time
                      Load   **Returned**
              AH = 2Ch   CH = Hour
                                                CL = Minutes
                                                DH = Seconds
                                                DL = Seconds/100

V1   Set Time
                      Load   **Returned**
              AH = 2Dh   AL = Status
              CH = Hour
              CL = Minutes
              DH = Seconds
              DL = Seconds/100

V2   Set Disk Transfer Verify Flag
                      Load   **Returned**
              AH = 2Eh   None
              AL = Flag control
              DH = 00h (Before V3)

V2   Read Disk Transfer Area Address
                      Load   **Returned**
              AH = 2Fh   ES:BX = Pointer to DTA buffer

V2   Determine DOS Version
                      Load   **Returned**
              AH = 30h   AL = Major version number
                                                AH = Minor version number

V2   Terminate Program, but Stay Resident (TSR)

Load	**Returned**
AH = 31h	None
AL = Return code	
DX = Memory size for TSR program	

V2   Read Address of Drive Parameter Block (DPB)

Load	**Returned**
AH = 32h	AL = Status
DL = Drive number	DS:BX = Pointer to DPB

V2   Read Ctrl-Break Flag Status

Load	**Returned**
AX = 3300h	DL = Flag status

V2   Set Ctrl-Break Flag

Load	**Returned**
AX = 3301h	None
DL = Flag control	

V2   Read Drive Boot Code

Load	**Returned**
AX = 3305h	DL = Boot drive code

V2   Read Interrupt Status Flag Address

Load	**Returned**
AH = 34h	ES:BX = Pointer to flag address

V2   Get Interrupt Type Vector

Load	**Returned**
AH = 35h	ES:BX = Pointer to interrupt handler
AL = Interrupt type number	

V2   Determine Free Disk Space

Load	**Returned**
AH = 36h	AX = Sectors per cluster
DL = Drive number	BX = Free clusters
	CX = Bytes per sector
	DX = Total clusters

V2   Read New Switch Character (Command Character)

Load	**Returned**
AX = 3700h	AL = Status
	DL = Old ASCII character

V2   Write New Switch Character (Command Character)

Load	**Returned**
AX = 3701h	AL = Status
DL = New ASCII character	

V2   Read Device Flag

Load	**Returned**
AX = 3702h	AL = Status
	DL = Flag control

V2   Set Device Flag

Load	**Returned**
AX = 3703h	AL = Status
DL = Flag control	DL = Flag control

V2   Read or Write Current Country Information

Load	Returned
AH = 38h	CF = 0:
AL = 00h:	BX = Country code read (V3 only)
Read information	DS:DX = Pointer to data
V3 and up:	CF = 1:
(00h < AL < FFh), AL = Code	AX = Errors
AL = FFh:	
BX = Code	
DX = Write command (V3 and up)	
DS:DX = Pointer to information buffer (V2)	

V2   Create a Subdirectory

Load	Returned
AH = 39h	CF = Status
DS:DX = Pointer to	AX = Errors
ASCIIZ path name	

V2   Remove an Empty Subdirectory

Load	Returned
AH = 3Ah	CF = Status
DS:DX = Pointer to	AX = Errors
ASCIIZ path name	

V2   Select Current Directory

Load	Returned
AH = 3Bh	CF = Status
DS:DX = Pointer to ASCIIZ	AX = Errors
path name	

V2   Create a New File or Erase an Old File

Load	Returned
AH = 3Ch	CF = 0:
CX = File attribute	AX = File handle
DS:DX = Pointer to ASCIIZ	CF = 1:
file path name	AX = Errors

V2   Open a File

Load	Returned
AH = 3Dh	CF = 0:
AL = Access mode	AX = File handle
DS:DX = Pointer to ASCIIZ	CF = 1:
file path name	AX = Errors

V2   Close a File

Load	Returned
AH = 3Eh	CF = Status
BX = File handle	AX = Errors

V2   Read Bytes From a File

Load	Returned
AH = 3Fh	CF = 1:
BX = File handle	AX = Number of bytes read
CX = Number of bytes	CF = 0:
DS:DX = Pointer to buffer	AX = Errors

V2   Write Bytes To a File

	Load	**Returned**
	AH = 40h	CF = 1:
	BX = File handle	AX = Number of bytes written
	CX = Number of bytes	CF = 0:
	DS:DX = Pointer to data	AX = Errors

V2   Delete File

	Load	**Returned**
	AH = 41h	CF = Status
	DS:DX = Pointer to ASCIIZ	AX = Errors
	file path	

V2   Change File Pointer

	Load	**Returned**
	AH = 42h	CF = 0:
	AL = Offset method code	DX:AX = New pointer address
	BX = File handle	CF = 1:
	CX:DX = Offset	AX = Errors

V2   Read File Attributes

	Load	**Returned**
	AX = 4300h	CF = 0:
	DS:DX = Pointer to ASCIIZ	CX = File attributes
	file path	CF = 1:
		AX = Errors

V2   Write File Attributes

	Load	**Returned**
	AX = 4301h	CF = Status
	CX = File attribute	AX = Errors
	DS:DX = Pointer to ASCIIZ	
	file path	

V2   Read Device or File Information Word

	Load	**Returned**
	AX = 4400h	CF = 1:
	BX = File handle	DX = Device information word
	AX = Errors	CF = 0:
		AX = Errors

V2   Write Character Device Information Word

	Load	**Returned**
	AX = 4401h	CF = Status
	BX = File handle	AX = Errors
	DX = Device information word	

V2   Read Device Control Information

	Load	**Returned**
	AX = 4402h	CF = 0:
	BX = File handle number	AX = Number of bytes read
	CX = Number of bytes to read	CF = 1:
	DS:DX = Pointer to buffer	AX = Errors

V2   Write Device Control Information

Load	**Returned**
AX = 4403h	CF = 0:
BX = File handle number	AX = Number of bytes written
CX = Number of bytes to read	CF = 1:
DS:DX = Pointer to data	AX = Errors

V2   Get Block Device (Disk Drive) Control Information

Load	**Returned**
AX = 4404h	CF = 0:
BL = Drive number	AX = Number of bytes read
CX = Number of bytes to read	CF = 1:
DS:DX = Pointer to buffer	AX = Errors

V2   Write Control Information to Block Device (Disk Drive)

Load	**Returned**
AX = 4405h	CF = 0:
BL = Drive number	AX = Number of bytes written
CX = Number of bytes to write	CF = 1:
DS:DX = Pointer to data	AX = Errors

V2   Get Input Status

Load	**Returned**
AX = 4406h	CF = 0:
BX = File handle	AX = Status
	CF = 1:
	AX = Errors

V2   Get Output Status

Load	**Returned**
AX = 4407h	CF = 0:
BX = File handle	AX = Status
	CF = 1:
	AX = Errors

V3   Get Block Device (Disk Drive) Removable Status

Load	**Returned**
AX = 4408h	CF = 0:
BL = Drive number	AX = Status
	CF = 1:
	AX = Errors

V3.1   Get Local or Remote Block Device (Disk Drive) Status

Load	**Returned**
AX = 4409h	CF = 0:
BL = Drive number	DX = Status
	CF = 1:
	AX = Errors

V3.1   Get Local or Remote File Handle Status

Load	**Returned**
AX = 440Ah	CF = 0:
BX = File handle number	DX = Status
	CF = 1:
	AX = Errors

V3   Write Sharing Retry Count

Load	**Returned**
AX = 440Bh	CF = Status
CX = Pause time	AX = Errors
DX = Number of tries	

V3.2   File Handle I/O Control

Load	**Returned**
AX = 440Ch	CF = Status
BX = File handle	AX = Errors
CH = Device code	
CL = Action code	
DS:DX = Pointer to parameters	

V3.2   Block Device (Disk Drive) I/O Control

Load	**Returned**
AX = 440Dh	CF = Status
BL = Drive number	AX = Errors
CH = Disk drive code	
CL = Action code	
DS:DX = Pointer to parameter block	

V3.2   Read Number of Logical Drives per Device

Load	**Returned**
AX = 440Eh	CF = 0:
BL = Logical drive number	AL = Last logical drive assigned
	CF = 1:
	AX = Errors

V3.2   Assign Logical Name to Drive

Load	**Returned**
AX = 440Fh	CF = 0:
	AL = Logical drive number
	CF = 1:
	AX = Errors

V2   Assign File a New Handle

Load	**Returned**
AH = 45h	CF = 0:
BX = Old handle	AX = New handle
	CF = 1:
	AX = Errors

V2   Assign File a Second Handle

Load	**Returned**
AH = 46h	CF = Status
BX = First handle for file	AX = Errors
CX = Second handle for file	

V2   Get Current Directory Name

Load	**Returned**
AH = 47h	CF = 0:
DL = Drive number	Buffer now contains ASCIIZ
DS:DI = Pointer to buffer	string name of directory
	CF = 1:
	AX = Errors

V2 Allocate Block of Memory
Load	**Returned**
AH = 48h	CF = 0:
BX = Memory paragraphs	AX = Initial memory segment
	CF = 1:
	AX = Errors

V2 Free Block of Memory
Load	**Returned**
AH = 49h	CF = Status
ES = Memory block segment	AX = Errors

V2 Change Size of Memory Block
Load	**Returned**
AH = 4Ah	CF = Status
BX = New size	AX = Errors
ES = Block segment	BX = Largest block available

V2 Execute Program
Load	**Returned**
AH = 4Bh	CF = Status
AL = 00, execute program	AX = Errors
AL = 03h, load overlay	
ES:BX = Pointer to parameter block	
DS:DX = Pointer to program specification	

V2 Exit Child Program and Return to Parent Program
Load	**Returned**
AH = 4Ch	None
AL = Return code	

V2 Get System and Child Return Exit Codes
Load	**Returned**
AH = 4Dh	AX = Exit codes

V2 Find First Matching File Name
Load	**Returned**
AH = 4Eh	CF = Status
CX = Attribute	AX = Errors
DS:DX = Pointer to ASCIIZ file path	

V2 Find Next File Match (Also see Function 4Eh)
Load	**Returned**
AH = 4Fh	CF = Status
	AX = Errors

V2 Write PSP Address
Load	**Returned**
AH = 50h	None
BX = PSP Address	

V2 Read PSP Address
Load	**Returned**
AH = 51h	BX = PSP Address

V2 Get Address of Disk Drive Parameters Block List Table
Load	**Returned**
AH = 52h	ES:BX = Pointer to list table

V2   Change BIOS Parameter Block to DOS Parameter Block

Load	**Returned**
AH = 53h	None
DS:SI = Pointer to BPB	
ES:BP = Pointer to DPB	

V2   Get Status of Read-after-Write (Verify) Flag

Load	**Returned**
AH = 54h	AL = Flag status

V2   Create PSP at Segment Address

Load	**Returned**
AH = 55h	None
DX = PSP segment address	

V2   Rename File

Load	**Returned**
AH = 56h	CF = Status
DS:DX = Pointer to old	AX = Errors
ASCIIZ file name	
ES:DI = Pointer to new	
ASCIIZ file name	

V2   Get File Date and Time

Load	**Returned**
AX = 5700h	CF = 0:
BX = Handle	CX,DX = Time, Date
	CF = 1:
	AX = Errors

V2   Set File Date and Time

Load	**Returned**
AX = 5701h	CF = Status
BX = File handle	AX = Errors
CX,DX = Time, Date	

V2   Read Memory Allocation Strategy

Load	**Returned**
AX = 5800h	CF = 0:
	AX = Strategy code
	CF = 1:
	AX = Errors

V2   Write Memory Allocation Strategy

Load	**Returned**
AX = 5801h	CF = Status
BX = Strategy code	AX = Errors

V3   Read Extended Error Code (Used after error status is returned)

Load	**Returned**
AH = 59h	AX = Extended error code (00h–5Ah)
BX = 00h	BH = Error class (nature of failure)
	BL = Recommended corrective action
	CH = Error locus (type of device)

V3   Create Unique Temporary File

Load	**Returned**
AH = 5Ah	CF = 0:
CX = Attribute	AX = File handle
DS:DX = Pointer to ASCIIZ	DS:DX = Pointer to ASCIIZ file path
path ending in \	CF = 1:
	AX = Errors

V3   Create a New File

Load	**Returned**
AH = 5Bh	CF = 0:
CX = Attribute	AX = File handle
DS:DX = Pointer to ASCIIZ	CF = 1:
file path string	AX = Errors

V3   Lock File Area for Multitasking

Load	**Returned**
AX = 5C00h	CF = Status
BX = File handle	AX = Errors
CX,DX = File region offset	
SI,DI = File region length	

V3   Remove Multitasking File Lock

Load	**Returned**
AX = 5C01h	CF = Status
BX = File handle	AX = Errors
CX,DX = File region offset	
SI,DI = File region length	

V3   Copy Data to DOS Save Area

Load	**Returned**
AX = 5D00h	DS:DI = Pointer to data

V3   Read Critical Error Flag Address

Load	**Returned**
AX = 5D06h	DS:DI = Pointer to flag address

V3   Change Extended Error Code Pointer Address

Load	**Returned**
AX = 5D0Ah	None
DS:SI = Pointer to error code table	

V3.1   Read Networked Machine Name

Load	**Returned**
AX = 5E00h	CF = 0:
DS:DX = Pointer to name buffer	CX = Status and name number
DS:SI = Pointer to setup	DS:DX = Pointer to name identifier
string	CF = 1:
	AX = Errors

V3.1   Write Networked Machine Name

Load	**Returned**
AX = 5E01h	CF = Status
DS:DX = Pointer to ASCIIZ	AX = Errors
string name	

V3.1   Setup Network Printer

Load	**Returned**
AH = 5E02h	CF = Status
BX = Redirection index	AX = Errors
CX = Length of printer	
setup string	
DS:SI = Pointer to setup string	

V3.1   Read Network Printer Setup String

Load	**Returned**
AX = 5E03h	CF = 0:
BX = Redirection index	CX = Printer setup string length
ES:DI = Pointer to setup	ES:DI = Pointer to setup string
string buffer	CF = 1:
	AX = Errors

V3.1   Read Network Device Redirection Entry

Load	**Returned**
AX = 5F02h	CF = 0:
BX = Entry index	BX = Device flag and type
DS:SI = Pointer to local	CX = Device parameter
device name buffer	DS:SI = Pointer to ASCIIZ local name
ES:DI = Pointer to network	ES:DI = Pointer to ASCIIZ network name
device name buffer	CF = 1:
	AX = Errors

V3.1   Write Network Device Redirection Entry

Load	**Returned**
AX = 5F03h	CF = Status
BL = Device number	AX = Errors
CX = Parameter	
DS:SI = Pointer to ASCIIZ local	
device name	
ES:DI = Pointer to ASCIIZ network	
device name and password	

V3.1   Cancel Network Device Redirection Entry

Load	**Returned**
AX = 5F04h	CX = Status
DS:SI = Pointer to ASCIIZ	AX = Errors
device name	

V3   Expand Path Name

Load	**Returned**
AH = 60h	CF = 0:
DS:SI = Pointer to ASCIIZ	ES:DI = Working buffer address
path name	CF = 1:
ES:DI = Working buffer address	AX = Errors

V3   Read PSP Segment Address

Load	**Returned**
AH = 62h	BX = PSP segment address

V2.25   Read Lead Byte Table Address

Load	**Returned**
AX = 6300h	DS:DI = Pointer to table address

V2.25 Only Set or Clear Console Flag

Load	**Returned**
AX = 6301h	None
DL = Set or clear code	

V2.25 Only Read Console Flag Status

Load	**Returned**
AX = 6302h	DL = Status

V3 Write DOS Current Country Byte

Load	**Returned**
AH = 64h	None
AL = Country code	

V3.3 Read Country Extended Data

Load	**Returned**
AH = 65h	CF = 0:
AL = Data ID	ES:DI = Pointer to storage buffer
BX = Code page	CF = 1:
CX = Amount of data	AX = Errors
DX = Country ID	
ES:DI = Pointer to data storage buffer	

V3.3 Read Current Country Code Pages

Load	**Returned**
AX = 6601h	CF = 0:
AL = 01h	BX = Active code page
	CX = System code page
	CF = 1:
	AX = Errors

V3.3 Write Country Global Pages

Load	**Returned**
AX = 6602h	CF = Status
BX = Active code page	AX = Errors
CX = System code page	

V3.3 Change Number of Allowed Handles

Load	**Returned**
AH = 67h	CF = Status
BX = Number of handles	AX = Errors

V3 Flush File Buffer

Load	**Returned**
AH = 68h	CF = Status
BX = File handle	AX = Errors

V4 Allocate a Block of Memory (See also Function 48h)

Load	**Returned**
AX = 6A00h	CF = Status
BX = File handle	AX = Errors

V4   Extended File Open/Create (See also Functions 3Ch, 3Dh, 5Bh)

Load	**Returned**
AX = 6C00h	CF = 0:
AL = 00h	AX = Handle
BX = Open File Mode	CX = Action
CX = File attributes	CF = 1:
DX = Action flags	AX = Errors
DS:SI = Pointer to ASCIIZ file path	

22h   V1 Terminate Current Program Return Address

Load	**Returned**
None	None

23h   V1 Ctrl-Break or Ctrl-C Key Combination Handler

Load	**Returned**
None	None

24h   V1 Critical-Error Device Handler

Load	**Returned**
AH = Error type	AL = Action code
DI = Error code	
BP:SI = Pointer to device driver	
STACK = CPU registers	

25h   V1 Read Logical Disk Sector

Load	**Returned**
AL = Drive number	CF = Status
CX = Number of sectors	AX = Errors
DX = Starting sector	SP = SP–2
DS:BX = Pointer to Data Transfer Area (DTA)	

26h   V1 Write Logical Disk Sector

Load	**Returned**
AL = Drive number	CF = Status
CX = Number of sectors	AX = Errors
DX = Starting sector	SP = SP–2
DS:BX = Pointer to Data Transfer Area (DTA)	

27h   V1 Terminate and Stay Resident (TSR)

Load	**Returned**
DX = PSP offset to end of TSR	None
CS = PSP Segment	

28h   V1 DOS Safe to Use Operations (Also see INT 21h, Function 34h)

Load	**Returned**
None	None

29h   V2 Write Character to Screen

Load	**Returned**
AL = ASCII character	None

2Ah   V3 Reserved for Microsoft Networks

Load	**Returned**
None	None

2Eh   V3 Shell Program Loader (Superseded by INT 21h, Function 4Bh)

Load	**Returned**
DS:SI = Pointer to command string	None

2Fh   V3, V4 Internal DOS System Services

Many interrupt 2Fh functions are DOS-specific (thus may only be used from within the DOS program kernel) and are not intended for application program use. Those functions that may be appropriate for user programs are listed next.

V3   Check for Print Spooler Program Installed

Load	**Returned**
AX = 0100h	CF = 0 if functional
	AL = Status:
	00h = May install PRINT.COM
	01h = May not install PRINT.COM
	FFh = Installed

V3   Print File via Print Spooler

Load	**Returned**
AX = 0101h	CF = 0 if functional, 1 if an error
DS:DX = Pointer to 5-byte	AL = Error code
field: priority byte	
and ASCIIZ file address	

V3   Remove File from Print Spooler Queue

Load	**Returned**
AX = 0102h	CF = 0 if functional, 1 if an error
DS:DX = Pointer to ASCIIZ	AL = Error code
file	

V3   Hold Print Spooler

Load	**Returned**
AX = 0104h	CF = 0 if successful
	DX = Error count
	DS:SI = Pointer to print queue
	CF = 1 if error
	AX = Error code

V3   Release Print Spooler

Load	**Returned**
AX = 0105h	CF = 0 if successful, 1 if error
	AX = Error code

V3   Determine Status of ASSIGN program

Load	**Returned**
AX = 0600h	CF = 0 if successful
	AL = Status:
	00h = May be installed
	01h = May not be installed
	FFh = Installed

V3   Determine Status of DRIVER program
Load   **Returned**
AX = 0800h   CF = 0 if successful
                                AL = Status:
                                    00h = May be installed
                                    01h = May not be installed
                                    FFh = Installed

V3   Determine Status of SHARE program
Load   **Returned**
AX = 1000h   CF = 0 if successful
                                AL = Status:
                                    00h = May be installed
                                    01h = May not be installed
                                    FFh = Installed

V3   Determine Status of NETWORK program
Load   **Returned**
AX = 1100h   CF = 0 if successful
                                AL = Status:
                                    00h = May be installed
                                    01h = May not be installed
                                    FFh = Installed

V3   Read DOS DS Value
Load   **Returned**
AX = 1203h   DS = Segment address

V3   Display Character from Stack
Load   **Returned**
AX = 1205h   None
SS:SP = Low byte on stack

V3   Standardize File Name
Load   **Returned**
AX = 1211h   ES:DI = Pointer to normalized file name
DS:SI = Pointer to original name   (all capitals and all / to \)
ES:DI = Pointer to standardized name

V3   Read ASCIIZ Name Length
Load   **Returned**
AX = 1212h   CX = Length, in bytes
ES:DI = Pointer to name

V3   Get Drive Number from Path Name
Load   **Returned**
AX = 121Ah   AL = Drive code (A = 1, B = 2, etc.)
DS:SI = Pointer to ASCIIZ   AL = 0, no drive letter in path name
    path name

V3   Compare ASCIIZ Strings
Load   **Returned**
AX = 121Eh   ZF = 1 if strings are equal
DS:SI = Pointer to first string
ES:DI = Pointer to second string

V4   Return ASCIIZ String Length

Load	**Returned**
AX = 1225h	CX = String length
DS:SI = Pointer to string	

33h   Mouse

A mouse-driver program (MOUSE.COM or MOUSE.SYS) may access interrupt 33h for mouse services. Mouse position is measured in rows from top (0) to bottom (199), and columns from left (0) to right (639).

V2   Initialize Mouse Hardware and Driver (See also Function 21h)

Load	**Returned**
AX = 0000h	AX = 0000h, mouse installed
	AX = FFFFh, mouse not installed
	BX = Number of mouse buttons

V2   Display Mouse Cursor

Load	**Returned**
AX = 0001h	None

V2   Blank Mouse Cursor

Load	**Returned**
AX = 0002h	None

V2   Read Mouse Position

Load	**Returned**
AX = 0003h	BL = Buttons:
	BL bit 2 1 0
	x x 1 Left button down
	x 1 x Right button down
	1 x x Center button down
	CX = Horizontal position (0–639)
	DX = Vertical position (0–199)

V2   Position Mouse Cursor

Load	**Returned**
AX = 0004h	None
CX = Horizontal position (0–639)	
DX = Vertical position (0–199)	

V2   Button Press Activity Status

Load	**Returned**
AX = 0005h	AX = Status (See BL of Function 0003h)
BX = Desired button:	BX = Button presses since last query
0000h = Left	At last press position:
0001h = Right	CX = Horizontal position (0–639)
0002h = Center	DX = Vertical position (0–199)

V2   Button Press Activity Status

Load	**Returned**
AX = 0006h	AX = Status (See BL of Function 0003h)
BX = Desired button:	BX = Button releases since last query
0000h = Left	At last release position:
0001h = Right	CX = Horizontal position (0–639)
0002h = Center	DX = Vertical position (0–199)

V2   Limit Horizontal Position

Load	**Returned**
AX = 0007h	None
CX = Left limit (0–639)	
DX = Right limit (0–639)	

V2   Limit Vertical Position

Load	**Returned**
AX = 0008h	None
CX = Upper limit (0–199)	
DX = Lower limit (0–199)	

V2   Graphics Mode Cursor Shape

Load	**Returned**
AX = 0009h	None
BX = Spot horizontal position	
CX = Spot vertical position	
ES:DS = Pointer to graphic masks	

V2   Text Mode Cursor Shape

Load	**Returned**
AX = 000Ah	None
BX = Shape	
CX = Screen mask	
DX = Cursor mask	

V2   Read Relative Motion

Believe it or not, relative cursor motion is measured in *mickeys* (undoubtedly after the famous cartoon character). One mickey is approximately 0.02 inches.

Load	**Returned**
AX = 000Bh	CX = Horizontal motion since last query
	DX = Vertical motion since last query

V2   Vector to User-Defined Handler

Load	**Returned**
AX = 000Ch	None
CX = Mask	
ES:DX = Vector to user handler	

V2   Start Light Pen Emulation

Load	**Returned**
AX = 000Dh	None

V2   Stop Light Pen Emulation

Load	**Returned**
AX = 000Eh	None

V2   Set Pixel to Mickey Ratio (See also Function 000Bh)

Load	**Returned**
AX = 000Fh	None
CX = Number of mickeys per 8-pixel horizontal motion	
DX = Number of mickeys per 8-pixel vertical motion	

V2   Hide Cursor Screen Area

	Load	**Returned**
	AX = 0010h	None
	CX = Upper screen column	
	DX = Upper screen row	
	SI = Lower screen column	
	DI = Lower screen row	

V2   Double Movement Rate Threshold

	Load	**Returned**
	AX = 0013h	None
	DX = Threshold, in mickeys	

V2   Swap User Handlers

	Load	**Returned**
	AX = 0014h	CX = Old user mask
	CX = New user mask	ES:DX = Old interrupt vector
	ES:DX = Old user interrupt vector	

V2   Read Mouse State Buffer Size

	Load	**Returned**
	AX = 0015h	BX = Buffer size, in bytes

V2   Save Mouse State

	Load	**Returned**
	AX = 0016h	None
	ES:DX = Pointer to buffer	

V2   Read Mouse State

	Load	**Returned**
	AX = 0017h	None
	ES:DX = Pointer to buffer	

V2   Set Special Handler

	Load	**Returned**
	AX = 0018h	AX = Status
	CX = Interrupt mask	
	ES:DX = Handler pointer	

V2   Get Special Handler

	Load	**Returned**
	AX = 0019h	BX:DX = Handler vector
	CX = Mask	CX = Mask

V2   Set Sensitivity (Also see Function 000Bh)

	Load	**Returned**
	AX = 001Ah	None
	BX = Horizontal mickeys	
	CX = Vertical mickeys	
	DX = Double speed threshold	

V2   Read Sensitivity (Also see Function 000Bh)

	Load	**Returned**
	AX = 001Bh	BX = Horizontal mickeys
		CX = Vertical mickeys
		DX = Double speed threshold

V2   Set Driver Interrupt Rate

Load	**Returned**
AX = 001Ch	None
BX = Rate	

V2   Set Cursor Page Number (Also see BIOS Interrupt 10h)

Load	**Returned**
AX = 001Dh	None
BX = Screen page number (0–7)	

V2   Get Cursor page Number (Also see BIOS Interrupt 10h)

Load	**Returned**
AX = 001Eh	BX = Page number (0–7)

V2   Disable Driver

Load	**Returned**
AX = 001Fh	AX = Status
	ES:BX = Vector to INT 33h

V2   Enable Driver

Load	**Returned**
AX = 0020h	None

V2   Reset Mouse Software (Also see Function 00h)

Load	**Returned**
AX = 0021h	AX = Driver status
	BX = Number of buttons (2 or 3)

V2   Set International Language

Load	**Returned**
AX = 0022h	None
BX = Country identification code	

V2   Read Current Language

Load	**Returned**
AX = 0023h	BX = Country identification code

V2   Read Mouse Driver Information and IRQ

Load	**Returned**
AX = 0024h	BX = Version number
	CH = Mouse type
	CL = IRQ number (2–7)

67h   Expanded Memory System

Additional memory may be added to a system using a hardware-software technique called *bank switching*. Bank switching involves enabling physical memory blocks under software control. Lotus, Intel, and Microsoft corporations established a bank switching standard, given the acronym *LIM*, to add additional memory to DOS-based systems. Interrupt 67h, and subsequent functions, deal with LIM expanded memory. Please refer to one of the DOS reference books, listed in Appendix H, for a discussion of expanded memory use.

D.6 Ten-Minute Tutorial on Using DOS Commands

Those who have not used DOS may wish to study the next section, which contains a brief description of how to use the most common DOS commands. Remember, the primary purpose of DOS is to find a file and load it into the computer for execution.

The DOS Prompt = >

When DOS gains control of the computer on boot-up you should see the common DOS *greater than prompt* >. You may see *more* than the prompt, but the prompt sign means that DOS is waiting for you to type a DOS command. Commands are actions you wish to have DOS perform for you. Commands must be typed exactly right or they will not work.

Prompts may appear as A:>, A:\WHAT>, A: Wed 07-22-1992>, but there usually is a > as the last character. If not, ask the system manager what *prompt* is in use.

Files, Files, and More Files

Everything on a disk is kept in a file identified by a file name. File names are assigned by the programmer who created the file in the first place. To find a file you must first find the drive that contains the file disk.

FILE NAMES

File names are written in the form of *myfile.any,* where myfile = any file *name* you wish to use up to eight alphabetic characters, and any = any *extension* you wish to use up to three alphabetic characters.

Disk drives also have names, such as a:, b:, c: and so on. Drives a: and b: are normally *floppy* disk drives, whereas those named c: and beyond are *hard* disk drives. Note that the *colon* (:) is part of a drive name. For instance, C is a file, but C: is a drive.

The *default* disk drive is the one currently in use. Usually, the default drive is printed just before the prompt, as, for instance, A:\>.

Changing to Another Disk Drive

If you wish to use a disk drive other than the one currently in use, just type the name of the drive next to the prompt. For instance, you are in drive a: and you want to use drive b:

■ Prompt = A:\> You type b: next to the prompt

■ Prompt = B:\> Means you have switched to drive b:

Directories

Groups of files can be organized on a disk by putting them all in a directory. Directories are given names by the programmer, just as file names are. Directories may have additional directories (called subdirectories) under them.

This tutorial assumes you are going to use only one directory level, with no subdirectories.

The symbol for a directory is the backslash \ symbol at the end of each directory name, so directory names do no get confused with file names. For example, assume there is a file, named myfile.any, placed in a directory named *misc*. Drive b: holds the disk with the *misc* directory on it. The way to get to myfile.any, or the *path* to the file, is:

b:\misc\myfile.any

The first \, after b:, is the root directory of the whole disk. Every drive has a root directory named, simply, \.

MAKING A DIRECTORY = MD

Type the command md anyname after the command prompt, where *anyname* is whatever name you wish to call the directory. The directory will be created on the default disk drive, *unless it already exists.* For instance, typing md what after the prompt will create a directory named *what* on drive a:.

CHANGE TO A DIRECTORY = CD

Once in the default drive that contains the disk with the directory on it, type cd anyname to get to a directory of any name. For instance, for the directory named what, type cd what after the prompt to get to directory what.

WHAT IS IN A DIRECTORY = DIR

The contents (file names) of a directory can be found by typing the dir command while in the directory. For instance, type dir to see every file (and directory) in the a: drive listed on the screen from *top to bottom* with date and time of last file change.

Type dir/w to see every file in the a: drive listed *across* the screen with no date or time of last change.

GET BACK TO ROOT DIRECTORY = CD \

Once in a directory, for example one named what, you may return to the root directory by typing the command cd \ to get back to the root directory named \. Or type cd . . .

REMOVE A DIRECTORY = RD

DOS will not remove a directory until it is completely empty, so use this command sequence when in the directory you wish to delete.

del *.* This will delete all files in the directory

You Sure (Y/N) DOS asks to make sure there is no mistake; type in a Y for yes, you're sure

cd \ Get back to the root directory named \

rd what   Removes directory named what

Delete Any File in a Directory = del

Get to the directory, for example directory what, that contains the file and type del myfile.any to get rid of the file named myfile.any.

Typing del *.* gets rid of all files in the directory, and DOS will ask you to type a Y(es) to make sure. Asterisks * can always be used for "wildcard" names, that is, an * can be any *name*. Question marks ? can be any *letter* of any name. The file myfile.any could also be found using m??ile.*, or *.any, or m?????.an?.

COPY A FILE = COPY

The source file (the file copied from) is copied to the destination file (the file copied to) using the copy command. Assume you are in drive a:, directory *here*, and you wish to copy the file with path *here\myfile.any* to drive b:, path *there*. Type as follows.

    copy a:\here\myfile.any b:\there

There are many copy command shortcuts, but typing the path of each file always works. You may also copy a file to the system printer (called prn:) using copy drive:\directory\myfile.any prn:.

Format a Disk = format

Formatting means wiping a disk clean. Type the format command and the drive you wish to format, as, for instance, format b: to format the disk in drive b:. Be sure you have a disk that matches the drive type you are using, such as a high-density disk for a high-density drive.

- WARNING: DO NOT FORMAT THE HARD DRIVE WITHOUT PERMISSION OF THE OWNER.
- When Things Don't Work:

    Did you type in a space, or leave out a space?

    Did you try to copy to a directory that does not exist?

    Are you in the correct drive and directory?
- When All Else Fails:

    Press Ctrl and Break keys or Ctrl and C keys at the same time.
- When Truly Desperate:

    Press Ctrl, Alt and Del keys at the same time. The computer will reboot. *Anything not saved on a disk is gone.*
- When Utterly Desperate:

    Turn power off to the computer, count to ten, then turn power on again. *Anything not saved on a disk is gone.*

Appendix

E

ASCII AND

KEYBOARD

CODES

E.0 The PC Keyboard

Keyboards are the most popular input devices for personal computers. Every PC keyboard contains a small microcontroller (a true single-chip computer) that operates the keyboard. The microcontroller generates a serial data *scan code* that is sent to the main computer whenever a key or key combination is pressed or released by the user.

BIOS hardware interrupt 09h receives each key scan code and pairs the scan code with a no-parity ASCII key code, *if* an ASCII code for the key is defined. The keyboard scan code and the ASCII code for each key are then stored in a 16-key keyboard buffer, which generally begins at low RAM location 0041Eh. Programs may retrieve key codes from the keyboard buffer, using BIOS or DOS interrupt service routines. Programs should not use BIOS interrupt 09h; it is a hardware-only interrupt. (See Appendix C for BIOS hardware and software interrupts.)

Note that each key generates a *single* scan code, *not* an ASCII code. For instance, key *L* on the keyboard generates one scan code. If the shift key is pressed while holding down the L key, then BIOS will convert the key to uppercase ASCII *L*. If the shift key is not held down when pressing the L key, then BIOS will generate lowercase ASCII for *l*. If *no* ASCII code exists for the key, such as for the F1 key for instance, then an ASCII *null code* of 00h is paired with the scan code. ASCII NUL code 00h has no printable symbol.

Three types of keyboards are in general use for PC applications: the 83-key PC keyboard, the 84-key AT keyboard, and the 101-key enhanced AT keyboard. All BIOS versions should recognize the original 83-key PC keyboard codes; thus the codes listed in this appendix are for PC keyboards. Please refer to references listed in Appendix H, "Sources," for information concerning AT keyboards.

The *complete* ASCII code is defined for hex numbers of 00h to 7Fh. ASCII codes above 7Fh display graphical symbols and non-English letters. Scan codes are shown in Table E.2.

Hex	Character[1]	Hex	Character	Hex	Character	Hex	Character
00	NUL (^@)	21	!	42	B	63	c
01	SOH (^A)	22	"	43	C	64	d
02	STX (^B)	23	#	44	D	65	e
03	ETX (^C)	24	$	45	E	66	f
04	EOT (^D)	25	%	46	F	67	g
05	ENQ (^E)	26	&	47	G	68	h
06	ACK (^F)	27	'	48	H	69	i
07	BEL (^G)	28	(	49	I	6A	j
08	BS (^H)	29	)	4A	J	6B	k
09	HT (^I)	2A	*	4B	K	6C	l
0A	LF (^J)	2B	+	4C	L	6D	m
0B	VT (^K)	2C	,	4D	M	6E	n
0C	FF (^L)	2D	–	4E	N	6F	o
0D	CR (^M)	2E	.	4F	O	70	p
0E	SO (^N)	2F	/	50	P	71	q
0F	SI (^O)	30	0	51	Q	72	r
10	DLE (^P)	31	1	52	R	73	s
11	DC1 (^Q)	32	2	53	S	74	t
12	DC2 (^R)	33	3	54	T	75	u
13	DC3 (^S)	34	4	55	U	76	v
14	DC4 (^T)	35	5	56	V	77	w
15	NAK (^U)	36	6	57	W	78	x
16	SYN (^V)	37	7	58	X	79	y
17	ETB (^W)	38	8	59	Y	7A	z
18	CAN (^X)	39	9	5A	Z	7B	{
19	EM (^Y)	3A	:	5B	[	7C		
1A	SUB (^Z)	3B	;	5C	\	7D	}
1B	ESC (^[)	3C	<	5D	]	7E	~
1C	FS (^\)	3D	=	5E	^	7F	(^BS)[2]
1D	GS (^])	3E	>	5F	_	80–FF[3]	
1E	RS (^^)	3F	?	60	`		
1F	US (^–)	40	@	61	a		
20	(SPACE)	41	A	62	b		

[1]ASCII control characters may be generated by pressing the Ctrl key, denoted by a circumflex (^), and the second key listed in each parentheses. For example, the DC2 control code character, ASCII 12h, may be generated by pressing the Ctrl-R keys simultaneously.

[2]Delete (7Fh) is generated by pressing the Ctrl-BackSpace (BS) keys. The symbol for 7Fh is a small triangle.

[3]Characters (happy faces and other graphic symbols) may be entered from the keyboard by pressing the Alt key and typing in the *decimal* equivalent for the symbol on the *keypad*.

Table E.1 ■ ASCII Keyboard Codes for Text and Control Characters, No Parity

Key	Code	Key	Code	Key	Code	Key	Code
A	1E	V	2F	F7	41	Enter	1C
B	30	W	11	F8	42	NumLk	45
C	2E	X	2D	F9	43	SclLk	46
D	20	Y	15	F10	44	CapLk	3A
E	12	Z	2C	Ctrl[1]	1D	BkSp	0E
F	21	0)	0B	Esc	01	End 1	4F
G	22	1 !	02	Alt[1]	38	Crsr Dwn 2	50
H	23	2 @	03	Tab	0F	Pg Dn 3	51
I	17	3 #	04	Ins	52	Crsr Lft 4	4B
J	24	4 $	05	Del	53	5	4C
K	25	5 %	06	; :	27	Crs Rgt 6	4D
L	26	6 ^	07	' "	28	Home 7	47
M	32	7 &	08	` ~	29	Crsr Up 8	48
N	31	8 *	09	– _	0C	Pg Up 9	49
O	18	9 (	0A	= +	0D	–	4A
P	19	F1	3B	, <	33	+	4E
Q	10	F2	3C	. >	34	Left Shft	2A
R	13	F3	3D	/ ?	35	Right Shft	36
S	1F	F4	3E	\ |	2B	Print Scrn	37
T	14	F5	3F	[{	1A	Space Bar	39
U	16	F6	40	] }	1B		

[1]The Ctrl key and the Alt key are on the *left* side of the PC keyboard, and on both sides of later keyboards.
[2]Scan code 00h is not assigned.

Table E.2 ■ PC Keyboard Hexadecimal Scan Codes

Appendix

8086
Assembly
Language
Instructions

The complete 8086 instruction set is shown in this appendix. Each instruction lists the instruction mnemonic, legal addressing modes for source and destination operands, the operation performed, and affected flags. Flags that are set equal to a question mark (?) are *undefined* for the instruction.

Instructions that are explained in some detail in the text are denoted by a *chapter number in brackets*. Instructions that are not covered in the text are discussed here.

ASCII ADJUST for ADDITION

Mnemonic	Destination	Source	Operation	Flags
AAA [8]	AX	AL	AXubcd[1] ← ALhex[2]	A, C O, P, S, Z = ?

ASCII ADJUST for DIVISION

Mnemonic	Destination	Source	Operation	Flags
AAD [8]	AL	AX	ALhex ← AXubcd	P, S, Z A, C, O = ?

ASCII ADJUST for MULTIPLICATION

Mnemonic	Destination	Source	Operation	Flags
AAM [8]	AX	AL	AXubcd ← ALhex	P, S, Z A, C, O = ?

ASCII ADJUST for SUBTRACTION

Mnemonic	Destination	Source	Operation	Flags
AAS [8]	AX	AL	AXubcd ← ALhex	A, C O, P, S, Z = ?

ADD with CARRY

Mnemonic	Destination	Source	Operation	Flags
ADC [5]	R/M	R/M/I	D ← D+S+C	A, C, O, P, S, Z

ADDITION

Mnemonic	Destination	Source	Operation	Flags
ADD [5]	R/M	R/M/I	D ← D+S	A, C, O, P, S, Z

AND logical

Mnemonic	Destination	Source	Operation	Flags
AND [5]	R/M	R/M/I	D ← D&S	C, O = 0; P, S, Z A = ?

CALL procedure

Mnemonic	Destination	Source	Operation	Flags
CALL [6]			CS:IP ← Add	None

CONVERT BYTE to WORD

Mnemonic	Destination	Source	Operation	Flags
CBW [8]	AX	AL	AH ← ALsb[3]	None

CLEAR CARRY flag

Mnemonic	Destination	Source	Operation	Flags
CLC	CF		CF = 0	C

Operation:
Clear the Carry flag to a binary 0 value.

CLEAR DIRECTION flag

Mnemonic	Destination	Source	Operation	Flags
CLD [8]	DF		DF = 0	D

CLEAR INTERRUPT flag

Mnemonic	Destination	Source	Operation	Flags
CLI [10]	IF		IF = 0	I

COMPLEMENT CARRY flag

Mnemonic	Destination	Source	Operation	Flags
CMC	CF		CF = /CF	C

Operation:
Complement the Carry flag.

COMPARE

Mnemonic	Destination	Source	Operation	Flags
CMP [5]	R/M	R/M/I	D – S	A, C, O, P, S, Z

COMPARE STRING

Mnemonic	Destination	Source	Operation	Flags
CMPS [8]	(DI)	(SI)	(SI) – (DI)	A, C, O, P, S, Z

CONVERT WORD to DOUBLE WORD

Mnemonic	Destination	Source	Operation	Flags
CWD [8]	DX	AX	DX ← AXsb[3]	None

DECIMAL ADJUST for ADDITION

Mnemonic	Destination	Source	Operation	Flags
DAA [8]	AL	AL	ALpbcd[4] ← ALhex	A, C, P, S, Z O = ?

DECIMAL ADJUST for SUBTRACTION

Mnemonic	Destination	Source	Operation	Flags
DAS [8]	AL	AL	ALpbcd ← ALhex	A,C,P,S,Z O = ?

DECREMENT

Mnemonic	Destination	Source	Operation	Flags
DEC [5]	R/M		D ← D–1	A, O, P, S, Z

DIVIDE

Mnemonic	Destination	Source	Operation	Flags
DIV [8]	DX,AX	R/M	DX,AX ← DXAX\S	A, C, O, P, S, Z

ESCAPE

Mnemonic	Destination	Source	Operation	Flags
ESC	I	M/R	Coprocessor	None

Operation:
Escape. Places the contents of the effective address on the data bus. Executes a NOP. Used to permit multiprocessing and coprocessing systems to be implemented. *Obsolete and not assembled by most assemblers.*

HALT

Mnemonic	Destination	Source	Operation	Flags
HLT			Processor Halt	None

Operation:

Halt all CPU program execution. Executes NOPs until NMI, Reset, or INTR signal received. Used to permit wait for external stimulus as a substitute for an endless software loop.

INTEGER DIVIDE

Mnemonic	Destination	Source	Operation	Flags
IDIV	AX DX,AX	M/R	Integer Divide	A, C, O, P = ? S, Z = ?

Operation:

Integer Divide. Performs signed number division.

If the source is a byte, then AX contains the dividend. The source is divided into AX, and the quotient is returned in AL, with the remainder in AH. The maximum permissible quotient is +127 (7Fh) or −128 (80h).

If the source is a word, then the dividend is in DX,AX. The remainder is returned in DX and the quotient returned in AX. The maximum permissible quotient is +32767 (7FFFh) or −32768 (8000h).

Division that results in a quotient larger than allowed (such as division by zero) results in an automatic INT Type 0, which normally stops program execution.

INTEGER MULTIPLY

Mnemonic	Destination	Source	Operation	Flags
IMUL	AX AX,DX	M/R	Integer Multiply	C, O A, P S, Z = ?

Operation:

Integer Multiply. Performs signed number multiplication.

If the source is a byte, then the source is multiplied by AL and the word result returned in AX. The maximum result ranges from −16256 (C080h) to +16384 (4000h).

If the source is a word, then the source is multiplied by AX, and the double word result returned in DX,AX. The maximum result ranges from −1,073,709,056 (C0008000h) to +1,073,741,824 (40008000h)

If the result returned in AH or DX is not the sign extension of AL or AX, then OF and CF are set indicating that the high-order register holds the most significant byte (or word) of the result. Overflow is not possible using signed number multiplication.

INPUT byte or word from port

Mnemonic	Destination	Source	Operation	Flags
IN [8]	AX	I/R	AX ← (S)port	None

INCREMENT

Mnemonic	Destination	Source	Operation	Flags
INC [5]	R/M		D ← D+1	A, O, P, S, Z

INTERRUPT

Mnemonic	Destination	Source	Operation	Flags
INT [7]	I		Call Interrupt	I, T

INTERRUPT on OVERFLOW

Mnemonic	Destination	Source	Operation	Flags
INTO			Call Int Type 4	None

Operation:
Interrupt on Overflow. Perform Interrupt Type 4 if OF is set to 1. Go to the next instruction if OF cleared to 0. Generally used to activate an overflow error routine while performing signed number arithmetic.

INTERRUPT RETURN

Mnemonic	Destination	Source	Operation	Flags
IRET [7]			Return from Int	Old flags popped

JUMP on ABOVE   JUMP on NOT BELOW or EQUAL

Mnemonic	Destination	Source	Operation	Flags
JA	IP		If (CF&ZF) = 0	None
JNBE				

Operation:
Jump if Above (Not Below or Equal). If a comparison type instruction has been previously executed, and CF and ZF are both 0, then the comparison shows that the operands are not equal (ZF = 0), and the destination is larger than the source (CF = 0). *Applies to unsigned numbers only.*

JUMP on ABOVE or EQUAL   JUMP on NOT BELOW   JUMP on NO CARRY

Mnemonic	Destination	Source	Operation	Flags
JAE [5]	IP		If CF = 0	None
JNB				
JNC				

JUMP on BELOW   JUMP on NOT ABOVE or EQUAL   JUMP on CARRY

Mnemonic	Destination	Source	Operation	Flags
JB [5]	IP		If CF = 1	None
JNAE				
JC				

JUMP on BELOW or EQUAL   JUMP on NOT ABOVE

Mnemonic	Destination	Source	Operation	Flags
JBE	IP		If (CF ^ ZF) = 1	None
JNA				

Operation:
Jump Below or Equal (Not Above). If a comparison type instruction has been previously executed, and CF or ZF are 1, then the comparison shows that the operands are either equal (ZF = 1), or the destination is larger than the source (CF = 1). *Applies to unsigned numbers only.*

JUMP on CX ZERO

Mnemonic	Destination	Source	Operation	Flags
JCXZ	IP		If CX = 0	None

Operation:
Jump if register CX is 0. May be useful to leave a loop that uses CX as a counter, or to skip a loop that uses CX as a counter, and CX happens to have an initial value of 0.

JUMP on EQUAL   JUMP on ZERO

Mnemonic	Destination	Source	Operation	Flags
JE [5]	IP		If ZF = 1	None
JZ				

JUMP on GREATER   JUMP on NOT LESS or EQUAL

Mnemonic	Destination	Source	Operation	Flags
JG	IP		If (SF = OF) ^ ZF = 0	None
JNLE				

Operation:
Jump if Greater (Not Less or Equal). If a comparison type instruction has been previously executed, and SF = OF or ZF = 0, then the comparison shows that the operands are not equal (ZF = 0), or the destination is larger than the source (SF = OF). *Applies to signed numbers only.* [Note that, for *signed* numbers, 01h (+1) is larger than FFh (−1).]

JUMP on GREATER or EQUAL   JUMP on NOT LESS

Mnemonic	Destination	Source	Operation	Flags
JGE	IP		If SF = OF	None
JNL				

Operation:
Jump if Greater or Equal (Not Less). If a comparison type instruction has been previously executed, and CF = OF, then the comparison shows that the destination is larger than the source (CF = OF). *Applies to signed numbers only.* Note that, for signed numbers, 01h (+1) is larger than FFh (−1).

JUMP LESS   JUMP on NOT GREATER or EQUAL

Mnemonic	Destination	Source	Operation	Flags
JL	IP		If SF<>OF	None
JNGE				

Operation:
Jump if Less (Not Greater or Equal). If a comparison type instruction has been previously executed, and SF <> OF, then the comparison shows that the destination is smaller than the source (SF <> OF). *Applies to signed numbers only.* [Note that, for *signed* numbers, 01h (+1) is larger than FFh (−1).]

JUMP on LESS or EQUAL   JUMP on NOT GREATER

Mnemonic	Destination	Source	Operation	Flags
JLE	IP		If (SF<>OF) ^ ZF = 1	None
JNG				

Operation:
Jump if Less or Equal (Not Greater). If a comparison type instruction has been previously executed, and SF <> OF, or ZF = 1, then the comparison shows that the operands are equal (ZF = 0), or the destination is smaller than the source (SF <> OF). *Applies to signed numbers only.* [Note that, for *signed* numbers, 01h (+1) is larger than FFh (−1).]

JUMP unconditionally

Mnemonic	Destination	Source	Operation	Flags
JMP [5]	CS:IP	I/M/R	Jump Always	None

JUMP on NOT EQUAL   JUMP on NOT ZERO

Mnemonic	Destination	Source	Operation	Flags
JNE [5]	IP		If ZF = 0	None
JNZ				

JUMP on NOT OVERFLOW

Mnemonic	Destination	Source	Operation	Flags
JNO	IP		If OF = 0	None

Operation:
Jump if Overflow flag is 0.

JUMP on NOT SIGN

Mnemonic	Destination	Source	Operation	Flags
JNS	IP		If SF = 0	None

Operation:
Jump if Sign flag is 0. Signifies that a signed number is positive.

JUMP on NOT PARITY   JUMP on PARITY ODD

Mnemonic	Destination	Source	Operation	Flags
JNP	IP		If PF = 0	None
JPO				

Operation:
Jump if Parity flag is 0 (parity odd). Signifies that parity is odd.

JUMP on OVERFLOW

Mnemonic	Destination	Source	Operation	Flags
JO	IP		If OF = 1	None

Operation:
Jump if Overflow flag is 1. Signifies that an overflow has occurred.

JUMP on PARITY   JUMP on PARITY EVEN

Mnemonic	Destination	Source	Operation	Flags
JP	IP		If PF = 1	None
JPE				

Operation:
Jump if Parity flag is 1 (even parity). Signifies even parity.

JUMP on SIGN

Mnemonic	Destination	Source	Operation	Flags
JS	IP		If SF = 1	None

Operation:
Jump if Sign flag is 1. Signifies a negative signed number.

LOAD AH from FLAGS

Mnemonic	Destination	Source	Operation	Flags
LAHF	AH	Flags	AH ← A,C,P,S,Z	None

Operation:
Load register AH from Flag register. Loads low byte of Flag registers SF, ZF, AF, PF, and CF into AH bits 7, 6, 4, 2, and 0. AH bits 5, 3, and 1 are undefined. May be used to inspect flags without using flag-dependent jumps.

LOAD DS and pointer

Mnemonic	Destination	Source	Operation	Flags
LDS [8]	DS:R	M	DS:R ← (M)w	None

LOAD EFFECTIVE ADDRESS

Mnemonic	Destination	Source	Operation	Flags
LEA [8]	R	M	R ← Mea[5]	None

LOAD ES and pointer

Mnemonic	Destination	Source	Operation	Flags
LES [8]	ES:R	M	ES:R < (M)w[6]	None

LOCK the bus

Mnemonic	Destination	Source	Operation	Flags
LOCK			Assert LOCK signal	None

Operation:
Assert LOCK output signal active-low (prefix). Forces bus LOCK signal low while instruction following the LOCK prefix executes. Applies only to CPU operating in maximum mode. Used in multiprocessor systems.

LOAD STRING

Mnemonic	Destination	Source	Operation	Flags
LODS [8]	AX	SI	AX ← (SI)	None

LOOP

Mnemonic	Destination	Source	Operation	Flags
LOOP	IP		Jump If CX <> 0	None

Operation:
Loop until CX = 0. A form of automatic decrement and jump instruction. Register CX is decremented by 1. If CX is not 0, then a jump is made to the label following the LOOP instruction.

LOOP while EQUAL   LOOP while ZERO

Mnemonic	Destination	Source	Operation	Flags
LOOPE			Jump If (ZF = 1)&CX<>0	None
LOOPZ				

Operation:
Loop IF CX > 0 AND ZF set to 1. A form of automatic decrement, test for equality, and jump instruction. Register CX is decremented by 1 and the ZF checked. If CX is not 0, and the ZF is set, then a jump is made to the label following the LOOPE/LOOPZ instruction. Loop will continue until CX has counted down to 0, or a mismatch is found in a loop containing a comparison type instruction.

LOOP while NOT EQUAL   LOOP while NOT ZERO

Mnemonic	Destination	Source	Operation	Flags
LOOPNE			Jump if (ZF = 0)&CX<>0	None
LOOPNZ				

Operation:
Loop IF CX > 0 AND ZF set to 0. A form of automatic decrement, test for inequality, and jump instruction. Register CX is decremented by 1 and the ZF checked. If CX is not 0, and the ZF is clear, then a jump is made to the label following the LOOPNE/LOOPNZ instruction. Loop will continue until CX has counted down to 0, or a match is found in a loop containing a comparison type instruction.

MOVE

Mnemonic	Destination	Source	Operation	Flags
MOV [5]	R/M	R/M/I	D ← S	None

MOVE STRING

Mnemonic	Destination	Source	Operation	Flags
MOVS [8]	DI	SI	(DI) ← (SI)	None

MULTIPLY

Mnemonic	Destination	Source	Operation	Flags
MUL [8]	DXAX	R/M	DXAX ← AX*S	C, O A, P, S, Z = ?

NEGATE

Mnemonic	Destination	Source	Operation	Flags
NEG	R/M		D ← 0–D	A, C, O, P, S, Z

Operation:
Negate Destination. Subtracts destination (byte or word) from 0 thus forming the two's complement of the destination. Negating a byte of 80h (–128) or a word of 8000h (–32,768) sets OF. Negating a 0 destination clears CF.

NO OPERATION

Mnemonic	Destination	Source	Operation	Flags
NOP			Do Nothing	None

Operation:
No Operation. Fetch the NOP instruction and do nothing.

NOT logical

Mnemonic	Destination	Source	Operation	Flags
NOT [5]	R/M		D ← /D	None

OR logical

Mnemonic	Destination	Source	Operation	Flags
OR [5]	R/M	R/M/I	D ← D ^ S	C, O = 0; P, S, Z A = ?

OUTPUT byte or word to port

Mnemonic	Destination	Source	Operation	Flags
OUT [8]	R/I	AX	(D)port ← AX	None

POP from stack

Mnemonic	Destination	Source	Operation	Flags
POP [6]	R/M	SP	D ← Stack	None

POP FLAGS from stack

Mnemonic	Destination	Source	Operation	Flags
POPF [6]	Flags	SP	Flags ← Stack	All

PUSH on stack

Mnemonic	Destination	Source	Operation	Flags
PUSH [6]	SP	R/M	Stack ← D	None

PUSH FLAGS on stack

Mnemonic	Destination	Source	Operation	Flags
PUSHF [6]	SP	Flags	Stack ← Flags	None

ROTATE through CARRY LEFT

Mnemonic	Destination	Source	Operation	Flags
RCL [5]	R/M		CF&D ← CF&Drotl[7]	C, O

ROTATE through CARRY RIGHT

Mnemonic	Destination	Source	Operation	Flags
RCR	R/M	CL/1	CF&D ← CF&Drotr[7]	C, O

Operation:
Rotate through Carry to the Right. Rotates destination to the right, using the carry as the most significant bit of the destination. If source = 1 then one rotate is done, else CL is counted down to 0, and a rotate done for each decrement of CL. See RCL in Chapter 5 for the rotate left equivalent.

REPEAT

Mnemonic	Destination	Source	Operation	Flags
REP [8]			Repeat MOVS, STOS Until CX = 0	None

REPEAT while EQUAL   **REPEAT while ZERO**

Mnemonic	Destination	Source	Operation	Flags
REPE [8]			Repeat CMPS, SCAS Until CX = 0 or ZF = 0	None
REPZ [8]				

REPEAT while NOT EQUAL   **REPEAT while NOT ZERO**

Mnemonic	Destination	Source	Operation	Flags
REPNE [8]			Repeat CMPS, SCAS Until CX = 0 or ZF = 1	None
REPNZ [8]				

RETURN

Mnemonic	Destination	Source	Operation	Flags
RET [6]	IP		Return From Call	None

ROTATE LEFT

Mnemonic	Destination	Source	Operation	Flags
ROL	R/M	CL/1	D ← Drotl	C, O

Operation:
Rotate to the Left. Rotates destination to the left. If source = 1 then one rotate is done, else CL is counted down to 0, and a rotate done for each decrement of CL.

ROTATE RIGHT

Mnemonic	Destination	Source	Operation	Flags
ROR	R/M	CL/1	D ← Drotr	C, O

Operation:
Rotate to the right. Rotates destination to the right. If source = 1 then one rotate is done, else CL is counted down to 0, and a rotate done for each decrement of CL.

STORE AH in FLAGS

Mnemonic	Destination	Source	Operation	Flags
SAHF	Flags	AH	A,C,P,S,Z ← AH	A, C, P, S, Z

Operation:
Load Flag register from AH. Loads SF, ZF, AF, PF, and CF into Flag register low byte from AH bits 7, 6, 4, 2, and 0. All other flags are unaffected.

SHIFT ARITHMETIC LEFT   SHIFT LEFT logical

Mnemonic	**Destination**	**Source**	**Operation**	**Flags**
SAL	R/M	CL/1	$D \leftarrow Dsal^8$	C, O, P, S, Z
SHL				A = ?

Operation:
Shift Arithmetic Left (Logical Left). Shifts each bit of the destination one place to the left. Replaces the LSB of the destination with a binary 0. The CF is set to the last bit shifted out of the MSB of the destination.

   If the count is 1, OF is set if the original sign bit and the new sign bit are different, else OF is cleared. If the count is in CL, then the state of OF is undefined.

SHIFT ARITHMETIC RIGHT

Mnemonic	**Destination**	**Source**	**Operation**	**Flags**
SAR	R/M	CL/1	$D \leftarrow Dsar^9$	C, O, P, S, Z

Operation:
Shift Arithmetic Right. Shifts each bit of the destination one place to the right. Replaces the MSB of the destination with itself. The CF is set to the last bit shifted out of the LSB of the destination.

   If the count is 1, OF is set if the original sign bit and the bit to its right are different, else OF is cleared. If the count is in CL, then OF is cleared.

SUBTRACT with BORROW

Mnemonic	**Destination**	**Source**	**Operation**	**Flags**
SBB [5]	R/M	R/M/I	$D \leftarrow D-S-CF$	A, C, O, P, S, Z

SCAN STRING

Mnemonic	**Destination**	**Source**	**Operation**	**Flags**
SCAS [8]	DI	AX	AX–(DI)	A, C, O, P, S, Z

SHIFT RIGHT logical

Mnemonic	**Destination**	**Source**	**Operation**	**Flags**
SHR	R/M	CL/1	$D \leftarrow Dslr^8$	C, O, P, S, Z
				A = ?

Operation:
Shift Logical Right. Shifts each bit of the destination one place to the right. Replaces the MSB of the destination with a binary 0. The CF is set to the last bit shifted out of the LSB of the destination.

   If the count is 1, OF is set if the original sign bit and old sign bit are different, else OF is cleared. If the count is in CL, then the state of OF is undefined.

SET CARRY

Mnemonic	**Destination**	**Source**	**Operation**	**Flags**
STC	CF		CF = 1	C

Operation:
Set the Carry flag.

SET DIRECTION

Mnemonic	**Destination**	**Source**	**Operation**	**Flags**
STD [8]	DF		DF = 1	D

SET INTERRUPT

Mnemonic	**Destination**	**Source**	**Operation**	**Flags**
STI [10]	IF		IF = 1	I

STORE STRING

Mnemonic	Destination	Source	Operation	Flags
STOS [8]	DI	AX	(DI) ← AX	None

SUBTRACT

Mnemonic	Destination	Source	Operation	Flags
SUB [5]	R/M	R/M/I	D ← D–S	A, C, O, P, S, Z

TEST

Mnemonic	Destination	Source	Operation	Flags
TEST	R/M	R/M/I	D&S	C, O, P, S, Z

Operation:
Test the Operands. The source and destination are logically ANDed together with no change to either, however the flags are set based on the AND operation. Useful for masking operations while leaving the test operand unchanged.

WAIT for test signal

Mnemonic	Destination	Source	Operation	Flags
WAIT				None

Operation:
Suspend CPU operation until TEST input signal is low. May be used to halt program operation until an external signal on the TEST pin goes active-high.

EXCHANGE

Mnemonic	Destination	Source	Operation	Flags
XCHG	R/M	R	(D) ↔ (S)	None

Operation:
Exchange Operands. Swaps the contents of the operands. Unlike a move instruction, which copies data from one location to another, XCHG exchanges data between two locations.

TRANSLATE

Mnemonic	Destination	Source	Operation	Flags
XLAT	AL	BX	AL ← (BX:AL)	None

Operation:
Translate Using AL. Replaces AL with the contents of the lookup table pointed to by BX:AL. Useful for small (256 byte) lookup tables.

EXCLUSIVE OR logical

Mnemonic	Destination	Source	Operation	Flags
XOR [5]	R/M	R/M/I	D ← D ⊕ S	C, O = 0; P, S, Z A = ?

NOTES

1. ubcd = Unpacked BCD number with a single BCD decimal digit in each half of AX.

2. hex = Hexadecimal number from 00 to 63h.

3. sb = The sign bit of the byte (bit 7) or word (bit 15) fills each bit of the extended byte or word.

4. pbcd = Packed BCD number, from 00 to 99.

5. ea = Effective address.

6. w = Word at that address.

7. rotl, rotr = Rotate left (right). A location is rotated when each bit moves to the left (right) one position. The most significant bit and the least significant bit are considered to be adjacent. If the Carry flag is included, then it is the most significant bit position.

8. slr, sal = Shift logic right/arithmetic left. A logical right shift or arithmetic left shift involves moving each bit of the location one place to the right (left). Binary 0 is shifted into the vacant bit position on the left (right).

9. sar = Shift arithmetic right. An arithmetic right shift maintains the sign bit of the location by keeping the most significant bit of the location at its original value while shifting it right into the next bit position.

SYMBOLS

←	= Moves Contents
( )	= Contents Pointed to by Register
&	= Logical AND
^	= Logical OR
⊕	= Logical Exclusive OR
+	= Addition
−	= Subtraction
*	= Multiplication
\	= Division
<>	= Not Equal
/	= Invert
R	= Register Address Mode
M	= Memory Address Mode
I	= Immediate Address Mode

Appendix

The Rest of the Family

G.0 The 8086 CPU Family

The 8086 (1978)

The 8086 microprocessor design was the first of a family of CPU architectures that persists to this day. I shall outline here the advances that have been made over the original 8086 CPU by newer family members called, collectively, the *X86 family*. One fact stands out: Buried in every X86 family member is the architecture of the original 8086 CPU.

The desire to maintain family compatibility for software written for the 8086 has driven the design of succeeding X86 CPUs. Programs that were originally written for the 8086 CPU can be run on any later X86 family member. The reverse is *not* true. Programs written for more advanced X86 family CPUs cannot be run on an 8086 unless the programmer takes great pains to *not use any advanced CPU software features.* 8086-compatible programs sell to a huge market of installed computers. Many programmers prefer to write application programs using the original 8086 software instructions so the programs can run on any X86 family member.

The 8088 (1979)

The 8088 CPU is an 8086 CPU (with a 4-byte BIU queue) that transfers only 1 *byte* of information from memory during each BIU cycle. Word transfers, therefore, use two BIU addressing cycles. (The 8086 can transfer a byte or a *word* of memory in a BIU cycle.)

Instructions for the 8088 and 8086 are identical. Except for speed of execution, the difference between byte and word transfers on the 8088 is transparent to the programmer.

The 80186 (1982)

The 80186 CPU is an 8086 CPU with several of the most common peripheral support chips integrated on the die with the 8086 CPU.

Chapters 11 and 12 discuss several peripheral support chips that are needed to construct a working computer using the 8088 CPU chip. Many of the separate support chips discussed in Chapters 11 and 12 are integrated *inside* the 80186 CPU design. The support functions that are included on the 80186 chip are listed in Table G.1 along with their 8086-equivalent part numbers.

Several other functions, including a programmable wait state generator and programmable chip-select logic, are also included on the 80186 die.

8086 Part Number	80186 Functional Equivalent
8237	DMA controller
8254	Programmable timers
8259	Interrupt controller
8284	Clock generation

Table G.1 ■ 80186 Integrated Functions

Because of its high level of functional integration, the 80186 has found wide application as an *embedded controller.* Embedded controllers are computers that are buried inside of machinery and "invisible" to the machine operator. Examples of embedded controllers include automotive applications such as engine management and anti-skid brakes, laser printers, chess games, and cameras. Appliances, such as microwave ovens and even steam irons, are now equipped with embedded controllers.

80186 SOFTWARE ADDITIONS

The 80186 adds the instructions listed in Table G.2 to the 8086 opcode set. The most notable 80186 opcode additions include a push and pop (that handle all eight CPU working registers at a time), high-level language procedures, and string moves involving I/O ports. Table G.2 also lists 8086 instructions that have been modified so that immediate numbers may take part in an operation.

Taken as a whole, the 80186 represents the culmination of 8086 hardware and software design. Significant advances in computational power are made possible by later X86 family CPU designs.

The 80286 (1982)

A significant hardware improvement of the 80286 over the 8086 is a reduction in the number of clock cycles used in an 80286 machine cycle. The 8086 uses four clock pulses (T states) to execute a machine cycle, whereas an 80286 uses two clock pulses per machine cycle. For example: An 80286, with a clock period of 200 ns, requires 400 ns for one machine cycle. An 8086 using the same clock requires 800 ns for one machine cycle. The 80286 thus

New	Action
BOUND	Check that array address is within array memory bounds.
ENTER	Execute calling sequence for high-level language procedure.
INS	Input *string* from I/O port.
LEAVE	Execute procedure return for high-level language.
OUTS	Output *string* to I/O port.
POPA	Pop CPU working *registers.*
PUSHA	Push CPU working *registers.*
Modified	
IMUL	8086 type IMUL that may use an immediate operand.
PUSH	8086 type PUSH that allows immediate *byte* or word push.
RCL	8086 type RCL that allows *n* immediate rotates.
RCR	8086 type RCR that allows *n* immediate rotates.
ROL	8086 type ROL that allows *n* immediate rotates.
ROR	8086 type ROR that allows *n* immediate rotates.
SAL/SHL	8086 type SAL that allows *n* immediate shifts.
SAR	8086 type SAR that allows *n* immediate shifts.
SHR	8086 type SHR that allows *n* immediate shifts.

Table G.2 ■ Additional 80186 Instructions

runs twice as fast as does an 8086 fed by the same clock frequency. Other improvements led Intel to claim that a 10 mHz 80286 was over three times as fast as a 10 mHz 8086.[1]

The 80286 can be considered a bridge between the 16-bit 8086 and the 32-bit 80386 CPU. The 80286 has 16-bit CPU registers that are *identical* to those of the 8086. In addition, however, the 80286 is equipped with memory addressing features, and *memory management* features, that are *not* found in the 8086. Memory management is a prominent feature, along with 32-bit registers, of the 80386.

Memory management is a term that is associated with *multi-user (multi-tasking)* operating systems, such as AT&T's UNIX operating system. Multi-tasking means that the computer switches between numerous application programs, running each for a brief instant. Multi-user means that many users, either humans or peripherals, have access to the CPU. Multi-user and multi-tasking operating systems are very concerned with keeping application programs separated from each other—and, even more important, with keeping the application programs from interfering with the operating system program! Beginning with the 80286, we begin to see CPU features and concepts that "make the operating system safe from the (application) programmer."

Many of the 27 new instructions added to the 80286, and most of the additional CPU registers, have to do with multi-user, multi-tasking *operating* system programs. Multi-user, multi-tasking operating system programs require *specialized instructions and hardware* just as application programs require standard instructions and hardware.

Most operating system issues are beyond the scope of this book, but I shall outline some features here that may be of importance to assembly language programmers.

80286 Real Mode

The 80286 can be operated in one of two modes of operation—*real* and *protected.* Real mode is exactly as if the 80286 were programmed as an 8086. When the 80286 is reset, it begins to operate in real mode. Real mode uses the segment registers in the same way as an 8086. Up to 1,000K of memory can be addressed by a combination of segment registers and offsets, as is the case with the 8086. Memory is addressed using 20 address bits formed by shifting the segment registers 4 bits to the left before adding the offset. In real mode, the 80286 is sort of a super 8086, adding additional instructions and running at higher CPU clock speeds than the 8086.

80286 Protected Mode

The second 80286 mode, the protected mode, is entirely different from 8086 operation. In protected mode, the 80286 can directly address up to 16M of physical memory, and up to 1G (1 gigabyte, or 1 billion bytes) of *virtual* memory. Physical memory is addressed, in the usual way, using 24 address lines that emanate from the CPU. (The upper 4 address bits are not used when the 80286 is in real operating mode.)

---

[1]*Introduction to the iAPX 286* Santa Clara, CA: Intel, 1982, pp. 2–3.

PROTECTED MODE VIRTUAL MEMORY

Virtual memory means that the CPU is capable of addressing 1G of memory by specifying up to 16,384 segments, each segment containing 64K. Virtual memory is conceptual, that is, 1G of segments are *addressable* by the CPU, but only 16M can be present in *physical* memory at any time. All other memory segments that are not present in physical memory are stored in some mass storage medium such as a hard disk drive. Any attempt by the CPU to access a memory segment that is *not* at the moment in the 16M semiconductor memory causes a signal to the operating system software to be generated by the CPU. The operating system reacts to the CPU signal by loading the desired memory segment from disk to physical semiconductor memory for access by the CPU.

To speed up the virtual memory process, the operating system tries to load as many virtual memory bytes as it can from disk to RAM. Programs that are written using virtual memory attempt to group program parts into large, contiguous groups of code and data segments in order to limit disk access by the operating system. Excessive disk access by the operating system results in very slow program execution as the physical RAM memory is constantly rewritten with new segments.

80286 PHYSICAL MEMORY ADDRESSING

Programs written for the 8086 can access 1M of memory space. The same 8086 program can also run in a protected mode 80286 16M memory space. The 80286 can address 16M of physical memory (from an 8086 program written to address 1M of memory) using *operating system hardware and software*. The operating system hardware is a set of registers that is *transparent* to application programs. The operating system hardware is manipulated by software instructions meant *only* for operating system use.

Figure G.1 shows the 80286 operating system hardware registers. The registers are listed in Table G.3.

THE MSW

A control register, named the Machine Status Word (MSW) register, is part of the 80286 architecture. One of the bits in the MSW is the Protected mode Enable (PE) bit, which places the 80286 in protected mode when *set*. The PE flag is initially cleared, on system reset, so that the 80286 always resets in *real* mode.

Protected mode is entered by setting the PE flag to 1. Once in protected

---

```
Machine Status Word (MSW)
Segment Selectors and Cache Registers
Task Register (TR) and Cache Register
Local Descriptor Table Register (LDTR) and Cache Register
Global Descriptor Table Register (GDTR) Cache Register
Interrupt Descriptor Table Register (IDTR) Cache Register
```

Table G.3 ■ 80286 Operating System Registers

Figure G.1 ■ 80286 Protected mode registers.

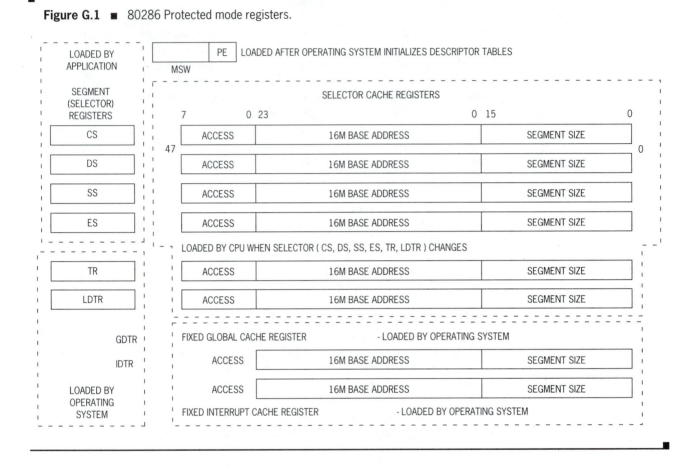

mode, however, the CPU will not allow the load MSW (LMSW) instruction to clear the PE flag. Reentering real mode from protected mode is very difficult because the computer must be reset (to clear PE) without booting (which clears memory). Before entering protected mode, the operating system must ensure that all of the operating system registers shown in Figure G.1, and all operating system data, are initialized.

PROGRAM PRIVILEGE

Protected mode embodies the philosophy of *program privilege.* Privilege implies that some programs, usually the operating system, have access to areas of memory that other, less privileged application programs do not. Application programs are named *tasks.* A task is a program that is loaded into memory and allowed to run by the operating system. Any attempt by a task to access a memory address not allowed by its privilege level results in corrective action by the operating system. Privilege levels vary from 0, the most privileged, to 3, the least privileged. The operating system is, typically, the only program that has a 0 privilege level. The operating system has access to all memory and directs the execution of all lower privileged programs. Next in privilege level are operating system service routines, then operating system extensions, and, last, application tasks.

The 80286 was intended to be used as a multi-tasking processor, that is, a CPU that can load many tasks into physical memory at one time. Each task is allowed to run, either until completion or for a short time, by a multi-tasking operating system. Many tasks, each running for a few milliseconds at a time, appear to be running at the same time to slower users, such as humans. Multi-tasking enables the CPU to service a number of users, each of whom thinks the computer is running only his or her application program. (Unfortunately, a DOS-compatible multi-tasking operating system was not developed concurrently with the 80286. Single-user [real mode] DOS continues to dominate X86 family personal computer operating systems.)

DESCRIPTORS

Part of the duties of a multi-tasking operating system is to allocate memory for application programs, load application programs into memory, and keep track of them as they run. Every task must be assigned physical memory segments for code, data, and stack bytes, as well as a privilege level.

The operating system assigns privilege levels and segment addresses for each task. Each task's privilege and segment address information is stored in operating system memory as data structures known as *descriptors*. Descriptors consist of 8 data bytes containing information that the operating system will use to load and run a task. Typically, a descriptor will specify a *24-bit physical memory base address,* as well as control and privilege data that is useful to the operating system. There are several types of descriptors, including segment and control descriptors.

An 8-byte segment descriptor is shown in Figure G.2. A segment descriptor consists of a *24-bit segment base address,* a 16-bit segment size limit, and a byte that describes the segment's privilege level and other operating system cues. (The last word of a segment descriptor is *not* used by the 80286 CPU, but is used in succeeding family members.)

A *complete set* of descriptors used *only* by a *single* task is stored in memory by the operating system as *local* descriptors for that task. There may be up to 8K local descriptors for each task. Descriptors that may be used by *all* tasks are stored as *global* descriptors. There may be up to 8K global descriptors.

Figure G.2 ■ 8-byte 80286 segment descriptor structure.

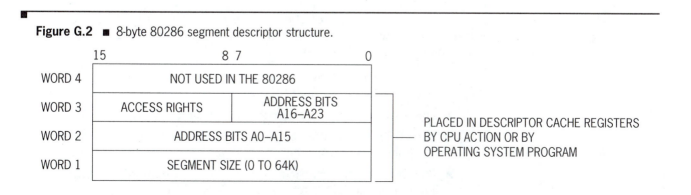

Figure G.3 ■ Selector register.

REAL MODE SEGMENT REGISTERS USED AS PROTECTED MODE SELECTORS

A real mode segment register used as a protected mode selector register is shown in Figure G.3. Register Bits 3–15 are used to specify 1 of the 8K possible descriptors, and Bit 2 identifies the Local or Global Cache register base address. Bits 0 and 1 identify the privilege level (RPL) of the segment as follows:

Selector Register Bits 1 and 0	Level	
0 0	0	(highest)
0 1	1	
1 0	2	
1 1	3	(lowest)

Privilege levels in the selector must agree with privilege levels in the descriptor, or the operating system will intervene.

Each task uses the *8086-type segment* registers as *pointers* (selectors) to its descriptors. The descriptors, in turn, *actually address physical* memory. By controlling the *contents* of the descriptors, the operating system can ensure that tasks do not interfere with each other. For instance, Task A is assigned descriptors for one area of physical memory, while Task B is assigned descriptors for a different area of physical memory.

DESCRIPTOR TABLES

Each task must have a means to access its own *set* of local descriptors from the array of local descriptors stored in operating system memory. There may be as many as 8K tasks and their associated local descriptor sets in an 80286 system. The operating system organizes the *base address* of each local task set into a *Local Descriptor Table* (LDT). The LDT is, in turn, made up of descriptors that contain the address (in physical memory) of each local set. The LDT is placed in the *global descriptor table*. The base address of a local task set is found using a combination of operating system program action and automatic CPU operation, as follows.

- A task is to be run by the CPU. The operating system finds a local task set selector number for the task.

- The *operating system loads* the CPU LDTR with the task selector.

- The LDTR now addresses one of the LDT descriptors contained in the global descriptor table.

- The CPU *automatically* loads the descriptor pointed to by the contents of the LDTR into the LDTR *Cache* register.

- The LDTR Cache register now holds the base address of the local descriptor set to be used by the task.
- All local descriptors used by the task are drawn from the local descriptor table using the contents of the LDTR Cache register base memory address. All global descriptors used by the task are drawn from the Global Descriptor Cache register base memory address.

SEGMENT DESCRIPTOR CACHE REGISTERS

Every segment selector register has an associated descriptor cache register, as shown in Figure G.1. Three descriptor bytes (24 bits) in a selector cache register address the 16M physical memory space. The other descriptor bytes include data on segment size and privilege and other memory access information for use by the operating system.

To find the particular descriptors that apply to a given task, the *application* program loads each selector register (segment register) with *offsets* into the local or global descriptor set. The segment selector numbers (loaded by the application program) are *automatically added* by the CPU to the LDTR or Global Cache register 16M *base address*. The resulting address in memory locates a descriptor for the selector. The CPU then *automatically* addresses the proper descriptor and loads it into the associated 80286 segment descriptor *cache* register. Program offsets are thereafter added to the *cache base address* to address physical memory.

To illustrate the difference between segment register use in a real mode program and selector register use in a multi-tasking program, consider the following example:

A *real* mode program might address physical memory data address 60070h by placing a 6007h in DS and using an address offset of 0. The CPU adds DS segment 6007h to offset 0000h to address physical address 60070h in memory.

In *protected* mode, the DS contents (6007h) act as a *pointer* to an entry in the *local* (selector Bit 2 = 1) descriptor table. A descriptor from the local descriptor set is *automatically* loaded by the CPU into the DS descriptor cache register.

The *DS cache* register contents then address a segment in the 16M 80286 memory. The *actual* physical address in the 16M memory space is determined by the DS descriptor assigned to the task by the operating system. Descriptor loading occurs *whenever* the contents of a segment selector are *changed* by the application program.

Once a descriptor is loaded into a cache register, it will be automatically used by the CPU to generate 24-bit addresses to the 16M physical memory. Addresses to the 16M physical RAM memory are formed, in a manner similar to the 8086, by adding an offset quantity to the 24-bit descriptor cache memory base address. Note that address formation involves the addition of a 16-bit offset to a *complete* 24-bit base address. No shifting of the cache register segment address is needed. 80286 protected mode segments may thus begin at any byte address in memory.

GDTR AND IDTR

There are no physical GDTR or IDTR registers because the 80286 does not use selectors to load the GDTR or IDTR cache registers. The global table may contain 8K global descriptors. The interrupt must contain 256 interrupt descriptors (2K). The operating system program loads the GDTR cache with a global table base address and the IDTR cache with the base address of descriptors used by system interrupts.

TR

The Task register, and its associated descriptor cache, address memory locations that save the *state* of every task. Multi-user or multi-tasking operating systems run each task for a fraction of a second. Whenever one task's operation is suspended, *all* CPU application register contents (the state of the task) must be saved in memory. When the task is resumed, the CPU register contents are restored from memory so the task resumes operation exactly at the point at which it was suspended. The TR cache descriptor addresses the memory location where the state of the task is saved (and retrieved) by the operating system.

Using Memory as Additional Registers

I have described a very complicated process for obtaining segment addresses for tasks run using an 80286 protected mode operating system. The essential features of the 80286 protected mode are that memory is used to hold segment address data and that an operating system is needed to manage memory.

The hardware and software features added to the 80286 were intended to make the 80286 an attractive CPU multi-user operating system. In actuality, the 80286 found its greatest application for single-user DOS running in real mode. The appeal of the 80286 in personal computers is its higher clock rates versus those of the 8086 (up to 25 MHz for the 80286 versus 10 MHz for the 8086) and 27 additional opcodes. The 80286 can be thought of as a very high speed 8086 and, when run in real mode, is absolutely compatible with programs written for the 8086.

The advent of the 80386 CPU eclipsed any potential the 80286 might have had for multi-user operating systems and defined a new level of performance for microprocessors.

80286 Software Additions

Not surprisingly, almost half of the 27 new 80286 instructions added to the 8086/80186 set have to do with managing the memory in protected mode. Table G.4 lists the 80286 additions to the 8086/80186 opcode set, with operating system (OS) codes marked with an x.

The 80386 (1985)

Application Program CPU Architecture

The 80386 CPU exhibits features of the 8086 register architecture, the 80286 memory management hardware, and additional control and testing hardware found on no earlier X86 family CPU. Similar to the 80286, the 80386 may be operated in a real or a protected mode.

You should keep in mind that, beginning with the 80286, an X86 CPU must be thought of as *two* computers. One part of the CPU is primarily for

Mnemonic	OS	Action
ARPL	x	Adjust RPL (requested privilege level) field of selector.
BOUND		Check that array pointer is within array memory limits.
CLTS	x	Clear TS (task switched) flag.
ENTER		Execute calling sequence for high-level language procedure.
INS		Input string from I/O port.
INSB		Input byte from I/O port.
INSW		Input word from I/O port.
LAR		Load access rights byte to destination register.
LEAVE		Execute procedure return for high-level language.
LGDT	x	Load global descriptor table cache register.
LIDT	x	Load interrupt descriptor table cache register.
LLDT	x	Load local descriptor table register.
LMSW	x	Load machine status word
LSL		Load segment limit.
LTR	x	Load task register.
OUTS		Output string to I/O port.
OUTSB		Output byte to I/O port.
OUTSW		Output word to I/O port.
POPA		Pop all CPU working registers.
PUSHA		Push all CPU working registers.
SGDT	x	Store global descriptor table cache register.
SIDT	x	Store interrupt descriptor table cache register.
SLDT	x	Store local descriptor table register.
SMSW	x	Store machine status word.
STR	x	Store task register
VERR		Verify a segment for reading.
VERW		Verify a segment for writing.

Table G.4 ■ Additional 80286 Opcodes

the use of application programs. The second part of the CPU is solely for the use of the operating system software. We are mainly concerned with application programming, and will investigate the 80386 from an application programmer's view. Those who write system software will be concerned with the operating system features of the 80386. Operating system programming is a subject that is beyond the scope of this book, thus operating system program CPU features will be discussed in a more general manner.

Figure G.4 shows the 80386 application program register architecture. The most prominent hardware change from the 80286 to the 80386 is the growth of internal working registers from 16 to 32 bits. Thirty-two bit working registers means that the 80386 can address physical memory as large as 4,294,967,296 (4 giga) bytes, from offset address 00000000h to FFFFFFFFh.

As may be seen in Figure G.4, the 8086/286 general purpose registers AX, BX, CX, and DX grow to 32 bits in length in the 80386. The names for the new 32-bit registers are formed by adding an *E* (Extended) to the older, 16-bit, register names. Register AX thus becomes register EAX; register BX becomes EBX, and so on.

Each 80386 general-purpose extended register may be thought of as four

Figure G.4 ■ 80386 program registers.

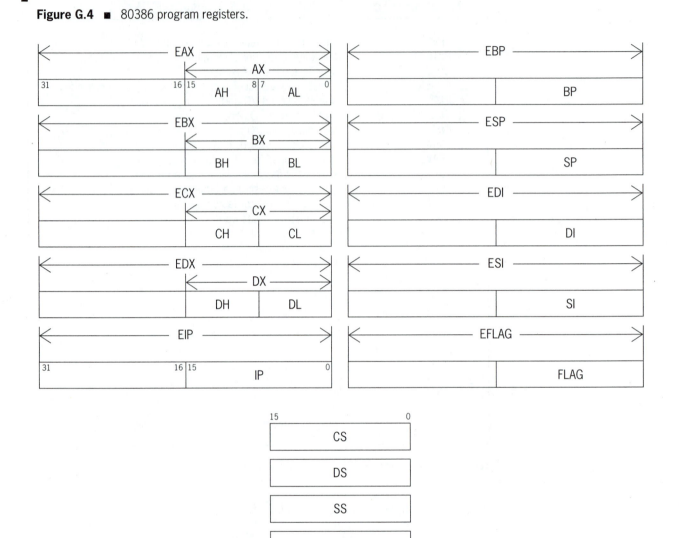

separate registers: a 32-bit register, a 16-bit register, and two 8-bit registers. For instance, using the A register, the complete 32-bit register is named EAX—the low-order 16-bit register is named AX, and each part of AX is named AH (the high byte of AX) and AL (the low byte of AX). There is no name for the high-order 16 bits of the extended register, and it may not be addressed.

Pointing registers IP, SI, DI, BP, and SP have also grown to 32-bit size becoming, respectively, EIP, ESI, EDI, EBP, and ESP. Each 80386 pointing register can be addressed as a 32-bit register, by including the E in its name, or as the lower, 16-bit register by omitting the E. The Source Index register, as an example, can be addressed as a full, 32-bit register as ESI, or just the low-order word addressed as SI.

The Flag register has been extended to 32 bits and can be addressed as a 32-bit register using the name EFLAG, or the low-order word of EFLAG can be addressed as FLAG.

APPLICATION PROGRAM SEGMENT REGISTERS: OPERATING SYSTEM DESCRIPTORS

The 80386 adds two additional data segment registers named FS and GS to the original 8086/80286 segment register set. The additional data segment registers add to the data-addressing versatility of the 80386. Interestingly, the 80386 segment registers have *not* grown from 16 to 32 bits. The reason the segment registers have not grown is clear: The descriptor method of segment selection first used in the 80286 is continued in the 80386. As was the case for the 80286, each segment register has an associated descriptor cache register. (Please refer to the 80286 CPU section for a discussion of descriptors and descriptor cache registers.) The 80386 segment registers act as pointers to descriptors located in tables maintained by the operating system. Descriptors are automatically loaded into 80386 descriptor cache registers from the operating-system controlled descriptor tables whenever a segment (selector) register's contents are changed by the application program.

The 80386 uses the same 64-bit descriptor format as the 80286, including employing descriptor Bits 48 to 63. The unused word of the 80286 descriptor data structure, shown in Figure G.2, is used by the 80386 as follows:

Bits	80386 Use
48–51	Segment size limit Bits 16–19
52	Available for operating system
53	Must be 0
54	Default segment size (0 = 64K, 1 = 4G)
55	Segment granularity (1 = Page, 0 = Byte)
56–63	Segment base address Bits 24–31

The term *granularity* associated with Bit 55 of an 80386 descriptor refers to how finely the data in a segment is divided. Note that segment size limits are expressed using a 20-bit number, or a size of 1M. If segment granularity is byte-sized, then the segment is limited to 1M. If the granularity is *page-size*, each page is defined as 4K, for a segment size of 4G.

80386 OPERATING MODES

The 80386 may be used in a real mode or a protected mode. Protected mode, in the 80386, has an additional specialized mode named *virtual* (V86) mode. Real mode is selected when the 80386 is reset.

80386 REAL MODE

When in the real mode of operation, the 80386 operates exactly as an 8086. One megabyte of physical memory may be addressed by the 80386 when in the real mode. Physical memory addresses are formed by shifting a segment register 4 bits to the left and adding a 16-bit offset quantity. Memory, in the real mode, is composed of 64K segments arranged on 16-byte boundaries. Although only 1M of memory is addressable, the programmer does have access to the 32-bit general purpose registers in the 80386, as well as the 8086/286 16- and 8-bit registers.

80386 PROTECTED MODE

The 80386 contains four 32-bit control registers, named CR0 to CR3. The 80386 Control register CR0 corresponds to the 80286 MSW register. Control register 1 is reserved for future use, and Control registers 3 and 4 are concerned with memory paging.

Programming the PE bit in Control register CR0 (see Figure G.5) will place the 80386 in protected, virtual memory, mode. Unlike the 80286, the 80386 can be returned to real mode from protected mode by resetting the PE bit. In protected mode, the physical address range of the 80386 is 4G, made possible by 32 address bits. Segment sizes, for the 80386 in protected mode, can be as large as the physical memory size, or 4G, because an offset quantity can be as large as a working register, which is 32 bits.

Program segments for a particular task are identified in a manner similar to the 80286 CPU. Descriptors for each task are arranged in memory as descriptor tables by the operating system and addressed by the application program selector (segment) registers. Segment descriptors are loaded into associated segment descriptor caches, and physical memory is addressed using the segment data contained in each segment descriptor. Up to 16K, 4G segments may be identified in the descriptor tables for a single task, making virtual memory as large as 64T (terabytes), or 70,368,744,000,000 bytes.

VIRTUAL (V86) MODE

Virtual mode, in essence, divides up memory space into a number of 8086 real mode pieces. The overall operating system is running the computer using protected mode mechanisms, but each program in memory is running as an 8086 real mode application, including its *own* copy of DOS.

Virtual mode allows many users to access a 386-based system as if each user had his or her own real-mode computer. The overall operating system can time-share the 80386 system among the many users. Most 386 and 486 computers are used by one person, however, usually operating in virtual mode. Single-user virtual mode programs can take advantage of the paging unit to access memory above the 1M DOS limit.

PAGING

The 80386 has a *paging* mechanism that allows a program address to be "translated" into a different physical memory address. For example, say the program uses address D0000000h. The page mechanism may translate this address to physical address C0000000h because physical address D0000000h is in use by another program. Pages are 4K of memory addresses that may be used by the paging translator for address translation. Paging may be done in real or protected modes of operation, allowing operating systems to access memory anywhere in the 4G address range.

Control registers CR0 to CR3, shown in Figure G.5, control the 80386 paging mechanism. The lower half of CR0 replaces the 80286 MSW register.

80386 TEST FEATURES

To aid in the testing and debugging of the inevitably complicated programs that run on the 80386, hardware testing registers are part of the 80386 design. The test and debug registers are shown in Figure G.5, and consist of breakpoint registers DR0 to DR5, debug control registers DR6 and DR7, and page mechanism test registers TR6 and TR7.

Code and data breakpoint addresses may be placed in registers DR0–DR3 (DR4 and DR5 are currently unused) to halt program execution whenever an address matching one of the breakpoint addresses is reached in a program. Registers DR6 and DR7 allow the programmer to control the debugging process. Registers TR6 and TR7 allow the paging mechanism to be tested.

Figure G.5 ■ 80386 control, test , and debug registers.

CR0

PAGING CONTROL	MSW	PE

CR1	NOT USED

CR2	PAGING MECHANISM

CR3	PAGING MECHANISM

DR0	BREAKPOINT	0

DR1	BREAKPOINT	1

DR2	BREAKPOINT	2

DR3	BREAKPOINT	3

DR4	NOT USED

DR5	NOT USED

DR6 DEBUG INTERRUPT

DR7 DEBUG INTERRUPT

TR6 PAGE MECHANISM TEST

TR7 PAGE MECHANISM TEST

Mnemonic	Action
BSF	Bit scan forward.
BSR	Bit scan reverse.
BT	Bit test.
BTC	Bit test and complement.
BTR	Bit test and clear (reset).
BTS	Bit test and set.
CDQ	Convert double word to quad word.
CMPSD	Compare double word strings.
CWDE	Convert word to extended double word.
INSD	Input double word of string from I/O port.
JECXZ	Jump if register ECX is 0.
LFS	Load FS register.
LGS	Load GS register.
LODSD	Load a double word from string into EAX.
LSS	Load SS register.
MOVSD	String move a double word.
MOVSX	Move data and extend sign.
MOVZX	Move data and zero high-order bits of destination.
OUTSD	Output double word of string to I/O port.
POPAD	Pop all general-purpose double word registers.
POPFD	Pop double word to extended stack register.
PUSHAD	Push all general-purpose double word registers.
PUSHFD	Push extended flag register.
SCASD	Scan string for double word.
SETA	Set byte if above.
SETAE	Set byte if above or equal.
SETB	Set byte if below.
SETBE	Set byte if below or equal.
SETC	Set byte on carry.
SETE	Set byte if equal.
SETG	Set byte if above.
SETGE	Set byte if greater or equal.
SETL	Set byte if less than.
SETLE	Set byte if less than or equal.
SETNA	Set byte if not above.
SETNAE	Set byte if not above or equal.
SETNB	Set byte if not below.
SETNBE	Set byte if not below or equal.
SETNC	Set byte if no carry.
SETNE	Set byte if not equal.
SETNG	Set byte if not greater.
SETNGE	Set byte if not greater or equal.
SETNL	Set byte if not less.
SETNLE	Set byte if not less than or equal.
SETNO	Set byte if no overflow.
SETNP	Set byte if parity odd.
SETNS	Set byte if not a negative number.
SETNZ	Set byte if not 0.

Table G.5 ■ Additional 80386 Instructions

Mnemonic	Action
SETO	Set byte if overflow.
SETP	Set byte if parity even.
SETPE	Set byte if parity even.
SETPO	Set byte if parity odd.
SETS	Set byte if a negative number.
SETZ	Set byte if 0.
SHLD	Shift left, double word.
SHRD	Shift right, double word.
STOSD	String Store EAX as a double word.

Table G.5 ■ *Continued*

Inclusion of the test registers in the 80386 design is an indication of the level of sophistication (and difficulty) expected for programs that run on the 80386.

80386 Software Additions

The 80386 instruction set has been enormously enhanced over the 80286. There are instructions that deal with the increased register size operands and the new segment registers. New bit scan and double-precision instructions are added. Most notably, new indirect addressing modes that use any of the working registers are added.

Fifty-seven new instructions are added to the 80286 instruction set by the 80386 CPU design, as listed in Table G.5. Many of the new instructions deal with the 32-bit nature of the 80386, and the new segment registers FS and GS. Other new instructions add bit and flag manipulation instructions to the 80386. Memory addressing capabilities have also been expanded by using all of the 32-bit working registers for indirect addressing.

The 80386SX (1988)

Intel followed the same data-size strategy with the 80386 CPU as was done for the 8086. Instead of using a different type number (there is no 80388), Intel uses the suffix *SX* and *DX* to denote differences between various models of the same CPU architecture. The 80386SX can transfer information from memory in byte and word lengths. The 80386DX can transfer information in byte, word, and double word lengths. Programming opcodes are identical for both variants.

Math Coprocessors

A coprocessor is exactly what the name implies—another processor that may be used in a multi-processor system. All X86 family member have hardware features (see Chapter 12) that permit many CPUs to share system resources. A math coprocessor is a unique type of processor in that it is optimized to perform floating-point, trigonometric, and transcendental math operations. Moreover, a math coprocessor *must* operate in conjunction with a general-purpose CPU. A math coprocessor is not intended for stand-alone operation.

It is entirely feasible to include a math coprocessor on the same chip with the CPU. Putting the CPU and math processor in the same package, unfortunately, adds tremendously to the cost of the pair as manufacturing testing costs mount up. Not all customers want, or need, extensive math capabilities. The higher cost of a combined-ability chip might drive them to a competitor's model.

Early X86 CPU designs did not have the physical space, on the silicon die, to include a math coprocessor. The CPU and the math coprocessor had to be built on separate pieces of silicon. For cost and space reasons, the 8086, 80186, 80286, and 80386 have separate CPU and coprocessor chips.

MATH COPROCESSOR OPERATION

A math coprocessor is a specialized CPU with its own set of mnemonics, internal architecture, and connections. In use, the math coprocessor is electrically connected in parallel with the main CPU, and synchronized with it. The coprocessor monitors the main CPU control and data signals and performs one of its special math opcodes when such a special opcode is detected being fetched by the main CPU. When the coprocessor has finished its calculations, it signals the main CPU to get the results. The coprocessor can use many system assets, such as memory, to obtain the operands it needs to take part in calculations.

Math coprocessors greatly speed up overall computer operations when a significant amount of calculations must be done, such as in CAD or spreadsheet programs. Math coprocessors have become very popular with customers, so much so that many companies now offer X86-compatible coprocessors.

The 80486 (1989)

The 80486 is essentially, from a programming standpoint, an 80386 CPU with an 80387 math coprocessor integrated on the same silicon die. By placing the math coprocessor on the same die as the CPU, considerable increases in program execution speed are obtained. Reliability increases also, because of the elimination of external connections. Unfortunately, cost goes up also. Manufacturing and testing a singe chip as complex as the 80486 is a costly and time-consuming process. A single flaw in the 80387 part of the chip results in scrapping a perfectly good 80386 CPU. And the reverse. As a result of the expense of testing the math coprocessor on an 80486, Intel introduced the low-cost 80486SX, which is an 80486DX with the coprocessor disabled!

From a hardware standpoint, the 80486 is designed to operate at an *internal* speed nearly twice that of an equivalent 80386. Many operations that require two clock cycles in an 80386 take one clock cycle in an 80486. For example, an 80386 uses two clock cycles for internal register operations, whereas a 80486 uses one clock cycle.

The 80486 also integrates a high-speed 8K *cache memory*. A cache memory is used to interface between a high-speed CPU and slower DRAM. The cache can be filled from the DRAM in a burst, and the CPU can then access the cache at full speed.

Dual-speed 80486 processors have also appeared on the personal computer scene. A dual-speed processor is designed to operate at two clock speeds, one internal and the other external. The internal clock runs all internal processing operations at twice the speed of the external clock frequency. The half-speed external clock is used to synchronize memory access and opcode fetches by the CPU.

Once inside the CPU, however, operations proceed at the full-speed internal clock frequency. Dual-speed CPUs use the slower external clock to access memory, because high-speed memory systems are very expensive to design and buy. High-speed internal chip operation is not nearly as expensive as high-speed memory, so overall system performance can be enhanced by using dual-speed processors. Dual-speed system performance is *not* doubled, however, over a single-speed CPU (internal and external clocks at the same speed) because memory access time still slows down the dual-speed system.

To differentiate between full-speed and half-speed models, Intel continues to use DX to indicate a single-speed 80486, and uses the suffix *DX2* to indicate a dual-speed 80486 CPU.

80486 SOFTWARE ADDITIONS

Very few instructions have been added to the X86 family by the 80486. The new instructions are listed in Table G.6.

Pentium (1993)

Intel announced the Pentium CPU during March, 1933. Major features of the Pentium are listed as follows.

Clock Speeds: 60 to 100 MHz

Internal 8K byte Data Cache

Internal 8K byte Code Cache

Dual Instruction Pipes

Dual Integer ALUs

Dynamic Branch (Jump) Prediction

Improved Floating-Point Math

64-bit Data Bus

Mnemonic	Action
BSWAP	Swap register bytes
CMPXCHG	Compare and exchange
INVD	Clear cache memory
INVLPG	Invalidate a TLB entry
WBINVD	Write to DRAM memory and clear cache memory
XADD	Exchange and ADD

Table G.6 ◼ Additional 80486 Instructions

Model	Data Bits	Address Bits	Register Bits	Multi-tasking Multi-programming	Transistors	Relative Speed*
8086	16	20	16	No	29K	1
80286	16	24	16	Yes	130K	12
80386	32	32	32	Yes	280K	38
80486	32	32	32	Yes	1,200K	125
Pentium	64	64	32	Yes	3,100K	250

*At the fastest clock frequency for the model.

Table G.7 ■ X86 Family Comparison

Internally, Pentium is a 32-bit machine with the same architectural features as the 80486. If one wished to forecast the future, however, the appearance of a 64-bit data bus on Pentium implies a 64-bit microprocessor on the horizon.

X86 Family Comparisons

Table G.7 compares some of the important features of the X86 family members. Table G.7 does not tell the whole story on speed. Faster memory and disk access speeds have also served to increase the overall performance of PC systems.

Appendix

Sources

H.0 References

No book of this type is written in a vacuum. One of the striking facts observed in writing this book was how many references were needed.

This textbook is not an exhaustive reference on any one subject, as most chapters' themes form the basis for entire books. Rather, a distillation of many pieces of information to those deemed to be essential to the student has been my intent. Once past the beginning stages, however, the advanced student and practitioner will need a collection of references. The following books are recommended as the nucleus of an 8086-DOS reference library.

DOS

Dettman, Terry. *DOS Programmer's Reference, 3rd Edition.* Carmel, IN: Que, 1992.

IBM Corporation. *Personal Computer Technical Reference.* 1983. IBM United Kingdom International, Products Limited North Harbour, Portsmouth, England.

Lai, Robert. *Writing MS-DOS Device Drivers.* New York: Addison-Wesley, 1987.

Microsoft Corporation. *MS-DOS Programmer's Reference.* Redmond, WA: Microsoft Press, 1991.

8086 Assembly Language Programming

Broquard, Victor, and Westley, William. *Fundamentals of Assembler Language Programming for the IBM PC and XT.* Columbus OH: Merrill, 1990.

Haskell, Richard. *Assembly Language Tutor.* Englewood Cliffs, NJ: Regents/Prentice Hall, 1993.

Intel Literature Department. *iAPX 86/88, 186/188 User's Manual Programmer's Reference.* Santa Clara, CA: Intel Corporation, 1983.

Irvine, Kip. *Assembly Language Programming for the IBM PC.* New York: Macmillan, 1990.

PC Interfacing

Choiser, John. *The XT-AT Handbook, 3rd Edition.* San Diego, CA: Annabooks, 1990.

Eggebrecht, Lewis. *Interfacing to the IBM Personal Computer, 2nd Edition.* Carmel IN: SAMS, 1990.

Hall, Douglas. *Microprocessors and Interfacing.* New York: McGraw-Hill, 1986.

Haskell, Richard. *IBM PC—8088 Assembly Language Programming.* Rochester Hills, MI: REHI Books, 1986.

McNamara, John. *Technical Aspects of Data Communication.* Bedford, MA: Digital Press, 1978.

Solari, Edward. *AT BUS Design.* San Diego, CA: Annabooks, 1990.

H.1 Sources of Equipment

There are many sources of integrated circuits and blank bus expansion board cards available to the customer who buys in small quantities. The following list is made up of those companies with whom I have had satisfactory personal experience.

Digikey
P.O. Box 677
Thief River Falls, MN 56701
Phone: (800) 344 4539

Jameco
1355 Shoreway Road
Belmont, CA 94002
Phone: (800) 831 4242

JDR Microdevices
2233 Samaritan Drive
San Jose, CA 95124
Phone: (800) 538 5000

M. P. Jones
P.O. Box 12685
Lake Park, FL 33403
Phone: (407) 844 8764

Servo Systems Co.
P.O. Box 97
Montville, NJ 07045
Phone: (800) 922 1103

INDEX

C

PLEASE READ BEFORE OPENING DISKETTE PACKAGE

The enclosed diskette contains the complete, fully-functional A86 assembler and D86 debugger packages, provided on a "shareware" basis. The author, Eric Isaacson, allows the software to be freely copied and distributed for evaluation purposes, subject to the terms of permission given in Chapter 1 of the documentaion files. West Publishing is providing you this diskette under those terms, and has not paid the author any fees or royalties for doing so. If you decide to use the assembler and/or the debugger, you are expected to license your use of the packages by sending a registration fee to the author.

Single-user registration fees range between $50 and $90, depending on what you order—see the documentation files on the diskettte for details. The maximum amount, $90, buys you the full license for both A86 and D86, the latest version of those programs at the time that you register, the printed manual covering both programs, several extra utility programs to enchance your usage of the packages, and the licensed current versions of the A386 assembler and D386 debugger, that support 32-bit operations on the 386, 486, and Pentium computers.

Terms for multi-user fees, such as for classroom use, are found in the file TERMS on the diskette.

INSTALLATION INSTRUCTIONS

To install this software, make sure you are at your DOS prompt, logged into your hard drive (usually the "C" drive). Insert the diskette into its drive. If it is the A drive, type A:INSTALLA to the DOS prompt. If it is the B drive, type B:INSTALLB to to the DOS prompt.